Aflatoxins: Progress and Challenges

Aflatoxins: Progress and Challenges

Edited by **Debbie Harms**

Callisto Reference

New York

Published by Callisto Reference,
106 Park Avenue, Suite 200,
New York, NY 10016, USA
www.callistoreference.com

Aflatoxins: Progress and Challenges
Edited by Debbie Harms

International Standard Book Number: 978-1-63239-056-1 (Hardback)

Printed in China.

Contents

Preface IX

Section 1 Molecular Genetics and Management Strategies 1

Chapter 1 **Terrestrial Bacteria from Agricultural Soils: Versatile Weapons against Aflatoxigenic Fungi** 3
Masoomeh Shams-Ghahfarokhi, Sanaz Kalantari and Mehdi Razzaghi-Abyaneh

Chapter 2 **Development of Maize Host Resistance to Aflatoxigenic Fungi** 20
Robert L. Brown, Deepak Bhatnagar, Thomas E. Cleveland, Zhi-Yuan Chen and Abebe Menkir

Chapter 3 **A New Approach in Aflatoxin Management in Africa: Targeting Aflatoxin/Sterigmatocystin Biosynthesis in Aspergillus Species by RNA Silencing Technique** 40
Amos Emitati Alakonya and Ethel Oranga Monda

Chapter 4 **Novel Methods for Preventing and Controlling Aflatoxins in Food: A Worldwide Daily Challenge** 57
Eva Guadalupe Lizárraga-Paulín, Susana Patricia Miranda-Castro, Ernesto Moreno-Martínez, Irineo Torres-Pacheco and Alma Virginia Lara-Sagahón

Chapter 5 **Recent Trends in Microbiological Decontamination of Aflatoxins in Foodstuffs** 93
Carlos Augusto Fernandes Oliveira, Fernanda Bovo, Carlos Humberto Corassin, Alessandra Vincenzi Jager and Kasa Ravindranadha Reddy

Chapter 6 **Recent Advances for Control, Counteraction and Amelioration of Potential Aflatoxins in Animal Feeds** 127
N.K.S. Gowda, H.V.L.N. Swamy and P. Mahajan

Section 2 Food and Agriculture **139**

Chapter 7 **Occurrence of Aflatoxins in Food** **141**
Ayhan Filazi and Ufuk Tansel Sireli

Chapter 8 **Aflatoxins Importance on Animal Nutrition** **169**
Vivian Feddern, Giniani C. Dors, Fernando de C. Tavernari, Helenice Mazzuco, Anildo Cunha Jr., Everton L. Krabbe and Gerson N. Scheuermann

Chapter 9 **Occurrence of Aflatoxin M_1 in Raw and Pasteurized Goat Milk in Thailand** **194**
Suthep Ruangwises, Piyawat Saipan and Nongluck Ruangwises

Chapter 10 **Aflatoxin in Fish Flour from the Amazon Region** **207**
Ariane M. Kluczkovski and Augusto Kluczkovski Junior

Section 3 Chemico-Biological Interactions and Human Health **217**

Chapter 11 **Review of the Biological and Health Effects of Aflatoxins on Body Organs and Body Systems** **219**
Godfrey S. Bbosa, David Kitya, A. Lubega, Jasper Ogwal-Okeng, William W. Anokbonggo and David B. Kyegombe

Chapter 12 **The Significance of Glutathione Conjugation in Aflatoxin Metabolism** **246**
Tahereh Ziglari and Abdolamir Allameh

Chapter 13 **Synergistic Interaction Between Aflatoxin and Hepatitis B Virus in Hepatocarcinogenesis** **266**
Michael C. Kew

Section 4 Detection and Analysis **282**

Chapter 14 **Characteristics of Mycotoxin Analysis Tools for Tomorrow** **284**
Luis Miguel Contreras-Medina, Alejandro Espinosa-Calderon, Carlos Duarte-Galvan, Arturo Alfonso Fernandez-Jaramillo, Rafae Francisco Muñoz-Huerta, Jesus Roberto Millan-Almaraz, Ramon Gerardo Guevara-Gonzalez and Irineo Torres-Pacheco

Chapter 15 **Lateral Flow Immunoassays for Aflatoxins B and G and for Aflatoxin M_1** **309**
Laura Anfossi, Claudio Baggiani, Cristina Giovannoli and Gianfranco Giraudi

Section 5 Risk Assessment, Economics and Trade **334**

Chapter 16 **Aflatoxin and Peanut Production Risk and Net Incomes** **336**
Cynthia Bley N'Dede, Curtis M. Jolly, Davo Simplice Vodouhe and Pauline E. Jolly

Chapter 17 **Aflatoxins: Risk, Exposure and Remediation** **355**
Antonello Santini and Alberto Ritieni

Permissions

List of Contributors

Preface

Challenges and progress in the field of aflatoxins are described extensively in this book. It features latest developments in aflatoxins research from epidemiology to molecular genomics and control measures, biocontrol processes, modern analytical techniques, economic concerns and underlying mechanisms of contamination processes. This book will update readers on various cutting-edge facets of aflatoxins research with updated information helpful for mycologists, toxicologists, microbiologists, agriculture scientists, plant pathologists and pharmacologists, who may be interested in understanding the influence, significance and current developments within the field of aflatoxins with a focus on control strategy.

After months of intensive research and writing, this book is the end result of all who devoted their time and efforts in the initiation and progress of this book. It will surely be a source of reference in enhancing the required knowledge of the new developments in the area. During the course of developing this book, certain measures such as accuracy, authenticity and research focused analytical studies were given preference in order to produce a comprehensive book in the area of study.

This book would not have been possible without the efforts of the authors and the publisher. I extend my sincere thanks to them. Secondly, I express my gratitude to my family and well-wishers. And most importantly, I thank my students for constantly expressing their willingness and curiosity in enhancing their knowledge in the field, which encourages me to take up further research projects for the advancement of the area.

Editor

Section 1

Molecular Genetics and Management Strategies

Chapter 1

Terrestrial Bacteria from Agricultural Soils: Versatile Weapons against Aflatoxigenic Fungi

Masoomeh Shams-Ghahfarokhi, Sanaz Kalantari and Mehdi Razzaghi-Abyaneh

Additional information is available at the end of the chapter

1. Introduction

Invasion of food, feed and agricultural crops with mycotoxigenic fungi from the genera *Aspergillus, Fusarium* and *Penicillium* is an economic problem that is not yet under adequate control despite modern food production technologies and the wide range of preservation techniques available (Bennett & Klich, 2003). A small number of characterized fungi are as important as the genus *Aspergillus,* a taxonomic group which encompasses members with pathogenic, agricultural, industrial and pharmaceutical importance (Jamali et al., 2012). Nearly all fungi that produce aflatoxins, the most potent naturally occurring hepatocarcinogens, are members of the genus *Aspergillus* classified into the section *Flavi.* Among 22 closely related species in *Aspergillus* section *Flavi,* the members frequently encountered in agricultural products i.e. *Aspergillus flavus* and *A. parasiticus* are responsible for the majority of aflatoxin (AF) contamination events, with *A. flavus* being by far the most common (Varga et al., 2011). Aflatoxigenic fungi are common soil habitants all over the world and they frequently contaminate agricultural crops, such as peanuts, cottonseed, maize, and tree nuts (Bennett & Klich, 2003; Hedayati et al., 2007; Razzaghi-Abyaneh et al., 2006; Sepahvand et al., 2011). The fungal community structure composed of several players, species, strains, isolates and vegetative compatibility groups (VCGs), in the soil and on the crop determines the final AF concentration (Jamali et al., 2012; Razzaghi-Abyaneh et al., 2006). The life cycle of *A. flavus* in a pistachio orchard is shown in Fig. 1. AF contamination of agricultural crops is a major concern due to economical losses resulting from inferior crop quality reduced animal productivity and impacts on trade and public health. In a global context, AF contamination is an everlasting concern between the 35N and 35S latitude. Most of the countries in the belt of concern are developing countries and this makes the situation even worse because in those countries people frequently rely on highly susceptible crops for their daily nutrition and income. It has also been evident that AF more and more becomes a problem in countries that previously did not have to worry about AF contamination.

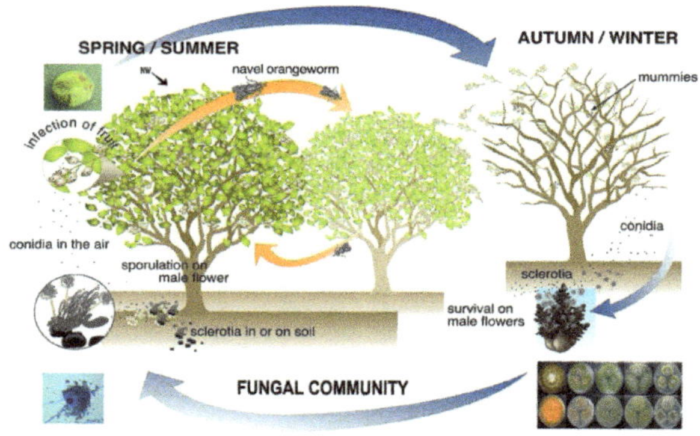

Figure 1. The life cycle of *A. flavus* is shown in a pistachio orchard. Infection of fruits with air-borne conidia occurs during Spring/Summer, while the fungus will survive by resistant structures named "sclerotia" during Autumn/Winter.

To ensure global safety on food and feed supplies, extensive researches have been carried out to effectively control and manage AF contamination of crops. The strategies for preventing AF contamination are generally divided into two categories including pre- and post-harvest controls (Kabak et al., 2006). Pre-harvest control strategies include appropriate field management practices (crop rotation, irrigation, soil cultivation, etc.), enhancing host resistance (transgenic or genetically modified crops), biological (application of antagonistic fungi and bacteria) and chemical control (fungicides, insecticides). Respect to biocontrol approaches, the rapid expansion in our knowledge about the role of microorganisms in inhibiting AF biosynthesis has enabled us to utilize them as potential AF biocontrol agents (Holmes et al., 2008; Raaijmakers et al., 2002). A large number of plants, mushrooms, bacteria, microalgae, fungi and actinomycetes have now been screened for the ability to inhibit toxigenic fungal growth and/or AF production (Alinezhad et al., 2011, Bagheri-Gavkosh et al., 2009; Ongena & Jacques, 2007; Razzaghi-Abyaneh & Shams-Ghahfarokhi, 2011; Razzaghi-Abyaneh et al., 2005, 2007, 2008, 2009, 2010, 2011). Substantial efforts have been carried out in identifying organisms inhibitory to AF biosynthesis through co-culture with aflatoxigenic fungi with the aim of finding potential biocontrol agents as well as novel inhibitory metabolites. The use of beneficial microorganisms is one of the most promising methods to the development of environmentally friendly alternatives to chemical pesticides in preventing the growth of aflatoxigenic fungi and subsequent AF contamination of susceptible crops. Among beneficial microorganisms, antagonistic bacteria are in the first line of investigation because of a much greater diversity than that of any other organism and possessing valuable pharmaceutically active molecules (Ongena & Jacques, 2007; Stein, 2005). Recent advances in analytical methods and enormous expanding of natural products libraries, cloning, and genetic engineering have provided a unique opportunity for isolation and structural elucidation of novel bioactive antifungal compounds from bacterial communities all over the world. It has been reported that, on average, two or three antibiotics derived from bacteria break into the market each year (Clark, 1996). Among an estimated number of 1.5 million bacterial species exists on our planet, only a little portion (less than 1%) has been identified yet of which a more little have tested for bioactive antifungal metabolites. Terrestrial bacteria are an interesting group of antagonistic microorganisms capable of efficiently inhibit toxigenic fungus growth and AF production. They mainly belong to the genera *Bacillus, Pseudomonas, Agrobacterium* and *Streptomyces* which have worldwide distribution (Holmes et al., 2008; Ongena & Jacques, 2007; Razzaghi-Abyaneh et al., 2011; Stein, 2005). Metabolites from *Bacillus subtilis* (Fengycins A and B, plipastatins A and B, iturin A, mycosubtilin, bacillomycin D), *Streptomyces* spp. (dioctatin A, aflastatin A, blasticidin A), and *Achromobacter xylosoxidans* [cyclo (L-leucyl-L-propyl)] are good examples of potent inhibitors of AF biosynthesis in laboratory conditions, crop model systems and also in the field (For review, see Razzaghi-Abyaneh et al., 2011). Since production of antifungal metabolites in bacteria is quite dependent to the strain and species, ongoing search on finding strange bacteria within the existing biodiversity to increase the chance of finding novel antifungals is currently done all over the world (Ranjbarian et al., 2011; Stein, 2005).

This chapter highlights comprehensive data on antagonistic bacteria isolated from agricultural soils of pistachio, peanuts and maize fields with an emphasis on their ability for inhibiting growth of aflatoxigenic fungi and AF production. We first describe how we can isolate and identify a large number of soil bacteria with antagonistic activity against toxigenic *A. parasiticus* by simple, efficient and low-cost screening methods. Next to be addressed will be a practical approach to isolation, purification and identification of antifungal metabolites from antagonistic bacteria by a combination of traditional and recent advanced technologies.

2. Biological control: a powerful management strategy

Biological control is defined as i) a method of managing pests by using natural enemies ii) an ecological method designed by man to lower a pest or parasite population to acceptable subclinical densities or iii) to keep parasite populations at a non-harmful level using natural living antagonists (Baker, 1987). The history of biological control dates back to an outstanding successful story, the biocontrol of the cottony-cushion scale (*Icerya purchasi*) on *Citrus* plant in California (Debach & Rosen, 1991). Biological control agents act against plant pathogens through different modes of action. Antagonistic interactions that can lead to biological control include antibiosis, competition and hyperparasitism (Bloom et al., 2003; Bull et al., 2002; Cook, 1993; Hoitink & Boehm, 1999). Competition occurs when two or more microorganisms require the same resources in excess of their supply. These resources can include space, nutrients, and oxygen. In a biological control system, the more efficient competitor, i.e., the biological control agent out-competes the less efficient one, i.e., the pathogen. Antibiosis occurs when antibiotics or toxic metabolites produced by one microorganism have direct inhibitory effect on another. Hyperparasitism or predation results from biotrophic or necrotrophic interactions that lead to parasitism of the plant pathogen by the biological control agent. Some microorganisms, particularly those in soil, can reduce damage from diseases by promoting plant growth or by inducing host resistance against a myriad of pathogens. Nowadays, atoxigenic *A. flavus* strains, biocompetitive bacteria and antagonistic yeasts has been effectively used to reduce AF contamination in field and laboratory conditions (Brown et al., 1991; Dorner et al., 1998, 1999; Hua et al., 1999; Palumbo et al., 2006). Commercial products from atoxigenic *A. flavus* under the names of AF36, AflaSafe and AflaGuard have been successfully used for biocontrol of aflatoxigenic fungi in maize, peanuts, cottonseed and pistachio fields in Southern US, Northern Mexico, Nigeria and West Africa (Atehnkeng et al., 2008; Donner et al., 2010).

3. Biocompetitive bacteria from agricultural soil

Regard to biocompetitive bacteria, *Bacillus subtilis* was first introduced as an inhibitor of growth and AF production of aflatoxigenic fungi by Kimura and Hirano (1988) and the effective compound, iturin A, had been patented for the control of AF in nuts and cereals (Kimura & Ono, 1988). Nowadays, ubiquitous inhabitants of agricultural soils i.e. the genera *Bacillus* and *Pseudomonas* are widely recognized as effective biocontrol agents of aflatoxigenic fungi. The broad host range, ability to form endospores and produce different biologically active compounds with a broad spectrum of activity made these bacteria as potentially useful biocontrol agents (Saharan & Nehra, 2011).

3.1. Soil sampling and bacterial isolation

One-hundred fifty soil samples were collected from pistachio, maize and peanut fields located in different regions of Damghan, Sari and Astaneh cities during June-July 2009. Sampling was done according to the latitude of each field. Each soil comprised from ten subsamples each of approximately 1000 mm^3 which were obtained using a sterile trowel at 10 m intervals. The subsamples were collected from the 50 mm top of the surface soil and then mixed thoroughly in a Nylon bag. The samples were air-dried in sterile Petri-dishes and stored at 4°C before use.

For bacteria isolation, 3 g of each soil sample was added to 10 ml of sterile normal saline solution (0.8 M), mixed vigorously by vortex for 2 min and centrifuge at 2500 rpm for 10 min. The amount of 10 µl aliquots of each sample supernatant was spread on to GY (Glucose 2%, Yeast extract 0.5%) agar and KB (King's B) agar plates and incubated for 3 days at 28°C. Discrete bacterial colonies were selected every 12 h and their purity was insured after transferring to master GY plate by tooth pick spot technique as shown in Fig. 2.

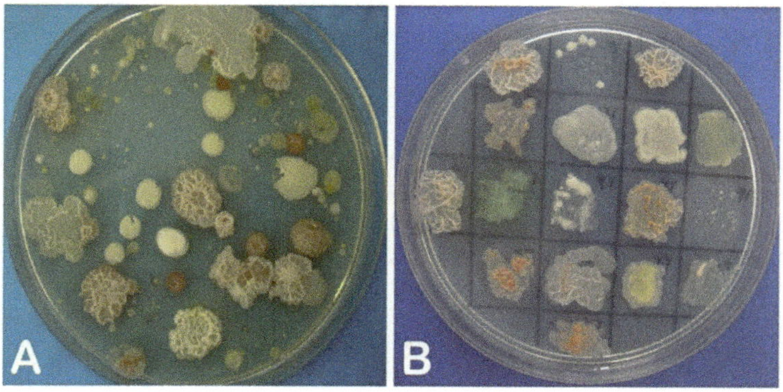

Figure 2. Various bacterial colonies appeared on GY agar after 3 days cultivation of soil suspensions (A). Separation and purification of colonies by using pick spot technique on GY agar master plates (B).

3.2. Screening for antifungal activity by visual agar plate assay

For selecting bacteria that inhibit either fungal growth or AF production, a visual agar plate assay was used as described by Hua et al. (1999) with some modifications. A 5 μl aliquot of a conidial suspension (200 conidia/μl) of a norsolorinic acid (NA)-accumulating mutant of *Aspergillus parasiticus* NRRL 2999 was streaked on the center of a Potato dextrose agar (PDA) plate. A single streak of 10 μl aliquots of isolated bacteria grown overnight in 0.5X Tryptic soy agar (TSA; Difco, Becton Dickinson, Franklin Lakes, NJ) at 28°C was inoculated in peripheral lines in distance of 1.5 cm from central line by tooth pick. Screen plates were incubated for 3-5 days at 28°C and assessed visually for antifungal phenotypes (Fig. 3). Antifungal activity was assessed by comparing the zone of fungal growth inhibition in fungus co-cultured with bacteria as tests, in comparison with control plates which were inoculated only with the fungus. The effect of bacteria on AF production was assessed from the underside of the fungus where a decrease in the red pigment (NA) in the mycelium indicated inhibition of AF production by the bacterium (Fig. 3).

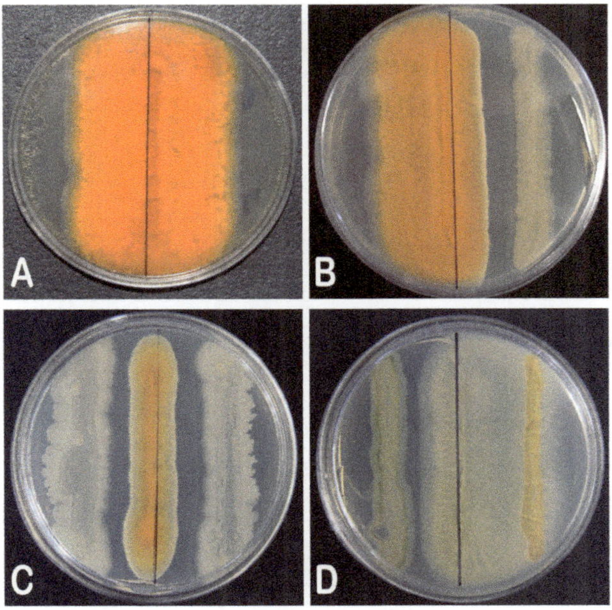

Figure 3. Visual agar plate assay shows screen identifying antagonistic bacteria with inhibitory activity against fungal (NA-accumulating mutant of *A. parasiticus* NRRL 2999) growth and/or NA accumulation (AF production):A) Control fungal culture against distilled water on both sides of GY agar.B) Control fungal culture against distilled water (left) and an antagonistic bacterium for fungal growth (right).C) Antagonistic bacteria for fungal growth with very weak inhibitory activity on NA accumulation on both sides.D) Antagonistic bacteria for both fungal growth and NA accumulation (left) and for only NA accumulation without affecting fungal growth (right).

Table 1 represents the results of antifungal phenotypes among soil bacteria isolated from pistachio, peanuts and maize fields. Different phenotypes were identified in all soils including NA and fungal growth inhibitors (type I), NA inhibitors (type II), growth inhibitors (type III) and finally non-inhibitors of NA and growth (type IV). The only exception were bacteria type II which was not isolated from peanuts field soils. In all fields, a pattern of type IV > type I > type III > type II were obtained regard to the number of antagonistic bacteria isolated. The phenotypes I and III are suitable candidates for biocontrol of AF-producing fungi in the field, while bacteria from type II are useful for elucidate AF biosynthesis pathway.

Fields of soil sampling	Total bacteria	Inhibitory bacteria	Inhibition of	
			NA	Fungal growth
Pistachio	290	37	+	+
		9	+	–
		22	–	+
		222	–	–
Maize	227	49	+	+
		6	+	–
		13	–	+
		159	–	–
Peanuts	87	19	+	+
		0	+	–
		16	–	+
		62	–	–

Table 1. Visual agar plate assay of antifungal phenotypes among soil bacteria isolated from pistachio, maize and peanuts field of Iran on PDA plates using a norsolorinic acid (NA) mutant of *A. parasiticus* NRRL 2999.

3.3. Identification of biocompetitive bacteria

The strongest antagonistic bacteria recognized from initial screening on PDA by visual agar plate assay were selected for identifying at genus and species level.

3.3.1. Biochemical identification

Selected bacteria were first determined to be either Gram-positive or Gram-negative using potassium hydroxide (Gregersen, 1978). Catalase and oxidase enzymatic activities were also determined (Barrow & Feltham, 1993). Gram-positive isolates were identified using GP2 MicroPlates (Biolog), whereas Gram-negative isolates were identified using GN2 MicroPlates (Biolog), according to the instructions of the manufacturer. Identification was based on the similarity index of carbon source utilization by each isolate relative to that of identified reference strains in the Biolog GP and GN databases.

3.3.2. *Molecular identification*

Fig. 4 illustrates all the steps for molecular identification of antagonistic bacteria. Overnight bacterial cultures on LB medium at 30°C were streaked on TSA plates. Single colonies from cultures grown on 0.5X TSA at 28°C were suspended in 2.0 ml sterile distilled water. Bacterial cells were pelleted by centrifugation at 12,000 × g for 10 min. and resuspended in 0.1 ml sterile distilled water. Total DNA from bacteria was prepared from single colonies grown on TSA according to the QIAGEN instruction. The 16s rRNA gene fragment was amplified in PCR using 1 to 5 μl of each cell suspension as template and universal primers 27F (5′-AGAGTTTGATCMTGGCTCAG-3′) and 1525R (5′AAGGAGGTGWTCCARCC-3′) (Lane, 1991). The PCRs were carried out using approximately 500 ng of total bacterial DNA, 10 μl of 10x PCR buffer, 8 μl of $MgCl_2$ (25 mM), 10 μl of deoxynucleoside triphosphates (dNTPs) (2 mM each), 3.3 μl of each primer (20 μM), 0.5 μl of *Taq* polymerase (5 U/μl), and enough Milli Q water so that the final volume of the mixture was 100 μl.

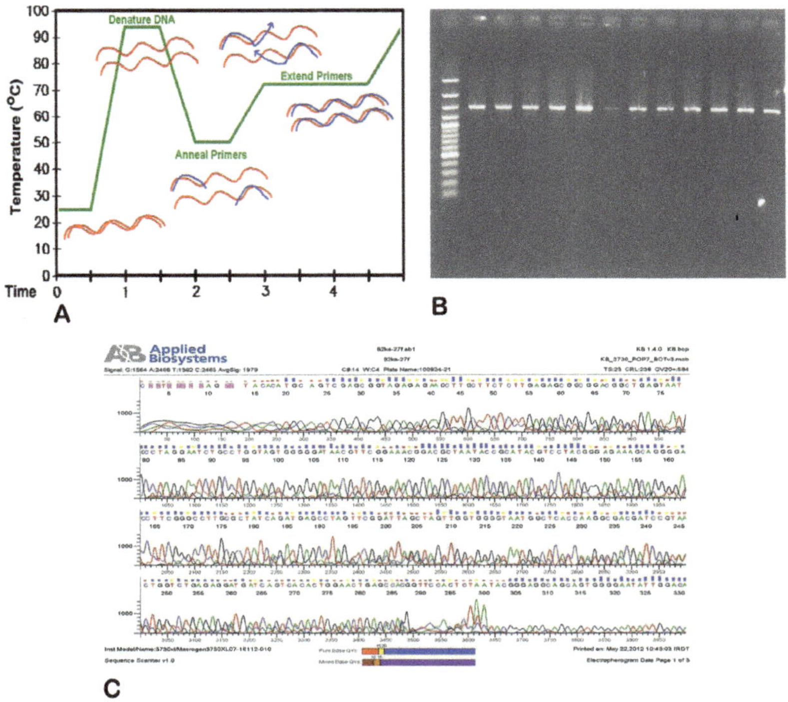

Figure 4. Molecular identification of antagonistic bacteria using PCR and DNA sequencing:A) PCR reaction temperature cycling; denaturing at 94°C, annealing at 55°C and extension at 72°C. Every cycle, DNA between primers is duplicated.B) An agarose gel stained with ethidium bromide shows PCR amplified bacterial DNAs (lines 2 to 13 from left). DNA molecular marker (100 bp DNA ladder) is shown in line 1 from left.C) Electroherogram data of purified DNA fragments of *Pseudomonas fluorescens* 82 which originated from sequence analysis by an ABI Prism Big Dye® Terminator v3.1 Cycle Sequencing Kit (Applied Biosystems).

The PCR mixtures were denatured at 95°C for 5 min, which was followed by 35 cycles of 94°C for 30s, 55°C for 30s, and 72°C for 90s and then a final extension at 72°C for 5 min. Amplification was checked for purity by electrophoresis on a 1.0% agarose gel. The bands of interest were excised from the gel, and the DNA was purified using QIAquick PCR purification columns (Qiagen, Inc., Valencia, CA). Purified DNA fragments were sequenced using the same sets of primers that were used for amplification by an ABI Prism Big Dye® Terminator v3.1 Cycle Sequencing Kit (Applied Biosystems). Bacteria were identified based on sequence similarities to homologous 16S rRNA gene fragments in the Ribosomal Database Project database (Cole et al., 2005) (accessed at http://rdp.cme.msu.edu/index.jsp).

3.4. Antagonistic activity against aflatoxigenic *A. parasiticus* NRRL 2999

Cell free culture supernatants of inhibitory bacteria were used in an antagonistic assay system. Table 2 represents the strongest antagonistic bacteria which were identified by a combination of biochemical and molecular methods in relation to their source of isolation.

Antagonistic bacteria	Strain number	Field	% of growth inhibition	% of AFB_1 inhibition	Surfactant production on blood agar
P. aeruginosa	320	Maize	63.9	95.3	+
	214	Maize	57.7	95.7	+
	155	Maize	48.9	78.2	–
	313	Peanuts	60.4	63.4	+
	257	Maize	64.5	85.3	+
	271	Maize	55.6	74.7	+
	293	Pistachio	55.7	73.6	+
	247	Maize	59.3	65.3	+
	287	Maize	59.0	87.6	+
	307	Peanuts	35.0	84.4	+
	168	Maize	62.6	96.9	+
	266	Maize	69.3	70.3	+
P. chlororaphis	236	Peanuts	15.3	65.9	–
P. fluorescens	82	Pistachio	72.7	91.1	–
B. subtilis	248	Maize	52.0	19.1	+
	298	Pistachio	70.6	18.7	+
	295	Pistachio	56.0	43.0	+
B. amyloliquefaciens	296	Maize	66.7	24.4	+

Table 2. Inhibitory effects of the strongest antagonistic bacteria selected from screening plates of visual agar plate assay on *A. parasiticus* NRRL 2999 growth and AF production in Potato dextrose broth. Control fungal culture had a growth rate of 51.17 mg and an AFB_1 amount of 697.78 ng/mg fungal dry weight.

Identified bacteria (0.1 ml of bacterial inoculums containing Ca. 10^7 CFU/ml) were inoculated on 20 ml of PDB prepared in 100 ml capacity flasks and incubated for 48 h at 28°C in shaking condition (100 rpm). Cell free supernatant fluids were prepared by centrifuging the cultures at 23990×g for 15 min. The supernatant was supplemented with PDB to compensate for the consumption of nutrient by bacterial growth the pH of supernatant fluid was adjusted to that of the original medium. Supernatant fluids were sterilized by filtration through a 0.45 μm pore size nylon membrane. Five ml aliquots of sterilized bacterial supernatant were aseptically dispensed in 25 ml Erlenmeyer flasks and inoculated with 0.1 ml of a spore suspension of *A. parasiticus* NRRL 2999 containing Ca 10^7 conidia/ml. Cultures were incubated for 96 h at 28°C and analyzed for fungal growth and AF production. At the end of incubation period, fungal mycelia were separated from culture medium using filter paper. Mycelia dry weight was determined as an index of fungal growth by incubating a known weight of fungal biomass at 80°C for 3 h and then until a constant weight was obtained. AF was extracted from the culture medium using chloroform. The chloroformic extracts were concentrated by a rotary evaporator (EYELA N-1000, Japan) to dryness. Quantitation of AFB_1 was carried out using HPLC (KNAUER D-14163 UV-VIS system, Germany) (Razzaghi-Abyaneh et al., 2007). Fifty ml of each sample (chloroformic extract) were injected into the HPLC column (TSKgel ODS-80TS; 4.6 mm ID × 150 mm, TOSOH BIOSCIENCE, Japan) and eluted at a flow rate of 1 ml/min. by water-acetonitrile-methanol (60:25:15, v/v/v) as mobile phase. AFB_1 was measured at wavelength of 365 nm. The elution time of the samples was compared with AFB_1 standards and quantified on the basis of the ratio of the peak area of samples to those of the standards. As shown in Table 2, secretory metabolites of all tested antagonistic bacteria including *Pseudomonas aeruginosa* (12 isolates), *Bacillus subtilis* (3 isolates), and one isolate of each *Pseudomonas chlororaphis, P. fluorescens* and *Bacillus amyloliquefaciens* inhibited both *A. parasiticus* growth and AFB_1 production by different extents. Fungal growth was inhibited in the range of 15.3 to 72.7%, while AFB_1 synthesis was suppressed by 18.7 to 96.9%. The highest inhibition of fungal growth and AFB_1 production was related to *P. fluorescens* 82 and *P. aeruginosa* 168, respectively. In contrast to *Pseudomonas, Bacillus* species strongly inhibited fungal growth with a weak suppressive effect on AF production. All antagonistic bacteria except *P. aeruginosa* 155 from maize, *P. chlororaphis* 236 from peanuts and *P. fluorescens* 82 from pistachio were capable of producing surfactants as a part of their pathogenesis system (Table 2).

4. Purification of antifungal metabolites from soil bacteria: A practical approach

4.1. Culture conditions for metabolite production

As the first step for production of bioactive antifungals, different culture conditions including medium, incubation time and aeration should be optimized. In order to initial purification of inhibitory metabolites, the selected bacterium with strongest antifungal activity in initial screening was cultured on suitable liquid media such as GY (2% glucose, 0.5% yeast extract), SCD (2% bacto dextrose, 20% potato infusion), PDB (potato dextrose broth) or even KB (King's B). The cultures were checked for optimal conditions of aeration (stationary cultures to shaking at different rpm from 100 to 250), incubation times (for at least 1 to maximum 7 days) and temperature (from 20 to 40°C). After culturing the bacterium at optimized condition, the whole culture as the main source of secretory metabolites was centrifuged at 8,000 x *g* for 30 min at room temperature. The cell free culture filtrate was then sterilized by filtration through a 0.22-µm-pore-size Millipore membrane (Millex-GV; Millipore) and kept at -20°C before use. The heat stability of the inhibitory metabolites can be examined by incubating the bacterial culture filtrate at 60, 80 and 100°C for 120 min or autoclaving at 121°C for 15 min. The acid and alkaline stabilities of the inhibitory metabolites can be checked by changing the pH of the culture medium to 1.5 and 11 by adding 1 M HCl or 1 M NaOH and incubating the solution at room temperature for 3 h.

4.2. Purification of antifungal metabolites

Consecutive steps of purification of bioactive metabolites from bacterial culture filtrate are summarized in Fig. 5. As the first step, the inhibitory bacterium should be cultured at optimized culture conditions from section 4.1. The next steps are Ion exchange column chromatography on Diaion HP20 resin, preparative thin layer chromatography on silica gel $60F_{254}$ and finally HPLC purification of bioactive metabolites.

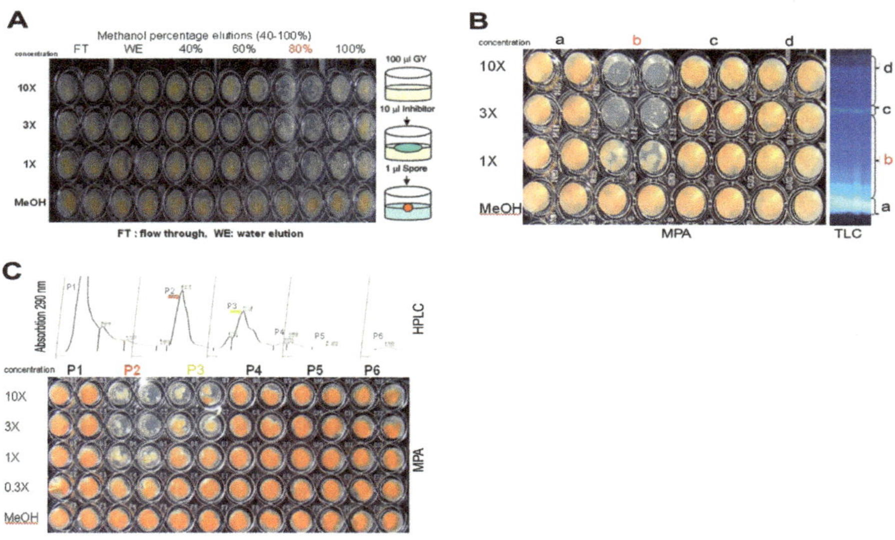

Figure 5. Sequential steps of purification of *A. parasiticus* growth inhibitory metabolites from bacterial culture filtrate:A) Stepwise elution of culture broth from a Diaion HP20 resin column using 40-100% aqueous MeOH. Fungal growth inhibition was reported for only 80% MeOH elution in microtiter agar plate assay (MPA).B) Further purification of fungal growth inhibitory metabolites from active Diaion HP20 column fraction (80% MeOH from step A) by thin layer chromatography (TLC). According to MPA result, section "b" was scrapped from TLC gel contained inhibitory compounds and thus, it was selected for further study.C) Final purification of inhibitory metabolites from section "b" of TLC in step B by normal-phase HPLC. Among 6 separated peaks shown (P1 to P6), two peaks i.e. P2 and P3 showed fungal growth inhibition in MPA.

4.2.1. Metabolite production at pre-optimized culture conditions

The selected bacterium with strongest antifungal activity was cultured in 1000 ml capacity flasks contained 250 ml GY as selected medium from section 4.1. The cultures were incubated at pre-optimized conditions (28°C for 5 days with shaking at 120 rpm). The whole culture (2 liters totally) was then centrifuged at 8,000 × g at room temperature for 30 min. The supernatant was used for purification of the inhibitory metabolites.

4.2.2. Ion exchange column chromatography

A glass column (2.5 × 60.0 cm) was equilibrated with MeOH. Five hundred grams of Diaion HP20 resin was suspended in MeOH and then packed onto the glass column. After removing of MeOH, the column was equilibrated with distilled water. The culture broth of selected bacterium (500 ml) was loaded onto the column. The resin was washed with 3 liters of distilled water, and the substances bound to the resin were then stepwise eluted by using 2 liters each of 40, 60, 80, and 100% methanol (MeOH) in water. Each elution was concentrated to dryness with a rotary evaporator and dissolves in desirable amounts of 100% MeOH. The 80% MeOH fraction which showed the highest growth and/or AF inhibitory activity against NA-mutant of *A. parasiticus* NRRL 2999 in microtiter agar plate assay (MPA), was selected for further purification (Fig. 5A).

4.2.3. Preparative thin layer chromatography

The 80% MeOH fraction from section 4.2.2 (an approximate of 250 mg dry weight) was applied to Silica gel $60F_{254}$ TLC plate and then developed with a mixture of chloroform/methanol/water (65:25:4, v/v/v) as mobile phase. Total area developed on the TLC plate was divided into at least 5 regions under 365 nm UV light, and the silica gel was scraped separately from each region. The substances presented in the silica gel were extracted with tenfold amounts of 100% MeOH. Each fraction was concentrated to dryness, dissolves in a small amount of MeOH, and subjected to the MPA on 96-well microplates. The fraction "b" (75.6 mg dry weight) which contained the strongest inhibitory activity against fungal growth and/or AF production was selected for further purification (Fig. 5B).

4.2.4. High performance liquid chromatography (HPLC)

The fraction "b" from section 4.2.3 was finally purified by HPLC equipped with a Cosmosil 5C18-AR column (4.6 × 150 mm; 5 μm). After injecting the sample, the column was washed with MeOH/water (50:50, v/v) for 80 min. The flow rate was adjusted at 1.0 ml/min, and elution was monitored at 290 nm wavelength. The number of 6 separated peaks (P1 to P6) were collected from the ODS column as shown in Fig. 5. Based on the MPA results, two peaks i.e. P2 and P3 were able to inhibit fungal growth and pigment production by *A. parasiticus* NRRL 2999 (Fig. 5C). These peaks were selected for further characterization by LC-MS and MALDI-TOF.

4.3. Structural elucidation of antifungal metabolites

With a combination of Liquid chromatography-Mass spectrometry (LC-MS) and Matrix-assisted laser desorption/ionization (MALDI-TOF), we will be able to elucidate the chemical structure of a protein or peptide in a best way. LC-MS spectrum determines retention time and an approximate mass of a purified compound, while complementary MALDI-TOF enable us to explain chemical formula and precise mass of the compound as the final step of identification. LC-MS and MALDI-TOF spectra of purified antifungal are shown in Fig. 6.

4.3.1. Liquid chromatography-Mass spectrometry (LC-MS)

The LC-MS system usually consists of a LC-10Avp separation module equipped with a SPD-M10Avp photodiode array detector and LC-MS2010A single quadruple mass spectrometer with atmospheric pressure photo ionization (APPI) source. The probe can be operated in the positive/negative mode under the condition of defined probe voltage, temperature of 300°C, CDL temperature of 200°C, nabulization gas (N2) flow 2.5 1/min, and scan range 900-1600 m/z (sec/scan). The amount of 2 µl of each inhibitory peak purified from HPLC separation was injected to an Ascentis C18 column (150 mm × 2.1 mm, 5 µm) and washed with MeOH (65% aqueous solution) acidified with 0.1% acetic acid in a flow rate of 0.2 ml/min. The column temperature should be maintained at 40°C during the operation. Approximate mass and retention time of the compound were recorded at the end of analysis.

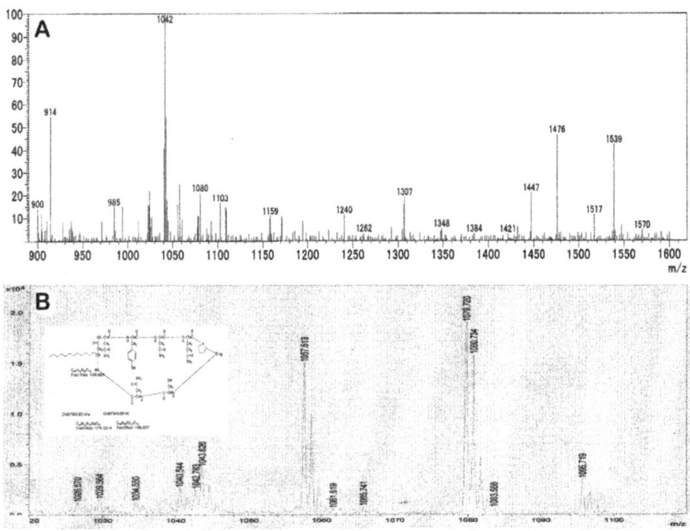

Figure 6. Liquid chromatography-Mass spectrometry (LC-MS) analysis of a HPLC purified inhibitory metabolite for *A. parasiticus* growth shows an approximate retention time of 17.0 min and a mass of 1042.0 m/z (A), while MALDI-TOF data indicates a structural formula of $C_{48}H_{76}N_{12}O_{14}$ and an exact mass of 1042.5447 m/z (B).

4.3.2. MALDI-TOF

Matrix-assisted laser desorption ionization-time of flight spectrometer (MALDI-TOF) is a soft ionization technique used in mass spectrometry, allowing the analysis of biomolecules (biopolymers such as DNA, proteins, peptides and sugars) and large organic molecules (polymers, dendrimers and other macromolecules), which tend to be fragile and fragment when ionized by more conventional ionization methods. The MALDI-TOF is a two step process. First, desorption is triggered by a UV laser beam. Matrix material heavily absorbs UV laser light, leading to the ablation of upper layer of the matrix material. The second step is ionization which takes place in the hot plume. Aside from peptide mass fingerprinting and useful application in identifying of microorganisms such as bacteria and fungi, MALDI-TOF is used for the rapid identification of proteins isolated by using gel electrophoresis: SDS-PAGE, size exclusion chromatography, affinity chromatography, strong/weak ion exchange, isotope coded protein labeling (ICPL), and two-dimensional gel electrophoresis. MALDI-TOF analysis of inhibitory compounds with defined retention time and an approximate mass from LC-MS step reveals valuable data about chemical formula and exact mass and provides finally identification of the absolute configuration of the purified inhibitory bacterial metabolite (Fig. 6).

5. Concluding remarks and future prospective

AF contamination of food and feed remains a major risk for human and animal health all over the world. Despite the long history of our knowledge about AF, little has been documented on how we can virtually combat the global distress of AF contamination of crops and agricultural commodities. AF-producing fungi can infect grains from pre-harvest conditions in the field through to post-harvest stages in the stores. Several pre- and post-harvest strategies have being tested to reduce risk of AF contamination. One of the management strategies being developed is biological control using various antagonistic microorganisms such as fungi, bacteria, and actinomycetes by a competitive exclusion mechanism. Biological control in conjunction with other management practices has potential to dramatically reduce AF contamination. Natural population of *A. flavus* consists of toxigenic strains that produce considerable amount of AF and atoxigenic strains that lack the capacity to produce AF. Nowadays, introducing atoxigenic strains has been successfully used to compete and exclude toxigenic strains in the field thereby reducing AF production in contaminated crops. However, there are some important limitations from the type of vegetative compatibility groups which shows the progeny of the fungus for AF-producing ability to geographic limitations in selection of atoxigenic strains. Considerable tolerance of *B. subtilis* and *P. chlororaphis* to environmental stresses, their large capacity for producing diverse array of beneficial antifungal metabolites and their readily producing by current fermentation technology make them promising tools for biocontrol of aflatoxigenic fungi in practice. Bacterial population from the genera *Bacillus* and *Pseudomonas* identified in pistachio, maize and peanut fields in the present study with potent antagonistic activity against aflatoxigenic *Aspergillus parasiticus* can potentially be developed into new biocontrol agents for combating AF contamination of crops in the field. These bacteria must be evaluated for a set of selection criteria for further use in biocontrol field experiments. Inability to produce toxic substances for biological systems and propensity to multiply, colonize and survive are the most important selection criteria to make sure that the selected antagonistic bacterial strains are safe and applicable when they introduced in to the environment. This endeavor shows biological control holds promise of offering a long-term solution for colonizing crops with aflatoxigenic fungi and thereby reducing AF contamination in the field.

Acknowledgements

This work was supported financially by Pasteur Institute of Iran.

Author details

Masoomeh Shams-Ghahfarokhi[1], Sanaz Kalantari[2] and Mehdi Razzaghi-Abyaneh[2*]

*Address all correspondence to: mrab442@yahoo.com, mrab442@pasteur.ac.ir

1 Department of Mycology, Faculty of Medical Sciences, Tarbiat Modares University, Tehran 14115-331 Iran

2 Department of Mycology, Pasteur Institute of Iran, Tehran 13164 Iran

References

[1] Alinezhad, S., Kamalzadeh, A., Shams-Ghahfarokhi, M., Rezaee, M. B., Jaimand, K., Kawachi, M., Zamani, Z., Tolouei, R., & Razzaghi-Abyaneh, M. (2011). Search for novel antifungals from 49 indigenous medicinal plants: Foeniculum vulgare and Platycladus orientalis as strong inhibitors of aflatoxin production by Aspergillus parasiticus. *Annals of Microbiology*, 61, 673-681.

[2] Atehnkeng, J., Ojiambo, P. S., Ikotun, T., Sikora, R. A., Cotty, P. J., & Bandyopadhyay, R. (2008). Evaluation of atoxigenic isolates of Aspergillus flavus as potential biocontrol agents for aflatoxin in maize. *Food Additives & Contaminants: Part A: Chemistry, Analysis, Control, Exposure & Risk Assessment*, 25, 1264-1271.

[3] Bagheri-Gavkosh, M., Bigdeli, M., Shams-Ghahfarokhi, M., & Razzaghi-Abyaneh, M. (2009). Inhibitory effects of Ephedra major Host on Aspergillus parasiticus growth and aflatoxin production. *Mycopathologia*, 168, 249-255.

[4] Baker, K. F. (1987). Evolving concepts of biological control of plant pathogens. *Annual Review of Phytopathology*, 25, 67-85.

[5] Barrow, G. I., & Feltham, R. K. A. (1993). Cowan and Steel's manual for the identification of medical bacteria, 3rd ed. Cambridge University Press, Cambridge, United Kingdom.

[6] Bennett, J. W., & Klich, M. (2003). Mycotoxins. *Clinical Microbiology Reviews*, 16, 497-516.

[7] Hedayati, M. T., Pasqualetto, A. C., Warn, P. A., Bowyer, P., & Denning, D. W. (2007). Aspergillus flavus: human pathogen, allergen and mycotoxin producer. *Microbiology*, 153, 1677-1692.

[8] Bloom, B., Ehlers, R., Haukeland-Salinas, S., Hoddanen, H., & Jung, K. (2003). Biological control agents: Safety and regulatory policy. *BioControl*, 48, 477-484.

[9] Brown, R. L., Cotty, P. J., & Cleveland, T. E. (1991). Reduction in aflatoxin content of maize by atoxigenic strains of Aspergillus flavus. *Journal of Food Protection*, 54, 623-626.

[10] Bull, C. T., Shetty, K. G., & Subbarao, K. V. (2002). Interactions between myxobacteria, plant pathogenic fungi and biocontrol agents. *Plant Disease*, 86, 889-896.

[11] Clark, A. M. (1996). Natural products as a resource for new drugs,. *Pharmaceutical Research*, 13, 1133-1141.

[12] Cole, J. R., Chai, B., Farris, R. J., Wang, Q., Kulam, S. A., McGarrell, D. M., Garrity, G. M., & Tiedje, J. M. (2005). The Ribosomal Database Project (RDP-II): sequences and tools for high-throughput rRNA analysis. *Nucleic Acids Research*, 33, D294-296.

[13] Cook, R. J. (1993). Making greater use of introduced microorganisms for biological control of plant pathogens. *Annual Review of Phytopathology*, 31, 53-80.

[14] DeBach, P., & Rosen, D. (1991). Biological control by natural enemies. Cambridge University Press, New York.

[15] Donner, M., Atehnkeng, J., Sikora, K. A., Bandyopadhyay, R., & Cotty, P. J. (2010). Molecular characterization of atoxigenic strains for biological control of aflatoxins in Nigeria. *Food Additives & Contaminants: Part A: Chemistry, Analysis, Control, Exposure & Risk Assessment*, 27, 576-590.

[16] Dorner, J. W., Cole, R. J., & Blankenship, P. D. (1998). Effect of inoculum rate of biocontrol agents on preharvest aflatoxin contamination of peanuts. *Biological Control*, 12, 171-176.

[17] Dorner, J. W., Cole, R. J., & Wicklow, D. T. (1999). Aflatoxin reduction in corn through field application of competitive fungi. *Journal of Food Protection*, 62, 650-656.

[18] Gregersen, T. (1978). Rapid method for distinction of Gram-negative from Gram-positive bacteria. *European Journal of Applied Microbiology and Biotechnology*, 5, 123-127.

[19] Hoitink, H., & Boehm, M. (1999). Biocontrol within the context of soil microbial communities: A substrate-dependent phenomenon. *Annual Review of Phytopathology*, 37, 427-446.

[20] Holmes, R. A., Boston, R. S., & Payne, G. A. (2008). Diverse inhibitors of aflatoxin biosynthesis. *Applied Microbiology and Biotechnology*, 78, 559-572.

[21] Hua, S. S., Baker, J. L., & Flores-Espiritu, M. (1999). Interactions of saprophytic yeasts with a nor mutant of Aspergillus flavus. *Applied and Environmental Microbiology*, 65, 2738-2740.

[22] Jamali, M., Ebrahimi, M. A., Karimipour, M., Shams-Ghahfarokhi, M., Dinparast-Djadid, N., Kalantari, S., Pilehvar-Soltanahmadi, Y., Amani, A., & Razzaghi-Abyaneh, M. (2012). An insight into the distribution, genetic diversity and mycotoxin production of Aspergillus section Flavi in soils of pistachio orchards. *Folia Microbiologica*, 57, 27-36.

[23] Kabak, B., Dobson, A. D. W., & Var, I. (2006). Strategies to prevent mycotoxin contamination of food and animal feed: A review. *Critical Reviews in Food Science and Nutrition*, 46, 593-619.

[24] Kimura, N., & Hirano, S. (1988). Inhibitory strains of Bacillus subtilis for growth and aflatoxin production of aflatoxigenic fungi. *Journal of Agricultural and Biological Chemistry*, 52, 1173-1179.

[25] Kimura, N., & Ono, M. (1988). Republique Française Patent No. 2644038.

[26] Lane, D. J. (1991). 16S/23S rRNA sequencing. *Nucleic acid techniques in bacterial systematic*, Stackebrandt, E. & Goodfellow, M. (ed.), 115-175, John Wiley & Sons, New York.

[27] Ongena, M., & Jacques, P. (2007). Bacillus lipopeptides: versatile weapons for plant disease biocontrol. *Trends in Microbiology*, 16, 115-125.

[28] Palumbo, J. D., Baker, J. L., & Mahoney, N. E. (2006). Isolation of bacterial antagonists of Aspergillus flavus from almonds. *Microbial Ecology*, 52, 45-52.

[29] Raaijmakers, J. M., Vlami, M., & de Souza, J. E. (2002). Antibiotic production by bacterial biocontrol agents. *Antonie van Leeuwenhoek*, 81, 537-547.

[30] Ranjbarian, A. R., Shams-Ghahfarokhi, M., Kalantari, S., & Razzaghi-Abyaneh, M. (2011). Molecular identification of antagonistic bacteria from Tehran soils and evaluation of their inhibitory activities toward pathogenic fungi. *Iranian Journal of Microbiology*, 3, 140-146.

[31] Razzaghi-Abyaneh, M., Allameh, A., Tiraihi, T., Shams-Ghahfarokhi, M., & Ghorbanian, M. (2005). Morphological alterations in toxigenic Aspergillus parasiticus exposed to neem (Azadirachta indica) leaf and seed aqueous extracts. *Mycopathologia*, 159, 565-570.

[32] Razzaghi-Abyaneh, M., & Shams-Ghahfarokhi, M. (2011). Natural Inhibitors of Food-borne Fungi from Plants and Microorganisms. *Natural Antimicrobials in Food Safety and Quality*, Rai, M. & Chikindas, M. (ed.), 182-203, CABI International, UK.

[33] Razzaghi-Abyaneh, M., Shams-Ghahfarokhi, M., Allameh, A., Kazeroon-Shiri, A. M., Ranjbar-Bahadori, S., Mirzahoseini, H., & Rezaee, M. B. (2006). A survey on distribution of Aspergillus section Flavi in corn field soils in Iran: Population patterns based on aflatoxins, cyclopiazonic acid and sclerotia production. *Mycopathologia*, 161, 183-192.

[34] Razzaghi-Abyaneh, M., Shams-Ghahfarokhi, M., Chang, P., & , K. (2011). Aflatoxins: Mechanisms of Inhibition by Antagonistic Plants and Microorganisms. *Aflatoxins: Biochemistry and Molecular Biology*, Ramon G. Guevara-Gonzalez (ed.), 285-304, INTECH Open Access Publisher, Croatia.

[35] Razzaghi-Abyaneh, M., Shams-Ghahfarokhi, M., Rezaee, M. B., Jaimand, S., Alinezhad, S., Saberi, R., & Yoshinari, T. (2009). Chemical composition and antiaflatoxigenic activity of Carum carvi L., Thymus vulgaris and Citrus aurantifolia essential oils. *Food Control*, 20, 1018-1024.

[36] Razzaghi-Abyaneh, M., Shams-Ghahfarokhi, M., Rezaee, M. B., & Sakuda, S. (2010). Natural Aflatoxin Inhibitors from Medicinal Plants. *Mycotoxins in Food, Feed and Bioweapons*, Rai, M. & Varma, A. (ed.), 329-352, Springer-Verlag, Berlin, Heidelberg.

[37] Razzaghi-Abyaneh, M., Shams-Ghahfarokhi, M., Yoshinari, T., Rezaee, M. B., Nagasawa, H., & Sakuda, S. (2008). Inhibitory effects of Satureja hortensis L. essential oil on growth and aflatoxin production by Aspergillus parasiticus. *International Journal of Food Microbiology*, 123, 228-233.

[38] Razzaghi-Abyaneh, M., Yoshinari, T., Shams-Ghahfarokhi, M., Rezaee, M. B., Nagasawa, H., & Sakuda, S. (2007). Dillapiol and apiol as specific inhibitors of the biosynthesis of aflatoxin G_1 in Aspergillus parasiticus. *Bioscience, Biotechnology and Biochemistry*, 71, 2329-2332.

[39] Saharan, B. S., & Nehra, V. (2011). Plant growth promoting rhizobacteria: A critical review. *Life Sciences and Medicine Research*, 21, 1-30.

[40] Sepahvand, A., Shams-Ghahfarokhi, M., Allameh, A., Jahanshiri, Z., Jamali, M., & Razzaghi-Abyaneh, M. (2011). A survey on distribution and toxigenicity of Aspergillus flavus from indoor and outdoor hospital environments. *Folia Microbiologica*, 56, 527-534.

[41] Stein, T. (2005). Bacillus subtilis antibiotics: structures, syntheses and specific functions. *Molecular Microbiology*, 56, 845-857.

[42] Varga, J., Frisvad, J. C., & Samson, R. A. (2011). Two new aflatoxin producing species and an overview of Aspergillus section Flavi. *Studies in Mycology*, 69, 57-80.

Chapter 2

Development of Maize Host Resistance to Aflatoxigenic Fungi

Robert L. Brown, Deepak Bhatnagar,
Thomas E. Cleveland, Zhi-Yuan Chen and
Abebe Menkir

Additional information is available at the end of the chapter

1. Introduction

Aflatoxins, the toxic and highly carcinogenic secondary metabolites of *Aspergillus flavus* and *A.parasiticus* are the most widely investigated of all mycotoxins because of their central role in establishing the significance of mycotoxins in animal diseases, and the regulation of their presence in food [1, 2]. Aflatoxins pose serious health hazards to humans and domestic animals, because they frequently contaminate agricultural commodities [3, 4]. Presently, numerous countries have established or proposed regulations for controlling aflatoxins in food and feeds [5]; the US Food and Drug Administration (FDA) has limits of 20 ppb, total aflatoxins, on interstate commerce of food and feed, and 0.5 ppb of aflatoxin M1 on the sale of milk. However, many countries, especially in the developing world, experience contamination of domestic-grown commodities at alarmingly greater levels than does the U.S. Evidence of this was shown in a study that revealed a strong association between exposure to aflatoxin and both stunting (a reflection of chronic malnutrition) and being underweight (a reflection of acute malnutrition) in West African children [6]. Also, a 2004 outbreak of acute aflatoxicosis in Kenya, due to the ingestion of contaminated maize, resulted in 125 deaths [7].

Recognition of the need to control aflatoxin contamination of food and feed grains has elicited responses outlining various approaches from researchers to eliminate aflatoxins from maize and other susceptible crops. The approach to enhance host resistance through breeding gained renewed attention following the discovery of natural resistance to *A. flavus* infection and aflatoxin production in Maize [8-12]. While several resistant maize genotypes have been identified through field screening, there is always a need to continually identify and utilize additional sources of maize genotypes with aflatoxin-resistance.

An important contribution to the identification/investigation of kernel aflatoxin-resistance has been the development of a rapid laboratory screening assay. The kernel screening assay (KSA), was developed and used to study resistance to aflatoxin production in GT-MAS:gk kernels [13, 14]. The KSA is designed to address the fact that aflatoxin buildup occurs in mature and not developing kernels. Although, other agronomic factors (e.g. husk tightness) are known to affect genetic resistance to aflatoxin accumulation in the field, the KSA measures seed-based genetic resistance. The seed, of course, is the primary target of aflatoxigenic fungi, and is the edible portion of the crop. Therefore, seed-based resistance represents the core objective of maize host resistance. Towards this aim, the KSA has demonstrated proficiency in separating susceptible from resistant seed [13, 14]. This assay has several advantages, as compared to traditional field screening techniques [14]: 1) it can be performed and repeated several times throughout the year and outside of the growing season; 2) it requires few kernels; 3) it can detect/identify different kernel resistance mechanisms; 4) it can dispute or confirm field evaluations (identify escapes); and 5) correlations between laboratory findings and inoculations in the field have been demonstrated. The KSA can, therefore, be a valuable complement to standard breeding practices for preliminary evaluation of germplasm. However, field trials are necessary for the final confirmation of resistance.

2. Discovery of aflatoxin-resistance

2.1. Traditional screening techniques

Screening maize for resistance to kernel infection by *Aspergillus flavus* or for resistance to aflatoxin production is a more difficult task than most disease screening. Successful screening in the past had been hindered [15] by the lack of 1) a resistant control; 2) inoculation methods that yield infection/aflatoxin levels high enough to differentiate among genotypes (natural infection is undependable); 3) repeatability across different locations and years; and, 4) rapid and inexpensive methods for assessment of fungal infection and aflatoxin levels. Several inoculation methods, including the pinbar inoculation technique (for inoculating kernels through husks), the silk inoculation technique, and infesting corn ears with insect larvae infected with *A. flavus* conidia have been tried with varying degrees of success [9, 16]. These methods can each be useful, however, clarity must exist as to the actual resistance trait to be measured (e.g. husk tightness; silk traits; the kernel pericarp barrier; wounded kernel resistance), before an appropriate technique can be employed. Silk inoculation, however, (possibly more dependent upon the plant's physiological stage and/or environmental conditions) has proven to be the most inconsistent of the inoculation methods [17].

Plating kernels to determine the frequency of kernel infection and examining kernels for emission of a bright greenish-yellow fluorescence (BGYF) are methods that have been used for assessing *A. flavus* infection [15]. While both methods can indicate the presence of *A. flavus* in seed, neither can provide the kind of accurate quantitative or tissue-localization data useful for effective resistance breeding. Several protocols have been developed and used for separation and relatively accurate quantification of aflatoxins [18].

2.2. Early identification of resistant maize lines

Two resistant inbreds (Mp420 and Mp313E) were discovered and tested in field trials at different locations and released as sources of resistant germplasm [11, 19]. The pinbar inoculation technique was one of the methods employed in the initial trials, and contributed towards the separation of resistant from susceptible lines [11]. Several other inbreds, demonstrating resistance to aflatoxin contamination in Illinois field trials (employing a modified pinbar technique) also were discovered [12]. Another source of resistance discovered was the maize breeding population, GT-MAS:gk. This population was derived from visibly classified segregating kernels, obtained from a single fungus-infected hybrid ear [10]. It tested resistant in trials conducted over a five year period, where a kernel knife inoculation technique was employed.

These discoveries of resistant germplasm may have been facilitated by the use of inoculation techniques capable of repeatedly providing high infection/aflatoxin levels for genotype separation to occur. While these maize lines do not generally possess commercially acceptable agronomic traits, they may be invaluable sources of resistance genes, and as such, provide a basis for the rapid development of host resistance strategies to eliminate aflatoxin contamination.

3. Investigations of resistance mechanisms/traits in maize lines

3.1. Molecular genetic investigations of aflatoxin-resistant lines

Chromosome regions associated with resistance to *A. flavus* and inhibition of aflatoxin production in maize have been identified through Restriction Fragment Length Polymorphism (RFLP) analysis in three "resistant" lines (R001, LB31, and Tex6) in an Illinois breeding program, after mapping populations were developed using B73 and/or Mo17 elite inbreds as the "susceptible" parents [20, 21]. Chromosome regions associated with inhibition of aflatoxin in studies considering all 3 resistant lines demonstrated that there are some regions in common. Regions on chromosome arms 2L, 3L, 4S, and 8S may prove promising for improving resistance through marker assisted breeding into commercial lines [21]. In some cases, chromosomal regions were associated with resistance to *Aspergillus* ear rot and not aflatoxin inhibition, and vice versa, whereas others were found to be associated with both traits. This suggests that these two traits may be at least partially under separate genetic control. QTL studies involving other populations have identified chromosome regions associated with low aflatoxin accumulation.

In a study involving 2 populations from Tex6 x B73, conducted in 1996 and 1997, promising QTLs for low aflatoxin were detected in bins 3.05-6, 4.07-8, 5.01-2, 5.05-5, and 10.05-10.07 [22]. Environment strongly influenced detection of QTLs for lower toxin in different years;

QTLs for lower aflatoxin were attributed to both parental sources. In a study involving a cross between B73 and resistant inbred Oh516, QTL associated with reduced aflatoxin were identified on chromosomes 2, 3 and 7 (bins 2.01 to 2.03, 2.08, 3.08, and 7.06) [23]. QTLs contributing resistance to aflatoxin accumulation were also identified using a population created by B73 and resistant inbred Mp313E, on chromosome 4 of Mp313E [24]. This confirmed the findings of an earlier study involving Mp313E and susceptible Va35 [25]. Another QTL in this study, which has similar effects to that on chromosome 4, was identified on chromosome 2 [24]. A recent study to identify aflatoxin-resistance QTL and linked markers for marker-assisted breeding was conducted using a population developed from Mp717, an aflatoxin-resistant maize inbred, and NC300, a susceptible inbred adapted to the southern U.S. QTL were identified on all chromosomes, except 4, 6, and 9; individual QTL accounted for up to 11% of phenotypic variance in aflatoxin accumulation [26]. Lastly, in a study of population of F2:3 families developed from resistant Mp715 and a southern-adapted susceptible, T173, QTL with phenotypic effects up to 18.5% were identified in multiple years on chromosomes 1, 3, 5, and 10 [27].

A number of genes corresponding to resistance-associated proteins (RAPs), that were identified in proteomics studies (see section 3.5.1 below) have been mapped to chromosomal location using the genetic sequence of B73 now available online (http://archive.maizesequence.org/index.html) [28]. Using the DNA sequence of the RAPs and blasting them against the B73 sequence allowed us to place each gene into a virtual bin, allowing us to pinpoint the chromosomal location to which each gene maps. The chromosomes involved include the above-mentioned chromosomes 1, 2, 3, 7, 8 and 10, some in bins closely located to those described above. Another study also mapped RAPs to bins on the above-chromosomes as well as chromosomes 4 and 9 [29].

3.2. Kernel pericarp wax

Kernel pericarp wax of maize breeding population GT-MAS:gk has been associated with resistance to *Aspergillus flavus* infection /aflatoxin production. Previously, kernel wax of GT-MAS:gk was compared to that of 3 susceptible genotypes. Thin layer chromatography (TLC) of wax from these genotypes showed a band unique to GT-MAS:gk and a band unique to the three susceptible lines [30]. GT-MAS:gk kernel wax also was shown to inhibit *A. flavus* growth. A later investigation compared GT-MAS:gk wax resistance-associated traits to that of twelve susceptible maize genotypes [31]. TLC results of wax from these lines confirmed findings of the previous investigation, demonstrating both the unique GT-MAS:gk TLC band and the unique 'susceptible' band. Gas chromatography/mass spectroscopy (GC/MS) analysis of the whole wax component showed a higher percentage of phenol-like compounds in the resistant genotype than in the susceptibles. Alkylresorcinol content was dramatically higher in GT-MAS:gk wax than in susceptible lines. An alkylresorcinol, 5-methylresorcinol, also inhibited *in vitro* growth of *A. flavus*. Further research is needed for a clear identification of the component(s) responsible for kernel wax resistance and to determine its expression level in other maize lines.

3.3. Two levels of resistance

The KSA employs a very simple and inexpensive apparatus involving bioassay trays, petri dishes, vial caps as seed containers, and chromatography paper for holding moisture [14]. Kernels screened by the KSA are maintained in 100% humidity, at a temperature favoring *A. flavus* (31° C) growth and aflatoxin production, and are usually incubated for seven days. Aflatoxin data from KSA experiments can be obtained two to three weeks after experiments are initiated. KSA experiments confirmed GT-MAS:gk resistance to aflatoxin production and demonstrated that it is maintained even when the pericarp barrier, in otherwise viable kernels, is breached [13]. Penetration through the pericarp barrier was achieved by wounding the kernel with a hypodermic needle down to the endosperm, prior to inoculation. Wounding facilitates differentiation between different resistance mechanisms in operation, and the manipulation of aflatoxin levels in kernels for comparison with other traits (e.g. fungal growth; protein induction). The results of this study indicate the presence of two levels of resistance: at the pericarp and at the subpericarp level. The former was supported by the above-studies which demonstrated a role for pericarp waxes in kernel resistance [30], and highlighted quantitative and qualitative differences in pericarp wax between GT-MAS:gk and susceptible genotypes [31, 32].

3.4. Comparing fungal growth to toxin production

When selected resistant Illinois maize inbreds (MI82, CI2, and T115) were examined by the KSA, modified to include an *A. flavus* GUS transformant (a strain genetically engineered with a gene construct consisting of a β-glucuronidase reporter gene linked to an *A. flavus* beta-tubulin gene promoter for monitoring fungal growth) [14], kernel resistance to fungal infection in nonwounded and wounded kernels was demonstrated both visually and quantitatively, as was a positive relationship between the degree of fungal infection and aflatoxin levels [14, 33]. This made it possible assess fungal infection levels and to determine if a correlation exists between infection and aflatoxin levels in the same kernels. *A. flavus* GUS transformants with the reporter gene linked to an aflatoxin biosynthetic pathway gene could also provide a way to indirectly measure aflatoxin levels [34-36], based on the extent of the expression of the pathway gene.

Recently, It was demonstrated, using the KSA and an *F. moniliforme* strain, genetically transformed with a GUS reporter gene linked to an *A. flavus* β-tubulin gene promoter, that the aflatoxin-resistant genotype, GT-MAS:gk, inhibits growth of *F. moniliforme* as well [37]. This indicates that some resistance mechanisms may be generic for ear rotting/ mycotoxigenic fungi.

A more recent use of reporter genes was performed on cotton using a green fluorescent protein reporter; a GFP-expressing *A. flavus* strain to successfully monitor fungal growth, mode of entry, colonization of cottonseeds, and production of aflatoxins [38]. This strain provides for an easy, potentially non-destructive, rapid and economical assay which can be done in real time, and may constitute an advance over GUS transformants.

3.5. Resistance-associated proteins

Developing resistance to fungal infection in wounded as well as intact kernels would go a long way toward solving the aflatoxin problem [17]. Studies demonstrating subpericarp (wounded-kernel) resistance in maize kernels have led to research for identification of subpericarp resistance mechanisms. Examinations of kernel proteins of several genotypes revealed differences between genotypes resistant and susceptible to aflatoxin contamination [39]. Imbibed susceptible kernels, for example, showed decreased aflatoxin levels and contained germination-induced ribosome inactivating protein (RIP) and zeamatin [40]. Both zeamatin and RIP have been shown to inhibit *A. flavus* growth *in vitro* [40]. In another study, two kernel proteins were identified from a resistant corn inbred (Tex6) which may contribute to resistance to aflatoxin contamination [41]. One protein, 28 kDa in size, inhibited *A. flavus* growth, while a second, over 100 kDa in size, primarily inhibited toxin formation. When a commercial corn hybrid was inoculated with aflatoxin and nonaflatoxin-producing strains of *A. flavus* at milk stage, one induced chitinase and one ß-1,3-glucanase isoform was detected in maturing infected kernels, while another isoform was detected in maturing uninfected kernels [42].

In another investigation, an examination of kernel protein profiles of 13 maize genotypes revealed that a 14 kDa trypsin inhibitor protein (TI) is present at relatively high concentrations in seven resistant maize lines, but at low concentrations or is absent in six susceptible lines [43]. The mode of action of TI against fungal growth may be partially due to its inhibition of fungal -amylase, limiting *A. flavus* access to simple sugars [44] required not only for fungal growth, but also for toxin production [45]. TI also demonstrated antifungal activity against other mycotoxigenic species [46]. The identification of these proteins may provide markers for plant breeders, and may facilitate the cloning and introduction of antifungal genes through genetic engineering into other aflatoxin-susceptible crops.

An investigation into maize kernel resistance [47] determined that both constitutive and induced proteins are required for resistance to aflatoxin production. It also showed that one major difference between resistant and susceptible genotypes is that resistant lines constitutively express higher levels of antifungal proteins compared to susceptible lines. The real function of these high levels of constitutive antifungal proteins may be to delay fungal invasion, and consequent aflatoxin formation, until other antifungal proteins can be synthesized to form an active defense system.

3.5.1. Proteomic analysis

Two-dimensional (2-D) gel electrophoresis, which sorts proteins according to two independent properties, isoelectric points and then molecular weights, has been recognized for a number of years as a powerful biochemical separation technique. Improvements in map resolution and reproducibility [48, 49], rapid analysis of proteins, analytical soft ware and computers, and the acquisition of genomic data for a number of organisms has given rise to another application of 2-D electrophoresis: proteome analysis. Proteome analysis or "proteomics" is the analysis of the protein complement of a genome [50, 51]. This involves the systematic separation, identification, and quantification of many proteins simultaneously. 2-D electrophoresis is also unique in its ability to detect post- and cotranslational modifications, which cannot be predicted from the genome sequence.

Through proteome analysis and the subtractive approach, it may be possible to identify important protein markers associated with resistance, as well as genes encoding these proteins. This could facilitate marker-assisted breeding and/or genetic engineering efforts. Endosperm and embryo proteins from several resistant and susceptible genotypes have been compared using large format 2-D gel electrophoresis, and over a dozen such protein spots, either unique or 5-fold upregulated in resistant maize lines (Mp420 and Mp313E), have been identified, isolated from preparative 2-D gels and analyzed using ESI-MS/MS after in-gel digestion with trypsin [52, 53]. These proteins, all constitutively expressed, can be grouped into three categories based on their peptide sequence homology: (1) storage proteins, such as globulins and late embryogenesis abundant proteins; (2) stress-responsive proteins, such as aldose reductase, a glyoxalase I protein and a 16.9 kDa heat shock protein, and (3) antifungal proteins, including the above-described TI.

During the screening of progeny developed through the IITA-USDA/ARS collaborative project, near-isogenic lines from the same backcross differing significantly in aflatoxin accumulation were identified, and proteome analysis of these lines is being conducted [54]. Investigating corn lines from the same cross with contrasting reaction to *A. flavus* should enhance the identification of RAPs clearly without the confounding effect of differences in the genetic backgrounds of the lines.

Heretofore, most RAPs identified have had antifungal activities. However, increased temperatures and drought, which often occur together, are major factors associated with aflatoxin contamination of maize kernels [55]. It has also been found that drought stress imposed during grain filling reduces dry matter accumulation in kernels [55]. This often leads to cracks in the seed and provides an easy entry site to fungi and insects. Possession of unique or of higher levels of hydrophilic storage or stress-related proteins, such as the aforementioned, may put resistant lines in an advantageous position over susceptible genotypes in the ability to synthesize proteins and defend against pathogens under stress conditions. Further studies including physiological and biochemical characterization, genetic mapping, plant transformation using RAP genes, and marker-assisted breeding should clarify the roles of stress-related RAPs in kernel resistance. RNAi gene silencing experiments involving RAPs may also contribute valuable information. [54].

3.5.2. Further characterization of RAPs

A literature review of the RAPs identified above indicates that storage and stress-related proteins may play important roles in enhancing stress tolerance of host plants. The expression of storage protein GLB1 and LEA3 has been reported to be stress-responsive and ABA-dependant [56]. Transgenic rice overexpressing a barley LEA3 protein HVA1 showed significantly increased tolerance to water deficit and salinity [57]. The role of GLX I in stress-tolerance was first highlighted in an earlier study using transgenic tobacco plants overexpressing a *Brassica juncea* glyoxalase I [58]. The substrate for glyoxalase I, methylglyoxal, is a potent cytotoxic compound produced spontaneously in all organisms under physiological conditions from glycolysis and photosynthesis intermediates, glyceraldehydes-3-phosphate and dihydroxyacetone phosphate. Methylglyoxal is an aflatoxin inducer even at low concentrations; experimental evidence indicates that induction is through upregulation of aflatoxin biosynthetic pathway transcripts including the *AFLR* regulatory gene [59]. Therefore, glyoxalase I may be directly affecting resistance by removing its aflatoxin-inducing substrate, methylglyoxal. PER1, a 1-cys peroxiredoxin antioxidant identified in a proteomics investigation [60], was demonstrated to be an abundant peroxidase, and may play a role in the removal of reactive oxygen species. The PER1 protein overexpressed in *Escherichia coli* demonstrated peroxidase activity *in vitro*. It is possibly involved in removing reactive oxygen species produced when maize is under stress conditions [60]. Another RAP that has been characterized further is the pathogenesis-related protein 10 (PR10). It showed high homology to PR10 from rice (85.6% identical) and sorghum (81.4% identical). It also shares 51.9% identity to intracellular pathogenesis-related proteins from lily (AAF21625) and asparagus (CAA10720), and low homology to a RNase from ginseng [61]. The PR10 overexpressed in *E. coli* exhibited ribonucleolytic and antifungal activities. In addition, an increase in the antifungal activity against *A. flavus* growth was observed in the leaf extracts of transgenic tobacco plants expressing maize *PR10* gene compared to the control leaf extract [61]. This evidence suggests that PR10 plays a role in kernel resistance by inhibiting fungal growth of *A. flavus*. Further, its expression during kernel development was induced in the resistant line GT-MAS:gk, but not in susceptible Mo17 in response to fungal inoculation [61]. Recently, a new *PR10* homologue was identified from maize (*PR10.1*) [62]. *PR10* was expressed at higher levels in all tissues compared to *PR10.1*, however, purified PR10.1 overexpressed in *E. coli* possessed 8-fold higher specific RNase activity than *PR10* [62]. This homologue may also play a role in resistance. Evidence supporting a role for *PR10* in host resistance is also accumulating in other plants. A barley *PR10* gene was found to be specifically induced in resistant cultivars upon infection by *Rhynchosporium secalis*, but not in near-isogenic susceptible plants [63]. In cowpea, a *PR10* homolog was specifically up-regulated in resistant epidermal cells inoculated with the rust fungus *Uromyces vignae* Barclay [64]. A *PR10* transcript was also induced in rice during infection by *Magnaporthe grisea* [65].

To directly demonstrate whether selected RAPs play a key role in host resistance against *A. flavus* infection, an RNA interference (RNAi) vector to silence the expression of endogenous RAP genes (such as *PR10, GLX I* and *TI*) in maize through genetic engineering was constructed [59, 66]. The degree of silencing using RNAi constructs is greater than that obtained using either co-suppression or antisense constructs, especially when an intron is included [67]. Interference of double-stranded RNA with expression of specific genes has been widely described [68, 69]. Although the mechanism is still not well understood, RNAi provides an extremely powerful tool to study functions of unknown genes in many organisms. This posttranscriptional gene silencing (PTGS) is a sequence-specific RNA degradation process triggered by a dsRNA, which propagates systemically throughout the plant, leading to the degradation of homologous RNA encoded by endogenous genes, and transgenes. Both particle bombardment and *Agrobacterium*-mediated transformation methods were used to introduce the RNAi vectors into immature maize embryos. The former was used to provide a quick assessment of the efficacy of the

RNAi vector in gene silencing. The latter, which can produce transgenic materials with fewer copies of foreign genes and is easier to regenerate, was chosen for generating transgenic kernels for evaluation of changes in aflatoxin-resistance. It was demonstrated using callus clones from particle bombardment that *PR10* expression was reduced by an average of over 90% after the introduction of the RNAi vector [66]. The transgenic kernels also showed a significant increase in susceptibility to *A. flavus* infection and aflatoxin production. The data from this RNAi study clearly demonstrated a direct role for PR10 in maize host resistance to *A. flavus* infection and aflatoxin contamination [66]. RNAi vectors to silence other RAP genes, such as *GLX I* and *TI*, have also been constructed, and introduced into immature maize embryos through both bombardment and *Agrobacterium* infection [70]. It will be very interesting to see the effect of silencing the expression of these genes in the transgenic kernels on host resistance to *A. flavus* infection and aflatoxin production.

ZmCORp, a protein with a sequence similar to cold-regulated protein and identified in the above-proteomic studies, was shown to exhibit lectin-like hemagglutination activity against fungal conidia and sheep erythrocytes [71]. When tested against *A. flavus*, ZmCORp inhibited germination of conidia by 80% and decreased mycelial growth by 50%, when germinated conidia were incubated with the protein. Quantitative real-time RT-PCR revealed *ZmCORp* to be expressed 50% more in kernels of a resistant maize line *versus* a susceptible.

ZmTIp, a 10 kDa trypsin inhibitor, had an impact on *A. flavus* growth, but not as great as the previously-mentioned 14 kDa TI [72].

3.5.3. Proteomic studies of rachis and silk tissue

A study was conducted to investigate the proteome of rachis tissue, maternal tissue that supplies nutrients to the kernels [75]. An interesting finding in this study is that after infection by *A. flavus*, rachis tissue of aflatoxin-resistant genotypes did not up-regulate PR proteins as these were already high in controls where they had strongly and constitutively accumulated during maturation. However, rachis tissue of aflatoxin-susceptible lines did not accumulate PR proteins to such an extent during maturation, but increased them in response to fungal infection. Given the relationship of the rachis to kernels, these results confirm findings of a previous investigation [47], which demonstrated levels of proteins in resistant *versus* susceptible kernels was a primary factor that determined kernel genetic resistance to aflatoxin contamination. Another study was conducted to identify proteins in maize silks that may be contributing to resistance against *A. flavus* infection/colonization [76]. Antifungal bioassays were performed using silk extracts from two aflatoxin-resistant and two–susceptible inbred lines. Silk extracts from resistant inbreds showed greater antifungal activity compared to susceptible inbreds. Comparative proteomic analysis of the two resistant and susceptible inbreds led to the identification of antifungal proteins including three chitinases that were differentially-expressed in resistant lines. When tested for chitinase activity, silk proteins from extracts of resistant lines also showed significantly higher chitinase activity than that from susceptible lines. Differential expression of chitinases in maize resistant and susceptible inbred silks suggests that these proteins may contribute to resistance.

3.5.4. Transcriptomic analyses

To investigate gene expression in response to *A. flavus'* infection and to more thoroughly identify factors potentially involved in the regulation of RAP genes, a transcriptomic profile was conducted on maize kernels of two inbred lines that were genetically closely-related [73]. Similar work had previously been performed using Tex6 as the resistant line and B73 as the susceptible [74], however, in the study using closely-related lines, imbibed mature kernels were used (for the first time) and proved to be a quicker and easier approach than traditional approaches. The involvement of certain stress-related and antifungal genes previously shown to be associated with constitutive resistance was demonstrated here; a kinase-binding protein, Xa21 was highly up-regulated in the resistant line compared to the susceptible, both constitutively and in the inducible state.

4. Current efforts to develop resistant lines

4.1. Closely-related lines

Recently, the screening of progeny generated through a collaborative breeding program between IITA-Nigeria (International Institute of Tropical Agriculture) and the Southern Regional Research Center of USDA-ARS in Center (SRRC) of USDA-ARS in New Orleans facilitated the identification of closely-related lines from the same backcross differing significantly in aflatoxin accumulation, and proteome analysis of these lines is being conducted [77, 78]. Investigating corn lines sharing close genetic backgrounds should enhance the identification of RAPs without the confounding effects experienced with lines of diverse genetic backgrounds. The IITA-SRRC collaboration has attempted to combine resistance traits of U.S. resistant inbred lines with those of African lines, originally selected for resistance to ear rot diseases and for potential aflatoxin-resistance (*via* KSA) [77, 78]. Five elite tropical inbred lines from IITA adapted to the Savanna and mid-altitude ecological zones of West and Central Africa were crossed with four U.S. resistant maize lines in Ibadan, Nigeria. The five African lines were originally selected for their resistance to ear rot caused by *Aspergillus, Botrydiplodia, Diplodia, Fusarium,* and *Macropomina* [77, 78]. The F1 crosses were backcrossed to their respective U.S. inbred lines and self-pollinated thereafter. The resulting lines were selected through the S4 generation for resistance to foliar diseases and desirable agronomic characteristics under conditions of severe natural infection in their respective areas of adaptation. Promising S5 lines were screened with the KSA (Table 1). In total, five pairs of closely-related lines were shown to be significantly different in aflatoxin resistance, while sharing as high as 97% genetic similarity [79]. Using these lines in proteomic comparisons to identify RAPs has advantages: (1) gel comparisons and analyses become easier; and (2) protein differences between resistant and susceptible lines as low as twofold can be identified with confidence. In addition, the likelihood of identifying proteins that are directly involved in host resistance is increased. In a preliminary proteomics comparison of constitutive protein differences between those African closely-related lines, a new category of resistance-associated proteins (putative regulatory proteins) was identified, including a serine/threonine protein kinase and a translation initiation factor 5A [29, 79]. The genes encoding these two resistance associated regulatory proteins are being cloned and their potential role in host resistance to *A. flavus* infection and aflatoxin production will be further investigated. Conducting proteomic analyses using lines from this program not only enhances chances of identifying genes important to resistance, but may have immediate practical value. The IITA-SRRC collaboration has registered and released six inbred lines with aflatoxin-resistance in good agronomic backgrounds, which also demonstrate good levels of resistance to southern corn blight and southern corn rust [80]. Resistance field trials for these lines on U.S. soil is being conducted; the ability to use resistance in these lines commercially will depend on having identified excellent markers, since seed companies desire insurance against the transfer of undesirable traits into their elite genetic backgrounds. The fact that this resistance is coming from good genetic backgrounds is also a safeguard against the transfer of undesirable traits.

Entry	Aflatoxin B_1 (ppb)
Susceptible control	10197 a
22*	1693 b
19	1284 bc
28	1605 bcd
27	1025 bcd
21	1072 bcd
26	793 bcde
20	574 cde
24	399 cde
GT-MAS:gk	338 de
25*	228 e
23	197 e
Resistant control	76 e

Table 1. KSA screening of IITA-SRRC maize breeding materials which identified 2 closely related lines (87.5% genetic similarity), #22 and #25, from parental cross (GT-MASgk x Ku1414SR) x GT-MAS:gk; these contrast significantly in aflatoxin accumulation. Values followed by the same letter are not significantly different by the least significant difference test (P = 0.05).

4.2. Recent breeding efforts

Recent breeding efforts towards the development of aflatoxin-resistant maize lines has resulted in a number of germplasm releases including the above-mentioned IITA-SRRC inbreds. In 2008, TZAR 101-106, derived from a combination of African and southern-adapted U.S. lines are being field-tested in different parts of the Southern U.S. (Figure 1) [80]. These have also exhibited resistance to lodging and common foliar diseases. GT-603 was released in 2011, after having been derived from GT-MAS:gk [81], while Mp-718 and Mp-719 were released as southern adapted resistant lines which are both shorter and earlier than previous Mp lines [82, 83]. These lines are also being tested as inbreds and in hybrid combinations in the southern U.S. [83].

Figure 1. Inoculation of maize ears with *Aspergillus flavus* spores using a 'side needle' wound technique for field evaluations of TZAR lines developed through IITA-SRRC program.

5. Conclusion

The host resistance approach to eliminating aflatoxin contamination of maize has been advanced forward by the identification/development of maize lines with resistance to aflatoxin accumulation. However, to fully exploit the resistance discovered in these lines, markers must be identified to transfer resistance to commercially useful backgrounds. Towards this goal numerous investigations have been undertaken to discover the factors that contribute to resistance, laying the basis for exploiting these discoveries as well.

These investigations include QTL analyses to locate regions of chromosomes associated with the resistant phenotype, and the discovery of kernel resistance-related traits. We now know that there are two levels of resistance in kernels, pericarp and subpericarp. Also, there is a two-phased kernel resistance response to fungal attack: constitutive at the time of fungal attack and that which is induced by the attack. Thus far, it's been demonstrated that natural resistance mechanisms discovered are antifungal in nature as opposed to inhibiting the aflatoxin biosynthetic pathway.

One of the most important discoveries, thus far, has been that of resistance-associated proteins or RAPs. Due to the significance of the constitutive response, constitutive RAPs were investigated first, although induced proteins are being studied as well. Investigations of other tissues such as rachis and silks begin to provide a more complete picture of the maize resistance response to aflatoxigenic fungi. RAP characterization studies provide greater evidence that these proteins are important to resistance, although clearly, more investigations are needed. Looking at data collectively that's been obtained from different types of studies may enhance the identification of markers for breeding. A good example of this may be the supporting evidence provided by QTL data to proteomic and RAP characterization data suggesting the involvement of 14 kDa TI, water stress inducible protein, zeamatin, heat shock, cold-regulated, glyoxalase I, cupin-domain and PR10 proteins in aflatoxin-resistance. It will be interesting to determine if this marker discovery approach can lead to the successful transfer of a multigene-based and quantitative phenomenon such as aflatoxin-resistance to commercially-useful genetic backgrounds.

Acknowledgements

Research discussed in this review received support from the USAID Linkage Program-IITA, Nigeria, and the USDA-ARS Office of International Research Programs (OIRP) -USAID Collaborative Support Program.

Author details

Robert L. Brown[1*], Deepak Bhatnagar[1], Thomas E. Cleveland[1], Zhi-Yuan Chen[2] and Abebe Menkir[3]

*Address all correspondence to: Robert.brown@ars.usda.gov

1 USDA-ARS, Southern Regional Research Center, New Orleans, LA, USA

2 Department of Plant Pathology and Crop Physiology, Louisiana State University Agricultural Center, Baton Rouge, LA, USA

3 International Institute of Tropical Agriculture, Ibadan, Nigeria

References

[1] Brown RL., Bhatnagar D., Cleveland TE., Cary JW. Recent Advances in Preventing Mycotoxin Contamination. In: Sinha KK., Bhatnagar D. (eds) Mycotoxins in Agriculture and Food Safety, Marcel Dekker: New York, NY, USA; 1998. p351-379.

[2] Dorner JW., Cole RJ., Wicklow DT. Aflatoxin Reduction in Maize through Field Application of Competitive Fungi. Journal of Food Protection 1999; 62 650–656.

[3] CAST. Aflatoxins and Other Mycotoxins: An Agricultural Perspective CAST Report No. 80; Council for Agricultural Science and Technology: Ames, IA; 1979.

[4] Diener UL., Cole RJ., Sanders TH., Payne GA., Lee LS. Epidemiology of Aflatoxin Formation by *Aspergillus flavus*. Annual Review of Phytopathology 1987; 25 249–270.

[5] Haumann F. Eradicating Mycotoxins in Food and Feeds. Inform 1995; 6 248–256.

[6] Gong YY., Cardwell K., Hounsa A., Egal S., Turner PC., Hall AJ.,Wild CP. Dietary Aflatoxin Exposure and Impaired Growth in Young Children from Benin and Togo: cross sectional study. British Medical Journal 2002; 325 20–21.

[7] Probst C., Njapau H., Cotty PJ. Outbreak of an Acute Aflatoxicosis in Kenya in 2004: Identification of the Causal Agent. Applied and Environmental Microbiology 2007; 73 2762–2764.

[8] Gardner CAC., Darrah LL., Zuber MS., Wallin JR. Genetic Control of Aflatoxin Production in Maize. Plant Disease 1987;71 426–429.

[9] King SB., Scott GE. Screening Maize Single Crosses for Resistance to Preharvest Infection of Kernels by *Aspergillus flavus*. Phytopathology 1982;72 942.

[10] Widstrom NW., McMillian WW., Wilson DM. Segregation for Resistance to Aflatoxin Contamination Among Seeds on an Ear of Hybrid Maize. Crop Science 1987;27 961–963.

[11] Scott GE., Zummo N. Sources of Resistance in Maize to Kernel Infection by *Aspergillus flavus* in the Field. Crop Science 1988;28 505–507.

[12] Campbell KW., White DG. (1995). Evaluation of Maize Genotypes for Resistance to Aspergillus Ear Rot, Kernel Infection, and Aflatoxin Production. Plant Disease 1995;79 1039–1045.

[13] Brown RL., Cotty PJ., Cleveland TE., Widstrom NW. (1993). Living Maize Embryo Influences Accumulation of Aflatoxin in Maize Kernels. Journal of Food Protection 1993; 56 967–971.

[14] Brown RL., Cleveland TE., Payne GA., Woloshuk CP., Campbell KW., White DG. Determination of Resistance to Aflatoxin Production in Maize Kernels and Detection of Fungal Colonization Using an *Aspergillus flavus* Transformant Expressing *Escherichia coli* β-glucuronidase. Phytopathology 1995; 85 983–989.

[15] King SB., Wallin JR. Methods for screening corn for resistance to kernel infection and aflatoxin production by Aspergillus flavus. In:Diener, UL., Asquith RL., Dickens JW. (eds) Aflatoxin and *Aspergillus flavus* in Corn (Southern Cooperative Bulletin 279). Auburn University, Alabama, 1983; p77.

[16] Tucker DH Jr., Trevathan LE., King SB., Scott GE. Effect of Four Inoculation Techniques on Infection and Aflatoxin Concentration of Resistant and Susceptible Corn Hybrids Inoculated with *Aspergillus flavus*. Phytopathology 1986; 76 290-293.

[17] Payne GA. Aflatoxin in Maize. Critical Reviews in Plant Science 1992; 10 423-440.

[18] Shotwell O. Aflatoxin detection and determination in corn. In: Diener UL.,Asquith RL., Dickens JW. (eds) Aflatoxin and *Aspergillus flavus* in Corn (Southern Cooperative Bulletin 279)., Auburn University, Alabama, 1983; p38.

[19] Windham GL, Williams WP. *Aspergillus flavus* Infection and Aflatoxin Accumulation in Resistant and Susceptible Maize Hybrids. Plant Disease 1998; 82 281-284.

[20] White DG., Rocheford TR., Kaufman B., Hamblin AM. Chromosome regions associated with resistance to *Aspergillus flavus* and inhibition of aflatoxin production in maize. In: Proceedings of USDA-ARS Aflatoxin Elimination Workshop, Atlanta, GA, USA, 1995; p.8.

[21] White DG., Rocheford TR., Naidoo G., Paul C., Hamblin AM. Inheritance of molecular markers associated with, and breeding for resistance to Aspergillus Ear Rot and aflatoxin production in corn using Tex6. In: Proceedings of USDA-ARS Aflatoxin Elimination Workshop, St. Louis, MO, USA, 1998; pp. 4–6.

[22] Paul C., Naidoo G., Forbes A., Mikkilineni V., White D., Rocheford T. Quantitative Trait Loci for Low Aflatoxin Production in Two Related Maize Populations. Theoretical and Applied Genetics 2003; 7 263–270.

[23] Busboom KN. White DG. Inheritance of Resistance to Aflatoxin Production and Aspergillus Ear Rot of Corn from the Cross of Inbreds B73 and Oh516. Phytopathology 2004; 94 1107–1115.

[24] Brooks TD., Williams WP., Windham GL., Wilcox MC., Abbas H. Quantitative Trait Loci Contributing Resistance to Aflatoxin Accumulation in Maize Inbred Mp313E. Crop Science 2005; 45 171–174.

[25] Davis GL., Windham GL., Williams WP. QTL for Aflatoxin Reduction in Maize. Maize Genetics Conference Abstracts 2000; 41 139.

[26] Warburton ML., Brooks TD., Krakowsky MD., Shan X., Windham GL., Williams WP. Identification and Mapping of New Sources of Resistance to Aflatoxin Accumulation in Maize. Crop Science 2009; 49 1403–1408.

[27] Warburton ML., Brooks TD., Windham GL., Williams WP. Identification of Novel QTL Contributing Resistance to Aflatoxin Accumulation in Maize. Molecular Breeding 2011; 27 491-499.

[28] Brown RL., Chen Z-Y., Luo M., Menkir A., Fakhoury A., Bhatnagar D. Discovery and Characterization of Proteins Associated with Aflatoxin-Resistance: Building a Case for Their Potential as Breeding Markers. Toxins 2010 2(4) 919-933.

[29] Chen Z-Y., Brown R L., Menkir A., Cleveland TE. Identification of Resistance-Associated Proteins in Closely-Rrelated Maize Lines Varying in Aflatoxin Accumulation. Molecular Breeding 2012; 30 53-68.

[30] Guo BZ., Russin JS., Cleveland TE., Brown RL., Widstrom NW. Wax and Cutin Layers in Maize Kernels Associated with Resistance to Aflatoxin Production by *Aspergillus flavus*. Journal of Food Protection 1995; 58 296-300.

[31] Gembeh SV., Brown RL., Grimm C., Cleveland TE. Identification of Chemical Components of Corn Kernel Pericarp Wax Associated with Resistance to *Aspergillus flavus* Infection and Aflatoxin Production. Journal of Agricultural and Food Chemistry 2001; 49 4635-4641.

[32] Russin JS., Guo BZ., Tubajika KM., Brown RL., Cleveland TE., Widstrom NW. Comparison of Kernel Wax from Corn Genotypes Resistant or Susceptible to *Aspergillus flavus*. Phytopathology 1997; 87 529-533.

[33] Brown RL., Cleveland TE., Payne GA., Woloshuk CP., White DG. Growth of an *Aspergillus flavus* Transformant Expressing *Escherichia coli* ß- glucuronidase in Maize Kernels Resistant to Aflatoxin Production. Journal of Food Protection 1997; 60 84-87.

[34] Brown RL., Brown-Jenco CS., Bhatnagar D., Payne GA. Construction and preliminary evaluation of an *Aspergillus flavus* reporter gene construct as a potential tool for screening aflatoxin resistance. Journal of Food Protection 2003a; 66 1927-1931.

[35] Brown-Jenco C S., Brown R L., Bhatnagar D., Payne G A. 1998, Use of an omtA(p)::GUS reporter construct to evaluate corn for resistance to aflatoxin accumulation. Proceedings of the USDA-ARS Aflatoxin Elimination Workshop, St. Louis, 1998; pp. 23.

[36] Payne GA. 1997. Characterization of inhibitors from corn seeds and the use of a new reporter construct to select corn genotypes resistant to aflatoxin accumulation. Proceedings of the USDA-ARS Aflatoxin Elimination Workshop, Memphis, TN, 1997; pp 66-67.

[37] Brown RL., Cleveland T E., Woloshuk CP., Payne GA., Bhatnagar D. Growth Inhibition of a *Fusarium verticillioides* GUS Strain in Corn Kernels of Aflatoxin -Resistant Genotypes. Applied Microbiology and Biotechnology 2001a; 57 708-711.

[38] Rajasekaran K., Cary JW., Cotty PJ., Cleveland TE. Development of a GFP-Expressing *Aspergillus flavus* Strain to Study Fungal Invasion, Colonization, and Resistance in Cottonseed. Mycopathologia 2008; 165 89-97.

[39] Guo BZ., Brown RL., Lax AR., Cleveland TE., Russin JS., Widstrom NW. Protein Profiles and Antifungal Activities of Kernel Extracts from Maize Genotypes Resistant and Susceptible to *Aspergillus flavus*. Journal of Food Protection 1998; 61 98–102.

[40] Guo BZ., Chen Z-Y., Brown RL., Lax AR., Cleveland TE., Russin JS., Mehta A.D., Selitrennikoff CP., Widstrom NW. Germination Induces Accumulation of Specific Proteins and Antifungal Activities in Maize Kernels. Phytopathology 1997; 87 1174–1178.

[41] Huang Z., White DG., Payne GA. Maize Seed Proteins Inhibitory to *Aspergillus flavus* and Aflatoxin Biosynthesis. Phytopathology 1997; 87 622–627.

[42] Ji C., Norton RA., Wicklow DT., Dowd PE. Isoform Patterns of Chitinase and β-1,3-glucanase in Maturing Maize Kernels (*Zea mays* L.) Associated with *Aspergillus flavus* Milk Stage Infection. Journal of Agricultural and Food Chemistry 2000; 48 507–511.

[43] Chen Z-Y., Brown RL., Lax AR., Guo BZ., Cleveland TE., Russin JS. Resistance to *Aspergillus flavus* in Maize Kernels is Associated with a 14 kDa Protein. Phytopathology 1998; 88 276–281.

[44] Chen Z-Y., Brown RL., Russin JS., Lax AR., Cleveland TE (1999b). A Maize Trypsin Inhibitor with Antifungal Activity Inhibits *Aspergillus flavus* α-amylase. Phytopathology 1996b; 89 902–907.

[45] Woloshuk CP., Cavaletto JR., Cleveland TE. Inducers of Aflatoxin Biosynthesis from Colonized Maize Kernels are Generated by an Amylase Activity from *Aspergillus flavus*. Phytopathology 1997; 87 164–169.

[46] Chen Z-Y., Brown RL., Lax AR., Cleveland TE., Russin JS. Inhibition Plant Pathogenic Fungi by a Maize Trypsin Inhibitor Over-Expressed in *Escherichia coli*. Applied and Environmental Microbiology 1999a; 65 1320–1324.

[47] Chen ZY., Brown RL., Cleveland TE., Damann KE., Russin JS. Comparison of Constitutive and Inducible Maize Kernel Proteins of Genotypes Resistant or Susceptible to Aflatoxin Production. Journal of Food Protection 2001; 64 1785–1792.

[48] Gorg A., Postel W., Gunther S., Weser J., Improved Horizontal Two-Dimensional Electrophoresis with Hybrid Isoelectric Focusing in Immobilized pH Gradients in the First Dimension and Laying-On Transfer to the Second Dimension. Electrophoresis 1985; 6 599-604.

[49] Gorg A., Postel W., Gunther S.,The Current State of Two-Dimensional Electrophoresis with Immobilized pH Gradients. Electrophoresis 1988; 9 531-546.

[50] Pennington SR., Wilkins MR., Hochstrasser DF., Dunn MJ. Proteome Analysis: From Protein Characterization to Biological Function. Trends in Cell Biology 1997; 7 168-173.

[51] Wilkins MR., Pasquali C., Appel RD., Ou K., Golaz O., Sanchez JC., Yan JX., Gooley AA., Hughes G., Humphrey-Smith I., Williams KL., Hochstrasser DF. From Proteins to Proteomes: Large-Scale Protein Identification by Two-Dimensional Electrophoresis and Amino Acid Analysis. Biotechnology 1996; 14 61-65.

[52] Chen Z-Y., Brown RL., Damann KE., Cleveland TE. 2000. Proteomics analysis of kernel embryo and endosperm proteins of corn genotypes resistant or susceptible to *As-*

pergillus flavus infection. In: Proceedings of the USDA-ARS Aflatoxin Elimination Workshop, 2000; pp. 88.

[53] Chen Z-Y., Brown R.L., Damann K.E., Cleveland T.E. Identification of Unique or Elevated Levels of Kernel Proteins in Aflatoxin-Resistant Maize Genotypes through Proteome Analysis. Phytopathology 2002; 92 1084–1094.

[54] Brown RL., Chen Z-Y., Menkir A., Cleveland TE. Using biotechnology to enhance host resistance to aflatoxin contamination of corn. African Journal of Biotechnology 2003b; 2 557-562.

[55] Payne GA. Process of Contamination by Aflatoxin-Producing Fungi and Their Impact on Crops. In: Sinha KK., Bhatnagar D. (eds), Mycotoxins in Agriculture and Food Safety. Marcel Dekker, New York, NY, USA, 1998; pp. 279-306.

[56] Thomann E.B., Sollinger J., White C., Rivin C.J. Accumulation of Group 3 Late Embryogenesis Abundant Proteins in *Zea mays* Embryos. Plant Physiology 1992; 99 607–614.

[57] Xu D., Duan X., Wang B., Hong B., Ho THD., Wu R. Expression of a Late Embryogenesis Abundant Protein Gene HVA1, from Barley Confers Tolerance to Water Deficit and Salt Stress in Transgenic Rice. Plant Physiology 1996; 110 249–257.

[58] Veena V., Reddy S., Sopory SK. Glyoxalase I from Brassica juncea: Molecular Cloning, Regulation and Its Over-expression Confer Tolerance in Transgenic Tobacco under Stress. Plant Journal 1999; 17 385–395.

[59] Chen ZY., Brown RL., Damann KE., Cleveland TE. Identification of a Maize Kernel Stress-Related Protein and Its Effect on Aflatoxin Accumulation. Phytopathology 2004; 94 938–945.

[60] Chen ZY., Brown RL., Damann KE., Cleveland TE. Identification of Maize Kernel Endosperm Proteins Associated with Resistance to Aflatoxin Contamination by *Aspergillus flavus*. Phytopathology 2007; 97 1094–1103.

[61] Chen Z.Y., Brown RL., Rajasekaran K., Damann KE., Cleveland TE. Evidence for Involvement of a Pathogenesis-Related Protein in Maize Resistance to *Aspergillus flavus* Infection/Aflatoxin Production. Phytopathology 2006; 96 87–95.

[62] Xie YR., Chen Z-Y., Brown RL., Bhatnagar D. Expression and Functional Characterization of Two Pathogenesis-Related Protein10 Genes from *Zea mays*. Journal of Plant Physiology 2010; 167 121–130.

[63] Steiner-Lange S., Fischer A., Boettcher A., Rouhara I., Liedgens H., Schmelzer E., Knogge W. Differential Defense Reactions in Leaf Tissues of Barley in Response to infection by *Rhynchosporium secalis* and to Treatment with a Fungal Avirulence Gene Product. Molecular and Plant-Microbe Interactions 2003; 16 893–902.

[64] Mould M.J., Xu T., Barbara M., Iscove NN., Heath MC. cDNAs Generated from Individual Eepidermal Cells Reveal that Differential Gene Expression Predicting Subse-

quent Resistance or Susceptibility to Rust Fungal Infection Occurs Prior to the Fungus Entering the Cell Lumen. Molecular and Plant-Microbe Interactions 2003; 16 835–845.

[65] McGee JD., Hamer JE., Hodges TK. Characterization of a PR-10 Pathogenesis-Related Gene Family Induced in Rice during Infection with *Magnaporthe grisea*. Molecular and Plant-Microbe Interactions 2001; 14 877–886.

[66] Chen Z-Y., Brown RL., Damann KE., Cleveland TE. PR10 Expression in Maize and Its Effect on Host Resistance against *Aspergillus flavusI*infection/Aflatoxin Production. Molecular Plant Pathology 2010; 11 69–81.

[67] Wesley SV., Helliwell CA., Smith NA., Wang MB., Rouse DT., Liu Q., Gooding PS., Singh SP., Abbott D., Stoutjesdijk PA. Robinson SP., Gleave AP., Green AG., Waterhouse PM. Construct design for efficient, effective and high-throughput gene silencing in plants. Plant Journal 2001; 27 581–590.

[68] Fire A. Xu S., Montgomery MK., Kostas SA., Driver SE., Mello CC. Potent and Specific Genetic Interference by Double Stranded RNA in *Caenorhabditis elegans*. Nature 1998; 391 806–811.

[69] Gura T. A silence that speaks volumes. Nature 2000; 404 804–808.

[70] Chen Z-Y., Damann KE. Louisiana State University, Baton Rouge, LA; Brown RL., Cleveland TE. USDA-ARS-SRRC New Orleans, LA., Unpublished work. 2007.

[71] Baker R., Brown RL., Cleveland TE., Chen Z-Y., Fakhoury A. COR, a Maize Lectin-like Protein with Antifungal Activity against *Aspergillus flavus*. Journal of Food Protection 2009a; 72 120–127.

[72] Baker R., Brown RL., Cleveland TE., Chen Z-Y., Fakhoury A. ZmTI, a Maize Trypsin Inhibitor with Limited Activity against *Aspergillus flavus*. Journal of Food Protection 2009b; 72 85–188.

[73] Luo M., Brown R L., Chen Z-Y., Menkir A., Yu J., Bhatnagar D. Transcriptional Profiles Uncover *Aspergillus flavus* Induced Resistance in Maize Kernels. Toxins 2011; 3 766-785.

[74] Luo M., Liu J., Lee RD., Scully BT., Guo B. Monitoring the Expression of Maize Genes in Developing Kernels under Drought Stress Using Oligo-Microarray. Journal of Integrative Plant Biology 2010; 52(12) 1059-1074.

[75] Pechanova O., Pechan T., Williams, WP., Luthe DS. Proteomic Analysis of Maize Rachis: Potential Roles of Constitutive and Induced Proteins in Resistance to Aspergillus flavus Infection and Aflatoxin Contamination. Proteomics 2011; 11 114-127.

[76] Peethambaran B., Hawkins L., Windham GL., Williams WP., Luthe DS. Antifungal Activity of Maize Silk Proteins and Role of Chitinases in *Aspergillus flavus* Resistance. Toxin Reviews 2010; 29 27–39.

[77] Brown RL., Chen Z-Y., Menkir A., Cleveland T.E., Cardwell K., Kling, J., White D.G. Resistance to Aflatoxin Accumulation in Kernels of Maize Inbreds Selected for Ear Rot Resistance in West and Central Africa. Journal of Food Protection 2001b; 64 396–400.

[78] Menkir A., Brown RL., Bandyopadhyay R.., Chen Z-Y., Cleveland TE. A USA-Africa Collaborative Strategy for Identifying, Characterizing, and Developing Maize Germplasm with Resistance to Aflatoxin Contamination. Mycopathologia 2006; 162 225–232.

[79] Chen ZY., Brown RL., Menkir A., Damann KE., Cleveland TE. Proteome Analysis of Near Isogenic Maize Lines Differing in the Level of Resistance against *Aspergillus flavus* Infection/Aflatoxin Production. Phytopathology 2005; 95 S19.

[80] Menkir A., Brown RL., Bandyopadhyay R., Cleveland TE. Registration of Six Tropical Maize Germplasm Lines with Resistance to Aflatoxin Contamination. Journal of Plant Registrations 2008; 2 246–250.

[81] Guo BZ., Krakowsky MD., Ni X., Scully BT., Lee RD., Coy AE., Widstrom NW. Registration of Maize Inbred Line GT603. Journal of Plant Registrations 2011; 5(2) 211-214.

[82] Williams WP.,Windham GL. Registration of Mp718 and Mp719 Germplasm Lines of Maize. Journal of Plant Registrations 2012; 6 200-202.

[83] Scully BT., Guo BZ., Ni X., Williams WP., Henry WB., Krakowsky MD., Brown RL. Development of Aflatoxin and Insect Resistant Corn Inbreds Adapted to the Southern U.S. Proceedings of the Corn Utilization and Technology Workshop, Abstract (in press).

Chapter 3

A New Approach in Aflatoxin Management in Africa: Targeting Aflatoxin/Sterigmatocystin Biosynthesis in Aspergillus Species by RNA Silencing Technique

Amos Emitati Alakonya and Ethel Oranga Monda

Additional information is available at the end of the chapter

1. Introduction

Africa is faced with the challenge of merging its food crop production with its ever-increasing population in order to ensure food security of its people. The effort to meet Africa's food demand is however hampered by drought, crop diseases, insect pests, suitable storage facilities for various agricultural products, markets, lack of fertilizers, flooding, suitable seeds for various agro-ecological zones and poor rural infrastructure. The most limiting aspect and also ahealth concern is infestation of grains by fungal pathogens that also produce toxic fungal metabolites called mycotoxins [77, 25]. Though the fungi produce various mycotoxins, aflatoxins are a major concern in Africa [77]. This is partly because of the conducive weather for their accumulation in Africa (wet and humid climates and dry regions), their lethality on ingestion and widespread occurrence in maize (Zea mays) a main stable food crop grown in Africa by small-scale farmers for local consumption [28, 7].What this implies is that the main fungal genera and mycotoxin contaminant of maize in Africa is therefore Aspergillus species and aflatoxins respectively. Aspergillus species and aflatoxins not only attract worldwide attention but also are of great significance in Africa due to their negative impact on yield, human health, animal productivity and trade [54, 7, 77, 79, 28]. To exacerbate the problem, Sub Sahara Africa (SSA) experiences high temperatures and high relative humidity that predisposes many crops to fungal pathogens. In addition, majority of farmers in Africa are small scale hence rely on the consumption of homegrown crops. Therefore, irrespective of the quality considerations normally applied by some African governments to control aflatoxin contamination in food supply, aflatoxicoses will frequently occur in the continent.

The high temperatures and high relative humidity predisposes many crops to fungal and other pathogens. There is a significant correlation in aflatoxin levels in products after long storage in Agro-ecological zones with wet and humid climates and dry regions [28]. Maize is a staple food throughout the African continent but is highly colonized by Aspergillus species that produce aflatoxins [7] and the fungal contamination is of great concern. Peanuts (Arachishypogaea) are also grown in many African countries by small-scale farmers for local consumption and also export if food safety regulations are followed. Aflatoxin in peanuts seeds hamper international trade and also adversely affects health of consumers [54]. There should be reduction in food losses and maintenance of food quality. Due to malnutrition, there are approximately over 5million deaths in children under the age of 5 years in developing countries every year and aflatoxin contamination is suspected to be a factor in infant under-nutrition [38]. Some of the factors that contribute to aflatoxin contamination include; contact of product with soil during drying, high kernel moisture during storage, time of harvest [33]. Aflatoxins are mainly classified in B1 B2 G1 G2 M1 M2 based on chromatographic and fluorescent characteristics [42]. They occur in maize and other cereal crops, peanuts, cotton and oil seed crops. When Dairy cattle feed on commodities contaminated with Aflatoxin B1, the toxin is excreted in milk as aflatoxin M1 and can cause DNA damage, gene mutation and chromosomal abnormalities. Aflatoxins particularly B1 is confirmed a potential carcinogen [32]. In Kenya Aflatoxin M1 has been reported in milk [37] and in Gambia, Aflatoxin M1 has been detected in breast milk [87]. This leads to maternal exposure of aflatoxin M1 in breast milk to young children.

The failure of aflatoxin regulatory systems is therefore partly due to existing weather conditions, poor harvesting, transportation, marketing and processing conditions that favour proliferation of aflatoxin producing fungi [6, 62, 30, 4, 28, 67]. In addition, the Aspergillusspp have multiple infection courts that include;i) Mycelial growth on the silk kernels and cobs, ii) Kernel wounds created by insects and/or birds, iii) Soil debris and iv) Infected seed which predispose future maize crops to infection which makes it even harder to control [66, 50, 29]. This review summarises the current work on aflatoxins and their management in Africa. Furthermore it presents an argument based on the current knowledge on host and parasite macro and micromoleculartrafficking that suggeststhe possibility to circumvent the aflatoxin problem by use of cross species RNA interference. The aim is to arm maize with molecules that would shut down the aflatoxin biosynthesis upon infection with toxigenic fungi hence thwarting aflatoxin accumulation.

2. Health effects associated with aflatoxins

Aspergillusflavus and Aspergillusparasiticus are of great concern due to production of aflatoxins and millions of people in Africa are chronically exposed to aflatoxins due to feeding on contaminated food. The aflatoxin problem is most serious in tropical and subtropical countries due to favorable climatic conditions for Aspergillusflavus and Aspergillusparasiticus. Human and animals are exposed to aflatoxin through diet [16, 7]. Animal feed is of concern due to contaminated animal feeds. It is estimated that about 25,200 – 155,000 people worldwide have aflatoxin induced liver cancer. Of this population 40% occur in Africa [45]. There are economic losses that result from contamination of crops and animal feeds with aflatoxins and also public health problems that result from ingestion of products contaminated with aflatoxins [54, 7]. In many developed countries, there are stringent government regulations on aflatoxins than any other mycotoxins with very low threshold for tolerance [20]. Maximum limit of contamination with aflatoxin in peanuts in Brazil and USA is 20μg/kg while Canada and European Union have imposed a limit of 15μg/kg [24]. For animal feeds, European Commission has maximum level for aflatoxins in animal feeds at 0.02mg/kg [21]. A number of African countries still have to put in place regulatory mechanisms for aflatoxins. However, Kenya's limit for aflatoxin in products for human consumption is 20ppb [39]. The two general forms of effects of aflatoxins are acute and chronic toxicity.

a. Acute toxicity is caused by ingestion of large amount of aflatoxins from heavily contaminated food. This causes decreased liver function and could lead to blood clotting mechanism, jaundice, a decrease in serum proteins that are synthesized by the liver, edema, abdominal pain, vomiting and death of affected person. This was the case in Kenya in 2004 where they were 317 cases and 125 deaths reported due to consumption of maize contaminated with aflatoxins [17, 58] identified the S strain of Aspergillusflavus as the causal agent of the outbreak. Epidemiological, clinical and experimental studies have indicated that exposure to large doses of aflatoxin causes acute toxicity but exposure to small doses for prolonged periods of time is carcinogenic. The liver is adversely affected by aflatoxins that cause necrosis of liver cells and death [15].

b. Chronic toxicity is due to long time exposure to low aflatoxin concentration. The main symptoms are decreased growth rate that leads to stunted growth [26]. In Togo and Benin, children who are underweight as a result of aflatoxins are also at higher risk for infections and diarrhea [26]. Aflatoxin-albumin adducts (32.8pg/mg) were detected in 99% of children between 9 months – 5 years. Exposure to aflatoxin in children increases at weaning and this contributes to reduced growth [26]. Exposureof children to aflatoxin can be through contaminated milk containing Aflatoxin M1 that is a metabolite of AFB1. In domestic animals,aflatoxins cause lowered milk or egg production and immune suppression that is caused by reactivity of aflatoxin with T-cell and a decrease in vitamin K activities including decrease in phagocytic in macrophages. [61]. It has been reported that there is a high risk among people with Hepatitis B and Hepatitis C carriers to develop cancer due to consumption of food contaminated with aflatoxins [75]. Aflatoxins have also been linked to immune suppression [70] and higher prevalence of hepatocellular cancer has been reported in Africa [68].

3. Management strategies against aflatoxins

Aspergillus infection increase with high temperature, high humidity, insect damage and nitrogen deficiency. Temperature and humidity are therefore important in aflatoxin management. A. flavus and A. parasiticus are unable to grow or produce aflatoxin at water activity of less than 0.7 (relative humidity below 70% or temperature below 100C, however under stress condition such as drought, aflatoxin contamination can be higher [17]. Various strategies have been suggested in management of aflatoxins. The strategies should adhere to the following: a) aflatoxin must be transformed to non-toxic products, b) fungal spores and mycelia should be destroyed to prevent formation of new toxins, c) the food or feed material should retain its nutritive value and palatability, d) the physical properties of raw material should not change significantly d) it must be cost efficient [5, 16, 64].

The Physical and chemical treatment of contaminated commodities include detoxification of aflatoxins using physical means such as removal of contaminated commodities or inactivation of the toxin in the commodity. These methods include mechanical sorting and separation, washing, density segregation, solvent extraction, irradiationand oxidation [5]. However,efficiency of these techniques will depend on level of contamination. Furthermore, results obtained are often uncertain and relatively costly and could remove or destroy essential nutrients in feed [41]. Also some of the methods have disadvantages such as nutritional loss, toxic, limited efficiency and high cost therefore limiting practical application. Various natural and synthetic agents could prevent growth of toxigenic fungi and formation of mycotoxins and these have been reviewed by Mahoneyet al. [47]. Chemical methods of deactivating mycotoxins in feeds and also clay products that could be used in deactivating mycotoxins have been extensively been reviewed by Kolosova and Stroka [41]. Management strategies can be divided into Pre-harvest and Post-harvest strategies.

4. Pre- harvest strategies

These include;

a. Good agricultural practices (GAP) that involve adequate fertilizer application and crop rotation with non-host.

b. Management of insect pests that predispose crops to fungal infection through availability of infection channels such as wounds and other entry points.

c. Optimal harvest time so that crops are not left in the field exposed to environmental factors that predispose crops to pathogen infection. Harvesting immediately after physiological maturity is recommended since aflatoxin level can increase with delayed harvest interval [35].

d. Suitable management of crop residues as they harbor pathogens that are able to survive saprophytically [5].

e. Management with fungicides has challenges due to environmental pollution and also emergence of resistant pathogen populations and also chemical residue in food products. Of fundamental valueare environmentally friendly strategies. Polysaccharides and glycoproteins particularly β-glucans from basidiomycetes Lentinulaedodes (edible mushroom) is known to promote health effects in animals and human and have ability to inhibit aflatoxin biosynthesis by stimulating the antioxidant defence of the toxigenic fungus. Oxidative stress induced using paraquatenhanced the expression of β-glucan synthase gene and stimulated effect of β-glucans production that leads to a higher aflatoxin inhibiting capacity. Efficient inhibition could be due to higher content of β-glucans [60]. Utilization of microorganisms or their enzymatic metabolites to detoxify mycotoxins in food and feed has advantages such as mild reaction conditions, target specificity, efficiency and environmental friendly.

f. Resistant hybrids could be very promising but commercial hybrids are not always available [1]. However, availability of resistant varieties is the best solution for farmers so long as they are available and affordable. Some high yielding yellow maize varieties with good resistance to Aspergillus have been identified. This includes AO901-25 that has a grain yield of 7115kg/h and low aflatoxin level (IITA). However there is still a lot to be done in order to consider consumer prevalence as most people in Africa have prevalence to white maize. Furthermore, reduction in aflatoxin level is still required. Menkiret al. [51] registered tropical maize germplasm with resistance to aflatoxins. These varieties have been distributed to National programs for the development of locally adapted hybrids.

g. Biological control is use of one microorganism to control another microorganism such as Pseudomonas strains [55]. It has been noted that Aspergillusflavus strains differ in aflatoxin production and this influences crop contamination. There are strains that produce a lot of aflatoxins and also produce numerous small sclerotia (<400μm). These are the 'S' strains (toxigenic strain). Another strain the 'L' strain produces low aflatoxin levels and a few large sclerotia that are about >400μm and are atoxigenic [18]. There is competitive exclusion when one strain competes to exclude another in the environment. This implies that a shift of strain profile from toxigenic to atoxigenic is a viable biological control strategy. Atoxigenic strains of A. flavus from Nigeria have been combined as a bio-control product and registered as AflaSafe. It is used on sorghum as a carrier at the rate of 10kg/ha applied 2-3 weeks before flowering. Native strains have been identified and are being used in African countries. In Diourbel (Senegal) peanuts treated with AflaSafe had aflatoxin level of 1.9ng/g while control had 29.7ng/g giving a reduction in aflatoxin level of 93%. In Ibadan (Nigeria), crops in treated plots had Aflatoxin level of 11ppb while control had 42ppb giving 73% in reduction of aflatoxin. Stored products had 105ppb in treated samples while untreated samples had 2408ppb giving a reduction of 96% in Aflatoxin level [7, 18]. Due to good performance of atoxigenic strains, peanut producers in Senegal and Gambia are willing to adopt competitive exclusion technology for aflatoxin control in peanuts.

5. Post-harvest management

Reduction of moisture in grains is very important. There are a number of technologies that could be used to dry maize fast. Such technologies have extensively been reviewed by Lutfyet al., [46]. These technologies are expensive and most African farmers may not be able to acquire them. However some of the post-harvest strategies that could be used in Africa include the following: Rapid and proper drying of maize to moisture level of 13% or below. This will halt growth of fungi in the product. Products stored with high moisture increase growth of fungi in the stored product and this leads to increase of aflatoxin in the product [27]. Post-harvest insect control can prevent damage to maize. Clays such as Novasil could bind to aflatoxin in animal feeds [36]. Other control strategies have been reviewed by Kerstin and Mutegi, [40]. Quality management systems for Hazard Analysis Critical Control Point (HACCP) should be employed for management of mycotoxins [65).

6. Cost effectiveness of aflatoxin reduction strategy in Africa

It is important to consider economic impacts of food contaminants such as aflatoxins as it imposes enormous socio-economic cost to human society. Wu and Khlangwiset [80] analyzed two potential aflatoxin control strategies in Africa, 1) pre-harvest control using atoxigenic strains of Aspergillusflavus competitively to exclude toxigenic strains in maize and 2) post-harvest intervention in a package to reduce aflatoxin contamination in peanuts in Guinea. Health benefit was gained from each intervention in terms of fewer aflatoxin-induced cases compared to cost of implementing the intervention. Both interventions were found to be cost-effective if applied widely in Africa. The monetary value of life saved and quality of life gained by reducing aflatoxin induced hepatocellular carcinoma exceeds the cost of either bio-control or post-harvest intervention package. The estimated cost-effectiveness ratio (CER: gross domestic product multiplied by disability adjusted life years saved per unit cost) for bio-control in Nigerian maize ranged from 5.10 - 24.8 while estimated CER for post-harvest intervention package in Guinea peanut ranged from 0.21 - 2.08. Any intervention with a CER >1 is considered by world Health Organization (WHO) to be very cost effective while intervention with CER > 0.33 is considered cost effective [80]. The way forward with toxigenic strains of Aspergillusflavusis therefore:-

1. Each African country should identify local non-toxigenic strains and develop a package for legal registration for use in aflatoxin management and develop capacity for manufacturing the strains.
2. There should be extensive awareness programmes in each country since some African countries exchange agricultural products across the border without strict control. Awareness of aflatoxin problem and management strategies should be extended to Medical Practitioners, religious leaders, herbalists and Private Sector.
3. Efficacy of non-toxigenic strains should be demonstrated through farmers Schools, Non-Governmental Organizations (NGO), extension staff, outreach programmes and Women groups involved in agricultural services. This will enhance adoption by farmers.
4. Government should provide incentives to resource poor farmers to access non-toxigenic strains that should be available in small packages.

5. There should be surveillance of aflatoxin testing in food and feed products
6. Government officials should be sensitized on aflatoxins and advantages of using local non-toxigenic strains. This will assist in formulation of appropriate policies.

7. Current status of aflatoxins in Africa.

In 2010 the level of aflatoxin in maize stored by farmers in Kenya were found to be 1776ppb while in the markets the concentration was 1632ppb [49]. These levels are likely to cause acute toxicity if contaminated products are consumed. In 2011, 40% of samples that were taken from farmers' fields in Eastern and Western Kenya were found with aflatoxin level of >10ppb. In Mali between 2009- 2010 aflatoxin level in peanuts were found to be >10ppb in 35-61 % of samples from farmers' fields and 39-91% samples from farmers stores [73]. Peanut paste in Mali had high aflatoxin level of >300ppb. Apparently the levels of aflatoxins in West Africa have been quite high. Maize in Benin had 4,000ng/g, In Ghana aflatoxin level in peanuts was reported to be 216ng/g while peanut paste had 3,278ng/g and peanut sauce 943ng/g, cashew paste, 366ng/g. In Nigeria Peanut oil had 500ng/g while yam flour had 7600ng/g [7]. This an indication that Ghana urgently needs intervention strategies to mitigate the aflatoxin challenges. In Kenya,aflatoxin M1 has been reported in milk [37]. There have been re-occurrence of outbreaks of acute aflatoxicoses in Eastern province that causes various deaths [57, 58].The S strain morphotype of A.flavus was identified as the cause of aflatoxicoses in 2004 and 2006 [57]. Apparently the high incidence of S strain of A. flavushighly correlated with acute aflatoxicosis in Eastern region of Kenya [56, 58, 57]. A simple test for Aflatoxin in maize kernels is the Bright greenish-yellow fluorescence (BGYF) or the black light test. Kernels are viewed under UV lamp (365 nm) for characteristic BGYF. This indicates a possible presence of aflatoxin producing fungi or mycotoxin itself [84] Laboratories in Africa should be able to perform these tests during surveillance survey.

8. RNA interference Strategy and its mechanisms

RNA interference (RNAi) refers to post-transcriptional gene silencing mediated by either degradation or translation arrest. This mechanism was first discovered in plants where transgene and viral RNAs guide DNA methylation [74, 34, 52]. The process is a naturally occurring biological process that is highly conserved among multicellular organisms including plants. The process is mediated by small interfering RNAs (siRNAs) that are produced from long dsRNA of exogenous or endogenous origin by an endonuclease (an enzyme) called a dicer. The resulting siRNAs are about 21-24 nucleotides long with 2 nucleotide single stranded 3′ end overhangs on each strand. The siRNAs are then incorporated into a nuclease complex called the RNA-induced silencing complex (RISC), which then targets and cleaves mRNA that is complementary to the siRNA [86].

In plants, RNAi plays a role in cellular defense, protecting the cell from inappropriate expression of repetitive sequences, transposable elements and viral infections [43]. RNAi has proved to have ability to regulate the expression of genes involved in a variety of cell processes such as proliferation, apoptosis and differentiation [2]. Moreover, it is thought to play a role in protecting the genome against damage caused by transposons [44]. More recently, these findings have been extended by the observations that siRNA-directed DNA methylation in plants is linked to histone modification [89]. In fission yeast, hetero-chromatin formation at centromere boundaries is associated with siRNAs [72].

Application of RNAi in crop improvement has been derived from targeted degradation of gene products with significant homology to the introduced sequence. Successful utilization of RNAi-based resistance effects rely on; (i) identification of a target gene (ii) dsRNA delivery, which includes in plantaexpression of dsRNA and (iii) delivery of sufficient amounts of intact dsRNA. RNAi can therefore be an important tool for crop improvement given that the RNAi signal can be both local (cell-cell) and systemic (spread through vascular system) [8, 71]. RNAi mediated silencing for agricultural traits offers the advantage of transmission across many cells and application in multigene family silencing.

9. Application of RNA interference in management of biotic challenges in agriculture

RNAi against crop parasites that include insects, nematodes, viruses and parasitic plants has been demonstrated [3, 9, 10, 11, Day et al., 1991, 22, 31, 82, 83, 88]. For example, the cotton bollworm (Helicoverpaarmiger a; Lepidoptera) and western corn rootworm (Diabroticavirgiferavirgiferaleconte) where dsRNA directed against a gene encoding V-type ATPase A, demonstrated rapid knockdown of endogenous mRNA within 24 hours of ingestion. In addition, dsRNAs directed against three target genes (β-tubulin, V-ATPase A and V-ATPase E) in western corn rootworm effectively resulted in high larval mortality [9].

Root-knot nematodes (Meloidogynespp) cause significant crop losses in Africa with the most damaging ones being M. incognita, M. javanica, M. arenariaand M. hapla. Since the discovery that RNAi is active in worms through oral uptake of dsRNA [22], intense studies on the control of parasitic nematodes through targeting essential parasite genes have been carried out [31, 82]. Yadav [82] described almost complete resistance to Meloidogynespp infection in transgenic tobacco. Geminiviruses, a major problem on crops in tropical and subtropical countries have been targeted via RNAi [Asadet al., 2003, 10, 11, 19, 83, 88]. With dsRNA and antisense RNA (as RNAi technologies), several regions of the viral genome can be targeted by plants expressing fused viral siRNA or hairpin dsRNA sequences. Multiple targeting of the viral genome provides stable and durable resistance considering that the viral genome is highly recombinogenic [12].

Currently it is understood that transcripts can be trafficked from host to parasitic plants [59]. Therefore when the RNAi transformed host plant is attacked by the parasite, the gene specific RNAi transcripts can be trafficked into the parasitic plant via the haustorial connection leading to gene silencing. Some studies have proposed the targeting of KNOX genes which are vital in plant development while others have suggested targeting genes that code for aquaporins that aid in loosening of the host plant cell wall during parasite infection [63]. This ability to tap into native pathways has yielded crucial breakthrough in parasitic plant management [23].

10. Prospects of applying RNAi in management of Aspergillus species and aflatoxins in grains

Early genetic studies have identifiedaflatoxin biosynthesis to becontrolled by a cluster of aflatoxin and sterigmatocystin gene in an ~70kb region [85, 13, 69, 76]. A critical analysis of the pathway indicates that their exists three enzymes that catalyse the two rate-limiting steps in aflatoxinbiosynthesis. Two enzymes stcJ and stcKcatalyse the first step that involves the conversion of Acetate and Malonyl-CoA into Hexanoyl-CoA. A further critical examination of the pathway identifies another enzyme stcA which catalyses the conversion of Hexanoyl-CoA to Norsolorinic acid. To addstrength to this observation, Brown et al.,[1996] reported that Aspergillus species with mutations in the stcJ and stcK grew normally but could not produce aflatoxin and sterigmatocystin. In the same study addition of Hexanoic acid to growth media restored aflatoxin and sterigmatocystin production. Recent studies have reported the trafficking of molecular cues between hosts and parasites including fungi. Among the molecules are small interfering RNA SiRNA. This targeted downregulation of gene expression bySiRNA has been used to engineer crops against virueses, nematodes and parasitic plants in cross species version. The key steps for this strategy to succeed are;i) identifying a key gene to a process, ii) cloning the target sequence of the gene from the parasite, iii) making an RNAi construct with the target sequence of the parasite in sense and antisense direction separated by an intron so as to allow formation of primary small intereferingRNAs (SiRNA) in host (maize), iv) transforming the host (maize) with the construct tailored for RNAi. In this case the rate-limiting steps in aflatoxin biosynthesis are known to be catalysed bystcJ, stcK and stcA [85,]. The strategy is therefore be to make an RNAi construct containing either combined partial or full sequences of the stcA, stcK and stcJ in sense and antisense orientation and transform it into maize. Upon colonization with aflatoxigenic fungi in the field, the primary SiRNA molecules will then cross from transgenic maize into Aspergillussppfungi through the haustoria connection at infection. The siRNAs will then cleave the stcJ, stcK and stcA mRNAs into 20 to 28bp long double molecules hence downregulating or inhibiting aflatoxin and sterigmatocystin biosynthesis.The transformed aflatoxigenic species in the field will be unable to synthesizeaflatoxins both in field and storage. RNAi will succeed in this case because it can be both local (cell–cell) and systemic (spread through the vascular system), hence all parts of the transgenic plant shall remain armed against aflatoxin biosynthesis. The stcJ, stcK and stcA do not exist in maize hence their silencing will not affect the crop.

11. Conclusion

Mycotoxins especially aflatoxins are believed to have caused harm to mankind since time immemorial. It is now almost 54 years after the discovery of the Turkeys X disease suspected to have been caused by aflatoxin contamination. Several major steps have been made towards the understanding of the aflatoxin biosynthetic pathway and its related genes. This book chapter not only emphasizes such work, but also focuses on Africa where due to the complex social economic dynamics, aflatoxins have greatly impacted negatively on the grain consuming population. This work goes on to describe the biosynthetic control of aflatoxinand further explores how AF/ST pathways could be altered via cross species RNA interference of key steps. If adopted, together with other existing aflatoxin control methods we believe researchers targeting mycotoxinswill realign their efforts in the development of practical methods for preventing not only aflatoxin contamination but alsoall the major mycotoxins in grains and nuts.

Acknowledgements

This work was funded by a grant from the Bill & Melinda Gates Foundation through the Grand Challenges Explorations initiative.

Author details

Amos Emitati Alakonya[1*] and Ethel Oranga Monda[2]

*Address all correspondence to: aealakonya@gmail.com

1 Institute of Biotechnology Research, Jomo Kenyatta University of Agriculture and Technology, P. O. Box 62000-00200, Nairobi

2 Department of Plant and Microbial Sciences, Kenyatta University, P. O Box 43844 -00100, Nairobi

References

[1] Abbas, H. K., Wilkinson, J. R., Zablotowicz, R. M., Accinelli, C., Abel, C. A., Bruns, H. A., & Weaver, M. A. (2009). Ecology of Aspergillusflavus, regulation of aflatoxin production and management strategies to reduce aflatoxin contamination of corn. *Toxin Reviews*, 28, 142-153.

[2] Ambros, V. (2001). MicroRNAs: Tiny regulators with great potential. *Cell*, 107, 823-826.

[3] Asad, S., Haris, W. A., Bashir, A., Zafar, Y., Malik, K. A., Malik, N. N., & Lichtenstein, C. P. (2003). Transgenic tobacco expressing geminiviral RNAs is resistant to the serious viral pathogen causing cotton leaf curl disease. *Archives of Virology*, 148, 2341-2352.

[4] Avantaggio, G., Quaranta, F., Desidero, E., & Visconti, A. (2002). Fumonisin contamination of maize hybrids visibly damaged by Sesamia. *Journal of Science, Food and Agriculture*, 83, 13-18.

[5] Awad, W. A., Ghareeb, K., Bhm, J., & Zentek, J. (2010). Decontamination and detoxification strategies for Fusariummycotoxindeoxynivalenol in animal feed and the effectiveness of microbial biodegradation. *Food Addidives and Contaminants Part A*, 27, 510-520.

[6] Azziz-Baumgartner, E., Lindblade, K., Gieseker, K., Rogers, S., Kieszak, S., Njapau, H., Schleicher, R., Mc Coy, L., Misore, A., De Cock, K., Rubin, C., & Slutsker, L. (2005). AflatoxinInvestive Group. Case-Control Study of an Acute Aflatoxicosis Outbreak, Kenya, 2004. *Environment Health Perspectives*, 113, 1779-1783.

[7] Bandyopadhyay, R., & Cotty, P. (2011). *Presentation at Annual Meeting of APS/IAPPS 6-10August 2011, Honolulu, Hawaii.*

[8] Bauchera, M., Montiesb, B., van Montagua, M., & Boerjana, W. (1998). Biosynthesis and genetic engineering of lignin. *Critical Reviews in Plant Sciences*, 17, 125-197.

[9] Baum, J. A. (2007). Control of coleopteran insect pests through RNA interference. *Nature Biotechnology*, 25, 1322-1326.

[10] Bejarano, E. R., & Lichtenstein, C. P. (1994). Expression of TGMV antisense RNA in transgenic tobacco inhibits replication of BCTV but not ACMV geminiviruses. *Plant Molecular Biology*, 24, 241-248.

[11] Bendahmane, M., & Gronenborn, B. (1997). Engineering resistance against tomato yellow leaf curl virus (TYLCV) using antisense RNA. *Plant Molecular Biology*, 33, 351-357.

[12] Briddon, R. W., & Stanley, J. (2006). Subviral agents associated with plant single-stranded DNA viruses. *Virology*, 344, 198-210.

[13] Brown, D. W., Yu, J. H., Kel, Kar. H., Fernandes, M., Nesbitt, T. C., Keller, N. P., Adams, T. H., & Leonard, T. J. (1996). Twenty five coagulated transcripts define Sterigmatocystin gene cluster in Aspergillusnidulans. *Proceeding of National Academy of Science USA.*, 93, 1418-1422.

[14] Center for Disease Control (CDC). (2004). *Outbreak of aflatoxin poisoning- Eastern and Central provinces, Kenya, January-July.Morbility and Mortality Weekly Report*, 53, 790-792.

[15] Chao, T. C., Maxwell, S. M., & Wong, S. Y. (1991). An outbreak of aflatoxicosis and boric acid poisoning in Malaysia: Aclinic pathological study. *Journal of Pathology*, 164, 225-333.

[16] Chulze, S. N. (2010). Strategies to reduce mycotoxin levels in maize during storage: a review. Food Additives and Contaminants Part A , 27, 651-657.

[17] Codex Alimentarius Commission (CAC). (2005). Code of practice for the prevention and reduction of aflatoxin contamination in tree nuts. www.codexalimentarius.net/download/standand/10221/CXP_--_059e.pdf.

[18] Cotty, P. J., & Cardwell, K. F. (1999). Divergence of West African and North American communities of Aspergillus section Flavi. *Applied and Environmental Microbiology*, 65, 2264-2266.

[19] Day, A. G., Bejarano, E. R., Buck, K. W., Burrell, M., & Lichtenstein, C. P. (1991). Expression of an antisense viral gene in transgenic tobacco confers resistance to the DNA virus tomato golden mosaic virus. *Proceedings of National academy of Sciences of USA*, 88, 6721-6725.

[20] Doster, M. A., Cotty, P. J., & Michailides, T. J. (2009). Description of a distinctive aflatoxin-producing strain of Aspergillusnomius that produces submerged sclerotia. *Mycopathologia*, 168, 193-201.

[21] European Commission. (2003). Commission Direcctive 2003/100/EC of 31 October 2003 ammendingAnnex 1 to Directive 2002/32/EC of the European Parliament and of the Council on undesirable substances in animal feed. *Official Journal of the European Union L*, 285, 33-37.

[22] Fire, A., Xu, S., Montgomery, M. K., Kostas, S. A., Driver, S. E., Mello, C. C., & 199, . (1998). Potent and specific genetic interference by double-stranded RNA in Caenorhabditiselegans. *Nature*, 391, 806-811.

[23] Fleming, A. (2006). Metabolic aspects of organogenesis in the shoot apical meristem. *Journal of Experimental Botany*, 57, 1863-1870.

[24] Fonseca, H. (2011). International legislation on mycotoxins. http://wwwmicotoxinas.comb.br/Legis-i.pdf.

[25] Food Nutrition and Agriculture (FAO). (1991). *Food for the Future. FAO 1.*

[26] Gong, Y. Y., Hounsal, A., Egal, S., Turner, P. C., Sutcliffe, A. E., Hall, A. J., Cardwell, K. F., & Wild, C. P. (2004). Postweaning exposure to aflatoxin results in impaired child growth. *A longitudinal study in Benin, West Africa.Environmental Health Perspective*, 112, 1334-1338.

[27] Hell, K., Fandohan, P., Bandyopadhyay, R., Cardwell, K., Kiewnick, S., Sikora, R., & Cotty, P. (2008). *Pre- and post-harvest management of aflatoxin in maize. In Leslie, J.F., Bandyopadhyay, R., Visconti, A., (eds) Mycotoxins: Detection Methods, Management, Public Health and Agricultural Trade*, CABI Publishing.

[28] Hell, K., Cardwell, K. F., Setamou, M., & Poehling, H. M. (2000). The influence of storage practices on aflatoxin contamination in maize in four agro-ecological zones of Benin, West Africa. *Journal of stored Products Research*, 36, 365-382.

[29] Hesseltine, C. W., & Bothast, R. J. (1977). Mold development in ears of corn from tasseling to harvest. *Mycologia*, 69, 328-330.

[30] Holbrook, C. C., Guo, B. Z., Wilson, D. M., & and, Kvien. C. (2004). Effect of drought tolerance on preharvestaflatoxin contamination in peanut. *Proceedings of the 4th International Crop Science Congress Brisbane, Australia, 26 Sep-1 Oct 2004.*

[31] Huang, G., Allen, R., Davis, E., Baum, T., Hussey, R., & 200, . (2006). Engineering broad root-knot resistance in transgenic plants by RNAi silencing of a conserved and essential root-knot nematode parasitism gene. Proceedings of the National Academy of Sciences, USA , 103, 14302-14306.

[32] IARC. (1993). Some naturally occurring substance: Food items and constituents. *IARC monographs on evaluation of carcinogenic risks to humans* [56].

[33] Jeurissen, S. M. F., Seyhan, F., Kandhai, M. C., Dekkers, S., Booij, C.J. H., Bos, P. M. J., & vander-Klerx-Fels., H. J. (2011). *An indicator based'traffic light' model to pro-actively assess the occurrence of mycotoxin in tree nuts.*

[34] Jones, L., Ratcliff, F., & Baulcombe, D. C. (2001). RNA-directed transcriptional gene silencing in plants can be inherited independently of the RNA trigger and requires Met1 for maintenance. *Current Biology*, 11, 747-757.

[35] Kaaya, A. N., & Kyamuhangire, W. (2006). The effect of storage time and agroecological zone on mould incidence and aflatoxin contamination of maize from traders in Uganda. *International Journal of Food Microbiology*, 110, 217-223.

[36] Kabak, B., & Dobson, A. D. (2009). Biological strategies to counteract the effects of mycotoxins. *Journal of Food Production*, 72, 2006-2016.

[37] Kang'ethe, E. K., & Langa, K. A. (2009). Aflatoxin B1 and aflatoxin M1 contamination of animal feeds and milk from urban Centers in Kenya. *Africa Health Science*, 9, 218-226.

[38] Katerere, R. D., Shephard, G. S., & Faber, M. (2008). Infant malnutrition and chronic aflatoxicosis in Southern Africa. *International Journal of Food Safety, Nutrition and Public Health*, 1, 127-136.

[39] Kenya Bureau of Standards. (1988). *Dry-milled maize Products. Kenya Standard 05158 Nairobi: Kenya Bureau of Standards.*

[40] Kerstin, H., & Mutegi, C. (2011). Aflatoxin control and preventive strategies in key crops of sub-Sahara Africa. *African Journal of Microbiology Research*, 5, 459-466.

[41] Kolosova, A., & Stroka, J. (2011). Substances for reduction of the contamination of feed by mycotoxins: a review. *World Mycotoxin Journal*, 4, 225-256.

[42] Lerda, D. (2010). Mycotoxin fact sheet. 3rd edition. JRC technical notes. IRRM, Geel Belgium

[43] Li, X., Song, Y., Century, K., Straight, S., Ronald, P., Dong, X., Lassner, M., & Zhang, Y. (2001). A fast neutron deletion mutagenesis-based reverse genetics system for plants. *Plant Journal*, 27, 235-42.

[44] Lippman, Z., & Martienssen, R. (2004). The role of RNA interference in heterochromatic silencing. *Nature*, 431, 364-370.

[45] Liu, Y., & Wu, F. (2010). Global Burden of Aflatoxin, A Risk Assessment.Aflatoxin-induced Hepatocellular Carcinoma. *Environmental Health Perspectives*, 118, 818-824.

[46] Lutfy, O. F., Mohd, Noor. S. B., Abbas, K. A., & Marhaban, M. H. (2008). Some control strategies in agricultural grain driers: A review. *Journal of Food, Agriculture and Environment*, 6, 74-85.

[47] Mahoney, N., Molyneux, R. J., Kim, J. H., Campbell, B. C., Waiss, A. C., & Gagerman, A. E. (2010). Aflatoxigenesis induced in Aspergillusflavus by oxidative stress and reduction by phenolic antioxidants from tree nuts. *World Mycotoxin Journal*, 3, 49-57.

[48] Mahuku, G., & Nzioki, H. S. (2011). Prevalence of aflatoxin along maize value chains in Kenya. *Report on workshop on prevalence and control of aflatoxin contamination along maize value chains organized by FAO and University of Nairobi, September 2011, Nairobi, Kenya.*

[49] Mahuku, G., & Sila, H. (2011). http://www.slideshare.net/pchenevixtrench/prevalence-of-aflatoxin-along-the-maize-value-chain-in-kenya-10402612.

[50] Marsh, S. F., & Payne, G. A. (1984). Scaning EM studies on the colonization of dent corn by Aspergillusflavus. Phytopathology ., 74, 557.

[51] Menkir, A., Brown, R. L., Bandyopadhyay, R., & Cleveland, T. E. (2008). Registration of six tropical maize germplasm lines with resistance to aflatoxin contamination. *Journal of Plant Registration*, 2, 246-250.

[52] Mette, M. F., Aufsatz, W., van der Winden, J., Matzke, M.A., & Matzke, A. J. (2001). Transcriptional silencing and promoter methylation triggered by double-stranded RNA. *EMBO Journal*, 19, 5194-5201.

[53] Munkvold, G. P., & Hellminch, R. L. (2000). Genetically modified, insect resistant maize, implications for management of ear and stalk diseases. Plant Health Program. http://www.planthealthprogress.org/current/reviews/maize/article.html.

[54] Nigam, S. N., Waliyar, F., Aruna, R., Reddy, S. V., Lava, Kumar. P., Craufurd, P. Q., Diallo, A. T., Ntare, B. R., & Upadhyaya, H. D. (2009). Breeding Peanut for Resistance to Aflatoxin Contamination at ICRISAT. *Peanut Science*, 36, 42-49.

[55] Palumbo, J. D., O'Keeffe, T. L., Kattan, A., Abbas, H. K., & Johnson, B. J. (2010). Inhibition of Aspergillusflavus in soil by antagonistic Pseudomonas strains reduces the potential for airborne spore dispersal. *Phytopathology*, 100, 532-538.

[56] Probst, C., Bandyopadhyay, R., Price, L. E., & Cotty, P. J. (2011). Identification of atoxigenicAspergillusflavus isolates to reduce aflatoxins contamination of maize in Kenya. *Plant Disease*, 95, 212-218.

[57] Probst, C., Njapau, H., & Cotty, P. J. (2007). Outbreak of an acute Aflatoxicosis in Kenya in 2004: Identification of the causal agent. *Applied Environmental Microbiology*, 73, 2762-2764.

[58] Probst, C., Schulthess, H., & Cotty, P. J. (2010). Impact of Aspergillus sect.Flavi community structure on the development of lethal levels of aflatoxins in Kenyan maize (Zea mays). *Journal of Applied Microbiology*, 108, 600-610.

[59] Rakefet, Runo. S., Towsley, B., Machuka, J., & Sinha, N. (2008). Long-distance transport of mRNA via parechyma cells and phloem across the host-parasite junction in Cuscuta. *New Phytologist*, 179, 1133-1141.

[60] Reverberi, M., Zjalic, S., Ricelli, A., Di Meo, C., Scarpari, M., Fanelli, C., & Fabbri, A. A. (2011). Mushrooms versus fungi: natural compounds from Lentinulaedodes inhibit aflatoxin biosynthesis by Aspergillusparasiticus. *World Mycotoxin Journal*, 4, 217-224.

[61] Robens, J. K., & Richard, J. L. (1992). Aflatoxins in animal and human health. *Rev Environmental Contamination and Toxicology*, 127, 69-94.

[62] Robertson, A. (2005). Risk of aflatoxin contamination increases with hot and dry growing conditions. Integrated Crop Management http:// www.ipm.iastate.edu/ipm/icm/2005/9-19/aflatoxin.html., 185-186.

[63] Rushton, P. J., & Somssich, I. E. (1998). Transcriptional control of plant genes responsive to pathogens. *Current Opinion in Plant Biology*, 1, 311-315.

[64] Rustom, I. Y. S. (1997). Aflatoxin in food and feed.Occurrence, legislation and inactivation by physical methods. *Food chemistry*, 59, 57-67.

[65] Schamale, D., & Munkvold, G. (2011). Mycotoxin in crops: A threat to human and domestic animal. healthhttp://www.apsnet.org/edcenter/intropp/topics/pages/ managementstrategies.aspx. accessed 12/06 2012

[66] Scheidegger, K. A., & Payne, G. A. (2003). Unlocking the secrets behind secondary metabolism: a review of Aspergillusflavus from pathogenicity to functional genomics. *Journal of Toxicology*, 22, 423-459.

[67] St, Leger. R. J., Screen, S. E., & Shams-Pirzadeh, B. (2000). Lack of host specialization in Aspergillusflavus. *Applied and Environmental Microbiology*, 66, 320-324.

[68] Strosnider, H., Azziz-Baumgartner, E., Banziger, M., Bhat, R. V., Breiman, R., & Brune, M. N. (2006). Workgroup report: Public Health strategies for reducing aflatoxin exposure in developing countries. *Environmental Health Perspective*, 114, 1898-1903.

[69] Trail, F., Mahanti, N., Rarick, M., Mehigh, R., Liang, S. H., Zhou, R., & Linz, J. E. (1995). Physical and transcriptional mapof an aflatoxin gene cluster in Aspergillus-

parasiticus and functional disruption of a gene involved early in aflatoxin pathway. *Applied Environmental Microbiology*, 61, 2665-2673.

[70] Turner, P. C., Sylla, A., Gong, Y. Y., Diallo, M. S., Sutcliffe, A., E., , hall, A. J., & Wild, C. P. (2005). Reduction in exposure to carcinogenic aflatoxins by post-harvest intervention measures in West Africa: a common-based intervention study. *The Lancet*, 365, 1950-1956.

[71] Voinnet, O., & Baulcombe, D. C. (1997). Systemic signalling in gene silencing. Nature , 389, 553-553.

[72] Volpe, T. A., Kidner, C., Hall, I. M., Teng, G., Grewal, S. I., & Martienssen, R. A. (2002). Regulation of heterochromatic silencing and histone H3 lysine-9 methylation by RNAi. *Science*, 297, 1833-1837.

[73] Waliyar, F., Kumar, L. P., Traore, A., Ntare, B. R., Diarra, , , B., & Kodio, O. (2008). Pre- and postharvest management of aflatoxin contamination in peanuts. *Leslie et al. (eds) Mycotoxins: Detection methods, management, public health and agricultural trade. CAB International*, Cromwell Press, Trowbridge, UK, 209-218.

[74] Wassenegger, M., Heimes, S., Riedel, L., & Sanger, H. L. (1994). RNA-directed de novo methylation of genomic sequences in plants. Cell , 76, 567-576.

[75] William, J., Philips, T. D., Jolly, P. E., Stiles, J. K., Jolly, C. M., & Aggarwal, D. (2003). Human aflatoxicoses in developing countries: a review of toxicology exposure, potential health consequences and interventions. *American Journal of Clinical Nutrition*, 80, 1106-1122.

[76] Woloshuk, C. P., & Prieto, R. (1998). *Genetic organization and fuction of the aflatoxin B1 biosynthetic genes, FEMS Microbiology Letters*, 160, 169-176.

[77] World Health Organization (WHO). (2006). Mycotoxins in African foods: implications to food safety and health. *AFRO Food Safety Newsletter. World.*

[78] Health Organization Food safety (FOS). (2006). www.afro.who.int/des.

[79] Wu, F., & 200, . (2006). Mycotoxin reduction in Bt corn: potential economic, health, and regulatory impacts. *ISB News Report, September 2006.*

[80] Wu, F., & Khlangwiset, M. (2010). Health economic impact and cost-effectiveness of aflatoxin reduction strategies in Africa: case studies in bio-control and post-harvest interventions. *Food Additives and Contaminants*, 27, 496-509.

[81] Xu, L., Xu, Y., Dong, A., Sun, Y., Pi, L., & Huang, H. (2003). Novel as1 and as2 defects in leaf adaxial-abaxial polarity reveal the requirement for ASYMMETRIC LEAVES1 and 2 and ERECTA functions in specifying leaf adaxial identity. *Development*, 130, 4097-4107.

[82] Yadav, B. C. (2006). Host-generated double-stranded RNA induces RNAi in plant-parasitic nematodes and protects the host from infection. *Molecular Biochemistry and Parasitology*, 148, 219-222.

[83] Yang, Y., Sherwood, T. A., Patte, Hiebert. E., & Polston, J. E. (2004). Use of tomato yellow leaf curl virus (TYLCV) Rep gene sequences to engineer TYLCV resistance in tomato. *Phytopathology*, 94, 490-496.

[84] Yao, H., Hruska, Z., Kincaid, R., Brown, R., Cleveland, T., & Bhatnagar, D. (2010). Correlation and classification of single flourescencehyperspectral data with aflatoxin concentration in corn kernels inoculated with Aspergillusflavus spores. *Food Additives and Contaminants Part A*, 27, 701-709.

[85] Yi, J., Chang, P. K., Cary, J. W., Wright, M., Bhatnagar, D., Cleveland, T. E., Payne, G. A., & Linz, J. E. (1995). Comparative mapping of Aflatoxin pathway gene clusters in Aspergillusparasiticus and Aspergillusflavus. *Applied and Environmental Microbiology*, 61, 2365-2371.

[86] Zamore, P. D., Tuschl, T., Sharp, P. A., & Bartel, D. P. (2000). RNAi: double-stranded RNA directs the ATP-dependent cleavage of mRNA at 21 to 23 nucleotide intervals. *Cell*, 101, 25-33.

[87] Zarba, A., Wild, C. P., Hall, A. J., Montesano, R., & Groopman, J. D. (1992). Aflatoxin M1 in human milk from the Gambia West Africa, quantified by monoclonal antibody immunoaffinitychromatohraphy and HPLC. *Carcinogenesis*, 13, 891-894.

[88] Zhang, H., Shen, W. B., Zhang, W., & Xu, L. L. (2005). A rapid response of beta-amylase to nitric oxide but not gibberellin in wheat seeds during the early stage of germination. *Planta*, 220, 708-16.

[89] Zilberman, D., Cao, X., & Jacobsen, S. E. (2003). ARGONAUTE4 control of locus specific siRNA accumulation and DNA and histone methylation. *Science*, 299, 716-719.

Chapter 4

Novel Methods for Preventing and Controlling Aflatoxins in Food: A Worldwide Daily Challenge

Eva Guadalupe Lizárraga-Paulín,
Susana Patricia Miranda-Castro,
Ernesto Moreno-Martínez,
Irineo Torres-Pacheco and
Alma Virginia Lara-Sagahón

Additional information is available at the end of the chapter

1. Introduction

Talking about Aflatoxins is not a new issue. Aflatoxins are a big problem that day by day turns more important due to their implication in crop production, food quality and human and animal health. Aflatoxins are also everywhere because those toxic secondary metabolites are mycotoxins produced by a large number of Aspergillus species, being *A. flavus*, and *A. parasiticus* the main producers; nevertheless, species like *A. nomius, A. pseudotamarii, A. parvisclerotigenus, A.bombycis, A. ochraceoroseus, A. rambellii, Emericella astellata* and *E. venezuelensis* are aflatoxin generators too [1,2]. Since those toxins have been recognized as a significant worldwide problem in 1960 (because of being isolated and identified as the causative toxins in "Turkey-X-disease" after 100,000 turkeys died in England from liver acute necrosis and bile duct hyperplasia after consuming groundnuts infected with *Aspergillus flavus*) [3-5], researchers have studied lots of ways to fight against this threat; however, after more than a half century, aflatoxins are still a big problem that has not been easy to deal with, because humans are not able to manipulate essential factors that affect aflatoxin contamination like the region weather, the crop genotype, the soil type, the minimum and maximum daily temperatures and the daily net evaporation [5].

Aflatoxins (AF) affect almost everything we eat: cereals (maize, wheat and rice principally) and their derivates; oilseeds (cotton, peanut, rapeseed, coconut, sunflowers and others), cassava, nuts, dry fruits, delicatessen products, spices, wines, legumes, fruits, milk and milk derivates [6,7], and even chocolates [8]. In order to find a solution for this problem, some organizations and institutions have purposed prevention strategies in order to reduce the risks given by this public problem especially in low-income countries, but those strategies are not enough to give a real solution to this worldwide daily problem.

2. The global problem of AF in crops and food

The prevalence of AF in crops and livestock is a serious problem in many parts of the world, undermining public health and development efforts. AF are highly toxic, cancer causing fungal metabolites known to cause immune-system suppression, growth retardation, liver disease, and death in both humans and domestic animals. According to the United Nations Food and Agriculture Organization (FAO), 25% of world food crops are affected, and countries that are situated between 40ºN and 40ºS are most at risk. Over 4.5 billion people in developing countries are at risk of chronic AF exposure [9]. Unless AF levels in crops and livestock are effectively managed, international development efforts to achieve greater agricultural development, food security and improve health will be undermined.

AF are very stable and persistent, so they are difficult to remove. Due to they are contained in many crops that are consumed by animals, AF have turned into a serious animal problem too. The most susceptible animals are rabbits, turkeys, chickens, pigs, cows and goats [10]. AF can be transmitted from animals to human food (by eggs, meat and dairy) with the consequent risk to human health.

Even non-mouldy foods or raw materials may contain AF. Spores can be transferred by insects (especially flies, wasps and bees) or by birds to foods where the spores germinate, produce mycelium, and AF are excreted. Seeds can contain AF by infection of the egg-cells of the flowering plants. The spores of *A. flavus* and *A. parasiticus* can germinate on the stigma surfaces of plants, then the germ tube penetrates to the developing embryo mimicking pollen germ tubes. The mycelium can establish an endotrophic relationship which is not harmful for the healthy plant. However, if the plant is under drought stress, then significant levels of AF may be produced in the plant tissue during growth in the field. Under these circumstances food commodities may already be contaminated at harvest and, although the concentrations are never as high as those formed in stored commodities, they can be economically significant [11, 12].

The danger of AF lies in their mode of action by inhibiting the incorporation of precursors for the synthesis of DNA, RNA and proteins; they also block the action of some enzymes that are responsible for the synthesis of nucleic acids, causing centrilobular necrosis in the liver, polymorphonuclear infiltration and fatty degeneration. AF toxicity depends on the dose, the exposure degree, the age, the nutritional status of the animal and the possible synergic effects of the chemical agents to which they are exposed [13]. Some secondary metabolites produced by Aspergillus species are harmful for animals too. That's the case of cyclopiazonic acid (CPA), which causes necrosis of liver or gastrointestinal tissue and necrotic changes in skeletal muscle and kidney [14, 15].

The economic impacts of AF contamination can vary greatly among affected food and feed commodities. These differences include the severity of the contamination problem, the geographic range of AF problems, the types of AF control methods available, and which sectors bear the burden of the cost of AF contamination. All of these factors affect whether AF control methods are adopted [16]. AF and mycotoxins in general have not been widely prioritized from a public health perspective in low-income countries. This is because knowledge of mycotoxins and the full range and scale of their adverse health effects is incomplete and the known risks are poorly communicated to governments in regions where the contamination is greatest [17]. Matters that have to be considered by government to avoid diseases from aflatoxicosis are: an opportune and nonexpensive analytic detection, unifying worldwide government regulations, deviation of AF-contaminated commodities from the food supply, improving research on the biosynthesis and molecular biology of AF, and designing new control strategies for the abolition of AF contamination of food crops, inter alia [10].

3. How to prevent pre-harvest AF?

It was established in about 1970 that fungal contamination could start in the field before harvest [9]. Although the highest levels of AF are undoubtedly associated with post-harvest spoilage of food commodities stored under inappropriate conditions of water activity and temperature, the aflatoxigenic fungi have more complex ecologies [12]. Factors that influence the incidence of fungal infection and subsequent toxin development include invertebrate vectors, grain damage, oxygen and carbon dioxide levels, inoculum load, substrate composition, fungal infection levels, prevalence of toxigenic strains and microbiological interactions. Insect damage on crops allows fungi to access in them, increasing the chances of AF contamination, especially when loose-husked maize hybrids are used [18, 19].

Controlling or reducing infection by regulating the factors that increase the risk of AF contamination in the field contributes extensively in managing AF. Management practices that reduce the incidence of AF contamination in the field include timely planting, maintaining optimal plant densities, proper plant nutrition, avoiding drought stress, controlling other plant pathogens, weeds and insect pests and proper harvesting [20]. Pre-harvest measures that are efficient in reducing AF levels are the same as those that will enhance yields. Crop rotation and management of crop residues also are important in controlling *A. flavus* infection in the field. Tillage practices, fertilizer application, weed control, late season rainfall, irrigation, wind and pest vectors affect the source and level of fungal inoculum, maintaining a disease cycle in crops like maize [19, 21]. Lime application, use of farm yard manure and cereal crop residues as soil amendments have shown to be effective in reducing *A. flavus* contamination as well as AF levels by 50-90%. Calcium, which is part of lime, thickens the cell wall and accelerates pod filling, while manure facilitates growth of microorganisms that suppress soil infections [21].

In order to minimize the levels of AF and mycotoxins in general, the National Institute of Agricultural Technology of Argentina (INTA), recommends to make early plantings, to plant resistant genotypes, to do good farming practices, to avoid stress conditions, to minimize insect damage, to harvest early in order to avoid delays, to avoid damaged kernels and to storage at less of 13% moisture in a clean, fresh and airy place with no insects [22]. As mentioned before, it is important to avoid product moisture, high temperatures (between 25 and 32°C) and high relative humidity in storage and seeds preservation. Weeds have to be removed and crop rotation should be done routinely. Prior to the preparation of the ground, dead organic matter has to be disabled or burned; product mechanical damage has to be avoided; crops have to be collected at full maturity; storage places should be dry and the entry of water has not to be allowed; storage health standards have to be fulfilled (pallets, proper humidity levels, adequate ventilation and lighting, etc.), and periodic inspection of the stored product should be done [23].

To avoid risks to human and animal health, INTA also suggests to avoid feeding animals with crops in poor condition (especially corn), not to use fractions of discarded corn fodder, and to make good manufacturing practices [22].

4. Traditional AF control methods

Since AF have been recognized as a significant worldwide problem, researches have proposed some ways of detoxification. AF detoxification refers to those post-harvest treatments directed to eliminate or diminish the toxic effects of toxins. Those strategies can be divided into three different groups: natural methods, physical methods and chemical methods, which are focused on destroying, modifying or adsorbing AF [24]. There is variety of tools such as post-harvest drying (which is economically accessible), adequate storage, shelling, dehulling, product sorting, early harvest, regionally adjusted planting dates, and insect control. However, even when storage conditions are generally good, AF frequently form prior to harvest while the crop is maturing and/or awaiting harvest, which can result in significant losses [5].

4.1. Natural methods

The natural methods used to avoid AF are principally: seed cleaning, sorting and seed division by screening and extrusion. Nevertheless, those techniques are neither practical nor efficient at all, and food micronutrients content get diminished [24]. Since 1989, the FAO has supported some decontamination processes like the UK-Thai Project (UTP) System, which showed to reliably produce low AF-content maize during the rainy season. With the UTP system, maize is first field dried on the stalk for one to two weeks before harvesting to reduce moisture content to 20%. It is next shelled within 24 to 48 hours of harvest, and loaded into a drier within 12 hours of shelling. Thus, within 48 hours, it is dried to 14% moisture content, with no part exceeding 15%. AF content is monitored rapidly by a special adaptation of the bright greenish-yellow fluorescence (BGYF) test. Maize dried to 14% moisture content by the UTP system can be safely stored for a minimum of two months with no increase in AF content [25].

By the other hand, cleaning of stores before loading in the new harvests has been correlated with reduction in AF levels. Separating heavily damaged ears (those having greater than 10% ear damage) also reduces AF levels in crops like maize. Wild hosts, which constitute a major source of infestation for storage pests, should also be removed from the vicinity of stores. For some crops like peanuts, the standard practice is drying of pods in the sun. Often pods are left in the field after uprooting for up to four weeks to partially dry prior to home drying [19].

AF are unevenly distributed in a seed lot and may be concentrated in a very small percentage of the product. Sorting out of physically damaged and infected grains (known from colorations, odd shapes and size) from the intact commodity can result in 40-80% reduction in AF levels [19]. The advantage of this method is that it reduces toxin concentrations to safe levels without the production of toxin degradation products or any reduction in the nutritional value of the food. This could be done manually or by using electronic sorters. Some studies have also looked at the use of local plant products for the control of fungi mostly proving their efficacy in-vitro but these products have not been sufficiently tested for their efficiency in controlling AF in stored crops [19, 26].

4.2. Physical methods

Although natural methods are cost-effective, the fungal contamination in grains is often unavoidable, so there is the need to apply a suitable process to inactivate the toxin. Sorting can remove a major part of AF contaminated units, but levels in contaminated commodities may also be reduced through physical food processing procedures like dehulling (which reduces AF contamination by 92%), roasting, baking, frying, X-radiation, extrusion cooking and nixtamalization, being the last two the most studied because of their effectiveness [27-29].

Roasting, baking and frying are three common methods used in some low-income countries, and all of them involve heath. Nevertheless, the heat used as the only factor for the mycotoxins destruction is ineffective because the temperatures reached during the detoxification process affect vitamins and food proteins. In contrast, heath can be used for increase the reactive capacity of some food molecules such as acids, alkalis and other chemical agents [30].

Radiation has also been used against AF. X-rays are capable of producing a high issuance of energy, which causes the breakdown of stable molecular structures. It has been established that AFB1 and AFG1 are the most sensitive to X-rays [30, 31].

Extrusion cooking is a processing technology that involves pushing a granular food material down a heated barrel and through an orifice by a rotating, tight fitting Archimedean screw. The shear forces created by the rotating action of the screws, together with frictional, compressive and pressure forces provide the necessary environment for rapidly cooking and transforming the food into visco-elastic melt. Extrusion cooking is an efficient high temperature short time process, and it is used to produce a wide variety of foods and ingredients. To destroy or inactivate AF, the extrusion cooking conditions need to be severe (high shear, high temperature, and the right pH) in order to provide the necessary environment in the barrel, but such treatments to destroy or inactivate AF in peanuts may affect essential nutrients and compromise the nutritional quality of the food product [32].

In 2011, Saalia & Philips reported that extrusion of artificially contaminated food degrade AF to varying degrees depending on the extrusion conditions without altering nutritional quality. They extruded naturally contaminated peanut meal by varying the moisture (20, 28, 35 g/ 100 g); pH (7.5, 9.5) and extruder die diameter (2.5, 3, 3.5, 4.0 mm). The highest AF reduction in naturally contaminated peanut meal was 59% at feed moisture content of 35 g/100 g. Higher (91%) reduction was achieved in the artificially contaminated peanut meal at moisture of 20 g/100 g. In-vitro protein digestibility and Fluorodinitrobenzene (FDNB)-available lysine of the extrudates were not significantly different from non-extruded peanut meal, and extrusion conditions for AF reduction did not adversely affect protein nutritional quality. Extrusion conditions that reduced throughput in the single screw extruder promoted greater AF reduction. Those conditions also marginally reduced the protein nutritional quality of the extrudates. High moisture conditions provided extrudates with the least in-vitro protein digestibility and lowest available lysine. Decontamination of naturally contaminated peanuts using extrusion cooking was less successful (59%) than artificially contaminated peanut meal (91%) [32].

Nixtamalization (TNP) is an alkaline cooking process original from ancient Mexico which is applied in corn tortillas. Alkalinity largely destroys AF in corn. TNP consists on the cooking of the grain in abundant water and lime (2–3 L of water/kg of maize processed, with 1–3% $CaOH_2$) at boiling temperatures for 35–70 min, with a steeping period of 8–16 h. After the steeping, the lime cooking solution (nejayote) is decanted, and the grain is thoroughly washed to leave the grain ready for milling to obtain the maize dough for making tortillas [33, 34]. It has been shown that traditional nixtamalization is capable of destroying 85% of the AF present in maize, and 15% of AF remaining in mass does not retain its fluorescence properties, but can be recognized by the monoclonal antibodies used for recent studies detection [35]. Mendez-Albores and collaborators reported that traditional nixtamalization can reduce AF concentrations in 94% even in highly contaminated maize, being more effective than extrusion cooking; nevertheless, this finding has been widely questioned because other authors suggest that AF lactone rings, which are opened during nixtamalization alkaline process, can be closed when tortillas are acidified in stomach [34, 35]. It is important to mention that some authors have reported nixtamalization as a chemical method [24].

4.3. Chemical methods

Chemical AF control methods are principally those which involve the use of chemical reagents for different purposes. Most investigators are looking for new sources of materials to control spoilage caused by fungi in food. However, the application of synthetic preservatives has led to a number of environmental and health problems because they are themselves carcinogenic, teratogenic, and highly toxic with long degradation periods [36, 37].

Insecticides and fumigants were the first chemicals to be used to deal with aflatoxigenic fungi. The DOA Division of Plant Pathology and Microbiology screened since several decades ago, seven reagents in the laboratory for effectiveness in preventing or reducing AF contamination of maize. Only three of the reagents were found to be effective: sodium bisulphite, ammonia, and propionic acid. Sodium bisulphite and ammonia treatments resulted in grain with a strong residual odor; the ammonia treatment also produced darker grain. The most promising regent was the propionic acid-based fungicide formulation, which effectively controlled both mould growth (*A. flavus*) and AF formation, while not adversely affecting the physical quality of the grain [25]. Nowadays, the use of insecticides for this purpose has been abandoned due to the toxic residues that they generate [19]. About fumigants, only two were in common use in the last decade: methyl bromide and phosphine. Methyl bromide has been identified as a major contributor to ozone depletion, which casts a doubt on its future use in pest control. There have been repeated indications that certain insects have developed resistance to phosphine, so its use is now doubtful [30, 38]. It has also been reported that propionic acid, sodium propionate, benzoic acid, ammonia, urea and citric acid are the best anti-fungal chemical compounds tested in feeds [39].

Organic solvents can be used to remove AF in food because mycotoxins have the physicochemical characteristic to be soluble in them. Combinations such as hexane-acetone-water or isopropanol-water, inter alia, have been reported to be effective mycotoxins draggers. Some acids such as hydrochloric acid, sulfuric acid and their derivatives have the capability to react with the lactone groups of AFB1, AFG1, and with non-aromatic double bonds present in AF. Toxicologically, the addition reaction of the acids with the double bonds structures appears to be most effective in terms of detoxification because the reaction products are polar substances that can be eliminated in the urine. Alkalis like monoethylmethylamine, hydroxide and calcium chloride, sodium hydroxide and ammonium carbonate, are reactive with the lactone group of AF. Oxidant agents such as ozone, peroxides and permanganates in alkaline solutions are reactive with non-conjugated double bonds of AF. The ozonolysis reaction leads to the creation of smaller molecules, but some of the obtained products could be toxic. The glycosylation reaction results in the creation of two hydroxyl groups that can subsequently form hydrogen bonds; nevertheless although this mechanism is effective for AF detoxification, it should be used in combination with polymers or silicates capable of adsorbing physically AF through hydrogen bonds [30].

Adsorption of mycotoxin molecules has been studied recently. It can be done by different inert chemicals, such as some complex indigestible carbohydrates (cellulose, polysaccharides in the cell walls of yeast and bacteria like glucomannans, peptidoglycans and others), synthetic polymers (such as cholestyramine and polyvinylpyrrolidone), humic acid and vegetable fibers, and clays or synthetic silicates, which can sequester mycotoxins. The pyrrolidone mechanism of action is due to both, physical adsorption effect and the bridges establishment of hydrogen and nitrogen in its structure [30, 40, 41]. The adsorptive capacity of the carbohydrate complexes in the yeast cell wall offers an interesting alternative to inorganic adsorbing agents. Modifications in manufacturing techniques have enabled the production of specifically modified yeast cell wall preparations with the ability to adsorb a range of mycotoxins. Several reports indicate the possibility of there being more than one target for mycotoxin binding in cell wall preparation. However, it is too early to interpret the mechanistic aspects and more basic studies are needed on the interaction of individual mycotoxins with different components of *S. cerevisiae* cell wall. More studies are needed on the chemistry of binding and stability of the complex, especially under the harsh conditions of the gastrointestinal tract. Moreover, several studies suggest that yeasts or esterified glucomannan products may not be effective in reducing AFM1 concentrations. Further *in vivo* studies are needed to confirm the effectiveness of yeasts and derivative products in suppressing absorption of AF in ruminants. Results on the efficacy of synthetic polymers or vegetable fibers in sequestering mycotoxins are highly promising, although this field is still in its infancy and further research is needed [40].

The aluminum silicates belong to clays, highlighting bentonite, sepiolite and zeolite. These compounds possess a three-dimensional structure formed by the junction core of SiO_4 tetrahedra, wherein some ions such as aluminum ions are intercalated. Nowadays, between of all the chemical methods of detoxification, silicates are the most used because they don't create waste problems, they don't destroy food vitamins and proteins, they don't generate partial reactions, they don't create toxic metabolites, and their prices are not elevated. Not only natural aluminum silicates but also Hydrated Sodium Calcium Aluminosilicates (HSCAS) are used, because the last ones have a greater adsorption capability because of being refined products. In its structure, not only aluminum ions, but also calcium and sodium ions are intercalated, increasing the distance between silicon ions and improving adsorption capacity. Since 1988 there are numerous publications that demonstrate the use of HSCAS as adsorbents for mycotoxins, at *in vivo* and *in vitro* level [30, 41]. HSCAS clay can adsorb AFB1 with high affinity and high capacity in aqueous solutions (including milk) and in the meantime it can markedly reduce the bioavailability of AF in poultry; it can greatly diminish the effects of AF in young animals, i.e., rats, chicks, poults, ducklings, lambs, and pigs; and it can decrease the level of AFM1 in milk from lactating cows and goats [40].

5. Novel AF control methods

Although there are a lot of methods that have been practiced in order to fight against aflatoxigenic fungi and their toxins, they have been criticized because of their low effectiveness or due to their contaminant nature as mentioned before. That is why in recent years researchers have chosen new ways to deal with this threat involving microbiological and biotechnological methods that are promising because of the good results that have been obtained with them.

5.1. Microbiological methods

The use of microorganisms is a strategy that has been used recently. There have been reported some processes such as the action that ruminal flora has over mycotoxins. It was found that it is capable of esterifying ochratoxin A, turning it into ochratoxin C. The isolated action of bacteria and fungi such as *Corynebacterium rubrum, Aspergillus niger, Trichoderma viride* and *Mucor ambiguus* in the modification of the structure of AFB1 has been studied too [30].

The most studied microbiological decontamination is the fermentation process, which is used during the production of bread from wheat kernels contaminated with deoxynivalenol. After fermentation, a reduction in toxins levels is observed, and this is attributed to fermentation *per se* and to the thermal process to which the product is subjected. Decontamination occurs because yeast adsorb toxins [42]. Some reviews report that experiments of alcoholic fermentation by *Saccharomyces cerevisiae* with contaminated must with deoxynivalenol (DON) and zearalenone, showed results where after 7 to 9 days of fermentation the DON was stable to the process, the initial content of zearalenone was converted to β-zearalenol (β-ZEL), and α-zearalenol; most of the metabolization of zearalenone occurred in the first and second days of fermentation, showing the instability of the toxin to this process [42]. Not only *Saccharomyces cerevisiae* but also some lactic bacteria and yeasts are used widely in food fermentation because they have wall structures which are capable to adhere mycotoxins. Mycotoxins can be degraded by specific enzymes, as the case of ochratoxin A, which peptidic group is attacked by proteases [30]. Other researches have shown good inhibition results in AF production using microorganisms such as Bacillus spp (98%), *A. flavus* (90%), *A. parasiticus* (90%) and Trichoderma spp (75%) [42].

5.2. Biotechnological methods: Biological Control

Biotechnological methods are those in which biological systems or their derivates are used in order to obtain better products. From among them, talking about AF control, we can highlight the biological control, the use of natural extracts and essential oils and genetic engineering to mention a few.

5.2.1. Biocompetition

An option to supplement, but not to supplant the traditional methods of AF control is biological control. Most AF biological control programs can truly be defined as biocompetition since they do not utilize parasites or diseases of the pest, but instead use atoxigenic Aspergillus species to competitively exclude toxigenic fungi [43]. Augmentative biological control is as a pest management tactic that utilizes the deliberate introduction of living natural enemies to low the population level of invasive pests. Biological control has been utilized for more than 100 years in efforts to control a wide number of agricultural pests including fungi, insects and weeds [44]. Biocontrol strategies have been implemented to control AF contamination in several important agricultural crops, such as peanut, cotton and corn [43, 45, 46]. Some authors have reviewed some biological methods using bacteria, yeasts and fungi as competitors for containment of *A. flavus* growth and/or toxin production [46, 47]. Natural population of fungi like *A. flavus,* consists of toxigenic strains that produce copious amount of AF and atoxigenic strains that lack the capacity to produce AF. In the competitive exclusion mechanism, introduced atoxigenic strains out compete and exclude toxigenic strains from colonizing grains thereby reducing AF production in contaminated grains [48]. The use of *A. flavus* atoxigenic strains (afla–) reduce AF contamination in many crops; nevertheless, the mechanism by which a non-aflatoxigenic strain interferes with AF accumulation of toxigenic strains has not been definitively elucidated [49, 50].

Since the last decade of the past century, some yeasts and bacteria have shown to be effective on controlling fruits and vegetables postharvest diseases. In the early nineties, biological control of grain fungi was studied only to a limited extent. Most of the studies had dealt mainly with the interaction between mycotoxigenic strains (mostly aflatoxigenic ones) and other fungi, occurring naturally on grains, grown in competition. A limited number of fungi (especially *Aspergillus niger* van Tieghem), yeasts and bacteria were found to inhibit, detoxify or metabolize AF; however, it was determined that their antagonistic effect was highly dependent on cultural and environmental conditions [51]. There has been found that the yeast *Pichia guilliermondii* is effective in controlling major citrus fruit rots [52]. Based in those studies, in 1993, Paster and collaborators evaluated the efficacy of *Pichia guilliermondii* Wickerham for the control of the common *Aspergillus flavus* storage fungus and the natural microflora of soya beans, obtaining good results. The ability of *Pichia guilliermondii* to inhibit growth of grain microflora was studied using naturally contaminated soya beans and sterilized soya beans artificially inoculated with *Aspergillus flavus*. When *A. flavus* (at a spore concentration of 10^2 spores ml^{-1}) and *P. guilliermondii* (at concentrations of 10^7 or 10^9 spores ml^{-1}) were applied simultaneously to sterilized soya beans, fungal proliferation was inhibited during 16 days of storage. Application of yeast cells 3 days prior to fungal inoculation resulted in decreased inhibitory activity. The inhibitory effect of the yeast was compared with that of propionic acid using naturally infested soya beans at two levels of moisture content (11 and 16%). At both levels the yeast prevented fungal proliferation on the grain for a limited period, but propionic acid showed better fungistatic activity [51].

During 1994 and 1995, studies were conducted in the environmental control plot facility at the National Peanut Research Laboratory in Georgia to determine the effect of different inoculum rates of biological control agents on preharvest AF contamination of Florunner peanuts. Biocontrol agents were nontoxigenic color mutants of *Aspergillus flavus* and *Aspergillus parasiticus* that were grown on rice for use as soil inoculum. Those results were published three years later [53]. Findings like these were the basis of further studies focused on the use of aflatoxigenic Aspergillus species that researchers are still investigating with more detail.

In recent years, some antagonists have been applied in biocontrol of postharvest diseases of agricultural products. Naturally occurring populations of atoxigenic strains are considered reservoirs from which to select strongest biocompetitors. The atoxigenic strains colonizing the environment where crops are affected by repeated AF outbreaks should have adapted to, and hence acquired, a superior fitness, for the relevant environment. Selecting biocontrol strains is not straightforward, as it is difficult to assess fitness for the task without expensive field trials. Reconstruction experiments have been generally performed under laboratory conditions to investigate the biological mechanisms underlying the efficacy of atoxigenic strains in preventing AF production and/or to give a preliminary indication of strain performance when released in the field [54]. The mechanisms by which afla– strains interfere with AF accumulation has not yet been definitively established. The prevalent opinion is that it depends on the competitive exclusion of AF producer (afla+) strains from the substrate as a result of (a successful) physical displacement and competition for nutrients by afla– strains. However, different hypotheses may still be taken into consideration [55].

Biological control is a promising approach for reducing both preharvest and postharvest AF contamination. There are some studies that report reductions in AF that are achieved by applying nontoxigenic strains of *A. flavus* and *A. parasiticus* to soil around developing plants, especially in peanuts. When late-season drought conditions make peanuts susceptible to invasion and growth by these fungi, the applied nontoxigenic strains competitively exclude toxigenic strains present in the soil and thereby reduce subsequent AF concentrations. Reductions in AF contamination with the use of nontoxigenic strains, has also been demonstrated in corn and cottonseed [56-59].

In 2003, Dorner and collaborators reported the results of a study that was conducted to evaluate the efficacy of three formulations of nontoxigenic strains of *Aspergillus flavus* and *Aspergillus parasiticus* to reduce preharvest AF contamination of peanuts during two years. Formulations included a solid-state fermented rice, fungal conidia encapsulated in an extrusion product termed Pesta and conidia encapsulated in pregelatinized corn flour granules. Analysis of soils for *A. flavus* and *A. parasiticus* showed that a large soil population of the nontoxigenic strains resulted from all formulations. In the first year, the percentage of kernels infected by wild-type *A. flavus* and *A. parasiticus* was significantly reduced in plots treated with rice and corn flour granules, but it was reduced only in the rice-treated plots in year two. There were no significant differences in total infection of kernels by all strains of A. *flavus* and *A. parasiticus* in either year. AF concentrations in peanuts were significantly reduced in year two by all formulation treatments with an average reduction of 92%. Reductions were also noted for all formulation treatments in year one (average 86%), but they were not statistically significant because of wide variation in the AF concentrations in the untreated controls. Each of the formulations tested, therefore, was effective in delivering competitive levels of nontoxigenic strains of *A. flavus* and *A. parasiticus* to soil and in reducing subsequent AF contamination of peanuts [59]. The maize endophyte *Acremonium zeae* is antagonistic to kernel rotting and mycotoxin producing fungi *Aspergillus flavus* and *Fusarium verticillioides* in cultural tests for antagonism, and interferes with *A. flavus* infection and AF contamination of preharvest maize kernels. In 2005, Wicklow, reported results of chemical studies of an organic extract from maize kernel fermentations of *Acremonium zeae* (NRRL 13540), which displayed significant antifungal activity against *Aspergillus flavus* and *F. verticillioides*, and revealed that the metabolites accounting for this activity were two newly reported antibiotics pyrrocidines A and B. Pyrrocidines were detected in fermentation extracts for 12 NRRL cultures of *Acremonium zeae* isolated from maize kernels harvested in different places. Pyrrocidine B was detected in whole symptomatic maize kernels removed at harvest from ears of a commercial hybrid that were wound-inoculated in the milk stage with *A. zeae* (NRRL 13540) or (NRRL 13541). The pyrrocidines were first reported from the fermentation broth of an unidentified filamentous fungus LL-Cyan426, isolated from a mixed Douglas Fir hardwood forest on Crane Island Preserve, Washington, in 1993. Pyrrocidine A exhibited potent activity against most Gram-positive bacteria, including drug-resistant strains, and was also active against the yeast Candida albicans. In an evaluation of cultural antagonism between 13 isolates of *A. zeae* in pairings with *A. flavus* (NRRL 6541) and *F. verticillioides* (NRRL 25457), A. zeae (NRRL 6415) and (NRRL 34556) produced the strongest reaction, inhibiting both organisms at a distance while continuing to grow through the resulting clear zone at an unchanged rate. [60].

In 2005, Bandyopadhyay reported a test of twenty-four atoxigenic *A. flavus* isolates under field conditions in Nigeria to identify a few effective strains that could exclude toxigenic strains. These atoxigenic strains were evaluated for a set of selection criteria to further narrow down the numbers to a few for further use in biocontrol field experiments. Good criteria of selection will ensure that the candidate atoxigenic strains belong to unique vegetative compatibility groups (for which testers have been developed) that are unable to produce toxigenic progenies in the natural environment. Propensity to multiply, colonize and survive are other selection criteria to make sure that few reapplications will be required once the atoxigenic strains are introduced in the environment [48].

In 2006, Palumbo and collaborators isolated bacteria from California almond orchard samples to evaluate their potential antifungal activity against AF-producing *Aspergillus flavus.* Fungal populations from the same samples were examined to determine the incidence of aflatoxigenic *Aspergillus species.* Antagonistic activities of the isolated bacterial strains were screened against a neither nonaflatoxigenic nor mutant of *A. flavus*, which accumulates the pigmented AF precursor norsolorinic acid (NOR) under conditions conducive to AF production. 171 bacteria isolated from almond flowers, immature nut fruits, and mature nut fruits showed inhibition of A. *flavus* growth and/or inhibition of NOR accumulation. Bacterial isolates were further characterized for production of extracellular enzymes capable of hydrolyzing chitin or yeast cell walls. Molecular and physiological identification of the bacterial strains indicated that the predominant genera isolated were Bacillus, Pseudomonas, Ralstonia, and Burkholderia, as well as several plant-associated enteric and nonenteric bacteria [61].

Chang & Hua in 2007, from screening subgroups of nonaflatoxigenic *A. flavus,* identified an *A. flavus* isolate, TX9-8, which competed well with three *A. flavus* isolates producing low, intermediate, and high levels of AF, respectively. TX9-8 has a defective polyketide synthase gene (pksA), which is necessary for AF biosynthesis. Co-inoculating TX9-8 at the same time with large sclerotial (L strain) *A. flavus* isolates at a ratio of 1:1 or 1:10 (TX9-8:toxigenic) prevented AF accumulation. The intervention of TX9-8 on small sclerotial (S strain) *A. flavus* isolates varied and depended on isolate and ratio of co-inoculation. At a ratio of 1:1 TX9-8 prevented AF accumulation by *A. flavus* CA28 and reduced AF accumulation 10-fold by *A. flavus* CA43. No decrease in AF accumulation was apparent when TX9-8 was inoculated 24 h after toxigenic L- or S strain *A. flavus* isolates started growing so the competitive effect likely is due to TX9-8 outgrowing toxigenic *A. flavus* isolates [62].

In 2009, it was reported that *Serratia plymuthica* 5-6, isolatedfromthe rhizosphere of pea reduced dry rot of potato caused by *Fusarium sambucinum* [63]. In 2009, a new strain of *Bacillus pumilus* isolated from Korean soybean sauce showed strong antifungal activity against the AF-producing fungi *A. flavus* and *A.parasiticus* [64].

In 2010, a strain of marine *Bacillus megaterium* isolated from the Yellow Sea of East China was evaluated by Kong and collaborators for its activity in reducing postharvest decay of peanut kernels caused by *Aspergillus flavus* in *in vitro* and *in vivo* tests, this, because microorganisms are capable of producing many unique bioactive substances, and therefore could be a rich resource for antagonists [65]. The results showed that the concentrations of antagonist had a significant effect on biocontrol effectiveness *in vivo*: when the concentration of the washed bacteria cell suspension was used at 1×109 CFU/ml, the percentage rate of rot of peanut kernels was 31.67%±2.89%, which was markedly lower than that treated with water (the control) after 7 days of incubation at 28 °C. The results also showed that unwashed cell culture of *B. megaterium* was as effective as the washed cell suspension, and better biocontrol was obtained when longer incubation time of *B. megaterium* was applied. When the incubation time of B. megaterium was 60 h, the rate of decay declined to 41.67%±2.89%. Furthermore, relative to the expression of 18S rRNA, the mRNA abundances of aflR gene and aflS gene in the experiment group were 0.28±0.03 and 0.024±0.005 respectively, indicating that this strain of *B. megaterium* could significantly reduce the biosynthesis of AF and expression of aflR gene and aflS gene [66].

In 2011, Degola and collaborators conducted a study in order to evaluate the potential of the different atoxigenic *A. flavus* strains, colonizing the corn fields of the Po Valley, in reducing AF accumulation when grown in mixed cultures together with atoxigenic strains; additionally, they developed a simple and inexpensive procedure that might be used to scale-up the screening process and to increase knowledge on the mechanisms interfering with mycotoxin production during co- infection [54].

Farzaneh and collaborators reported in this year, an investigation in which *Bacillus subtilis* strain UTBSP1 was isolated from pistachio nuts and studied for the degradation of AFB1. The results indicated *B. subtilis* UTBSP1 could considerably remediate AFB1 from nutrient broth culture and pistachio nut by 85.66% and 95%, respectively. Cell free supernatant fluid caused an apparent 78.39% decrease in AFB1 content. The optimal conditions for AFB1 degradation by cell free supernatant appeared at 35 and 40°C, during 24 h. Furthermore, the results indicated that AFB1 degradation is enzymatic and responsible enzymes are extracellular and constitutively produced. They found that destructive AFB1 differed from standard AFB1 chemically, and lost a fluorescence property [67].

It was found that *A. flavus* K49 produces neither AFs nor cyclopiazonic acid (CPA) and is currently being tested in corn-growing fields in Mississippi. Its lack of production of AF and CPA results from single nucleotide mutations in the polyketide synthase gene and hybrid polyketide nonribosomal peptide synthase gene, respectively. Furthermore, based on single nucleotide polymorphisms of the AF biosynthesis omtA gene and the CPA biosynthesis dmaT gene, it is known that K49, AF36 and TX9-8 form a biocontrol group, appear to be derived from recombinants of typical large and small sclerotial morphotype strains [50].

Not only Aspergillus, but also other pathogens have been faced to biocontrol. For example, it is known that the plant pathogen *Fusarium solani* causes a disease root rot of common bean (*Phaseolus vulgaris*) resulting in great losses of yield in irrigated areas. Species of the genus Trichoderma have been used in the biological control of this pathogen as an alternative to chemical control. To gain new insights into the biocontrol mechanism used by *Trichoderma harzianum* against the phytopathogenic fungus, *Fusarium solani,* it was performed a transcriptome analysis using expressed sequence tags (ESTs) and quantitative real-time PCR

(RT-qPCR) approaches. A cDNA library from *T. harzianum* mycelium (isolate ALL42) grown on cell walls of *F. solani* (CWFS) was constructed and analyzed. A total of 2927 high quality sequences were selected from 3845 and 37.7% were identified as unique genes. The Gene Ontology analysis revealed that the majority of the annotated genes are involved in metabolic processes (80.9%), followed by cellular process (73.7%). Genes that encode proteins with potential role in biological control have been tested. RT-qPCR analysis showed that none of these genes were expressed when *T. harzianum* was challenged with itself. These genes showed different patterns of expression during in vitro interaction between *T. harzianum* and *F. solani* [68].

It is a fact that several papers have been published about AFB1 reduction by some bacterial isolates. Lactic acid bacteria such as Lactobacillus, Bifidobacterium, Propionibacterium and Lactococcus were found to be active in removing AFB1 primarily by the adhesion method. In addition, some bacteria such as *Rhodococcus erythropolis,* Bacillus sp., *Stenotrophomonas maltophilia, Mycobacterium fluoranthenivorans* and *Nocardia corynebacterioides* were reported to degrade AFB1 [67].

5.2.2. Natural products and essential oils

Plants produce lots of secondary metabolites as part of their normal growth and development in order to fight against environmental stress, pathogen attack or other adversities. One of the most important secondary metabolites are essential oils (EOs), which are extracted from plants, commonly by a distillation process [69] and then used as natural additives in different foods to reduce the proliferation of microorganisms and their toxins production due to their antifungal, antiviral, antibacterial, antioxidant and anticarcinogenic properties [70-72]. They have received major consideration in regard to their relatively safe status and enrichment by a wide range of structurally different useful constituents [73]. Until 1989, more than 1340 plants were known to be potential sources of antimicrobial compounds, which are safe for the environment and consumers, and are useful to control postharvest diseases, being an excellent alternative to reduce the use of synthetic chemicals in agriculture. The majorities of the essential oils are classified as Generally Recognized As Safe (GRAS) and have low risk for developing resistance to pathogenic microorganisms [74, 75].

There is a large number of different groups of chemical compounds present in EOs, that is why antimicrobial activity is not attributable to one specific mechanism but to the existence of several targets in the cell [76, 77]. There is a relationship between the chemical structures of the most abundant compounds in the EOs and the anitimicrobial activity; minor components have a critical part to play in antimicrobial activity, possibly by producing a synergic effect between other components [78]. Not only EOs but also alkaloids, phenols, glycosides, steroids, coumarins and tannins have been found to have antimicrobial properties [79]. Generally, the extent of the inhibition of the oils could be attributed to the presence of an aromatic nucleus containing a polar functional group [80], being phenols the majority group. For example, in 2008, Bluma and Etcheverry, based in the principle that phenolics are secondary metabolites synthesized via phenylpropanoid biosynthetic pathway which build blocks for cell wall structures serving as defense against pathogens, found that phenolic compounds such as acetocyringone, syringaldehyde and sinapinic acid inhibit AFB1 biosynthesis by *A. flavus* in PDA and reduce norsolinic acid production, because the presence of phenolic OH groups are able to form hydrogen bonds with the active sites of target enzymes increasing antimicrobial activity [69].

There is a wide list of natural products from the entire world (summarized in Table 1) used in the last decade to diminish Aspergillus populations to counteract the effect of AFs in food or to test fumigant activity in feed at specific inhibitory concentrations [81]. It has been demonstrated that the antifungal capability of those EOs depend on the concentration in which they are applied and the conditions around them. In 2001, Varma and Dubey reported that EOs from plants like *Caesulia axillaris* and *Mentha arvensis* have fumigant activity in the management of biodeterioration of stored wheat samples by *A. flavus* showing the same efficacy as postharvest fungicides used for this purpose [38]. In 2002, Soliman and Badeaa tested inhibitory activity of essential oils from 12 medicinal plants against *A. flavus, A. parasiticus, A. ochraceus* and *Fusarium moniliforme,* finding that the oils of thyme and cinnamon (at a 4500 ppm concentration), marigold (42000 ppm), spearmint, basil and quyssum (3000 ppm) completely inhibit all the test fungi. Caraway was inhibitory at 2000 ppm against A. *flavus, A. parasiticus* and 3000 ppm against *A. ochraceaus and F. moniliforme. A. flavus, A. ochraceus, A. parasiticus* and *F. moniliforme* were completely inhibited by anise at 4500 ppm, being chamomile and hazanbul essential oils just partially effective against the test toxigenic fungi [71].

NATURAL PRODUCT	COMMON NAME	PRINCIPAL METABOLITE	PATHOGEN INHIBITED	INHIBITORY CONCENTRATION	REFERENCE
Achillea fragrantissima	Qyssum	Polyphenolic compounds	A. flavus, A. parasiticus, A. ochraceus	3,000 ppm	[71]
Agave asperrima	Maguey Cenizo	Polyphenolic compounds	A. flavus A. parasiticus	< 2 mg ml^{-1}	[15]
Agave striata	Maguey Espadín	Polyphenolic compounds	A. flavus A. parasiticus	< 2 mg ml^{-1}	[15]
Ageratum conyzoides	Goatweed	Precocene, Cumarine, trans-Caryophyllene	A.flavus	0.10 μg ml^{-1}	[91]
Azadirachta indica A. Juss	Neem	Aromatic compounds	*A. parasiticus*	"/ 10% (v/v)	[81]
Caesulia axillaris	Pink Node Flower	Aromatic compounds	*A. flavus*	nd	[38]

NATURAL PRODUCT	COMMON NAME	PRINCIPAL METABOLITE	PATHOGEN INHIBITED	INHIBITORY CONCENTRATION	REFERENCE
Calendula ofricinalis L.	Marigold	Carfone	A. flavus, A. parasiticus, A. ochraceus	< 2,000 ppm	[71]
Carum carvi L.	Caraway	Carfone	A. flavus, A. parasiticus, A. ochraceus	2,000 – 3,000 ppm	[71]
Cicuta virosa L. var. latisecta Celak	Umbelliferae	γ-Terpinene p-Cymene Cumin Aldehyde	A. flavus	5 µl ml^{-1}	[75]
Cinnamomum cassia	Cassia	Aromatic compounds	*A. parasiticus*	2.5 % (v/v)	[79]
Cinnamomum zeylanicum L.	Cinnamon	Cinnamic aldehyde O-methoxy-cinnamaldehyde Carfone	A. flavus, A. parasiticus, A. ochraceus	200 – 250 ppm, < 500 ppm	[71, 83, 85]
Citrus limon	Lemon	Limomene	A. flavus	2, 000 ppm	[13]
Cymbopogon citratus	Lemongrass	Citral, geraniol, eugenol, α-pinene, linalool	A.flavus	1 – 5%, 1,200 ppm	[81, 83]
Eucalyptus globulus	Blue Gum	1,8-cineole	A.flavus A. parasiticus	nd	[86]
Hedeoma multiflora Benth	Mountain Thyme	α-Terpinene ∂-Terpinene ρ-Cimeno o-Cimeno Borneol Thymol Carvacrol	A. flavus, A. parasiticus	2,000 – 3,000 µg g^{-1}	[69]
Laurus nobilis	Bay Leaf	Aromatic compounds	*A. parasiticus*	1 – 5 % (v/v)	[79]

NATURAL PRODUCT	COMMON NAME	PRINCIPAL METABOLITE	PATHOGEN INHIBITED	INHIBITORY CONCENTRATION	REFERENCE
Lippia turbinate var. integrifolia (griseb)	Poleo	β-Cariofilene α-Humulene Camfene Sabinene	A. flavus, A. parasiticus,	2,000 – 3,000 μg g^{-1}, 2500 μl l^{-1}	[69, 95]
Mentha arvensis	Wild Mint	Menthone Menthol	A. flavus	nd	[38]
Mentha viridis	Spearmint	Menthone Menthol β-pinene α-pinene	A. flavus, A. parasiticus, A. ochraceus	3,000 ppm	[71]
Ocimum basilicum	Sweet Basil	β-pinene α-pinene Ocimene Methyl Chavecol	A. parasiticus	5% (v/v)	[71, 79]
Ocimum basilicum L	Basil	β-pinene α-pinene Ocimene Methyl Chavecol	A. flavus, A. parasiticus, A. ochraceus	3,000 ppm	[71]
Ocimum gratissimum	Clove Basil	γ-terpinene Methyl cinnamate	F.moniliforme, A.flavus A. fumigatus	800 ppm	[83, 93]
Origanum vulgare	Oregano	γ-terpinene p-cimeme Linalool Cariophyllene	A. flavus	500 μg g^{-1}, 100 – 2,000 ppm	[81, 85]
Pëumus boldus	Boldo	α-Pinene β-Pinene α-Terperpine ρ-Cimene Terpinen-4-ol α-Terpinolene	A. flavus, A. parasiticus	2,000 – 3,000 μg g^{-1}, 2500 μl l^{-1}	[69, 95]
Pimpinella anisum L.	Anise	Metilchavicol Anethol	A. flavus, A. parasiticus, A. ochraceus	< 500 ppm	[71]
Satureja hortensis L.	Winter Savory	Carvacrol Thymol	*A. parasiticus*	~0.5 mM	[81, 87]

NATURAL PRODUCT	COMMON NAME	PRINCIPAL METABOLITE	PATHOGEN INHIBITED	INHIBITORY CONCENTRATION	REFERENCE
Syzygium aromaticum	Clove	Humulene Cariophyllene Eugenol	A. flavus, A. parasiticus	1500 μl l^{-1}	[95]
Thymus eriocalyx	Avishan	Thymol β-phellandrene cis-sabinene hydroxide 1,8-cineole β-pinene	A. parasiticus	250 ppm	[84]
Thymus vulgaris L.	Thyme	β-pinene α-pinene Thymol p- cymene	A. flavus, A. parasiticus, A. ochraceus	< 500 ppm, 1000 ppm	[71, 83]
Thymus X-porlock	Thyme	Thymol β-phellandrene cis-sabinene hydroxide 1,8-cineole β-pinene	A. parasiticus	250 ppm	[84]
Trachyspermum ammi (L.)	Ajowan	Aromatic compounds	A. flavus	1 g ml^{-1}	[92]
Zingiber officinale	Ginger	Polyphenolic compounds	A.flavus	800 – 2,500 ppm	[83]

Table 1. Metabolites obtained from some natural products which are used to diminish fungal populations and AF production (nd= no data).

EOs and other natural products have been tested not only against Aspergillus species but also Fusarium species, which most of the times are developed in parallel. In 2003, Vellutti and collaborators reported the effect of cinnamon, clove, oregano, palmarose and lemongrass oils on fumonisin B1 growth and production by three different isolates of *F. proliferatum* in irradiated maize grain at 0.995 and 0.950 aw and at 20 and 30°C. The five essential oils inhibited growth of *F. proliferatum* isolates at 0.995 aw at both temperatures, while at 0.950 aw only cinnamon, clove and oregano oils were effective in inhibiting growth of *F. proliferatum* at 20°C and none of them at 30°C. Cinnamon, oregano and palmarose oils had significant inhibitory effect on FB1 production by the three strains of *F. proliferatum* at 0.995 aw and both temperatures, while clove and lemongrass oils had only significant inhibitory effect at 30°C [81]. In 2004, Nguefack and his group of researchers tested the inhibitory effect of EOs extracted from *Cymbopogon citratus, Monodora myristica, Ocimum gratissimum, Thymus vulgaris* and *Zingiber officinale* against *F. moniliforme*, being *O. gratissimum, T. vulgaris* and *C. citratus* the most effective over conidial germination and fungal growth at 800, 1000 and

1200 ppm, respectively. Moderate activity was observed for the EO from Z. *officinale* between 800 and 2500 ppm, while the EO from *M. myristica* was less inhibitory. These effects against food spoilage and mycotoxin producing fungi indicated the possible ability of each EO as a food preservative [83].

In 2005, Sánchez and collaborators prepared ethanolic, methanolic and aqueous extracts of flowers from mexican *Agave asperrima* and *Agave striata,* in order to diminish growth and production of AF from *A. flavus* and *A. parasiticus* at in vitro and in vivo level. All extracts, but specifically the methanolic one, showed an effective inhibition growth (99%) [15]. In the same year, Rasooli & Owlia extracted the EOS from *Thymus eriocalyx* and *Thymus X-porlock* in order to test antifungal activity against *A. parasiticus* growth and AF production. *T. eriocalyx* showed lethal effects at 250 ppm while *T. X-porlock* was lethal at 500 ppm [84].

EOs from common spices have been also investigated, that is the case of cinnamon (*Cinnamomum zeylanicum*) and oregano (*Origanum vulgare*) which shows antifungal activity against *A. flavus* at 2000 ppm and 1000 ppm respectively in a malt-agar medium and a fungistatic activity at 100 ppm. [85]. Eucalyptus (*Eucalyptus globules)* is effective against the storage fungi *A. flavus* and *A. parasiticus* [86]. Lemon EO (*Citrus limon*), applied in food AF-contaminated samples, results in a strong antiaflatoxigenic and antifungal substance, reducing AF concentrations in food samples for broilers up to 73.6% [13]. Sweet basil (*Ocimum basilicum*), cassia (*Cinnamomum cassia*), coriander (*Coriandrum sativum*) and bay leaf (*Laurus nobilis*) at 1–5% (v/v) concentration were studied in palm kernel over the aflatoxigenic fungus *A. parasiticus* CFR 223 and AF production. Sweet basil oil at optimal protective dosage of 5% (v/v) was fungistatic on *A. parasiticus*; in contrast, oils of cassia and bay leaf stimulated the mycelia growth of the fungus in vitro but reduced the AF concentration (AFB1+AFG1) of the fungus by 97.92% and 55.21% respectively, while coriander oil did not have any effect on both the mycelia growth and AF content of the fungus. The combination of cassia and sweet basil oils at half their optimal protective dosages (2.5% v/v) completely inhibited the growth of the fungus. It was found that the addition of whole and ground basil leaves markedly reduced AF contamination; however, 10% (w/w) of whole leaves was more effective as the reduction in AF was between 89.05% and 91% [79].

In 2008, Bluma and Etcheverry found that *Pimpinella anisum* L. (anise), *Pëumus boldus* Mol (boldus), *Hedeoma multiflora* Benth (mountain thyme), *Syzygium aromaticum* L. (clove), and *Lippia turbinate* var. *integrifolia* (griseb) (poleo) had an inhibitory effect on Aspergillus section Flavi growth rate, and their efficacy depended mainly on the water activity and EOs concentration. Boldus, poleo, and mountain thyme EOs completely inhibited AFB1 at 2000 and 3000 $\mu g\ g^{-1}$ [69]. *Satureja hortensis* L. has been also reported as a potent inhibitor of AFB1 and AFG1 produced by *A. parasiticus* at concentrations from 0.041 to 1.32 mM [87]. In 2009, Kumar and collaborators found that *Cymbopogon flexuosus* EO and its components were efficient in checking fungal growth and AF production, inhibiting absolutely inhibited the growth of *A. flavus* and AFB1 production at 1.3 $\mu l ml^{-1}$ and 1.0 $\mu l ml^{-1}$ respectively, due to the principal component: eugenol [88]. Razzaghi-Abyaneh and his investigation group found that *Thymus vulgari* and *Citrus aurantifolia* inhibit both *A. parasiticus* and AF production. The EOs from *Mentha spicata* L., *Foeniculum miller*, *Azadirachta indica* A. Juss, *Conium maculatum* and *Artemisia dracunculus* only inhibited fungal growth, while *Carum carvi* L. effectively inhibited AF production without any obvious effect on fungal growth. *Ferula gummosa, Citrus sinensis, Mentha longifolia* and *Eucalyptus camaldulensis* had no effect on *A. parasiticus* growth and AF production at all concentrations used [89]. There are other investigations of the potential use of antifungal component eugenol for the reduction of AFB1. Komala and collaborators reported some findings in stored sorghum grain due to fungal infestation of sorghum results in deterioration of varied biochemical composition of the grain. In this study, three genotypes (M35-1; C-43; LPJ) were inoculated with two highly toxigenic strains of *Aspergillus flavus* with three different eugenol treatments in order to evaluate the AFB1 production. From this study it was found that at 8.025 mg/g concentration, eugenol completely inhibited the AFB1 production. The lowest amount of AFB1 was observed in genotype M35-1, whereas higher amount AFB1 was observed in LPJ followed by C-43. In all sorghum genotypes there was a significant positive correlation existing between protein content and AF produced, the r values being 0.789 and 0.653, respectively. Starch in three genotypes was found to have a significant negative correlation with AF produced. The starch content decreased whereas the protein content in all sorghum varieties increased during infection [90].

Ageratum conyzoides EO is other specie that has been studied recently. It acts directly on the mycelial growth and AFB1 production by *A. flavus*, inhibiting fungal growth to different extents depending on the concentration, and completely inhibiting AF production at concentrations above 0.10 μg/mL, because this EO acts affecting mainly the fungal mitochondria [91]. This EO acts similarly than Ajowan extract (*Trachyspermum ammi* L., which acts directly over AFB1, AFB2 and AFG2 [92]. In 2011, it was found that *Ocimum gratissimum* EO acts a nontoxic antimicrobial and antiaflatoxigenic agent against fungal and AF contamination of spices infected with *A. flavus* isolated from *Piper nigrum* and *Myristica fragrans* respectively at 0.6 μl/ml and 0.5 μl/ml, as well as a shelf life enhancer in view of its antioxidant activity, playing a prominent role in the development of an ideal plant based food additive [93]. It was found too that EOs extracted from the fruits of *Cicuta virosa* L. var. latisecta Celak acts against *A. flavus, A, oryzae, A. niger,* and *Alternaria alternata*, having a strong inhibitory effect on spore production and germination in all tested fungi proportional to concentration. The oil exhibited noticeable inhibition on dry mycelium weight and synthesis of AFB1 by *A. flavus*, completely inhibiting AFB1 production at 4 μL/mL [75].

Because of the great results obtained with this kind of AFs biocontrol, researchers are still investigating new natural products and their active compounds in order to deal with those toxins ad the fungi which produce them, and avoiding the use of fumigants that are toxic for plants and for plant consumers. In this year, EOs from plants like *Zanthoxylum alatum* Roxb have been studied, because it has been proved that its two major constituents (linalool and methyl cinnamate) inhibit the growth of a toxigenic strain of *A. flavus* (LHP-10) as well as AFB1 secretion at different concentrations. *Zanthoxylum alatum* Roxb EO has also showed strong antioxidant activity with an IC50 value at 5.6 μl/ml [94]. EOs from boldo, clove, anise and thyme are still studied against aflatoxigenic Aspergillus strains in specific cultures like peanut-based medium, finding that those EOs have influence on lag phase, growth rate, and AFB1 accumulation [95]. The EO extracted from the bark of *Cinnamomum jensenianum* Hand.

Mazz has been tested for antifungal activity against *A. flavus*. Mycelial growth and spore germination was inhibited by the oil in a dose-dependent manner. The oil also exhibited a noticeable inhibition on the dry mycelium weight and the synthesis of AFB1 by *A. flavus*, completely restraining AFB1 production at 6 μl/ml. The possible mode of action of the oil against *A. flavus* is discussed based on changes in the mycelial ultrastructure [37]. Nevertheless, most research is needed in order to understand the mechanisms of action of the essential oils over aflatoxigenic fungi, turning them into potential sources for food preservation.

5.3. Genetic Engineering: Molecular biology and genetics proposals

The genome of plants has significant influence on fungal contamination and the subsequent biosynthesis of mycotoxins, hence, the importance of developing new varieties through genetic engineering, capable of withstanding the fungal attack or inhibiting toxin production. Several researchers have found some seed varieties with significant differences in regard to contamination by *Aspergillus flavus* and its subsequent AF production. These differences may be due to different factors, and the plant genome can influence the expression of the mycotoxin biosynthesis [95]. Various approaches have been suggested for genetic control of preharvest AF contamination including the development and use of crops with resistance to insects and resistance to plant stress (especially for tolerance to drought and high temperatures). Several sources of resistant germplasm have been identified and released for crop genetic improvement [95]. Using a combination of genetic, genomic and proteomic approaches to elucidate crop defense mechanisms and their genetic regulation will significantly improve the efficiency of genetic breeding for better crop cultivars [98].

One of the most important challenges in AFs genetic engineering has been the identification of the genes that are present in aflatoxigenic strains but not in the non-toxigenic ones, in order to design in the laboratory non-toxigenic strains by manipulating the genes of toxigenic strains. The AF pathway genes are found to be clustered in the genome of *A. flavus* and *A. parasiticus*. These genes are expressed concurrently except for the regulatory gene *aflR*. In this gene cluster, a positive-acting regulatory gene, *aflR*, is located in the middle of the gene cluster. Adjacent to *aflR* a divergently transcribed gene, *aflS* (*aflJ*), was also found to be involved in the regulation of transcription. Other physically unrelated genes, such as *laeA* and *veA*, also have been shown to exhibit a "global" regulatory role on AF biosynthesis [98]; nevertheless, although the basis of the toxigenic activity of AF are being well investigated, more research is still needed in order to get more information about how to manipulate genes in the different strains present in different crops and foods.

AF are synthesized by enzymes encoded within a large gene cluster. The initial step in the generation of the polyketide backbone of AF is proposed to involve polymerization of acetate and nine malonate units (with a loss of CO2) by a polyketide synthetase in a manner analogous to fatty acid biosynthesis. AF synthesis is controlled by different enzymes which are expressed through gene expression processes. Genetic studies on AF biosynthesis in *Aspergillus flavus* and *Aspergillus parasiticus* led to the cloning of 25 clustered genes within a 70 kb DNA region responsible for the enzymatic conversions in the AF biosynthetic pathway.

Regulatory elements such as *aflR* and *aflS (aflJ)* genes, nutritional and environmental factors, fungal developmental and sporulation were also found to affect AF formation [31].

Aflatoxigenic *Aspergillus flavus* isolates show four DNA fragments specific for *aflR, nor-1, ver-1, and omt-A* genes. Non-aflatoxigenic A. flavus strains give variable DNA banding pattern lacking one, two, three or four of these genes. Recently, it has been found that some AF non-producing *A. flavus* strains show a complete set of genes. Some studies suggest that 36.5% of non-aflatoxigenic *A. flavus* strains show DNA fragments that correspond to the complete set of genes (quadruplet pattern) as in aflatoxigenic *A. flavus*; 32% shows three DNA banding patterns grouped in four profiles where *nor-1, ver-1 and omt-A* are the most frequent profile; 18.7% of non-aflatoxigenic *A. flavus* strains yield two DNA banding pattern whereas 12% of the strains show one DNA banding pattern [99]. The *aflR* gene, encoding a 47 kDa sequence-specific zinc-finger DNA-binding protein is required for transcriptional activation of most, if not all, the structural genes of the AF gene cluster. Like other Gal4-type regulatory proteins that bind to palindromic sequences, functional AflR probably binds as a dimer. It binds to the palindromic sequence 5'-TCGN5CGR-3' in the promoter regions of the structural genes. The AflR-binding motifs are found to be located from 80 to 600 bp, with the majority at the 100 to 200 bp, relative to the translation start site. AflR binds, in some cases, to a deviated sequence rather than the typical motif such as in the case of *aflG (avnA)*. When there is more than one binding motif, only one of them is the preferred binding site such as in the case of *aflC (pksA)*. Deletion of *aflR* in *A. parasiticus* abolishes the expression of other AF pathway genes. Overexpression of *aflR* in *A. flavus* up-regulates AF pathway gene transcription and AF accumulation. AflR is specifically involved in the regulation of AF biosynthesis [98].

The *aflS* (*aflJ*) gene, although not demonstrating significant homology with any other encoded proteins found in databases, is necessary for AF formation. In the *A. parasiticus aflR* transformants, the production of AF pathway intermediates was significantly enhanced in transformants that contained an additional *aflR* plus *aflS*. Quantitative PCR showed that in the *aflS* knockout mutants, the lack of *aflS* transcript is associated with 5- to 20-fold reduction of expression of some AF pathway genes such as *aflC* (*pksA*), *aflD* (*nor-1*), *aflM* (*ver-1*), and *aflP* (*omtA*). The mutants lost the ability to synthesize AF intermediates and no AFs were produced. However, deletion of *aflS* (*aflJ*) did not have a discernible effect on *aflR* transcription, and vice versa. Overexpression of *A. flavus aflS* (*aflJ*) does not result in elevated transcription of *aflM* (*ver-1*), *aflP* (*omtA*), or *aflR*, but it appears to have some effect on *aflC* (*pksA*), *aflD* (*nor-1*), *aflA* (*fas-1*), and *aflB* (*fas-2*), which are required for the biosynthesis of the early AF pathway intermediate, averantin [98, 100, 101].

The global regulatory gene, *laeA* (for lack of *aflR* expression), is well conserved in fungi as shown by its presence in the genomes of all fungi so far sequenced. LaeA is a nuclear protein which contains an S-adenosylmethionine (SAM) binding motif and activates transcription of several other secondary metabolism gene clusters in addition to the AF cluster. It also regulates some genes not associated with secondary metabolite clusters, but this mechanism is not known yet. One proposed regulatory mechanism is that LaeA differentially methylates histone protein and it alters the chromatin structure for gene expression [98]. Recent analyses of nonaflatoxigenic *A. parasiticus sec-* (for secondary metabolism negative) variants generated through serial transfer of mycelia of the *sec+* parents show that *laeA* is expressed in both *sec+* and *sec-* strains, suggesting that LaeA only exerts its effect on AF biosynthesis at a certain level and is independent of other regulatory pathways that are involved in fungal development [102].

The *veA* gene is initially found to be crucial for light-dependent conidiation. The light dependence is abolished by a mutation (*veA1*) which allows conidiation of *A. nidulans* to occur in the dark. A comparison of the light effect on sterigmatocystin production by *A. nidulans veA+* and *veA1* strains showed that both strains produced sterigmatocystin but the highest amount was produced by the *veA+* strain grown in darkness. However, *veA*-deleted *A. flavus* and *A. parasiticus* strains completely lost the ability to produce AF regardless of the illumination conditions [103, 104]. Under normal growth conditions, some *A. flavus* and all *A. parasiticus* strains produce conidia in both dark and light conditions. VeA contains a bipartite nuclear localization signal (NLS) motif and its migration to the nucleus is light-dependent and requires the importing α carrier protein. In the dark VeA is located mainly in the nucleus; under light it is located both in cytoplasm and nucleus. VeA has no recognizable DNA-binding sequences and likely exerts its effect on sterigmatosyctin and AF production through proteinprotein interactions with other regulatory factors. Post- translational modifications such as phosphylation and dephosphorylation may modulate its activity. Lack of VeA production in the *veA*-deleted *A. flavus* and *A. parasiticus* strains consequently abolishes AF production because a threshold concentration of nuclear VeA might be necessary to initiate AF biosynthesis [98, 104]. One of the approaches in the field of AF research with regard to proteomics is to study the AF resistance proteins in host plants such as corn. The investigation on proteins associated with host resistance has been shown to be a possible strategy for controlling AF contamination of plants [105, 106].

An important factor affecting the agricultural commodities is the drought stress. Pre-harvest contamination of maize, peanuts and other products with AFs has been observed to be higher especially in the drought years, having devastating economical [106]. Guo and collaborators reviewed the potential of genetics, genomics and proteomics in understanding the relationship between drought stress and preharvest AF contamination in agricultural products. Different proteomic approaches revealed that resistant lines have elevated levels of stress-related proteins, antifungal and storage proteins in comparison to susceptible lines [95]. The use of proteomic tools has made possible to find different categories of resistance associated proteins which can be divided into 3 groups: stress-responsive proteins, storage proteins and antifungal proteins indicating that storage and stress-responsive proteins may play an important role in enhancing stress-tolerance of host plant [106, 107]. The use of proteomics is still a new tool to understand plant resistance against fungal contamination, so it promises to become an important field for understanding fungal genetic behavior.

5.4. Storage and packing technologies

As mentioned before, it is well known that AF contamination of foods increase with storage period. That is why proper selection of packaging materials is necessary to prevent absorption of moisture and AF formation which will influence the overall product quality and safety [19, 108]. Postharvest contamination of grain can also take place during transportation, so grains need to be well covered and/or aerated during transportation [19]. Storage prior and during marketing has to be done in appropriate bagging, preferably sisal bags, because this kind of material facilitates aeration in transit. The use of containers made from plant materials (wood, bamboo, thatch) or mud placed on raised platforms and covered with thatch or metal roofing sheet is another way to prevention. The stores should be constructed to prevent insect and rodent infestation and to prevent moisture from getting into the grains. While new storage technologies such as the use of metal or cement bins by small-scale farmers would serve better, their uptake has been slow due to their high cost. Many farmers nowadays store their grains in bags, especially polypropylene which are not airtight, but there is evidence that this method facilitates fungal contamination and AF development [19, 109, 110]. Presently there are efforts to market improved hermetic storage bags in Africa, based on triple bagging developed for cowpea which has been or is being tested for other commodities [19].

Not only optimal storage plastic bagging and container materials have been proposed. Shakerardekani and Karim reported in 2012 a short communication in which they studied the effect of five different types of flexible packaging films (low density polyethylene (LDPE) which served as the control, food-grade polyvinyl chloride (PVC), nylon (LDPE/PA), polyamide/polypropylene (PA/PP) and polyethylene terephthalate (PET)) on the moisture and AF contents of pistachio nuts during storage at room temperature (22–28 °C) and relative humidity of 85–100%. Samples were analyzed at 0, 2, 4, 6, 8 and 10 months during the storage period. Results showed that there was an increase in moisture content with the increase in storage time of pistachio nuts. The increase in moisture content was associated with the AF level of pistachio nuts during storage time. All the packaging materials except LDPE delayed the moisture absorption and AF formation of the product. The most suitable packaging materials for maintaining the quality and safety of pistachio nuts were PET films followed by nylon, PA/ PP and PVC. The shelf-life of pistachio showed to be extended from 2 months (Control) to 5 months when PET was used as the packaging material [108].

In the market, there are some products that have been proved recently on grain shelf-life extension. This is the case of Mater-Bi® (MB), a bioplastic product composed of starch, polycaprolactone (e-caprolactone) and a minor amount of a natural plasticizer, being a reliable and readily adaptable product currently used for making shopping bags, biofillers, agricultural films and a number of other commercial products [111]. Moreover, MB is completely biodegradable, having a rate of breakdown similar to that of cellulose, having a highly favorable low environmental impact profile [112]. Based in MB properties and reviewing previous research that demonstrated that AF contamination in corn is reduced by field application of wheat grains pre-inoculated with the non-aflatoxigenic *Aspergillus flavus* strain NRRL 30797, Accinelli and collaborators in 2009 conducted a series of laboratory studies on the reliability and efficiency of replacing wheat grains with the novel bioplastic formulation Mater-Bi® to serve as a carrier matrix to formulate this fungus. Mater-Bi® granules were inoculated with a conidial suspension of NRRL 30797 to achieve a final cell density of approximately log 7 conidia / granule. Incubation of 20-g soil samples receiving a single Mater-Bi® granule for 60-days resulted in log 4.2–5.3 propagules of *A. flavus* / g soil in microbiologically active and sterilized soil, respectively. Increasing the number of granules had no effect on the degree of soil colonization by the biocontrol fungus. In addition to the maintenance of rapid vegetative growth and colonization of soil samples, the bioplastic formulation was highly stable, indicating that Mater-Bi® is a suitable substitute for biocontrol applications of *A. flavus* NRRL 30797 [43].

Nowadays, the use of biopolymer covers on seeds has been a successful and economic biocontrol method. The most used is chitosan, a biopolymer which is found naturally in cell walls of certain fungi, but which primary production source is the hydrolysis of chitin in alkaline medium at high temperatures [113]. Chitosan is known for its antifungal and antimicrobial properties, and it can be used in solution, films, spheres, hydrogels, nanoparticles, fibers and coatings, which makes it useful for a variety of applications in different areas [114]. Since the nineties, chitosan has been used to coat fruits and vegetables because of its bactericidal and fungicidal properties, and its ability to form films favoring the preservation of products due to the modification of the internal atmosphere and reduced transpiration losses. In addition, the coating gives the fruit more firmness and promotes the reduction of microbial development [113, 115, 116]. Due to the success of the results obtained using chitosan as a biocide, a large number of researchers all over the world have applied chitosan in seeds under storage conditions, reporting a favorable decrease on storage fungi even under high humidity conditions and thereby decreasing the amount of mycotoxins developed in the grain [116, 117].

In 2011, Lizárraga-Paulín and collaborators reported their findings about the use of chitosan in maize against *Aspergillus flavus* and *Fusarium moniliforme*. The objective of this research was to determine the protective effect of chitosan in maize seedlings subjected to the fungi mentioned above. In order to achieve the aim, after some quality tests, three groups of seeds were separately subjected to attacks by *Aspergillus flavus* and *Fusarium moniliforme*. A first group was considered as a positive control, another was coated with chitosan solution and, a final group was mechanically damaged before application of the biopolymer. In the fifth week of growth, leaf structures of the seedlings were planted in agar PDA in order to determine the presence of stressful-fungi. It was found that leaves from the seeds treated with chitosan developed no fungal burden, suggesting that chitosan acts as an activator of defense mechanisms in maize seedlings, preventing infection by the pathogenic fungi and turning chitosan recovering into a good method to storage maize seeds under adverse conditions [118]. More research is needed in order to determine if not only *A. flavus* and *F. moniliforme* but also AF and fumonisins development can be prevented since seed level.

6. Conclusions

The use of biotechnological methods is a promising tool based on the use of biological systems, living organisms or their derivatives, and focused not only on increasing agricultural products quality, but also on the development of new approaches for fighting against AF and avoiding diseases caused by this threat. The use of new materials like biopolymers and biodegradable plastics on crops seems to be more effective against toxins, and moreover, they have the capability to replace substances that are harmful for health, avoiding contamination and offering the consumer better and uncontaminated products.

Acknowledgements

To PAPIIT program No. IT220411 "Estudio fitopatológico, bioquímico y molecular de la respuesta contra estreses bióticos y abióticos en plántulas de maíz" for financial support.

Author details

Eva Guadalupe Lizárraga-Paulín[1*], Susana Patricia Miranda-Castro[2], Ernesto Moreno-Martínez[2], Irineo Torres-Pacheco[3] and Alma Virginia Lara-Sagahón[1]

*Address all correspondence to: evee_123@hotmail.com

1 Laboratorio de Biotecnología, Facultad de Estudios Superiores Cuautitlán, Campo 1, UNAM. Cuautitlán Izcalli, Estado de México, México

2 Unidad de Investigación en Granos y Semillas, Facultad de Estudios Superiores Cuautitlán, CAT, UNAM. Cuautitlán Izcalli, Estado de México, México

3 CA Ingeniería de Biosistemas, División de Investigación y Posgrado, Facultad de Ingeniería, Universidad Autónoma de Querétaro, Cerro de las Campanas s/n, 76010, Querétaro, Qro., México

References

[1] IARC. (2002). Some traditional herbal medicines, some mycotoxins, naphthalene and styrene. 8 Summary of data reported and evaluation. IARC Monographs on the evaluation of the carcinogenic risk to humans. *International Agency for Research on Cancer, Lyon, France.*

[2] Frisvad, J. C., Smedsgaard, J., Larsen, T. O., & Samson, R. A. (2004). Mycotoxins, drugs and other extrolites produced by species in Penicillium subgenus Penicillium. *Studies in Mycology*, 49, 201-241.

[3] Asao, T., Buchi, G. H., Abdel-Kader, M. M., Chang, S. B., Wick, E. L., & Wogan, G. N. (1965). The structures of aflatoxins B1 and G1. *Journal of the American Chemical Society*, 87-882.

[4] D'Mello, J. P. F., & Macdonald, A. M. C. (1997). Mycotoxins. *Animal Feed Science and Technology*, 69(1997), 155-166.

[5] Strosnider, H., Azziz-Baumgartner, E., Banziger, M., Bhat, R. V., Breiman, R., & Wilson, D. (2006). Public Health Strategies for Reducing Aflatoxin Exposure in Developing Countries: A Workgroup Report. *Environmental Health Perspectives*, 114(12), 1898-1903.

[6] Gimeno. (2004). Aflatoxina M1 no leite. Riscos para a saúde pública, prevenção e controlo Alimentação Animal. *Revista de la Associação Portuguesa dos Industriais de Alimentos Compostos para Animais*, 49, 32-44.

[7] Wild, C. P., & Gong, Y. Y. (2010). Mycotoxins and human disease: A largely ignored global health issue. *Carcinogenesis*, 31(1), 71-82.

[8] Copetti, M. V., Iamanaka, B. T., Pereira, J. L., Lemes, D. P., Nakano, F., & Taniwaki, M. H. (2012). Co-occurrence of ochratoxin A and aflatoxins in chocolate marketed in Brazil. *Food Control*, 26(1), 36-41.

[9] Williams, J. H., Phillips, T. D., Jolly, P. E., Stiles, J. K., Jolly, C. M., & Aggarwal, D. (2004). Human aflatoxicosis in developing countries: a review of toxicology, exposure, potential health consequences, and interventions. *American Journal of Clinical Nutrition*, 80(5), 1106-1122.

[10] Lizárraga-Paulín, E. G., Moreno-Martínez, E., & Miranda-Castro, S. P. (2011). Aflatoxins and Their Impact on Human and Animal Health: An Emerging Problem. *Aflatoxins- Biochemistry and Molecular Biology*, Rijeka, InTech, 255-282.

[11] Hansen, E., & Jung, M. (1973). Control of Aflatoxins in the food industry. *Pure and Applied Chemistry*, 35-239.

[12] Moss, M.O. (2002). Risk assessment for aflatoxins in foodstuffs. *International Biodeterioration and Biodegradation*, 50(3-4), 137-142.

[13] Bejarano, R. R. J., & Centeno, B. S. J. (2009). Extracto de Citrus limon para el control de aflatoxinas y hongos aflatoxigénicos en alimentos concentrados para pollos de engorde producidos en Venezuela. *Revista de la Sociedad Venezolana de Microbiología*, 29(1), 57-61.

[14] Dorner, J. W., Cole, R. J., & Diener, U. L. (1984). The relationship of Aspergillus flavus and Aspergillus parasiticus with reference to production of aflatoxins and cyclopiazonic acid. *Mycopathologia*, 87(1-2), 13-15.

[15] Sánchez, E., Heredia, N., & García, S. (2005). Inhibition of growth and mycotoxin production of Aspergillus flavus and Aspergillus parasiticus by extracts of Agave species. *International Journal of Food Microbiology*, 98(3), 271-279.

[16] Wu, F., Liu, Y., & Bhatnagar, D. (2008). Cost-effectiveness of aflatoxin control methods: Economic incentives. *Toxin Reviews*, 27-203.

[17] Wild, C. P., & Gong, Y. Y. (2010). Mycotoxins and human disease: a largely ignored global health issue. *Carcinogenesis*, 31(1), 71-82.

[18] Dowd, P.F. (2003). Insect management to facilitate preharvest mycotoxin management. *Journal of Toxicology- Toxins Reviews*, 22(2-3), 327-350.

[19] Hell, K., & Mutegi, C. (2011). Aflatoxin control and prevention strategies in key crops of Sub-Saharan Africa. *African Journal of Microbiology Research*, 5(5), 459-466.

[20] Bruns, H.A. (2003). Controlling aflatoxin and fumonisin in maize by crop management. *Journal of Toxicology- Toxin Reviews*, 22(2-3), 153-173.

[21] Diener, U. L., Cole, R. J., Sanders, T. H., Payne, G. A., Lee, L. S., & Klich, M. A. (1987). Epidemiology of aflatoxin formation by Aspergillus flavus. *Annual Review of Phytopathology*, 25-240.

[22] Iglesias, J., Presello, D. A., Fauguel, C. M., & Botta, G. L. (2008). Micotoxinas: Debemos preocuparnos? *Presented at the 3 Jornada de Actualización Técnica de Maíz organized by INTA, AIANBA, MAIZAR, Pergamino, Buenos Aires, Argentina.*

[23] Bolet, A. M. B., & Socarrás, S. M. M. (2005). Micotoxinas y cáncer. *Revista Cubana de Investigaciones Biomédicas*, 24(1), 54-59.

[24] Denli, M., & Pérez, J. F. (2006). Contaminación por micotoxinas en los piensos: Efectos, tratamiento y prevención. *XXII Curso de especialización FEDNA, Barcelona, España.*

[25] FAO. (1989). Control of aflatoxin in maize. In: Agriculture and Consumer Protection (ed.) Mycotoxin prevention and control in foodgrains. Rome:. *FAO Corporate Document Repository*, http://www.fao.org/docrep/X5036E/X5036E00.htm.

[26] Hsieh, P. C., Mau, J. L., & Huang, S. H. (2001). Antimicrobial effect of various combinations of plant extracts. *Food Microbiology*, 18(1), 35-43.

[27] Fandohan, P., Zoumenou, D., Hounhouigan, D. J., Marasas, W. F., Wingfield, M. J., & Hell, K. (2005). Fate of aflatoxins and fumonisins during the processing of maize into food products in Benin. *International Journal of Food Microbiology*, 98(3), 249-259.

[28] Castells, M., Marin, S., Sanchis, V., & Ramos, A. J. (2005). Fate of mycotoxins in cereals during extrusion cooking: A review. *Food Additives and Contaminants*, 22(2), 150-157.

[29] Siwela, A. H., Siwela, M., Matindi, G., Dube, S., & Nziramasanga, N. (2005). Decontamination of aflatoxin-contaminated maize by dehulling. *Journal of the Science of Food and Agriculture*, 85(15), 2535-2538.

[30] Borrell, J., & Gimeno, G. (2002). Micotoxinas en los alimentos: medidas de prevención y detoxificación. *Selecciones Avícolas*, 1, 567-571.

[31] Allameh, A., Ziglari, T., & Rasooli, I. (2011). Phytoinhibition of Growth and Aflatoxin Biosynthesis. *in Toxigenic Fungi. In: Torres-Pacheco I. Aflatoxins- Detection, Measurement and Control*, Rijeka, InTech, 283-305.

[32] Saalia, F. K., & Phillips, R. D. (2011). Degradation of aflatoxins by extrusion cooking: Effects on nutritional quality of extrudates. *LWT - Food Science and Technology*, 44(6), 1496-1501.

[33] Serna-Saldívar, S. O., Gómez, M. H., & Rooney, L. W. (1990). Technology, chemistry and nutritional value of alkaline cooked corn products. *Pomeranz Y. (ed.) Advances in Cereal Science and Technology, St. Paul: American Association of Cereal Chemists*, 10, 243-307.

[34] Méndez-Albores, J. A., Arámbula-Villa, G., Loarca-Piña, M. G., González-Hernández, J., Castaño-Tostado, E., & Moreno-Martínez, E. (2004). Aflatoxins' fate during the nixtamalization of contaminated maize by two tortilla-making processes. *Journal of Stored Products Research*, 40(1), 87-94.

[35] Anguiano-Ruvalcaba, G. L., Vargas-Cortina, A. V., & Guzmán De Peña., D. (2005). Inactivación de aflatoxina B1 y aflatoxicol por nixtamalización tradicional del maíz y su regeneración por acidificación de la masa. *Salud Pública de México*, 47(5), 369-375, http://dialnet.unirioja.es/servlet/articulo?codigo=1389007.

[36] Lingk, W. (1991). Health risk evaluation of pesticide contamination. *Drinking water. Gesunde Pflangen*, 43-21.

[37] Tian, J., Ban, X., Zeng, H., He, J., Huang, B., & Wang, Y. (2011). Chemical composition and antifungal activity of essential oil from Cicuta virosa L. var. latisecta Celak. *International Journal of Food Microbiology*, 145(2-3), 464-470.

[38] Varma, J., & Dubey, N. (2001). Efficacy of essential oils of Caesulia axillaris and Mentha arensis against some storage pests causing biodeterioration of food commodities. *International Journal of Food Microbiology*, 68(3), 207-210.

[39] Gowda, N. K. S., Malathi, V., & Suganthi, R. U. (2004). Effect of some chemical and herbal compounds on growth of Aspergillus parasiticus and aflatoxin production. *Animal Feed Science and Technology*, 116(3-4), 281-291.

[40] Boudergue, C., Burel, C., Drsgacci, S., et al. (2009). Review of mycotoxin-detoxifying agents used as feed additives: mode of action, efficacy and feed / food safety. *CFP/EFSA/FEEDAP/2009/01. Scientific Report submitted to EFSA, France.*

[41] Kolosova, A., & Stroka, J. (2011). Substances for reduction of the contamination of feed by mycotoxins: A Review. *World Mycotoxin Journal*, 4(3), 225-256.

[42] Mallman, C. A., Rauber, R. H., & Giacomini, L. (2007). Factores de formación de las micotoxinas y sus formas de control. *Engormix*, http://www.engormix.com/MA-micotoxinas/foros/articulo-factores-formacion-micotoxinas-t13355/251-0 htm.

[43] Accinelli, C., Saccà, M. L., Abbas, H. K., Zablotowicz, R. M., & Wilkinson, J. R. (2009). Use of a granular bioplastic formulation for carrying conidia of a non-aflatoxigenic strain of Aspergillus flavus. *Bioresource Technology*, 100(17), 3997-4004.

[44] Stiling, P., & Cornelissen, T. (2005). What makes a successful biocontrol agent? A meta-analysis of biological control agent performance. *Biological Control,* 34(3 SPEC. ISS), 236-246.

[45] Abbas, H. K., Zablotowicz, R., Bruns, H. A., & Abel, C. (2006). Biocontrol of aflatoxin in corn by inoculation with non-aflatoxigenic Aspergillus flavus isolates. *Biocontrol Science and Technology,* 16(5), 437-449.

[46] Dorner, J. W., Cole, R. J., & Blankenship, P. D. (1992). Use of a biocompetitive agent to control preharvest aflatoxins in drought stressed peanuts. *Journal of Food Protection,* 55(11), 888-892.

[47] Yin, Y. N., Yan, L. Y., Jiang, J. H., & Z. H., Ma. (2008). Biological control of aflatoxin contamination of crops. *Journal of Zhejiang University: Science B,* 9(10), 787-792.

[48] Bandyopadhyay, R., Kiewnick, S., Atehnkeng, J., Donner, M., Cotty, P. J., & Hell, K. (2005). Biological control of aflatoxin contamination in maize in Africa. *Presented in the Conference on International Agricultural Research for Development, Tropentag.*

[49] Huang, C., Jha, A., Sweany, R., Derobertis, C., & Damann, K. E. Jr. (2011). Intraspecific aflatoxin inhibition. *Aspergillus flavus is thigmoregulated, independent of vegetative compatibility group and is strain dependent. PLoS One,* 6(8), e23470.

[50] Chang, P. K., Abbas, H. K., Weaver, M. A., Ehrlich, K. C., Scharfenstein, L. L., & Cotty, P. J. (2012). Identification of genetic defects in the atoxigenic biocontrol strain Aspergillus flavus K49 reveals the presence of a competitive recombinant group in field populations. *International Journal of Food Microbiology,* 154(3), 192-196.

[51] Paster, N., Drobyz, S., Chalutz, E., Mensaherov, M., Nitzan, T., & Wilson, C. L. (1993). Evaluation of the potential of the yeast Pichia guilliermondii as a biocontrol agent against Aspergillus flavus and fungi of stored soya beans. *Mycological Research,* 97(10), 1201-1206.

[52] Chalutz, E., & Wilson, C. L. (1990). Postharvest biocontrol of green and blue mold and sour rot of citrus fruit by Debaryomyces hansenii. *Plant Disease,* 74-134.

[53] Dorner, J. W., Cole, R. J., & Blankenship, P. D. (1998). Effect of inoculum rate of biological control agents on preharvest aflatoxin contamination of peanuts. *Biological Control,* 12(3), 171-176.

[54] Degola, F., Berni, E., & Restivo, F. M. (2011). Laboratory tests for assessing efficacy of atoxigenic Aspergillus flavus strains as biocontrol agents. *International Journal of Food Microbiology,* 146(3), 235-243.

[55] Mehl, H.L., & Cotty, P.J. (2010). Variation in competitive ability among isolates of Aspergillus flavus from different vegetative compatibility groups during maize infection. *Phytopathology,* 100(2), 150-159.

[56] Brown, R. L., Cotty, P. J., & Cleveland, T. E. (1991). Reduction in aflatoxin content of maize by atoxigenic strains of Aspergillus flavus. *Journal of Food Protection*, 54(8), 623-626.

[57] Cotty, P.J. (1994). Influence of field application of an atoxigenic strain of Aspergillus flavus on the populations of A. flavus infecting cotton bolls and on the aflatoxin content of cottonseed. *Phytopathology*, 84(11), 1270-1277.

[58] Dorner, J. W., Cole, R. J., & Wicklow, D. T. (1999). Aflatoxin reduction in corn through field application of competitive fungi. *Journal of Food Protection*, 62(6), 650-656.

[59] Dorner, J. W., Cole, R. J., Connick, W. J., Daigle, D. J., Mc Guire, M. R., & Shasha, B. S. (2003). Evaluation of biological control formulations to reduce aflatoxin contamination in peanuts. *Biological Control*, 26(3), 318-324.

[60] Wicklow, D. T., Roth, S., Deyrup, S. T., & Gloer, J. B. (2005). A protective endophyte of maize: Acremonium zeae antibiotics inhibitory to Aspergillus flavus and Fusarium verticillioides. *Mycological Research*, 109(5), 610-618.

[61] Palumbo, J. D., Baker, J. L., & Mahoney, N. E. (2006). Isolation of Bacterial Antagonists of Aspergillus flavus from Almonds. *Microbial Ecology*, 52(1), 45-52.

[62] Chang, P. K., & Hua, S. S. T. (2007). Nonaflatoxigenic Aspergillus flavus TX9-8 competitively prevents aflatoxin accumulation by A. flavus isolates of large and small sclerotial morphotypes. *International Journal of Food Microbiology*, 114(3), 275-279.

[63] Gould, M., Nelson, L. M., Waterer, D., & Hynes, R. K. (2008). Biocontrol of Fusarium sambucinum, dry rot of potato, by Serratia plymuthica 5-6. *Biocontrol Science and Technology*, 18(10), 1005-1016.

[64] Cho, K. M., Math, R. K., Hong, S. Y., Asraful, S. M. D., Mandanna, D. K., Cho, J. J., Yun, M. G., Kim, J. M., & Yun, H. D. (2009). Iturin produced by Bacillus pumilus HY1 from Korean soybean sauce (kanjang) inhibits growth of aflatoxin producing fungi. *Food Control*, 20(4), 402-406.

[65] Blunt, J. W., Copp, B. R., Hu, W. P., Munro, M. H. G., Northcote, P. T., & Prinsep, M. R. (2008). Marine natural products. *Natural Product Reports*, 25(1), 35-94.

[66] Kong, Q., Shan, S., Liu, Q., Wang, X., & Yu, F. (2010). Biocontrol of Aspergillus flavus on peanut kernels by use of a strain of marine Bacillus megaterium. *International Journal of Food Microbiology*, 139(1-2), 31-35.

[67] Farzaneh, M., Shi, Z. Q., Ghassempour, A., Sedaghat, S., Ahmadzadeh, M., Mirabolfathy, M., & Javan-Nikkhah, M. (2012). Aflatoxin B1 degradation by Bacillus subtilis UTBSP1 isolated from pistachio nuts of Iran. *Food Control*, 23(1), 100-106.

[68] Steindorff, A. S., Silva, R. D. N., Coelho, A. S. G., Nagata, T., Noronha, E. F., & Ulhoa, C. J. (2012). Trichoderma harzianum expressed sequence tags for identification of

genes with putative roles in mycoparasitism against Fusarium solani. *Biological Control*, 61(2), 134-140.

[69] Bluma, R. V., & Etcheverry, M. G. (2008). Application of essential oils in maize grain: Impact on Aspergillus section Flavi growth parameters and aflatoxin accumulation. *Food Microbiology*, 25(2), 324-334.

[70] Teissedre, P. L., & Waterhouse, A. L. (2000). Inhibition of Oxidation of Human Low-Density Lipoproteins by Phenolic Substances in Different Essential Oils Varieties. *Journal of Agricultural and Food Chemistry*, 48(9), 3801-3805.

[71] Soliman, K. M., & Badeaa, R. I. (2002). Effect of oil extracted from some medicinal plants on different mycotoxigenic fungi. *Food and Chemical Toxicology*, 40(11), 1669-1675.

[72] Lang, G., & Buchbauer, G. (2012). A review on recent research results (2008-2010) on essential oils as antimicrobials and antifungals. *Flavour and Fragance Journal A review*, 27(1), 13-39.

[73] Bruneton, J. (1995). Pharmacognosy. *Phytochemistry of Medicinal Plants. Andover, UK: Intercept Limited Anodover. UK.*

[74] Cardile, V., Russo, A., Formisano, C., Rigano, D., Senatore, F., Arnold, N. A., & Piozzi, F. (2009). Essential oils of Salvia bracteata and Salvia rubifolia from Lebanon: Chemical composition, antimicrobial activity and inhibitory effect on human melanoma cells. *Journal of Ethnopharmacology*, 126(2), 265-272.

[75] Tian, J., Ban, X., Zeng, H., He, J., Huang, B., & Wang, Y. (2011). Chemical composition and antifungal activity of essential oil from Cicuta virosa L. var. latisecta Celak. *International Journal of Food Microbiology*, 145(2-3), 464-470.

[76] Skandamis, P. N., & Nychas, G. J. E. (2000). Development and evaluation of a model predicting to survival of Escherichia coli O157H7 NCTC 12900 in homemade eggplant salad at various temperatures, pHs, and oregano essential oils concentration. *Applied and Environmental Microbiology*, 66(4), 1646-1653.

[77] Carson, C. F., Mee, B. J., & Riley, T. V. (2002). Mechanism of action of Melaleuca alternifolia (tea tree) oil on Staphyclococcus aureus determined by time-kill, lysis, leakage, and salt tolerance assay and electron microscopy. *Antimicrobial Agents and Chemotherapy*, 46(6), 1914-1920.

[78] Burt, S. (2004). Essential oils: their antimicrobial properties and potential applications in foods-a review. *International Journal of Food Microbiology*, 94(3), 223-253.

[79] Atanda, O. O., Akpan, I., & Oluwafemi, F. (2007). The potential of some spice essential oils in the control of A. parasiticus CFR 223 and aflatoxin production. *Food Control*, 18(5), 601-607.

[80] Farag, R. S., Daw, Z. Y., & Abo-Raya, S. H. (1989). Influence of some spice essential oils on Aspergillus parasiticus growth and production of aflatoxins in a synthetic medium. *Journal of Food Science*, 54(1), 74-76.

[81] Razzaghi-Abyaneh, M., Shams-Ghahfarokhi, M., Rezaee, M. B., & Sakuda, S. (2010). Natural Aflatoxin Inhibitors from Medicinal Plants. *Rai M. and Varma A. (ed.) Mycotoxins in Food, Feed and Bioweapons*, Springer-Verlag Publication, 329-354.

[82] Velluti, A., Sanchis, V., Ramos, A. J., Egido, J., & Marín, S. (2003). Inhibitory effect of cinnamon, clove, lemongrass, oregano and palmarose essential oils on growth and fumonisin B1 production by Fusarium proliferatum in maize grain. *International Journal of Food Microbiology*, 89(2-3), 145-154.

[83] Nguefack, J., Leth, V., Amvam Zollo, P. H., & Mathur, S. B. (2004). Evaluation of five essential oils from aromatic plants of Cameroon for controlling food spoilage and mycotoxin producing fungi. *International Journal of Food Microbiology*, 94(3), 329-334.

[84] Rasooli, I., & Owlia, P. Chemoprevention by thyme oils of Aspergillus parasiticus growth and aflatoxin production. 2005, *Phytochemistry*, 66(24), 2851-2856.

[85] García-Camarillo, E. A., Quezada-Viay, M. Y., Moreno-Lara, J., Sánchez-Hernández, G., Moreno-Martínez, E., & Pérez-Reyes, M. C. J. (2006). Actividad Antifúngica de Aceites Esenciales de Canela (Cinnamomum zeylanicum Blume) y Orégano (Origanum vulgare L.) y su Efecto sobre la Producción de aflatoxinas en Nuez. *Revista Mexicana de Fitopatología*, 24-8.

[86] Vilela, G. R., de Almeida, G. S., D'Arce, M. A. B. R., Moraes, M. H. D., Brito, J. O., da Silva, M. F. D. G. F., Silva, S. C., de Stefano, P. S. M., Calori-Domingues, M. A., & da Gloria, E. M. (2009). Activity of essential oil and its major compound, 1, 8-cineole, from Eucalyptus globulus Labill, against the storage fungi Aspergillus flavus Link and Aspergillus parasiticus Speare. *Journal of Stored Products Research*, 45(2), 108-111.

[87] Razzaghi-Abyaneh, M., Shams-Ghahfarokhi, M., Yoshinari, T., Rezaee, M. B., Jaimand, K., Nagasawa, H., & Sakuda, S. (2008). Inhibitory effects of Satureja hortensis L. essential oil on growth and aflatoxin production by Aspergillus parasiticus. *International Journal of Food Microbiology*, 123(3), 228-233.

[88] Kumar, A., Shukla, R., Singh, P., & Dubey, N. K. (2009). Biodeterioration of some herbal raw materials by storage fungi and aflatoxin and assessment of Cymbopogon flexuosus essential oil and its components as antifungal. *International Biodeterioration and Biodegradation*, 63(6), 712-716.

[89] Razzaghi-Abyaneh, M., Shams-Ghahfarokhi, M., Rezaee, M. B., Jaimand, K., Alinezhad, S., Saberi, R., & Yoshinari, T. (2009). Chemical composition and antiaflatoxigenic activity of Carum carvi L., Thymus vulgaris and Citrus aurantifolia essential oils. *Food Control*, 20(11), 1018-1024.

[90] Komala, V. V., Ratnavathi, C. V., Vijay, Kumar. B. S., & Das, I. K. (2012). Inhibition of aflatoxin B1 production by an antifungal component, eugenol in stored sorghum grains. *Food Control*, 26(1), 139-146.

[91] Nogueira, J. H. C., Gonçalez, E., Galleti, S. R., Facanali, R., Marques, M. O. M., & Felício, J.D. (2010). Ageratum conyzoides essential oil as aflatoxin suppressor of Aspergillus flavus. *International Journal of Food Microbiology*, 137(1), 55-60.

[92] Velazhahan, R., Vijayanandraj, S., Vijayasamundeeswari, A., Paranidharan, V., Samiyappana, R., Iwamotob, T., Friebe, B., & Muthukrishnan, S. (2010). Detoxification of aflatoxins by seed extracts of the medicinal plant, Trachyspermum ammi (L.) Sprague ex Turrill- Structural analysis and biological toxicity of degradation product of aflatoxin G1. *Food Control*, 21(5), 719-725.

[93] Prakash, B., Shukla, R., Singh, P., Kumar-Mishra, P., Dubey, N. K., & Nath-Kharawar, R. (2011). Efficacy of chemically characterized Ocimum gratissimum L. essential oil as an antioxidant and a safe plant based antimicrobial against fungal and aflatoxin B1 contamination of spices. *Food Research International*, 44(1), 385-390.

[94] Prakash, B., Singh, P., Mishra, P. K., & Dubey, N. K. (2012). Safety assessment of Zanthoxylum alatum Roxb. essential oil, its antifungal, antiaflatoxin, antioxidant activity and efficacy as antimicrobial in preservation of Piper nigrum L. fruits. *International Journal of Food Microbiology*, 153(1-2), 183-191.

[95] Passone, M. A., Girardi, N. S., Ferrand, C. A., & Etcheverry, M.A. (2012). Girardi N.S., Ferrand C.A., Etcheverry M. 2012. Invitro evaluation of five essential oils as botanical fungitoxicants for the protection of stored peanuts from Aspergillus flavus and A. parasiticus contamination. *International Biodeterioration and Biodegradation*, 70-82.

[96] Mallmann, C. A., Hummes, R. R., & Giacomini, L. (2007). Factores de formación de las micotoxinas y sus formas de control. *Universidad Federal de Santa María, Brasil*, 1-8.

[97] Guo, B., Chen, Z. Y., Dewey, L. R., & Scully, B. T. (2008). Drought stress and preharvest aflatoxin contamination in agricultural commodity: Genetics, genomics and proteomics. *Journal of Integrative Plant Biology*, 50(10), 1281-1291.

[98] Yu, J., & Ehrlich, K. C. (2011). AF Biosynthetic Pathway and Pathway Genes. *Guevara-González RG. (ed.) Aflatoxins - Biochemistry and Molecular Biology*, Rijeka, InTech, 41-66.

[99] Criseo, G., Racco, C., & Romeo, O. (2008). High genetic variability in non-aflatoxigenic A. flavus strains by using Quadruplex PCR-based assay. *International Journal of Food Microbiology*, 125(3), 341-343.

[100] Chang, P. K., Ehrlich, K. C., Yu, J., Bhatnagar, D., & Cleveland, T. E. (1995). Increased expression of Aspergillus parasiticus aflR, encoding a sequence-specific DNA binding protein, relieves nitrate inhibition of aflatoxin biosynthesis. *Applied and Environmental Microbiology*, 61(6), 2372-2377.

[101] Du, W., Obrian, G. R., & Payne, G. A. (2007). Function and regulation of aflJ in the accumulation of aflatoxin early pathway intermediate in Aspergillus flavus. *Food Additives and Contaminants*, 24(10), 1043-1050.

[102] Kale, S. P., Cary, J. W., Hollis, N., Wilkinson, J. R., Bhatnagar, D., Yu, J., Cleveland, T. E., & Bennett, J. W. (2007). Analysis of aflatoxin regulatory factors in serial transfer-induced non-aflatoxigenic Aspergillus parasiticus. *Food Additives and Contaminants*, 24(10), 1061-1069.

[103] Duran, R. M., Cary, J. W., & Calvo, A. M. (2007). Production of cyclopiazonic acid, aflatrem, and aflatoxin by Aspergillus flavus is regulated by veA, a gene necessary for sclerotial formation. *Applied Microbiology and Biotechnology*, 73(5), 1158-1168.

[104] Stinnett, S. M., Espeso, E. A., Cobeno, L., Araujo-Bazan, L., & Calvo, A. M. (2007). Aspergillus nidulans VeA subcellular localization is dependent on the importin alpha carrier and on light. *Molecular Microbiology*, 63(1), 242-255.

[105] Brown, R. L., Chen, Z. Y., Warburton, M., Luo, M., Menkir, A., Fakhoury, A., & Bhatnagar, D. (2010). Discovery and characterization of proteins associated with aflatoxin-resistance: Evaluating their potential as breeding markers. *Toxins*, 2(4), 919-933.

[106] Razzazi-Fazeli, E., Rizwan, M., Mayrhofer, C., & Nöbauer, . (2011). The Use of Proteomics as a Novel Tool in Aflatoxin Research. *Guevara-González RG. (ed.) Aflatoxins-Biochemistry and Molecular Biology*, Rijeka, InTech, 127-148.

[107] Chen, Z. Y., Brown, R. L., Damann, K. E., & Cleveland, T. E. (2002). Identification of unique or elevated levels of kernel proteins in aflatoxin-resistant maize genotypes through proteome analysis. *Phytopathology*, 92(10), 1084-1094.

[108] Shakerardekani, A., & Karim, R. (2012). Effect of different types of plastic packaging films on the moisture and aflatoxin contents of pistachio nuts during storage. *Journal of Food Science and Technology In Press* , 1-3.

[109] Hell, K., Cardwell, K. F., Setamou, M., & Poehling, H. M. (2000). The influence of storage practices on aflatoxin contamination in maize in four agro- ecological zones of Benin, West Africa. *Journal of Stored Products Research*, 36(4), 365-382.

[110] Udoh, J. M., Cardwell, K. F., & Ikotun, T. (2000). Storage structures and aflatoxin content of maize in five agroecological zones of Nigeria. *Journal of Stored Products Research*, 36(2), 187-201.

[111] Bastioli, C. (2001). Global status of the production of biobased packaging materials. *Starch/Starke*, 53(8), 351-355.

[112] Bastioli, C. (1998). Properties and applications of Mater-Bi starch-based materials. *Polymer Degradation and Stability*, 59(1-3), 263-272.

[113] Devlieghere, F., Vermeulen, A., & Debevere, J. (2004). Chitosan: antimicrobial activity, interactions with food components and applicability as a coating on fruit and vegetables. Food Microbiology ., 21(6), 703-714.

[114] Miranda-Castro, S.P. (2000). La quitina y su potencial aplicación industrial. Revista Investigación y Desarrollo. *Periodismo de Ciencia y tecnología,* http:// www.invdes.com.mx.

[115] Miranda-Castro, S.P. (2004). Películas compuestas de quitosano y esponjas para usos biomédicos y alimentarios. (A Chitosan and/or porous solids film for Biomedical and/or Food Uses, a process for the obtention thereof and use of the same). *Instituto Mexicano de la Propiedad Industrial, México. File Number: PA/a/2004/001347, Folio PA/E/ 2004/007243.*

[116] Miranda, S. P., Miranda, E., & Serrano, J. (2007). Chitosan and gamma irradiated chitosan against Aspergillus flavus in maize. *Asian Chitin Journal,* 3-37.

[117] Rivero, D., Cruz, A., Martínez, B., Ramírez, M. A., Rodríguez, A. T., & Cárdenas, R. M. (2004). Efecto protector de la quitosana en semillas de arroz frente a Fusarium sp. *Revista de Protección Vegetal,* 19(2), 140-144.

[118] Lizárraga-Paulín, E. G., Torres-Pacheco, I., Moreno-Martínez, E., & Miranda-Castro, S. P. (2011). Protección contra estrés biótico inducida por quitosán en plántulas de maíz (Zea mays). *Revista Mexicana de Ciencias Agrícolas,* 2(6), 813-827.

Chapter 5

Recent Trends in Microbiological Decontamination of Aflatoxins in Foodstuffs

Carlos Augusto Fernandes Oliveira, Fernanda Bovo,
Carlos Humberto Corassin,
Alessandra Vincenzi Jager and
Kasa Ravindranadha Reddy

Additional information is available at the end of the chapter

1. Introduction

Nowadays, about 100,000 fungi have already been identified. From these, more than 400 may be considered potentially toxigenic, and about 5% are known to produce toxic compounds or classes of compounds that cause adverse effects in animals and humans in several parts of the world [1]. These compounds, called mycotoxins, are secondary metabolites of low molecular weight produced by mycelia or spores of filamentous fungi [2]. It is suggested that mycotoxin production is generally limited to a relatively small number of mold species, and that toxin may be produced by the whole species or just one specific strain [3]. The more complex the synthesis pathway of a mycotoxin, the lesser the number of mold species that produce it.

The term "mycotoxin" originates from the Greek word "Mykes", meaning fungus, and from the Latin word "Toxicum", meaning poison or toxin [2]. Mycotoxins are classified as the most important chronic and noninfectious foodborne risk factor, more important than synthetic contaminants, plant toxins, food additives, and pesticide residues. Both humans and animals may show acute or chronic intoxication caused by mycotoxin ingestion, and the pathological condition that results from this ingestion is called mycotoxicosis [4]. Some factors affect the magnitude of toxicity in humans or animals, including the animal species, mechanism of action, metabolism and defense mechanisms [5].

About 400 types of mycotoxins have been already discovered, and they are generally divided into groups based on structural similarities and most important toxic effects [6]. From all mycotoxins that have been isolated, aflatoxin is one of the most well-known and widely distributed in foodstuffs, with proven and marked toxic properties. Aflatoxins are predominantly produced by *Aspergillus flavus* and *A. parasiticus*, but may also be produced by other strains, such as *A. nomius, A. tamari*, and *A. pseudotamarii* [7]. Contamination of foodstuff with aflatoxigenic fungi may occur at any moment during production, harvesting, processing, transportation, and storage [8]. The most different kinds of foods may be affected, such as corn, peanuts, cotton seeds, rice, pistachio, almonds, chestnuts, Brazil nuts, and pumpkin seeds, as well as other oily seeds, such as sunflower and coconut [9].

Aflatoxins are distributed worldwide. *Aspergillus* species are able to grow in a wide variety of substrates and under different environmental conditions. Toxin formation in agricultural products occurs in hot and humid weather, and in inadequate or deficient storage facilities. The most important factors that influence growth and aflatoxin production are relative humidity, ranging from 88 to 95% in most of the cases [8], and temperature, ranging from 36 to 38 C for mold growth, and 25 to 27 C for maximum toxin production [10].

Other factors may also influence aflatoxin production: substrate composition, water activity, pH, atmosphere (concentration of oxygen and carbon dioxide), microbial competition, mechanical damage to the seeds, mold lineage, strain specificity and variation, instability of toxigenic properties, plant stress, insect infestation, and use of fungicides or fertilizers [2, 5, 11]. It is important to remember that aflatoxin contamination is cumulative, and the moment of harvesting and drying, and storage conditions may also play an important role in aflatoxin production [12].

Concerns related to the negative impacts of aflatoxins on health led to the study of strategies to prevent toxin formation in foodstuffs, as well as to eliminate, inactivate or reduce toxin bioavailability in contaminated products [13]. Contamination may be prevented by improved agricultural practices, antifungal agents, genetic engineering, and control of storage conditions [2]. Bioavailability may be reduced by enterosorption, which is done by adding nutritionally inert adsorbent compounds to the diet. These compounds are mycotoxin sequestrants, and prevent the toxin from being absorbed in the gastrointestinal tract of the animals, making its distribution to the target organs impossible [14]. This method has limited practical use, due to the safety of the adsorbent agents used, and the difficulty in applying them to human foods [15]. Elimination or inactivation, that is, decontamination, may be achieved by physical, chemical, and biological methods, which have to present the following characteristics: complete inactivation; destruction or removal of the toxin; no production or toxic residues in foods or no remainders of them; preservation of nutritional value and palatability of the food; destruction of fungal spores and mycelia to prevent production or reappearance of the toxin; no significant changes in the physical properties of the food; low cost and ease of use [1,11].

Physical methods for mycotoxin decontamination involve procedures such as thermal inactivation, ultraviolet light, ionizing radiation, or extraction with solvents. Chemical methods are based on agents that break mycotoxin structure, such as chlorine treatment (sodium hypochlorite or chlorine gas), oxidizing agents (hydrogen peroxide, ozone and sodium disulfide), or hydrolytic agents (acids, alkalis and ammonia). However, both chemical and physical methods have disadvantages, either because removal is not efficient, or because of high costs or nutritional losses to the product [16,17]. Biological methods are based on the action of microorganisms on mycotoxins. These microorganisms may be yeasts, filamentous fungi, bacteria, algae, among others, and their mechanisms of action is based on competition by nutrients and space, interactions, and antibiosis, among others [18].

Biodegradation of aflatoxins by microorganisms offers an attractive alternative for the control or elimination of aflatoxins in foods and animal feed, preserving their quality and safety [19]. Besides, their use have a more "natural" appeal, given the ever-growing resistance of the consumer to chemical treatments [1]. Biological decontamination methods are being widely studied and may be a very promising choice, provided they show to be efficient, specific, cost-effective, and are environmentally friendly [20]. Among the types of microorganisms available and that may be used to remove aflatoxins from a contaminated medium, lactic acid bacteria (LAB) and yeasts are the most studied ones, showing the most promising results.

Therefore, the objective of this chapter was to present results of studies on microbiological methods for aflatoxin decontamination, more specifically on the ability of LAB and yeasts to degrade or sequestrate this mycotoxin.

2. Toxicological Properties of Aflatoxins

Nowadays, there are 18 similar compounds called aflatoxins. However, the most important types in terms of health and medical interest are identified based on their fluorescence under ultraviolet light (B = Blue and G = Green), such as aflatoxin B_1 (AFB_1), B_2 (AFB_2), G_1 (AFG_1) and G_2 (AFG_2). From these compounds, AFB_1 is the most prevalent and toxic one [21]. When AFB_1 is ingested by domestic animals in contaminated feed or foodstuffs, such as by dairy cows, the toxin undergoes liver biotransformation and is converted into aflatoxin M_1 (AFM_1), becoming the hydroxilated form of AFB_1, which is excreted in milk, tissues and biological fluids of these animals [22-24]. It was reported that of all AFB_1 ingested in feed, about 0.3% to 6.2% is transformed in AFM_1 in milk and that there is a linear relationship between the concentration of AFM_1 in milk and the concentration of AFB_1 in contaminated feeds consumed by the animals [25,26].

Chronic exposure to low levels of aflatoxins represents a serious risk to economy, and mainly to health [21]. Economic losses are related to decreased efficiency in industrial or agricultural production, with loss in quality, lower yield, and defective product [27]. It was also reported that in some states of the USA, economic losses to agriculture amount to 100 million dollars [19]. On the other hand, these losses caused by mold contamination and mycotoxins are greater than 1.6 billion dollars in the US, and African feeds lose about 670 billion dollars a year due to barriers to the trade of aflatoxin-contaminated foodstuffs [28].

As for human and animal health, biological effects of aflatoxins may be carcinogenic, mutagenic, teratogenic, hepatotoxic, and immunosuppressive [29]. The International Agency for

Research on Cancer classifies AFB_1 and AFM_1 as Group 1 human carcinogens, even though AFM_1 is about 10 times less carcinogenic than AFB_1 [30]. All these aflatoxin effects are influenced by variations according to the animal species, sex, age, nutritional status, and effects of other chemical products, besides the dose of toxin and the length of exposure of the organism to it [31].

Aflatoxicosis is the poisoning caused by the ingestion of moderate to high levels of alfatoxin in contaminated foods. Acute aflatoxicosis causes quick and progressive jaundice, edema of the limbs, pain, vomiting, necrosis, cirrhosis and, in severe cases, acute liver failure and death, caused by the ingestion of about 10 to 20 mg of aflatoxin in adults. Aflatoxin LD50 shows the following order of toxicity: $AFB_1 > AFM_1 > AFG_1 > AFB_2 > AFG_2$ [4, 32]. Chronic aflatoxicosis causes cancer, immunosuppression and other pathological conditions, having the liver as the primary target organ [4].

The greatest risk presented by aflatoxins for human beings is chronic exposure causing hepatocellular carcinoma, which may be made worse by hepatitis A virus [5]. It was also report that aflatoxins were found in the tissues of children affected by Reye syndrome (encephalopathy with serious lesions in liver and kidneys after influenza or chickenpox), and Kwashiorkor (protein-energy malnutrition). Aflatoxicosis is considered, then, a contributing factor to these diseases.

AFB_1 is metabolized in the liver by the cytochrome P450 system, generating its most carcinogenic metabolite, AFB_1-8,9-epoxide (AFBO), or other less mutagenic forms, such as AFM_1, Q_1 or P_1. There are several pathways for AFBO after it is metabolized, with one of them leading to cancer, another to toxicity and another one, to excretion. AFBO exo-form easily binds to cell macromolecules, including genetic material such as DNA proteins, producing adducts. Formation of these DNA adducts leads to genetic mutations and cancer, and their excretion in the urine of infected people is not only a proof that humans have the necessary biochemical pathways for carcinogenesis, but also offers a reliable biomarker for AFB_1 exposure [24].

Potential risk to human health caused by aflatoxins has led to surveillance programs for the toxin in different raw materials, as well as regulations determined by almost every country in the world [9]. A study carried out by the Food and Agriculture Organization of the United Nations (FAO) in 2002 pointed out that about 100 countries had specific regulations for the presence of aflatoxin in foods, dairy products and animal feed, and that the total population of these countries amounted to 90% of the world population. The same study showed that regulations for aflatoxin are getting more diverse and detailed, including sampling methods and methods of analysis [33].

In countries where a regulation for aflatoxin exists, tolerance levels for the total aflatoxin (sum of aflatoxins B_1, B_2, G_1 and G_2) ranges from 1 to 35 μg/kg for foods, with an average of 10 g/kg; and from zero to 50 μg/kg for animal feed, with an average of 20 μg/kg. For AFM_1 in milk, tolerance levels are between 0.05 and 0.5 μg/kg, with most countries adopting a threshold of 0.05 μg/kg [10].

3. Decontamination of Aflatoxins by Lactic Acid Bacteria

LAB is a large group of genetically different bacteria that, besides producing lactic acid as the main product of their metabolism, have similar characteristics: they are all gram-positive, non-sporoformers, non-motile, and catalase, and oxidase negative. They are, therefore, aerotolerant anaerobes. Besides, they mandatorily ferment sugars and tend to be nutritionally fastidious, frequently requiring specific amino acids and B-complex vitamins as growth factors [34]. Several LAB genera, such as *Lactobacillus*, *Bifidobacterium* and *Lactococcus* are known for they ability to act as preserving agents in fermented foods, such as vegetables, cereals, dairy and meat products, actively inhibiting spoilage and growth of pathogenic bacteria, besides increasing shelf life and sensory properties of these foods [23].

Fermentation enables longer shelf life and improves sensory and nutritional properties of the product, as sugar fermentation lowers pH and inhibits growth of spoilage and pathogenic microorganisms. Fermentation is also responsible for other reactions, such as proteins hydrolysis, improving texture and flavor; synthesis of aromatic components and texturizers, affecting the consistency of the product; and production of inhibitory components [35,36]. This inhibition is, in part, caused by the final products of fermentation, such as lactic acid, diacetyl, acetaldehyde and acetic acid, which may accumulate in inhibitory concentrations in certain foods and drinks. In other cases, inhibition may also be caused by secondary by-products of metabolism, such as hydrogen peroxide or bacteriocins [37].

Therefore, two aspects may be considered when LAB are used: fermentation and antibiosis ability. In the first case, the starter culture added to the food acts on the substrate, causing advantages to the food. In the second case, the starter culture has to inhibit the development of undesirable microorganisms that may spoil the product or be hazardous to human health. In reference [38], authors state that one of the effects that were identified in LAB was protection against toxins found in foods, such as heterocyclic amines, polycyclic aromatic hydrocarbons, reactive oxygen species, and mycotoxins. In the latter case, studies have demonstrated that LAB have the ability to inhibit aflatoxin biosynthesis, or that they have the ability to remove mycotoxins from the medium, reducing their effects.

It should be emphasized that with increased interest in probiotic food production all over the world, selection of LAB cultures with probiotic characteristics and greater ability to remove mycotoxins may help to reduce risk of exposure to these toxins in foodstuffs, which is a very promising line of research in mycotoxicology. Yeast and LAB strains have great ability to remove mycotoxins, and may be used as part of starter cultures in the fermentation of foods and drinks [39]. These microorganisms have, thus, ability to ferment and decontaminate the medium, and purified components of these strains may be used in small amounts as food additives without compromising the characteristics of the final product.

One of the first studies in this area was carried out in the 1960s, when these authors evaluated the ability of about 1,000 types of microorganisms to degrade aflatoxins [40]. Yeasts, filamentous fungi, bacteria, actinomycetes, algae, and fungal spores were among the organisms studied. From these, only the bacterium *Flavobacterium aurantiacum* B-184 (known today as *Nocardia corynebacterioides*) was able to irreversibly remove aflatoxins from the solution.

After this study, many others followed. However, the most significant ones started to appear after the 1990s. Table 1 presents the most relevant studies carried out with bacteria for aflatoxin decontamination. The action of 7 different types of bacteria on AFB_1 was evaluated and it was found that some strains of *Lactobacillus* (*L. rhamnosus* GG and *L. rhamnosus* LC-705) were able to efficiently remove most mycotoxin from the medium, up to about 80% [17]. In reference [27] authors analyzed 9 strains of *Lactobacillus* and achieved the same result, that *L. rhamnosus* GG and *L. rhamnosus* LC-705 were the most efficient strains in removing AFB_1, with removal rates of 78.9% and 76.5%, respectively. Fifteen types of LAB, among them *Lactobacillus* and *Lactococcus*, and 5 types of bifidobacteria, were studied and it was observed that removal of AFB_1 ranged from 5.6% to 59.7% [23]. Strains of *Lactobacillus amylovorus* (CSCC 5160 and CSCC 5197) and *L. rhamnosus* LC 1/3 showed the best results: 59.7%, 57.8%, and 54.6%, respectively. It was also observed that different strains of bifidobacteria removed from 37% to 46% AFB_1, and that *Staphylococcus aureus* and *Escherichia coli* removed 46% and 37%, respectively [22]. It may be observed that among a given genus, and even a given species, not all the strains show equivalent toxin removal rates. On the contrary, the ability to remove aflatoxin is a characteristic of specific lineages, and efficiency varies widely [41].

Most assays on aflatoxin removal in the studies cited above were carried out in phosphate-buffered saline (PBS). In reference [42], besides testing the ability of 27 strains of *Lactococcus* spp. and 15 strains of *Streptococcus* spp. isolated from yogurt, raw milk, and Karish cheese to remove AFB_1 in buffered solution, observed that *Lactococcus L. lactis* and *Streptococcus thermophilus* presented the greatest rates of toxin removal (54.85% and 81.0%, respectively). They also tested the ability of viable and non-viable cells to remove AFB_1 in different vegetable oils, and observed that viable *L. lactis* cells removed from 71% to 86.7% AFB_1, whereas non-viable cells removed 100% of the toxin in all the oils. Moreover, viable *S. thermophilus* cells removed from 66.5% to 91.5% of the toxin, and non-viable ones, from 81.7% to 96.8%.

AFB_1 was added to yogurt and acidified milk in concentrations ranging from 1,000 to 1,400 g/kg, and a reduction of AFB_1 in yogurt (pH 4.0), ranging from 97.8% to 90% was obtained [43]. Maximum decrease in AFB_1 was observed during milk fermentation. As for milk acidified with citric, lactic, and acetic acid (pH 4.0) AFB_1 reduction (concentration of 1,000 μg/Kg) was 90%, 84% and 73%, respectively. The ability of probiotic bacteria (*L. paracasei, L. casei, L. brevis* and *L. plantarum*) and the yeast *Saccharomyces cerevisiae* to remove a sum of aflatoxins (B_1, B_2, G_1 and G_2) during fermentation of dough made up of 50% barley flour, 45% wheat flour, and 5% corn flour was evaluated [44]. They observed that after 6 hours of fermentation, the amount of aflatoxin had decreased 18% and 33% for dough added of 4 and 40 μg of aflatoxin, respectively, and after 24 hours, the amount of aflatoxin decreased 27% and 50%, respectively.

Toxin polarity has an important role in the binding mechanism. The percentage of aflatoxin removed by LAB decreases in the following order: $AFB_1 > AFB_2 > AFG_1 > AFG_2$. This observation correlates with the decrease in the polarity of these toxins, and is consistent with hydrophobic reactions, which may also have a role in the binding mechanism [45]. AFM_1 is less efficiently removed than AFB_1. However, scientific literature has few studies on the ability of LAB to remove AFM_1.

In reference [46], authors examined the ability of 4 strains of *Lactobacillus* spp. and 2 strains of *Bifidobacterium* spp. to remove AFM_1 in PBS and reconstituted skim milk. In PBS, viable cells of 6 strains were able to remove from 10.22 to 26.65% AFM_1 in solution, depending on the level of contamination and the length of incubation, whereas non-viable cells removed from 14.04 to 28.97% of the toxin. In reconstituted skim milk incubated for 4 hours, 7.85 to 25.94% AFM_1 were removed by viable cells, and 12.85 to 27.31% for cells rendered non-viable by heat treatment. These researchers concluded that the removal process was fast, with no differences between 0, 4, and 24 hours of contact, different from what was observed in [47] for strains of *Lactobacillus* spp., *Lactobacillus* spp. and *Bifidobacterium* spp., which showed removal rates ranging from 0 to 14.6% after 24 hours of contact, and from 4.5 to 73.1% after 96 hours of contact.

The ability of *L. rhamnosus* GG to remove AFM_1 from reconstituted skim and whole milk was investigated and it was observed rates of 18.8% and 26.0%, respectively [29]. The authors concluded that the decrease in removal efficiency may be explained by the fact that AFM_1 is possibly not accessible in milk, that is, it is associated with casein, and the interference of proteins in toxin removal may be the greatest responsible factor for the difference between skim milk and whole milk (approximately 10% lower), once powdered skim milk used in the study contained 37g of protein / 100 g, whereas protein content in powdered whole milk was 25g /100g. In the same study, AFM_1 removal in buffered solution (50.7%) was compared with AFB_1 removal by the same bacterial strain in the same solution (75.3%). It was concluded that AFM_1 removal was less effective possibly due to the presence of an -OH group in the molecule, increasing its polarity and making it less hydrophilic, what increases the tendency of the molecule to be retained in aqueous solutions.

Some physical, chemical, and enzymatic treatments may increase the ability of LAB to bind to aflatoxin in the medium. In reference [48] authors studied the ability of *L. rhamnosus* GG to bind to AFB_1, observing little difference between aflatoxin removal by heat-treated and acid-treated cells (85% and 91%, respectively), compared with viable bacterial cells (86%). The use of physical and chemical treatments (chloric acid, and heat treatment in autoclave or boiling at 100 °C) on *L. rhamnosus* GG and LC-705 caused a significant increase in AFB_1 removal, showing that metabolic degradation caused by viable bacterial cells may be ruled out as a possible mechanism of action [15-17].

Comparing the ability of viable and heat-treated bifidobacteria cells, it was observed that viable cells removed 4 to 56% AFB_1 from the medium, whereas non-viable cells removed 12 to 82% [23]. Evaluating the influence of the inactivation treatment on the ability of 4 types of *Lactobacillus* spp. to remove AFB_1, it was observed that acid treatment (58.6 to 87.0%) and heat treatment (33.5 to 71.9%) increased the ability to remove the toxin, compared with viable cells in PBS (16.3 to 56.6%) [49]. On the other hand, alkali treatment (8.3 to 27.4%) and ethanol treatment (15.9 to 46.5%) decreased the amount of aflatoxin removed from the medium.

Removal of AFM_1 with 8 LAB strains showed that heat-treated cells bound more efficiently (25.5 to 61.5%) to the toxin than viable bacterial cells (18.1 to 53.8%) [29]. In reference [50] it was observed that heat-treated cells removed greater percentages of AFM_1 (12.4% to 45.7%) in PBS compared with viable cells (5.6% to 33.5%), with no significant differences between 15 minutes or 24 hours of contact. Similar results were found in [51], because viable cells of *Lactobacillus delbrueckii* spp. *bulgaricus* CH-2 removed 29.42% AFM_1 in PBS after 4 hours of contact at 37 C. These authors also analyzed the ability of *Streptococcus thermophilus* ST-36, observing that 18.70% AFM_1 was removed from the medium. Until today, only one bacterium, *Flavobacterium aurantiacum* NRRL B-184, was able to remove 100% of AFM_1 from contaminated liquid medium, at a cell concentration of 5×10^{10} CFU/mL and 4 hours of contact [52].

In [53] authors observed that *B. subtilis* UTBSP1 presented significant removal of AFB_1 from a medium contaminated with 2.5 μg/g (52.67% and 80.53%, after 24 and 48 hours, respectively). After 72 and 96 hours, there was no significant increase in the amount of toxin removed from the medium. Strains of *B. subtilis* were analyzed and it was concluded that strain ANSB060 was the one that best removed AFB_1, AFM_1, and AFG_1 from the medium (81.5%, 60%, and 80.7%, respectively) [54]. Results of this study also demonstrated that aflatoxin degradation is mainly observed in the supernatant culture, compared with cells or cell extracts. Besides, in assays that simulated the gastrointestinal environment (pH 2.0, and 0.3% of biliary salts), viable cells of the same strain were able to survive for 24 hours of incubation, and presented antimicrobial activity against *E. coli*, *S. typhimurium*, and *S. aureus*.

These examples show that both viable and non-viable cells are able to remove aflatoxin from aqueous solutions. As non-viable cells are also able to remove the toxin, it is supposed that cells are physically bound to the toxin, that is, components of the bacterial cell wall adhere to it, mainly polysaccharides and peptidoglycans, taking into account the possibility of a covalent bond or degradation caused by bacterial metabolism [1, 55, 56].

Both polysaccharides and peptidoglycans of the bacterial cell wall may be extremely affected by heat and acid treatment, once heat may denature proteins or form Maillard reaction products. Besides, acid treatment may break glycosidic bonds of polysaccharides, releasing monomers that may be further broken into aldehydes, also degrading proteins to smaller components, such as peptides and amino acids. Thus, acid treatment may break the peptidoglycan structure, compromising its structural integrity, that is, decreasing the thickness of this layer, reducing cross links and increasing the size of the pores. These changes caused by the treatments cited above enable AFB_1 to bind to the bacterial cell wall and to the components of the plasmatic membrane that were not available when the bacterial cell was intact [27].

In reference [57] authors explained that the integrity of the bacterial cell wall is important in the process of toxin removal by both viable and non-viable cells. In their study of AFB_1, they observed that both the bacterial cell wall and its purified fragments were able to remove aflatoxin from the medium. However, when the cell wall was lost or destroyed (totally or partially) by enzymatic treatment, there was a significant decrease in the ability to remove the toxin. It was observed, using atomic force microscopy, that the bond between AFB_1 and *Lactobacillus casei* Shirota produced structural changes that modified the surface of the bacterial cell [58]. Before the toxin was bound to it, the surface was well-defined, smooth and homogenous, and after AFB_1 adsorption, there were changes in shape. These changes were probably caused by the bond between the toxin and the surface of the cell wall, which became very irregular and rough, with undefined edges. The authors suggest that changes in the shape of teichoic acids are responsible for these alterations, once these molecules are found inside the cell wall in such a way that they produce no differences in the texture of the surface before the toxin was bound to it.

The ability of *L. rhamnosus* GG to bind to AFB_1 was studied, observing that the addition of urea - an anti-hydrophobic agent - to the medium, significantly decreased removal of the toxin by non-viable cells, from 85-91% to 50-60%, showing that hydrophobic interactions have a relevant role in the process [48]. Besides, addition of different concentrations of NaCl and $CaCl_2$ (from 0.01 to 1 M), and pH variations from 2.5 to 8.5 had practically no effect on AFB_1 removal by the bacterium, suggesting that hydrogen bonds and electrostatic interactions are not important in this process.

In the use of pronase E, lipase and periodate, treatment with periodate led to significant reduction in the ability to remove the toxin, both by viable and non-viable cells, once it oxidizes the -OH cis groups in aldehyde and carboxylic acid groups, suggesting that the bonds involve predominantly bacterial polysaccharides. Treatment with pronase E caused the same significant reduction in AFB_1 removal, evidencing that proteins may also be involved in the process. Thus, the fact that pronase E and periodate both have a significant reduction on AFB_1 removal indicates that binding sites are made of protein. Treatment with lipase, on its turn, did not cause any significant reduction in AFB_1, showing that lipids, such as lypoteichoic acid probably do not have a role in the process. Although the treatments decreased AFB_1 removal, it was still substantial in all cases, possibly showing the involvement of multiple components in the bond with mycotoxin [48].

However, not only the type of bacterial strain and the inactivation treatment used may influence formation and stability of the LAB/aflatoxin complex, but also of other factors, such as bacterial counts, specificity of the bacteria, pH, incubation temperature, addition of nutrients, and the solvents used, among others [23, 27, 48].

As for the number of bacterial cells in the medium, it has been concluded that there was a significant decrease in the amount of AFM_1 removed when cell counts changed from 10^7 CFU/mL (0 to 5.02%) to 10^8 CFU/mL (10.22 to 26.65%), indicating that bacterial counts are critical factors in the removal of AFM_1 by LAB [46]. In reference [59] authors observed that no less than 5 x 10^9 CFU/mL of *Lactobacillus acidophilus* or *Bifidobacterium longum* are necessary to remove only 13% AFB_1 in about one hour.

In reference [17] authors reported that, for *Lactobacillus rhamnosus* (strains GG and LC705), minimum counts of 2 x 10^9 CFU/mL were required to remove 50% AFB_1, and greater removal rates were obtained when LAB concentration was increased to 10^{10} CFU/mL. In this same study, the authors observed that the process depended on the temperature, once the efficiency in aflatoxin removal was greater at 37 °C than at 4 and 25 °C. Besides, the authors observed that Gram-positive bacteria are better aflatoxin sequestrants than Gram-negative bacteria, with removal rates of 80% and 20%, respectively, suggesting the ability to remove the toxin depends on the structure of the cell wall. It has also been stated that aflatoxin concentration in the medium also influences adsorption rates, leading to the conclusion that the greater its concentration in the medium, the greater the removal rate, both for viable and non-viable cells [60].

Assays with AFB_1 and *L. rhamnosus* GG and LC-705 at different incubation temperatures was also carried out, but it was not observed significant differences in the stability of the LAB/AFB_1 complex formed in the temperatures range between 4 °C and 37 °C [27]. When pH of the medium was changed from 2 to 10, a range that includes the pH switch in the gastrointestinal tract, only 10% AFB_1 removed was released back into the solution, different from what happened when organic solvents were used. In this case, almost all AFB_1 that was removed by the bacterial strains was released back into the medium, providing extra evidence that the process is based on a non-covalent bond. In this study, the release efficiency by solvents presented the following order: methanol < acetonitrile = benzene < chloroform, which does not coincide with the order of decreasing polarity. This may be explained by the fact that AFB_1 hydrophobicity is similar to that of the chloroform molecule. These results show once more that hydrophobic interactions have an important role in the binding mechanism between LAB and the toxin.

The effect of washing on the stability of the LAB/aflatoxin complex was analyzed [47]. They observed that after the first washing of bacterial pellets with PBS, the proportion of AFM_1 released by the bacteria was 87.3% for *Lactobacillus* spp. strains; 85.7% for *Lactococcus* spp. strains, and 85.7% for strains of *Bifidobacterium* spp. They also observed that after the third washing, practically all bacteria had released adsorbed AFM_1 back into the medium (92.0 a 100%). In reference [46] they concluded that AFM_1 removal by bacteria was reversible, and that small amounts of toxin were released back to the PBS solution (5.62 to 8.54%). This finding is consistent with those observations of reference [27], who reported that *L. rhamnosus* GG, *L. rhamnosus* LC-705, and *Lactobacillus casei* Shirota released, respectively, 3.7%, 3.0% and 2.4% AFB_1 back into the solution. Differently, in [23] authors showed that release of AFB_1 back into the solution in the first washing was 48.6%, 30.7% and 26.5% for *L. amylovorus* (strains CSCC 5160 and CSCC 5197) and *L. rhamnosus* Lc 1/3, respectively. After 5 washings, AFB_1 adsorbed by *L. amylovorus* CSCC 5160 was almost completely released (94.4%), whereas *L. amylovorus* CSCC 5197 and *L. rhamnosus* Lc 1/3 retained, respectively, only 17.4% and 32.2% AFB_1 found in the original solution.

Thus, the LAB/aflatoxin complex seems to be unstable, once part of the aflatoxin, both for AFB_1 and AFM_1, is released from the complex after washing, and gradually returns to the aqueous solution. Therefore, the greater the number of washings, the greater the amount of aflatoxin released back into the solution. This shows that the bond is not a strong one, suggesting it is a weak non-covalent bond and an association with hydrophobic sites on the surface of the bacteria [23, 48].

Different from this hypothesis, in reference [61], performing the same washings on a complex between *Flavobacterium aurantiacum* and AFB_1, authors observed that aflatoxin was not released into the aqueous solution. Analyzing the stability of the complex formed between AFB_1 and 8 strains of *Lactobacillus casei* after the washings, it was demonstrated that the amount of aflatoxin released ranged from practically zero and 9.2% [13]. Possible explanations for this variation in aflatoxin release include the differences in binding sites found in the different strains, or more probably, that these biding sites are similar, but that they present minimal differences depending on the strain. Authors explained that lower rate of toxin release into the medium after the washings may be attributed to the interactions between aflatoxin molecules retained on the cell wall of a bacterium and molecules retained on the cell wall of the adjacent bacterium, forming a kind of reticulated matrix that prevents aflatoxin release. It has also been suggested that the greater the number of molecules that are removed by the bacterial cells, the longer these molecules remain adsorbed on the cell surface [60].

The stability of the LAB/aflatoxin complex in a wide range of pH is an important factor in the use of these microorganisms to remove aflatoxin from foods, once gastric release of the toxin would have negative health implications. Therefore, the complex formed has to resist environmental stress caused by the gastrointestinal tract, such as low pH and presence of bile. When the influence of the presence of bile on the LAB/aflatoxin complex was analyzed, it was observed that *Lactobacillus casei* removed more AFB_1 when exposed to bile, suggesting that this exposure causes changes in the structure and composition of the bacterial cell wall, probably inducing the formation of new biding sites for aflatoxin, or increasing the size of the sites available [13].

The ability of *L. rhamnosus* (strains GG and LC705) and *Propionibacterium freudenreichii* spp. *shermanii* JS to remove AFB_1 from intestinal liquid medium extracted from the duodenum of chicks was investigated, and it was observed that AFB_1 concentration was reduced in 54% in only 1 minute in the presence of *L. rhamnosus* GG, whereas it was reduced in only 44% in the presence of *L. rhamnosus* LC705, and 36% in the presence of *P. freudenreichii* spp. *shermanii* JS [62]. The authors observed that the accumulation of AFB_1 in the intestinal tissue was reduced in 74%, 63%, and 37%, respectively, for *L. rhamnosus* (strains GG and LC705) and *P. freudenreichii* spp. *shermanii* JS, showing that these bacteria may affect aflatoxin bioavailability and be used to reduce its toxicity to humans and animals.

Rats treated with feed added of aflatoxin (3 mg/kg of feed) presented a significant decrease in the feed intake compared with the control group, different from the animals fed diets containing *Lactobacillus casei* and *Lactobacillus reuteri* (10 mL/kg of feed, with 1×10^{11} CFU/mL) and aflatoxin [63]. The second group did not show reduced feed intake. Consequently, animals treated with the diet containing only aflatoxin presented lower body weight, significant increase in serum levels of transaminase, alkaline phosphatase, cholesterol, triglycerides, total lipids, creatinine, uric acid, and nitric oxide; and in lipid peroxidation in the liver and kidneys, followed by a significant decrease in total antioxidant capacity. Treatment with bacteria was able to induce a significant improvement in all biochemical parameters and in the histological condition of the liver, with *L. reuteri* being more efficient than *L. casei*.

In Egypt, a pilot study investigated the effect of the addition of *L. rhamnosus* LC-705 and *P. freudenreichii* spp. *shermanii* JS in human diet on the levels of aflatoxin in feces samples. It was observed that from 11 of 20 volunteers, AFB_1 ranged from 1.8 to 6 μg AFB_1/kg feces, and after two weeks of supplementation with probiotic bacteria, there was a significant reduction in the excretion rate, showing that these strains have the ability to influence the concentration of AFB_1 in feces [64].

Microorganism	AF	Bound (%)	Conditions	Ref.
L. rhamnosus GG	B1			
Viable cells	5 µg/mL	78.4	2 x 10^{10} cfu/mL, 0h, 37 °C, PBS	[17]
Freeze-dried cells		65	4h, 37 C, PBS	
Heat-treated cells		81	4h, 37 °C, PBS	
L. rhamnosus LC-705				
Viable cells		78.8	2 x 10^{10} cfu/mL, 0h, 37 °C, PBS	
Freeze-dried cells		50	4h, 37 °C, PBS	
Heat-treated cells		82	4h, 37 °C, PBS	
L. gasseri		58.1	2 x 1010 cfu/mL, 0h, 37 C, PBS	
L. acidophilus		67.4	7 x 109 cfu/mL, 0h, 37 °C, PBS	
L. casei Shirota		33.2	1 x 1010 cfu/mL, 0h, 37 C, PBS	
E. coli		16.3	5 x 1010 cfu/mL, 0h, 37 °C, PBS	
L. paracasei, L. casei, L. brevis, L. plantarum and *S. cerevisiae*	B_1, B_2, G_1, G2 4 or 40 µg/kg	18-33	6h, 37 °C , barley flour (50%), wheat flour (45%) and corn flour (5%) mixed with water in 1:1.5 proportion	[44]
		27-50	24h, 37 °C , barley flour (50%), wheat flour (45%) and corn flour (5%) mixed with water in 1:1.5 proportion	
Lc. lactis ssp. cremoris	B_1	5.6	1 x 10^{10} cfu/mL, 24h, 37 °C, PBS	[23]
Lactobacillus delbrueckii	5 µg/mL	17.3		
Lb. acidophilus		18.2		
Lb. rhamnosus		22.7		
Lb. plantarum		28.4		
Lc. lactis ssp. lactis		31.6		
Bifidobacterium lactis		18.0		
Lb. helveticus		34.2		
Lc. lactis ssp. *cremoris*		41.1		
Lb. rhamnosus Lc		54.6		
Lb. acidophilus		20.7		
Lb. fermentum		22.6		
Lb. johnsonii		30.1		
Lb. rhamnosus		33.1		
Lb. amylovorus		57.8		
Lb. amylovorus		59.7		
Bb. lactis		34.7		
Bb. longum		37.5		
Bb. animalis		45.7		
Bb. lactis		48.7		

Microorganism	AF	Bound (%)	Conditions	Ref.
L. rhamnosus GG	B1			[27]
Viable cells	5 µg/mL	78.9	1 x 10^{10} cfu/mL, 1h, 37 °C, PBS	
Heat-treated cells		84.1		
Acid-treated cells		86.7		
L. rhamnosus LC-705				
Viable cells		76.5	1 x 10^{10} cfu/mL, 1h, 37°C, PBS	
Heat-treated cells		87.8		
Acid-treated cells		88.3		
L. acidophilus LC1				
Viable cells		59.7	1 x 10^{10} cfu/mL, 1h, 37 °C, PBS	
Heat-treated cells		74.7		
Acid-treated cells		84.2		
L. lactis subsp. lactis				
Viable cells		59.0	1 x 10^{10} cfu/mL, 1h, 37 °C, PBS	
Heat-treated cells		58.1		
Acid-treated cells		69.5		
L. acidophilus ATCC 4356				
Viable cells		48.3	1 x 10^{10} cfu/mL, 1h, 37 °C, PBS	
Heat-treated cells		69.7		
Acid-treated cells		81.3		
L. plantarum				
Viable cells		29.9	1 x 10^{10} cfu/mL, 1h, 37 °C, PBS	
Heat-treated cells		35.5		
Acid-treated cells		62.7		
L. casei Shirota				
Viable cells		21.8	1 x 10^{10} cfu/mL, 1h, 37 °C, PBS	
Heat-treated cells		41.5		
Acid-treated cells		32.3		
L. delbrueckii subsp. *bulgaricus*			1 x 10^{10} cfu/mL, 1h, 37 °C, PBS	
Viable cells		15.6		
Heat-treated cells		33.7		
Acid-treated cells		75.8		
L. helveticus				
Viable cells		17.5	1 x 10^{10} cfu/mL, 1h, 37 °C, PBS	
Heat-treated cells		29.8		
Acid-treated cells		58.1		
P. freudenreichii subsp. *shermanii* JS				
Viable cells		22.3	1 x 10^{10} cfu/mL, 1h, 37 °C, PBS	
Heat-treated cells		67.3		

Microorganism	AF	Bound (%)	Conditions	Ref.
Acid-treated cells		82.5		
Lc. lactis subsp. *cremoris*				
Viable cells		26.9	1 x 10^{10} cfu/mL, 1h, 37 °C, PBS	
Heat-treated cells		40.1		
Acid-treated cells		43.7		
S. thermophilus				
Viable cells		32.7	1 x 10^{10} cfu/mL, 1h, 37 °C, PBS	
Heat-treated cells		42.0		
Acid-treated cells		63.8		
E. Coli	B1	37	30 min, 37 °C, PBS	[22]
L. rhamnosus GG	2 µg/mL	37		
S. aureus		46		
Bifidobacterium sp. Bf6		25		
B. adolescentis 14		31		
B. bifidum BGN4		46		
Bifidobacterium sp. CH4		37		
B. longum JR20		37		
Bifidobacterium sp. JO3		41		
Lc. lactis	B1		10^7-10^8 cfu/mL,30 min, 37 °C, in:	[42]
Living cells	0.5 µg/mL	54.8	PBS	
		86.7	maize oil	
		82.3	sunflower oil	
		71.0	soybean oil	
Dead cells by boiling		81.0	PBS	
		100	maize, sunflower or soybean oil	
Dead cells by autoclaving		80.0		
			PBS	
S. thermophilus				
Living cells		81.0	PBS	
		91.5	maize oil	
		90.7	sunflower oil	
		66.5	soybean oil	
Dead cells by boiling		100.0	PBS	
		96.8	maize oil	
		81.7	sunflower oil	
		96.0	soybean oil	
Dead cells by autoclaving		83.0	PBS	

Microorganism	AF	Bound (%)	Conditions	Ref.
Yoghurt Culture	B1			[43]
	0.6mg/kg	97	42 °C/3h, pH 4.0, overnight, milk	
	1 mg/kg	91		
	1.4mg/kg	90		
	1 mg/kg	90	milk acidified with citric acid	
		84	milk acidified with latic acid	
		73	milk acidified with acetic acid	
L. acidophilus NCC12	M1			[46]
Living cells	5, 10 and	14.9-20.2	10^8 cfu/mL, 0, 4, 24 h, 37 °C, PBS	
	20	14.4-15.4	10^8 cfu/mL, 4h, 37 °C, milk	
Heated cells	ng/mL	17.0-24.9	0, 4 , 24 h, 37 °C, PBS	
		16.6-19.0	4h, 37 °C, milk	
L. acidophilus NCC36				
Living cells		20.4-25.3	10^8 cfu/mL, 0, 4, 24 h, 37 °C, PBS	
		21.8-22.7	10^8 cfu/mL, 4h, 37 °C, milk	
Heated cells		22.1-26.8	0, 4, 24 h, 37 °C, PBS	
		23.7-25.1	4h, 37 °C, milk	
L. acidophilus NCC68				
Living cells		10.2-16.0	10^8 cfu/mL, 0, 4, 24 h, 37 °C, PBS	
		7.8-10.5	10^8 cfu/mL, 4h, 37 °C, milk	
Heated cells		14.0-21.8	0, 4, 24 h, 37 °C, PBS	
		12.8-15.9	4h, 37 °C, milk	
B. bifidum Bb13				
Living cells		23.5-26.6	10^8 cfu/mL, 0, 4, 24 h, 37 °C, PBS	
		24.0-25.9	10^8 cfu/mL, 4h, 37 °C, milk	
Heated cells		24.3-28.9	0, 4,24 h, 37 °C, PBS	
		25.4-27.4	4h, 37 °C, milk	
B. bifidum NCC 381				
Living cells		16.6-22.1	10^8 cfu/mL, 0, 4,24 h, 37 °C, PBS	
		15.5-18.3	10^8 cfu/mL, 4h, 37 °C, milk	
Heated cells		17.4-23.5	0, 4 and 24 h, 37 °C, PBS	
		17.1-22.2	4h, 37 °C, milk	
L. rhamnosus		20.1-24.0	10^8 cfu/mL, 0, 4, 24 h, 37 °C, PBS	
Living cells		20.4-22.2	10^8 cfu/mL, 4h, 37 °C, milk	
Heated cells		23.4-27.8	0, 4 and 24 h, 37 °C, PBS	
		22.9-26.3	4h, 37 °C, milk	
Lactobacillus strains	AFM1	9.4-73.1	96 h, 37 °C , PBS	[47]

Microorganism	AF	Bound (%)	Conditions	Ref.
Lactococcus strains		4.5-38.3	96 h, 37 °C , PBS	
Bifidobacterium strains		7.8-41.6	96 h, 37 °C , PBS	
L. plantarum		73	96 h, 37 °C , PBS	
B. adolescentes		41.6	96 h, 37 °C , PBS	
Lactobacillus strains		64-80.5	96 h, 37 °C ,milk	
Lactococcus strains		46.0-68.5	96 h, 37 °C , milk	
Bifidobacterium strains		67.0-72.5	96 h, 37 °C , milk	
L. bulgaricus		80.5	96 h, 37 °C , milk	
B. adolescentes		73	96 h, 37 °C , milk	
L. rhamnosus strain GG	M1		5.3 x 10^8, 15 - 16h, , 37 °C, in:	[29]
(pre-cultured)	0.15			
Viable cells	µg/ml	50.7	PBS	
		18.8	skim milk	
		26.0	full cream milk	
Heat-killed cells		57.8	PBS	
		26.6	skim milk	
		36.6	full cream milk	
*L. rhamnosus*strain				
LC-705				
(pre-cultured)				
Viable cells		46.3	PBS	
		69.6	skim milk	
		27.4	full cream milk	
Heat-killed cells		51.6	PBS	
		63.6	skim milk	
		30.1	full cream milk	
L. rhamnosus strain GG				
(lyophilized)				
Viable cells		53.8	1.0 x 10^{10}, 15-16h, 37 °C, PBS	
Heat-killed cells		56.2		
L. rhamnosus strain				
LC-705				
(lyophilized)				
Viable cells		45.7	1.0 x 10^{10}, 15-16h, 37 °C, PBS	
Heat-killed cells		57.4		
L. lactis ssp. *cremoris*				
strain ARH74				
Viable cells		40.4	2.9 x 10^8, 15-16h, , 37 °C, PBS	

Microorganism	AF	Bound (%)	Conditions	Ref.
Heat-killed cells		38.9		
L. gasseri (ATCC 33323)				
Viable cells		30.8	3.9 x 10^8, 15-16h, , 37 °C, PBS	
Heat-killed cells		61,5		
L. acidophilus strain LA1				
Viable cells		18,3	1.7 x 10^9, 15-16h, , 37 °C, PBS	
Heat-killed cells		25,5		
L. rhamnosus strain 1/3				
Viable cells		18,1	3.9 x 10^8, 15-16h, , 37 °C, PBS	
Heat-killed cells		39,9		
L. rhamnosus strain GG	B_1			[48]
Pre-treatment:	5 µg/mL			
Pronase E				
Viable cells		66	1h, 37 °C, 5% CO_2, PBS	
Heat-treated cells		72	Boiled for 1h, PBS	
Acid-treated cells		85	2 mol/L HCl, 1h, 37 °C, 5% CO2	
Lipase				
Viable cells		76	1h, 37 °C, 5% CO2, PBS	
Heat-treated cells		74	Boiled for 1h, PBS	
Acid-treated cells		89	2 mol/L HCl, 1h, 37 °C, 5% CO2	
Phosphate Buffer				
Viable cells		86	1h, 37 °C, 5% CO2, PBS	
Heat-treated cells		85	Boiled for 1h, PBS	
Acid-treated cells		91	2 mol/L HCl, 1h, 37 °C, 5% CO2	
m-Periodater				
Viable cells		60	1h, 37 °C, 5% CO2, PBS	
Heat-treated cells		49	Boiled for 1h, PBS	
Acid-treated cells		36	2 mol/L HCl, 1h, 37 °C, 5% CO2	
Iodate				
Viable cells		83	1h, 37 °C, 5% CO2, PBS	
Heat-treated cells		84	Boiled for 1h, PBS	
Acid-treated cells		80	2 mol/L HCl, 1h, 37 °C, 5% CO2	
Urea				
Viable cells		64	1h, 37 °C, 5% CO2, PBS	
Heat-treated cells		60	Boiled for 1h, PBS	
Acid-treated cells		50	2 mol/L HCl, 1h, 37 °C, 5% CO2	
Water (Milli Q)				
Viable cells		76	1h, 37 °C, 5% CO2, PBS	
Heat-treated cells		83	Boiled for 1h, PBS	
Acid-treated cells		84	2 mol/L HCl, 1h, 37 °C, 5% CO2	

Microorganism	AF	Bound (%)	Conditions	Ref.
L. acidophilus	B1			[49]
Pre-treatment:	5 µg/mL			
None		56.6	4h, 37 °C, PBS	
Heat		71.9		
Ethanol		46.5		
Acid		87.0		
Alkaline		27.4		
L. casi				
None		22.4		
Heat		41.8		
Ethanol		21.8		
Acid		43.1		
Alkaline		12.0		
L. helveticus				
None		17.8		
Heat		28.5		
Ethanol		18.0		
Acid		56.3		
Alkaline		9.1		
L. bulgaricus				
None		16.3		
Heat		33.5		
Ethanol		15.9		
Acid		586		
Alkaline		8.3		
L. plantarum	M1		10^{10} cfu/mL, 15 min, 37°C, in:	[50]
Viable cells	0.15	5.6	PBS	
Heat-killed cells	µg/mL	8.1	PBS	
E. avium	PBS			
Viable cells	solution	7.4	PBS	
Heat-killed cells	0.5	6.6	PBS	
P. pentosaceus	µg/mL			
Viable cells	skimmed	8.7	PBS	
Heat-killed cells	milk	7.8	PBS	
L. gasseri				
Viable cells		21.4	PBS	
Heat-killed cells		22.8	PBS	
L. bulgaricus				
Viable cells		30.2	PBS	
Heat-killed cells		33.5	PBS	

Microorganism	AF	Bound (%)	Conditions	Ref.
		33.5	skimmed milk	
L. rhamnosus				
Viable cells		17.1		
Heat-killed cells		27.8	PBS	
		24.5	PBS	
			skimmed milk	
B. lactis				
Viable cells		16.9	PBS	
Heat-killed cells		23.6	PBS	
		32.5	skimmed milk	
L. delbrueckii subsp. *bulgaricus* CH-2	M1 10 ng/mL	18.7	4h, 37 °C, PBS	[51]
		27.6	4h, 42 °C, milk	
S. thermophilus ST-36		29.4	4h, 37 °C, PBS	
		39.2	4h, 42 °C, milk	
		14.8	Yoghurt	
F. aurantiacum NRRL B-184	M1 10 μg/mL	100	5 x 10^{10} cfu/mL, 30 °C, 4h, PBS and milk	[52]
B. subtilis UTBSP1	B1			[53]
Viable cells	2.5 μg/mL	85.7	96h, 30 °C, nutrient broth culture	
Cell Free Supernatant		95	10^8 cfu/ml, 120 h, 30 °C, pistachio nuts	
			120 h, 35 °C, nutrient broth culture	
		78.4		
B. subtilis ANSB060	B_1			[54]
"Inocula" suspension	G_1	81.5	72h, 37 °C, Luria-Bertani medium	
	M1	80.7		
	(0.5 μg/mL)	60		
Cell	B1	10.5	72h, 37 °C, PBS	
Cell extract		9.6		
Culture Supernatant		78.7		
L. rhamnosus strain GG	B1		10^{10}cfu/mL:	[62]
"In vivo"	3 μg/mL	51	1 min, duodenum of chicks	
		92		
"In vitro"		80	1 h, duodenum of chicks	
L. rhamnosus strain LC-705			37 °C, 1h , pH 7.3	

Microorganism	AF	Bound (%)	Conditions	Ref.
"In vivo"		36	1 min, duodenum of chicks	
		71	1 h, duodenum of chicks	
"In vitro"		77	37 °C, 1h , pH 7.3	
P. freudenreichii subsp. *shermanii* JS				
"In vivo"		37	1 min, duodenum of chicks	
		82	1 h, duodenum of chicks	
"In vitro"		22	37 °C, 1h , pH 7.3	

Table 1. Aflatoxin binding / absorption by microorganisms. Note: PBS, Phosphate-Buffered Saline; cfu, colony formingunit.

4. Decontamination of Aflatoxins by Yeasts

Yeasts are non-photosynthetic organisms with a separate nucleus and complex life cycle. They are larger than bacteria, normally spherical, non-motile, and reproduce by budding. Although their main function is alcoholic fermentation, these organisms are also capable of producing enzymes and vitamins. The primary substrates for yeasts are fermentable sugars, which are mainly transformed in ethanol, carbon dioxide, and biomass under oxygen-limited conditions. Under adequate oxygen supply, yeast produces carbon dioxide, water, and biomass [65]. *Saccharomyces cerevisiae* (SC) is the most well-known and commercially important species of yeast, and SC strains are widely used in the production of alcoholic drinks and in the baking industry.

As it occurs with LAB, SC cells have been studied to evaluate their ability to remove aflatoxins from contaminated media. The most important results obtained until now are summarized in Table 2. Products based on SC (cell wall from baker and brewer yeasts, inactivated baker yeast, or alcohol yeast) was studied, and it was observed that in pH 3, 37 °C and 15 minutes of contact, AFB_1 removal ranged from 2.5% to 49.3%, depending on the concentration of the toxin in the medium, and on the yeast-based products used [66]. These authors also observed a decrease in toxin adsorption as the initial concentration increased, and concluded that adsorption is not a linear phenomenon. Similar results with a SC strain and AFB_1 concentration ranging from 1 to 20 μg/mL was also reported [56]. At the 1 μg/mL concentration, 69.1% AFB_1 was removed; at 5 g/mL, removal rate was 41%; and at 20 μg/mL, 34%. *S. cerevisiae* strains were isolated from animal feed, feces and intestines, and tested for their ability to tolerate gastrointestinal conditions and remove AFB_1 from a contaminated medium [67]. These researchers observed that all strains isolated were able to survive in gastrointestinal conditions, and that the percentage of toxin removed ranged among SC strains (10^7 CFU/mL), and with AFB_1 concentration used (16.4% to 82% of adsorption for 50 ng/mL AFB_1; 21.3% to 48.7% for 100 ng/mL AFB_1; and 20.2% to 65.5% for 500 ng/mL AFB_1).

The ability of SC (0.1%, 0.2%, and 0.3%) to adsorb AFB_1 in contaminated corn (150, 300, 450 and 800 μg/kg corn was analyzed [68]. The adsorption process showed an inversely proportional relationship with the concentration, that is, the greater the AFB_1 concentration in the medium, the lower the efficiency of AFB_1 removal by SC (16% to 66% for 800 μg/kg AFB_1 vs. 40% to 93% for 150 μg/kg AFB_1). The authors concluded, using densitogram analysis, that the adsorption process did not change the molecular structure of the mycotoxin, and that the decreased AFB_1 adsorption rates observed as the toxin concentration increased may possibly be caused by saturation of the adsorption sites on the SC cell. Other factors, such as length of incubation, pH, method of biomass purification, and methods of analysis, may also influence this process.

Immobilized SC cells (ATTC 9763) was investigated for their ability to remove AFB_1 from pistachio seeds, and it was observed that the amount of toxin removed was dependent on its concentration in the medium (40% and 70% of removal for concentrations of 10 ng/mL and 20 ng/mL AFB_1, respectively) [69]. The authors also concluded that this ability to remove the toxin was greater in SC exponential growth phase, and that the process was a quick one, being saturated after 3 hours of contact. Besides, the ability of SC cells to remove toxin was increased after treatment with acid (60% and 73% for 10 ng/mL and 20 ng/mL AFB_1, respectively) and heat (55% and 75%, respectively). In another study, authors also concluded that the treatment of SC cells with heat at 60 °C and 120 °C, and with chloric acid (2 mol/L) increased their ability to remove AFB_1 from the medium to 68.8%, 79.3%, and 72.1%, respectively, against 38.7% when viable yeast cells were used [56].

Heat treatment may increase the permeability of the external layer of the cell wall due to the suspension of some mannanes on the cell surface, leading to increased availability of previously hidden binding sites. Besides, countless physical-chemical changes take place on the cell wall during heat treatment, leading to more exposed binding sites. On the other hand, acid conditions may affect polysaccharides by releasing monomers, which are further fragmented in aldehydes after glycosidic bonds are broken. Continuous removal of aflatoxin, even after use of acid and heat treatments, confirms that yeast cell viability is not a significant factor for the removal of aflatoxin from the medium [69].

During the fermentation of broiler feed using LAB (3 strains of *Lactobacillus*) and SC strains resistant to gastric juices and bile, 55% AFB_1 was removed when AFB_1 concentration in the medium was 1 mg/kg, and 39% when concentration was 5 mg/kg AFB_1, after 6 hours [70]. This tendency for removal was maintained as incubation continued, and after 24 hours, the amount of AFB_1 removed was 73% and 53%, respectively, for the two concentrations of the toxin. The authors considered that, from a practical point of view, the most important factor was the 6-hour fermentation period, once the passage of feed through the gastrointestinal tract of broilers lasts from 4 to 8 hours. In reference [71], authors analyzed the ability of SC to remove AFB_1 from a contaminated medium at different pH values (3.0, 6.0, and 8.0), and observed that the three strains analyzed showed great ability to remove the toxin (41.6% to 94.5%), and that after washing, only a small amount of AFB_1 was released back into the medium. *In vitro* studies are not always good indications of the *in vivo* responses, as *in vivo* studies are affected by physiological parameters, such as pH, peristaltic movements, and gastric and intestinal secretions.

In vivo studies using SC are not as rare as those with LAB, mainly in poultry science. Generally, SC is added to the feed as a growth promoter. However, the addition of yeasts has also presented beneficial effects against the exposure to AFB_1. It was observed that the addition of 1% SC to feed contaminated with 5 g/g of AFB_1 prevented loss of weight; liver and heart hyperplasia; and decreased serum albumin and total protein concentrations in broilers [72]. The addition of SC in feed containing aflatoxin decreased the deleterious effects on feed intake, weight gain, and feed conversion in Japanese quails [73]. Compared with control animals, weight gain was 37% lower in birds fed a diet added only of aflatoxin, and was 15% greater than the control in the group that received feed containing aflatoxin and SC. The authors concluded that the diet containing with only SC significantly improved all growth parameters investigated (about 40%), compared with the control group.

In a study with mice, it was observed that the addition of AFB_1 to the diet (0.4 and 0.8 mg/kg) caused a significant reduction in weight gain, and an increase of 85% (0.8 mg/kg) in the rate of micronucleated normochromatic erythrocytes (MNE) after 3 weeks of ingestion, compared with the control group [68]. When diets containing AFB_1 and SC (0.3%) were administered, weight gain was twice greater than in diets that contained only the toxin, and the rate of MNE increased only 46% (0.8 mg/kg) The authors stated that reduced body weight is one of the most common consequences of AFB_1 ingestion, because the toxin alters the activity of several digestive enzymes, giving rise to a malabsorption syndrome characterized by steatorrhea, hypovitaminosis A and a decrease in the levels of bile, pancreatic lipase, trypsin, and amylase. Besides, biotransformation of AFB_1 gives rise to several metabolites, particularly AFB_1-8,9-epoxide, which may bind covalently to DNA and proteins, changing enzymatic processes such as gluconeogenesis, Krebs cycle, and fatty acid synthesis [74]. MNE rate is used to determine the genotoxicity of $AFB_{1,}$ because it quantifies broken chromosomes and whole chromosomes that are abnormally distributed to daughter cells, showing thus, that AFB_1 is a potent mutagenic agent.

A diet containing 5 g/g of aflatoxin (82.06% AFB_{1}, 12.98% AFB_{2}, 2.84% AFG_{1}, and 1.12% AFG_2) by female quails (49 to 84 days of age) led to decreased egg production, feed intake, and feed conversion (31%, 28%, and 47%, respectively) [75]. However, addition of SC (2 g/kg) significantly increased these parameters (16%, 4%, and 14%, respectively). They also observed that the diet with aflatoxins caused a marked decrease in weight gain and egg weight, besides increasing animal mortality (39%, 7%, and 50%, respectively), whereas addition of SC reverted the negative effect on these parameters (65%, 8%, and 50%, respectively). The authors stated that these negative effects of aflatoxins in egg production, feed intake, and feed conversion may have been caused by anorexia, apathy, and inhibition of protein synthesis and lipogenesis. Besides, affected liver function and mechanisms of use of protein and lipids may have affected performance criteria and the general health of the animals. In reference [76] authors reported that the components of the cells wall of SC are able to adsorb mycotoxins, stimulate the immune system, and compete for binding sites in the enterocytes, inhibiting intestinal colonization by pathogens.

SC cell wall is mainly made up of polysaccharides (80-90%), and its mechanical resistance is due to an inside layer composed of β-D-glucans, which are formed by a complex network of highly polymerized β-(1,3)-D-glucans, branched off as β-(1,6)-D-glucans, that have a low level of polymerization. This inside layer is firmly bound to the plasmatic membrane by linear chains of chitin, which have a significant role in the insolubility of the overall structure and packing of the branched β-D-glucans. Both chitin chains and β-D-glucans affect the plasticity of the cell wall. The external layer of the yeast cell wall is formed by mannoproteins, which have an important role in the exchanges with the external environment. This whole structure is highly dynamic and may vary according to the yeast strain, phase of the cell cycle, and culture conditions, such as pH, temperature, oxygenation rate, nature of the medium, concentration and nature of the carbon source. Thus, these differences in the composition of the cell wall among yeast strains are related with their ability to bind to the mycotoxin [77].

Studies have shown that the components of SC cell wall, called oligomannanes, after esterification, are able to bind more than 95% AFB_1 [78]. Addition of 0.05% glucomannanes in the basal diet improved broiler performance [79].

The possible binding mechanisms between yeast cell wall and mycotoxins were studied, and authors suggested that β-D-glucans are the components of the cell wall that are responsible for forming the complex with the toxin, and that the reticular organization of β-D-glucans and their distribution in β-(1,3)-D-glucans and β-(1,6)-D-glucans have an important role in the efficiency of the bond [77]. Besides, studies have shown that weak hydrogen and van der Waals bonds are involved in the complex chemical connection between the mycotoxins and β-D-glucans, a chemical interaction that is much more "adsorption" than "bond". As for AFB_1, they observed that the aromatic ring, the lactone and ketone groups of the polar form of AFB_1, or chemical bonds with glucose units in the single helix of the β-D-glucans, are what keep the toxin bound to the glucans.

It was demonstrated that yeast strains isolated from environments were animals are raised are able to bind to AFB_1 in saline solution (PBS, pH 7) [67]. These strains presented other properties that were beneficial to the host, such as the inhibition of pathogenic bacteria. Therefore, SC strains acted both as probiotics (co-aggregation and inhibition of pathogenic bacteria), and as mycotoxin adsorbents.

In reference [72], SC was able to reduce the deleterious effects of AFB_1 in the diet of broilers and in [68] authors replicated these findings in rats. Protective effect against aflatoxins produced by yeasts was confirmed in rats. However, when yeast cells were inactivated by heat, they were inefficient [80] but when glucomannanes extracted from the cell wall of yeasts were used, there was an increase in the efficiency of the bond with AFB_1, OTA and T-2 toxin [81-84], individually or in combination [75, 79, 85, 86]. The addition of SC in the diet reduced AFB_1 toxic effects in chickens [72, 87]. The ability of SC to reduce AFB_1 toxic effects in quails was demonstrated, and this effect was apparently more efficient with the increase in inclusion rates [88].

In [89] authors obtained a significant reduction in AFB_1 concentration during beer production, probably due to the bond between mycotoxins and SC cell. This hypothesis was supported by other studies [39, 90]. A 19% reduction in AFB_1 during dough fermentation in bread production was observed [91].

Microorganism	AF	Bound (%)	Conditions	Ref.
S. cerevisiae	B_1		15 min, 37 °C:	[66]
	0.0058-	7.6-49.3	YCW from brewer's yeast	
	6.35 μg/mL	7.6-29	YCW from brewer's yeast	
		10-24	Inactivated baker's yeast	
		4-29	YCW from baker's yeast	
		17-44	Inactivated baker's yeast	
		3-44	YCW from baker's yeast	
		23-35	YCW from baker's yeast	
		27-44	Alcohol yeast	
S. cerevisiae	B1			[56]
Strain A18	1 μg/mL	69.1	3h, 25 C, PBS	
	5 μg/mL	41		
	10 μg/mL	33		
	20 μg/mL	34.2		
Strain 26.1.11	1 μg/mL	65.1	3h, 25 C, PBS	
	5 μg/mL	37.2		
	10 μg/mL	31		
	20 μg/mL	32.6		
Pre-treatment:				
Heated cells 52°C	5 μg/mL	58.8	3h, 25 C, PBS	
Strain A18		56.5		
Strain 26.1.11				
Heated cells 55 °C		64.5		
Strain A18		64		
Strain 26.1.11				
Heated cells at 60 °C		68.8		
Strain A18		67		
Strain 26.1.11				
Heat cells at 120 °C		79.3		
Strain A18		77.7		
Strain 26.1.11				
2 mol/L HCl / 1h		72.1		
Strain A18		69.3		
Strain 26.1.11				
S. cerevisiae	B_1 (ng/mL)			[67]

Microorganism	AF	Bound (%)	Conditions	Ref.
	50 ng/mL			
Strain RC008	100	67.6	10^7 cells/mL, 1h, 37 °C, PBS	
	500	43.5		
	50	38.2		
Strain RC009	100	16.4		
	500	21.3		
	50	31.8		
Strain RC012	100	29.6		
	500	20.6		
	50	20.2		
Strain RC016	100	82.0		
	500	48.7		
		65.5		
S. cerevisiae	B1			[68]
Yeast concentration:	(μg/kg)			
0.1 %	150	40	37 °C, 24 h, corn	
	300	25		
	450	17		
	800	16		
0.2 %	150	88	37 °C, 24 h, corn	
	300	76		
	450	64		
	800	51		
0.3 %	150	93	37 °C, 24 h, corn	
	300	86		
	450	81		
	800	66		
S. cerevisiae ATTC 9763	B1			[69]
Pre-treatment:	(ng/mL)			
None	10	40	3 h, 25 °C, pistachio nuts	
	20	70		
Acid treated cells	10	60		
(2 mol/L / 90 min)	20	73		
Heat-treated cells	10	55		
(120 °C / 20 min)	20	75		
L. paracasei LOCK 0920, *L. brevis* LOCK 0944, *L. plantarum* LOCK 0945, and *S. cerevisiae* LOCK 0140	B_1			[70]
	1 mg/kg	55	37 °C, 6h fermentation in broiler feed	
	5 mg/kg	39		

Microorganism	AF	Bound (%)	Conditions	Ref.
S. cerevisiae	B_1			[89]
	1 μg /g	86	12 °C, 8 days, brewing process	
	10 μg /g	72		

Table 2. Aflatoxin binding by yeasts. YCW, Yeast Cell Wall

5. Concluding Remarks

Considering the data from several studies carried out until now, it may be observed that microorganisms, among them lactic acid bacteria and yeasts, have a huge potential application in aflatoxin degradation in foodstuffs. However, new studies are necessary to identify bacterial species with greater binding potential with aflatoxins, once there are differences in sensitivity and selectivity, besides the influence of factors that are intrinsic and extrinsic to the bacteria in the decontamination process. After this step of choosing species with greater efficiency has been overcome, new production technologies that are economically viable to be applied to human and animal foods may be developed.

Several studies have demonstrated that the cell wall of SC and LAB and their components are responsible for binding with aflatoxins. However, the mechanisms by which this bond occurs remain unclear. Cell walls with glucomannanes and manno-oligosaccharides have been pointed out as the responsible elements for AFB_1 bond with yeasts. The great advantage in the commercial use of these microorganisms as binding agents is that these strains are approved and already used in a wide range of fermented food products, being recognized as safe. However, aflatoxin may be released from the cell-aflatoxin complex with changes in the pH and temperature conditions. Therefore, further studies are necessary to determine the behavior of yeasts in the different environmental conditions before they are used commercially.

Acknowledgements

The authors wish to acknowledge the Fundação de Amparo à Pesquisa do Estado de São Paulo (FAPESP) and the Conselho Nacional de Desenvolvimento Científico e Tecnológico (CNPq) of Brazil for financial support and fellowships.

Author details

Carlos Augusto Fernandes Oliveira[1*], Fernanda Bovo[1], Carlos Humberto Corassin[1], Alessandra Vincenzi Jager[1] and Kasa Ravindranadha Reddy

*Address all correspondence to: carlosaf@usp.br

1 Faculty of Animal Science and Food Engineering, University of São Paulo, Brazil

2 Research and Development Department, SRI Biotech Laboratories Ltd, India

References

[1] Bata, A., & Lasztity, R. (1999). Detoxification of mycotoxin-contaminated food and feed by microorganisms. *Trends in Food Science & Technology*, 10-223.

[2] Gonçalez, E., Pinto, M. M., & Felicio, J. D. (2001). Análise de micotoxinas no Instituto Biológico de 1989 a 1999. *Biológico*, 63(1/2), 15-19.

[3] D'Mello, J. P. F., & Mac, Donald. A. M. C. (1997). Mycotoxins. *Animal Feed Science and Technology*, 69-155.

[4] Nierman, W. C., Cleveland, T. E., Payne, G. A., Keller, N. P., Campbell, B. C., Bennett, J. W., Guo, B., Yu, J., & Robens, J. F. (2008). Mycotoxin Production and Prevention of Aflatoxin Contamination in Food and Feed. In: Goldman GH, Osmani SA. The Aspergilli: Genomics, Medical Aspects, Biotechnology, and Research Methods. Boca Raton: CRC Press , 457-472.

[5] Hussein, H. S., & Brasel, J. M. (2001). Toxicity, metabolism, and impact of mycotoxins on humans and animals. *Toxicology*, 167-101.

[6] Bhat, R., Rai, R. V., & Karim, A. A. (2010). Mycotoxins in Food and Feed: Present Status and Future Concerns. *Comprensive Reviews in Food Science and Food Safety*, 9-57.

[7] Alberts, J. F., Engelbrecht, Y., Steyn, P. S., Holzapfel, W. H., & Van Zyl, W. H. (2006). Biological degradation of aflatoxin B1 by Rhodococcus erythropolis cultures. *International Journal of Food Microbiology*, 109-121.

[8] Park, D. L., & Liang, B. (1993). Perspectives on aflatoxin control for human food and animal feed. *Trends in Food Science & Technology*, 41-334.

[9] Chu, F. S. (1991). Mycotoxins: food contamination, mechanism, carcinogenic potential and preventive measures. *Mutation Research*, 259-291.

[10] Abbas, H. K. (2005). Aflatoxin and food safety. Boca Raton, CRC Press.

[11] Magan, N., & Olsen, M. (2006). Mycotoxins in food: Detection and Control. Boca Raton, CRC Press.

[12] Prandini, A., Tansini, G., Sigolo, S., Filippi, L., Laporta, M., & Piva, G. (2009). On the occurrence of aflatoxin M_1 in milk and dairy products. *Food and Chemical Toxicology*, 47-984

[13] Hernandez-Mendoza, A., Garcia, H. S., & Steele, J. L. (2009). Screening of Lactobacillus casei strains for their ability to bind aflatoxin B_1. *Food and Chemical Toxicology*, 47-1064.

[14] Gratz , S., Mykkänem, H. , & El-Nezami, H. (2005). Aflatoxin B1 binding by a mixture of Lactobacillus and Propionibacterium: in vitro versus ex vivo. *Journal of Food Protection*, 68(11), 2470-2474.

[15] El -Nezami, H., Kankaanpaa, P., Salminen, S., & Ahokas, J. (1998). Physicochemical alterations enhance the ability of dairy strains of lactic acid bacteria to remove aflatoxin from contaminated media. *Journal of Food Protection*, 61(4), 466-468.

[16] Line, J. E., & Brackett, R. E. (1995). Factors affecting aflatoxin B1 removal by Flavobacterium aurantiacum. *Journal of Food Protection*, 58(1), 91-94.

[17] El -Nezami, H., Kankaanpaa, P., Salminen, S., & Ahokas, J. (1998). Ability of dairy strains of lactic acid bacteria to bind a common food carcinogen, aflatoxin B_1. *Food and Chemical Toxicology*, 36-321.

[18] Fazeli, M. R., Hajimohammadali, M., Moshkani, A., Samadi, N., Jamalifar, H., Khoshayand, M. R., et al. (2009). Aflatoxin B1 binding capacity of autochthonous strains of lactic acid bacteria. *Journal of Food Protection*, 72(1), 189-192.

[19] Alberts, J. F., Gelderblomb, W. C. A., Botha, A., & Van Zyl, W. H. (2009). Degradation of aflatoxin B1 by fungal laccase enzymes. *International Journal of Food Microbiology*, 135-47.

[20] Wu, Q., Jezkova, A., Yuan, Z., Pavlikova, L., Dohnal, V., & Kuca, K. (2009). Biological degradation of aflatoxins. *Drug Metabolism Reviews*, 41(1), 1-7.

[21] Oliveira, C. A. F., & Germano, P. M. L. (1997). Aflatoxinas: conceitos sobre mecanismos de toxicidade e seu envolvimento na etiologia do câncer hepático celular. *Revista de Saúde Pública*, 31(4), 417-424.

[22] Oatley, J. T., Rarick, M. D., Ji, G. E., & Linz, J. E. (2000). Binding of aflatoxin B1 to bifidobacteria in vitro. *Journal of Food Protection*, 63(8), 1133-36.

[23] Peltonem, K., El-Nezami, H., Haskard, C., Ahokas, J., & Salminen, S. (2001). Aflatoxin B1 binding by dairy strains of lactic acid bacteria and bifidobacteria. *Journal of Dairy Science*, 84, 2152-2156.

[24] Murphy, P. A., Hendrich, S., Landgren, C., & Bryant, C. M. (2006). Food Mycotoxins: An Update. *Journal of Food Science*, 71(5), 51-65.

[25] Creppy, E. E. (2002). Update of survey, regulation and toxic effects of mycotoxins in Europe. *Toxicology Letters*, 127-19.

[26] Bakirci, I. (2001). A study on the occurrence of aflatoxin M1 in milk and milk products produced in Van province of Turkey. *Food Control*, 12-47.

[27] Haskard, C. A., El -Nezami, H. S., Kankaanpaa, P. E., Salminen, S., & Ahokas, J. T. (2001). Surface binding of aflatoxin B1 by lactic acid bacteria. *Applied and Environmental Microbiology*, 67(7), 3086-3091.

[28] Oluwafemi, F., & Silva, F. A. (2009). Removal of aflatoxins by viable and heat-killed Lactobacillus species isolated from fermented maize. *Journal of Applied Biosciences.*, 16-871.

[29] Pierides, M., El -Nezami, H., Peltonem, K., Salminen, S., & Ahokas, J. (2000). Ability of dairy strains of lactic acid bacteria to bind aflatoxin M1 in a food model. *Journal of Food Protection*, 63(5), 645-650.

[30] IARC International Agency for Research on Cancer- World Health Organization. (2002). *IARC Monograph on the Evaluation of Carcinogenic Risk to Humans.*, 82, 171.

[31] Mishra, H. N., & Das, C. (2003). A Review on biological control and metabolism of aflatoxin. *Critical Reviews in Food Science and Nutrition*, 43(3), 245-264.

[32] Laciaková, A., Cicoòová, P., Máté, D., & Laciak, V. (2008). Aflatoxins and possibilities for their biological detoxification. *Medycyna Weterynaryjna*, 64(3), 276-279.

[33] Van Egmond, H. P., & Jonker, M. A. (2004). Worldwide regulations on aflatoxins- The situation in 2002. *Toxin Reviews*, 23, 273-293.

[34] Walstra, P., Wouters, J. T. M., & Geurts, T. J. (2006). Dairy Science and Technology. Boca Raton, CRC Press.

[35] Salminen, S., Von, Wright. A., & Ouwehand, A. C. (2004). Lactic acid bacteria: microbiological and functional aspects. 3.ed. New York Marcel Dekker, Inc.

[36] Sybesma, W. F., & Hugenholtz, J. (2004). Food fermentation by lactic acid bacteria for the prevention of cardiovascular disease. In: Arnoldi A. Functional Foods, Cardiovascular Disease and Diabetes. Boca Raton CRC Press

[37] Onilude, A. A., Fagade, O. E., Bello, M. M., & Fadahunsi, I. F. (2005). Inhibition of aflatoxin-producing aspergilli by lactic acid bacteria isolates from indigenously fermented cereal gruels. *African Journal of Biotechnology*, 4(12), 1404-1408.

[38] Fuchs, S., Sontag, G., Stidl, R., Ehrlich, V., Kundi, M., & Knasmüller, S. (2008). Detoxification of patulin and ochratoxin A, two abundant mycotoxins, by lactic acid bacteria. *Food and Chemical Toxicology*, 46-1398.

[39] Shetty, P. H., & Jespersen, L. (2006). Saccharomyces cerevisiae and lactic acid bacteria as potential mycotoxin decontaminating agents. *Trends in Food Science & Technology*, 17-48.

[40] Ciegler, A., Lillehoj, E. B., Peterson, R. E., & Hall, H. H. (1966). Microbial detoxification of aflatoxin. *Applied Environmental Microbiology*, 14(6), 934-939.

[41] El -Nezami, H., Mykkänen, H., Haskard, C., Salminen, S., & Salminen, E. (2004). Lactic acid bacteria as a tool for enhancing food safety by removal of dietary toxins. In:

Salminen S, Von Wright A, Ouwehand A. Lactic Acid Bacteria: Microbiological and Functional Aspects. 3.ed. Nova Iorque Marcel Dekker , 397-406.

[42] Shahin, A. A. M. (2007). Removal of aflatoxin B1 from contaminated liquid media by dairy lactic acid bacteria. *International Journal of Agriculture and Biology*, 9(1), 71-75.

[43] Rasic, J. L., Skrinjar, M., & Markov, S. (1991). Decrease of aflatoxins B1 in yoghurt and acidified milks. *Mycopathologia*, 113-117.

[44] Biernasiak, J., Piotrowska, M., & Libudzisz, Z. (2006). Detoxification of mycotoxins by probiotic preparation for broiler chickens. Mycotoxin Research; , 22(4), 230-235.

[45] Kabak, B., & Dobson, A. D. W. (2009). Biological strategies to counteract the effects of mycotoxins. *Journal of Food Protection*, 72(9), 2006-2016.

[46] Kabak, B., & Var, I. (2008). Factors affecting the removal of aflatoxin M_1 from food model by Lactobacillus and Bifidobacterium strains. *Journal of Environmental Science and Health, Part B.*, 43-617.

[47] Elgerbi, A. M., Aidoo, K. E., Candlish, A. A. G., & Williams, A. G. (2006). Effects of lactic acid bacteria and bifidobacteria on levels of aflatoxin M_1 in milk and phosphate buffer. *Milchwissenschaft*, 61(2), 197-199.

[48] Haskard, C., Binnion, C., & Ahokas, J. (2000). Factors affecting the sequestration of aflatoxin by Lactobacillus rhamnosus strain GG. *Chemico-Biological Interactions*, 128-39.

[49] Azab, R. M., Tawakkol, W. M., Hamad, A. R. M., Abou-Elmagd, M. K., El -Agrab, H. M., & Refai, M. K. (2005). Detection and estimation of aflatoxin B_1 in feeds and its biodegradation by bacteria and fungi. *Egypt Journal of Natural Toxins*, 2-39.

[50] Bovo, F., Corassin, C. H., Rosim, R. E., & Oliveira, C. A. F. (2012). Efficiency of lactic acid bacteria strains for decontamination of aflatoxin M_1 in phosphate buffer saline solution and in skimmed milk. *Food and Bioprocess Technology*.

[51] Sarimehmetoglu, B., & Küplülü, Ö. (2004). Binding ability of aflatoxin M1 to yoghurt bacteria. Ankara Üniversitesi Veteriner Fakültesi Dergisi; , 51-195.

[52] Lillehoj, E. B., Stubblefield, R. D., Shannon, G. M., & Shotwell, O. L. (1971). Aflatoxin M_1 removal from aqueous solutions by Flavobacterium aurantiacum. *Mycopathologia et mycologia applicata.*, 45-259.

[53] Farzaneh, M., Shi, Z. Q., Ghassempour, A., Sedaghat, N., Ahmadzadeh, M., Mirabolfathy, M., & Javan-Nikkhah, M. (2012). Aflatoxin B_1 degradation by Bacillus subtilis UTBSP1 isolated from pistachio nuts of Iran. *Food Control*, 23-100.

[54] Gao, X., Ma, Q., Zhao, L., Lei, Y., Shan, Y., & Ji, C. (2011). Isolation of Bacillus subtilis: screening for aflatoxins B_1, M_1 and G_1 detoxification. *European Food Research and Technology*, 232-957.

[55] Lahtinen, S. J., Haskard, C. A., Ouwehand, A. C., Salminen, S. J., & Ahokas, J. T. (2004). Binding of aflatoxin B1 to cell wall components of Lactobacillus rhamnosus strain GG. *Food Additives and Contaminants*, 21(2), 158-164.

[56] Shetty, P. H., Hald, B., & Jespersen, L. (2007). Surface binding of aflatoxin B_1 by Saccharomyces cerevisiae strains with potential decontaminating abilities in indigenous fermented foods. *International Journal of Food Microbiology*, 113-41.

[57] Hernandez-Mendoza, A., Guzman-de-Peña, D., & Garcia, H. S. (2009). Key role of teichoic acids on aflatoxin B_1 binding by probiotic bacteria. *Journal of Applied Microbiology*, 107-395.

[58] Hernandez-Mendoza, A., Guzman-de-Peña, D., González-Córdova, A. F., Vallejo-Córdoba, B., & Garcia, H. S. (2010). In vivo assessment of the potential protective effect of Lactobacillus casei Shirota against aflatoxin B_1. *Dairy Science and Technology*, 90-729.

[59] Bolognani, F., Rummey, C. J., & Rowland, I. R. (1997). Influence of carcinogen binding by lactic acid-producing bacteria on tissue distribution and in vivo mutagenicity of dietary carcinogens. *Food and Chemical Toxicology*, 35-535.

[60] Lee, Y. K., El -Nezami, H., Haskard, C. A., Gratz, S., Puong, K. Y., Salminen, S., & Mykkänen, H. (2003). Kinetics of adsorption and desorption of aflatoxin B_1 by viable and nonviable bacteria. *Journal of Food Protection*, 66(3), 426-430.

[61] D'Souza, D. H., & Brackett, R. E. (2001). Aflatoxin B1 degradation by Flavobacterium aurantiacum in the presence of reducing conditions and seryl and sulfhydryl group inhibitors. *Journal of Food Protection*, 64(2), 268-271.

[62] El -Nezami, H., Mykkänen, H., Kankaanpaa, P., Salminen, S., & Ahokas, J. (2000). Ability of Lactobacillus and Propionibacterium strains to remove aflatoxin B_1 from the chicken duodenum. *Journal of Food Protection*, 63(4), 549-552.

[63] Hathout, A. S., Mohamed, S. R., El -Nekeety, A. A., Hassan, N. S., Aly, S. E., & Abdel-Wahhab, M. A. (2011). Ability of Lactobacillus casei and Lactobacillus reuteri to protect against oxidative stress in rats fed aflatoxins-contaminated diet. *Toxicon*, 58-179.

[64] El -Nezami, H. S., Mykkänen, H., Kankaanpää, P., Suomalainen, T., Ahokas, J. T., & Salminen, S. (2000). The ability of a mixture of Lactobacillus and Propionibacterium to influence the faecal recovery of aflatoxins in healthy Egyptian volunteers: a pilot clinical study. *Bioscience and Microflora*, 19-41.

[65] Mc Kinney, R. E. (2004). Environmental pollution control microbiology: a fifty-year perspective. Boca Raton, CRC Press.

[66] Joannis-Cassan, C., Tozlovanu, M., Hadjeba-Medjdoub, K., Ballet, N., & Pfohl-Leszkowicz, A. (2011). Binding of zearalenone, aflatoxin B_1, and ochratoxin A by yeast-based products: a method for quantification of adsorption performance. Journal of Food Protection;, 74(7), 1175-1185.

[67] Armando, M. R., Dogi, C. A., Pizzolitto, R. P., Escobar, F., Peirano, M. S., Salvano, M. A., Sabini, L. I., Combina, M., Dalcero, A. M., & Cavaglieri, L. R. (2011). Saccharomyces cerevisiae strains from animal environment with in vitro aflatoxin B_1 binding ability and anti-pathogenic bacterial influence. *World Mycotoxin Journal*, 4(1), 59-68.

[68] Madrigal-Santillán, E., Madrigal-Bujaidar, E., Márquez-Márquez, R., & Reyes, A. (2006). Antigenotoxic effect of Saccharomyces cerevisiae on the damage produced in mice fed with aflatoxin B1 contaminated corn. *Food and Chemical Toxicology*, 44, 2058-2063.

[69] Rahaie, S., Emam-Djomeh, Z., Razavi, S. H., & Mazaheri, M. (2010). Immobilized Saccharomyces cerevisiae as a potential aflatoxin decontaminating agent in pistachio nuts. *Brazilian Journal of Microbiology*, 41-82.

[70] Slizewska, K., & Smulikowska, S. (2011). Detoxification of aflatoxin B_1 and change in microflora pattern by probiotic in vitro fermentation of broiler feed. *Journal of Animal and Feed Sciences;*, 20-300.

[71] Dogi, C. A., Armando, R., Ludueña, R., Le Blanc, A. M., Rosa, C. A. R., Dalcero, A., & Cavaglieri, L. (2011). Saccharomyces cerevisiae strains retain their viability and aflatoxin B_1 binding ability under gastrointestinal conditions and improve ruminal fermentation. *Food Additives and Contamiantes, Part A.*, 28(12), 1705-1711.

[72] Stanley, V. G., Ojo, R., Woldesenbet, S., Hutchinson, D. H., & Kubena, L. F. (1993). The use of Saccharomyces cerevisiae to suppress the effects of aflatoxicosis in broiler chicks. *Poultry Science*, 72(10), 1867-1872.

[73] Parlat, S. S., Özcan, M., & Oguz, M. (2001). Biological suppression of aflatoxicosis in Japanese quail (Coturnix coturnix japonica) by dietary addition of yeast (Saccharomyces cerevisiae). *Research in Veterinary Science*, 7-207.

[74] Lesson, S., Diaz, G., & Summers, J. (1995). Poultry metabolic disorders and mycotoxins. Montreal (Canada), University Books.

[75] Yildiz, A. O., Parlat, S. S., & Yildirim, I. (2004). Effect of dietary addition of live yeast (Saccharomyces cerevisiae) on some performance parameters of adult Japanese quail (Coturnix coturnix japonica) induced by aflatoxicosis. *Revue Médecine Vétérinaire*, 155(1), 38-41.

[76] Raju, M. V. L. N., & Devegowda, G. (2000). Influence of esterified-glucomannan on performance and organ morphology, serum biochemistry and haematology in broilers exposed to individual and combined mycotoxicosis (aflatoxin, ochratoxin and T-2 toxin). *British Poultry Science*, 41(5), 640-650.

[77] Jouany, J. P., Yiannikouris, A., & Bertin, G. (2005). The chemical bonds between mycotoxins and cell wall components of Saccharomyces cerevisiae have been identified. *Archiva Zootechnica*, 8-26.

[78] Devegowda, G., Aravind, B. I. R., & Morton, M. G. (1996). Saccharomyces cerevisiae and mannanoligosaccharides to counteract aflatoxicoses in broilers. Proceedings of Australian Poultry Science Symposium, , 103-106.

[79] Aravind, K. L., Patil, V. S., Devegowda, G., Umakhanta, B., & Ganpule, S. P. (2003). Efficacy of esterified glucomannan to counteract mycotoxicosis in naturally contaminated feed on performance and serum biochemical, hematological parameters in broilers. *Poultry Science*, 82-571.

[80] Baptista, A. S., Horii, J., Calori-domingues, M. A., Gloria, E. M., Salgado, J. M., & Vizioli, M. R. (2002). Thermolysed and active yeast to reduce the toxicity of aflatoxin. *ScientiaAgricola,*, 59(2), 257-260.

[81] Bejaoui, H., Mathieu, F., Taillandier, P., & Lebrihi, A. (2004). Ochratoxin A removal in synthetic and natural grape juices by selected oenological Saccharomyces strains. *Journal of Applied Microbiology,*, 97-1038.

[82] Cecchini, F., Morassut, M., Moruno, E., & Di Stefano, R. (2007). Influence of yeast strain on ochratoxin A content during fermentation of white and red must. *Food Microbiology*, 23-411.

[83] Angioni, A., Caboni, P., Garau, A., Farris, A., Orro, D., Budroni, M., & Cabras, P. (2007). In vitro interaction between ochratoxin A and different strains of Saccharomyces cerevisiae and Kloeckera apiculata. *Journal of Agricultural and Food Chemistry*, 55, 2043-2048.

[84] Freimund, S., Sauter, M., & Rys, P. (2003). Efficient adsorption of the mycotoxins zearalenone and T-2 toxin on a modified yeast glucan. *Journal of Environmental Science and Health Part B*, 38-243.

[85] Yiannikouris, A., François, J., Poughon, L., Dussap, C. G., Bertin, G., Jeminet, G., & Jouany, J. P. (2004). Alkali extraction of β-D-gluvcans from Saccharomyces cerevisiae cell wall and study of their adsorptive properties toward zearalenone. *Journal of Agriculture and Food Chemistry*, 52-3666.

[86] Karaman, M., Basmacioglu, H., Ortatatli, M., & Oguz, H. (2005). Evaluation of the detoxifying effect of yeast glucomannan on aflatoxicosis in broilers as assessed by gross examination and histopathology. *British Poultry Science*, 46-394.

[87] Santin, E., Paulillo, A. C., Nakagui, L. S. O., Alessi, A. C., Polveiro, W. J. C., & Maiorka, A. (2003). Evaluation of the efficacy of Saccharomyces cerevisiae cell wall to ameliorate the toxic effects of aflatoxin in broilers. *International Journal of Poultry Science*, 2(6), 465-468.

[88] Eshak, M. G., Khalil, W. K. B., Hegazy, E. M., Ibrahim, M. F., Fadel, M., & Stino, K. R. D. (2010). Effect of SCE on reduction of aflatoxicosis, enhancement of growth performance and expression of neural and gonadal genes in Japanese quail. *Journal of American Science*, 6-824.

[89] Chu, F. S., Chang, C. C., Ashoor, S. H., & Prentice, N. (1975). Stability of Aflatoxin B_1 and Ochratoxin A in Brewing. *Applied Microbiology*, 29(3), 313-316.

[90] Celyk, K., Denly, M., & Savas, T. (2003). Reduction of toxic effects of aflatoxin by using baker yeast (Saccharomyces cerevisiae): in growing broiler chicken diets. *Revista Brasileira de Zootecnia*, 32-615.

[91] El -Banna, A. A., Lau, P. Y., & Scott, P. M. (1983). Fate of mycotoxins during processing of foodstuffs. II- Deoxynivalenol (vomitoxin) during making of Egyptian bread. *Journal of Food Protection*, 46-484.

Chapter 6

Recent Advances for Control, Counteraction and Amelioration of Potential Aflatoxins in Animal Feeds

N.K.S. Gowda, H.V.L.N. Swamy and P. Mahajan

Additional information is available at the end of the chapter

1. Introduction

Mycotoxins are the toxic products of fungal metabolism occuring in a wide variety of commodities like animal feeds and human food products. Mycotoxins on ingestion can cause health hazards both in livestock and human beings and hence there is a greater economic and public health implication. The severity of mycotoxin contamination is determined by environmental factors like excessive moisture in the field as well as in storage, hot and humid climate and insect infestation. Mycotoxin contamination of feed affects practically all livestock but greater information is available on dairy cattle, poultry, and swine. In these animals mycotoxins reduce production efficiency, impairs resistance to infection and compromise reproduction. Economic losses due to mycotoxicosis are derived directly from livestock morbidity, mortality and wastage of contaminated feed. On a global scale, it is estimated that around 25% of the world's crops are affected by mycotoxins annually and in addition to the above losses costs involved in monitoring the level of mycotoxins should also be considered. The recent mycotoxin surveys have indicated that the percent contamination is much higher than the perceived 25%. The mycotoxins that are of significance in animal feed are: Aflatoxins, Ochratoxins and Fusarial toxins (Fumonisins, Zearalenone, Trichothecenes including Deoxynivalenol and T-2 toxin).

1.1. Aflatoxins and biological action

The aflatoxins are highly toxic and carcinogenic compounds produced by *Aspergillus* fungi at an optimum temperature of 25-32^0 C, moisture of greater than 12-16% and a relative humidity of 85%. Commonly affected feeds are maize, groundnut cake, cottonseed cake and copra cake and causes toxicity in poultry, cattle, sheep and swine. Animal consuming aflatoxin contaminated feed display poor performance, reduced immunity, liver damage, kidney and intestinal haemorrhage and liver tumors. Among the afltoxins B_1 is more prevalent and toxigenic. This is metabolized to Aflatoxin M_1 in liver and is excreted in milk of dairy cattle and also as residue in egg / meat.

Epoxide derivative of aflatoxin B_1 binds with DNA and disrupts transcription and translation activities, thus initiating carcinogenesis. Oxidative nature of the toxic derivative releases free radicals and cause cell damage (Fig.1). Advancement in molecular techniques like microarray and PCR has helped to understand the precise mechanism of action of aflatoxin. Recent gene expression studies have shown that down regulation of mitochondrial carnitine palmitoyltransferase (CPT) system, down regulation of fatty acid metabolism pathway, up-regulation of cell proliferation pathway and down regulation of B cell activation are respectively responsible for decreased body weight gain, fatty liver / increased liver weight, carcinoma and lowered immunity in birds fed aflatoxin. Supplementation of curcumin through turmeric powder ameliorated most of the ill effects induced by aflatoxin. Adverse effects of aflatoxicosis are much severe when there is a concurrent contamination with other toxins like ochratoxin and T-2 toxin.

1.2. Limits of aflatoxin

The presence of Aflatoxin M_1 in food products meant for human consumption is not desirable and the residual concentration should not exceed 0.5 ppb as per FDA regulations. Such regulations are much more stringent in European Union where the level should not exceed 0.05ppb. Aflatoxin B_1 level of 20 ppb in the diet of dairy cattle is appropriate for reducing the risk of aflatoxin M_1 in milk. In many countries there are strict guidelines for maximum tolerable limits of aflatoxins, beyond which the commodity is unsafe and not accepted (Table 1).

Limits
Cattle 20 ppb, Broiler chicken 20 ppb, Finisher pig 200 ppb
Beef cattle 300 ppb, Layer poultry 100 ppb
India : 60 ppb (B_1) for groundnut cake, 120 ppb (B_1) for groundnut cake (export)
UK & Spain : Complete feeds 10-20 ppb(B_1). Groundnut 50 ppb (B_1)
Other feed ingredients 200 ppb (B_1)
EEU : 500 ppb (B_1) for feed ingredients ; France : 300 ppb (B_1) for feed ingredients; Japan : 1000 ppb (B_1) for raw materials, 50 ppb (B_1) for complete feeds of cattle, 20 ppb (B_1) for complete feeds of pigs and poultry
USA : 300 ppb (B_1) for cottonseed meal; 20 ppb (B_1) for other feed ingredients, milk for human consumption 0.5 ppb.
Canada : 20 ppb (total aflatoxins) for livestock feeds
South Africa : 10 ppb(total), Australia : 15 ppb (B_1) for groundnut

Table 1. Suggested limits for aflatoxin.

2. Control and counteraction of aflatoxins

2.1. Preventive measures

Aflatoxins affect mainly liver and kidney and are also carcinogenic and mutagenic (Fig 1). Therefore effective control and detoxification measures need to be undertaken. Toxin producing fungi may invade at pre-harvesting period, harvest-time, during post harvest handling and in storage. According to the site and time of infestation, the fungi can be divided into three groups: (a) Field fungi (b) Storage fungi (c) Advanced deterioration fungi. Field fungi are generally plant pathogenic fungi; namely *Fusarium*. The storage fungi are *Aspergillus* and *Penicillium*. The advanced deterioration fungi, normally do not infest intact grains but easily attack damaged grains and requires high moisture content, that include *Aspergillus clavatus, Aspergillus fumigatus*.

Prevention and effective plan for reducing fungal growth and toxin production is very important. The recommended practices include 1. Development of fungal resistant varieties of plants, 2. Suitable pre-harvest, harvest and post harvest techniques, 3. Store commodities at low temperature as for as possible, 4. Use fungicides and preservatives against fungal growth and 5. Control of insect damage in grain storage with approved insecticides.

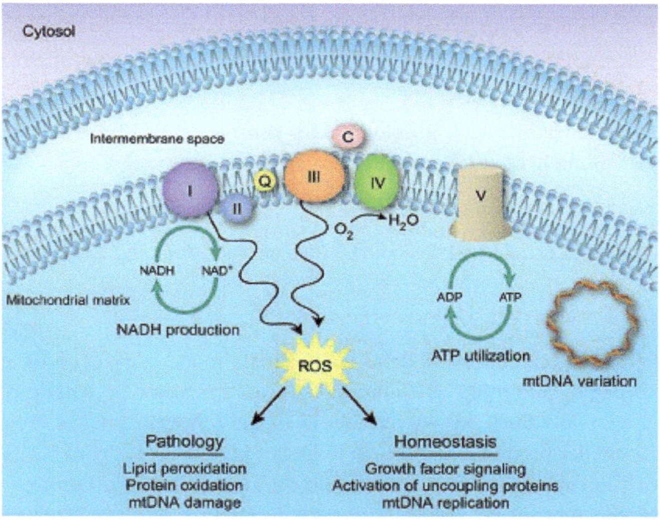

Figure 1. Mechanism of cell damage in mycotoxin toxicity.

(adopted from Joshua M Baughman and Vamsi K Mootha, 2006) [6]

The secondary prevention of fungal growth include limiting the growth of infested fungi by re-drying the product, removal of contaminated seeds. The tertiary measures could be to prevent the transfer of fungi and their health hazardous toxins into the food/feed and to the environment. This include complete destruction of the contaminated product or diversion for fermentation to produce ethanol or detoxification / destruction of mycotoxins to the minimum level. Among the mycotoxins, aflatoxin is the most well-known and thoroughly studied and its prevention and control has been most successfully practiced in many countries.

2.2. Fungal growth inhibition

The inhibition of fungal growth can be achieved by physical, chemical and biological treatments. After the crop is harvested, drying and proper storage and suitable transportation of the commodities are of prime importance. Factors contribute to the growth of fungi and toxin production includes high moisture content, humidity, warm temperature (25-40 °C), insect infestation and grain damage.

2.2.1. Physical methods

- Drying seeds and commodities to the safe moisture level (< 9-11%).
- Maintenance of the container or store house at low temperature and humidity.
- Keep out insects and pests from the storage.
- Gamma-irradiation of large-scale commodities.
- Dilution of the contaminated feed with safe feed.

2.2.2. Chemical methods

- Use of fungicides (acetic acid, propionic acid, benzoic acid, citric acid and their sodium salts, copper sulfate): 0.2–0.4 % in feed.
- Use of fumigants – ammonia: 0.2-0.4%
- Addition of herbal extracts (garlic, onion, clove oil, turmeric powder, thyme) : 0.25-0.5%

2.2.3. Biological methods

Anti-fungal enzymes, chitinase and Beta -1,3 glucanase found in plant seeds, may act as defense against pathogenic fungi as chitin and glucan are major polymeric components of many fungal cell walls. Such polysaccharides in fungal cell wall could be enzymatically hydrolysed into smaller products resulting in killing of mycelia or spore of fungi. It is foreseen that seeds rich in such anti-fungal enzymes likely to resist the infestation of fungi. Use of non-toxigenic biocompetitive *Aspergillus* strains to out-compete the toxigenic isolates has been found effective in reducing pre-harvest contamination with aflatoxin in peanut and cotton. However, the aflatoxin contamination process is so compelx that a combination of approaches will be required to eliminate toxin production.

Application of non-toxigenic strains of *Aspergillus flavus* and *Aspergillus parasiticus* to soil in maize plots, favoured the reduction in colonization of toxigenic fungi in subsequent years. When the weather conditions were suitable for fungal growth and resulted in 65-80% decline in aflatoxin production as compared to control. Inoculation of chitosan, *Bacillus subtilis* and *Trichoderma harzianum* to pre-harvest maize along with *Aspergillus flavus* inhibits aflatoxin production. Many anti-fungal metabolites (cyclic dipeptides, phenylactic acid, caproic acid, reuterin, lactic acid, acetic acid, fungicin) have been isolated from different cultures of lactic acid bacteria. Aflastin A, an anti-microbial compound produced by *Streptomyces* Spp,. MRI 142 strain of bacteria is known to inhibit aflatoxin production by *Aspergillus parasiticus*. Iturin, an anti-fungal peptide produced by *Bacillus subtilis* had inhibitory effect an *Aspergillus parasiticus*.

2.2.4. *Plant breeding, genetic engineering and microarray*

Genetic modification of mold susceptible plants holds some promise in ensuring food safety. This involves increasing production of compounds like anti-fungal proteins, hydroxamic acids, and phenolics that reduce fungal contamination. This may be accomplished by introducing a novel gene to express the target compound, or enhancing the expression of such compounds by the existing genes, thereby capitalizing on the plant's own defense mechanisms. Enzymes that catalyze production of anti-fungals could be targeted for their expression and such an approach is being actively pursued by researchers. Enhanced expression of an alpha-amylase inhibitor in *Aspergillus* could result in reduced aflatoxin synthesis. Hybrid varieties of cereals with Bt (*Bacillus thermophilus*) genes have shown reduced aflatoxin production, probably due to higher resistance of plants against pest and insects.

A cluster of genes are responsible for aflatoxin production through pathway-specific transcriptional regulator. A total of 20 genes in the aflatoxin biosynthetic cluster and 3 additional genes outside the aflatoxin biosynthetic cluster responsible for aflatoxin production have been identified. Identification of critical genes governing aflatoxin formation could lead to use of non-aflatoxigenic bio-competitive strains of *Aspergillus flavus* through use of gene disruption techniques. The advances in molecular biology could aid in early detection of mycotoxin production in food/feed material. DNA-chip with microarray system containing oligonucleotide primers that are homologues to genes of several fungal species responsible for the expression of mycotoxins can be employed to forecast the mycotoxin production in advance and accordingly critical anti-fungal strategies can be employed. Such PCR based molecular techniques are of value in assessing the potential for mycotoxin production. The time gap between expression of a set of genes and actual mycotoxin production is about 4-5 days. This early forecasting of extent of mycotoxin production will help in adopting immediate preventive anti-fungal measures.

2.3. Counteraction / Detoxification of aflatoxins

Aflatoxins in foods and feeds can be removed, inactivated or detoxified by physical, chemical and biological means. The treated products should be health safe from the chemicals and their essential nutritive value should not be deteriorated.

2.3.1. *Physical methods*

Physically, aflatoxin contaminated seeds can be removed by hand picking or photoelectric detecting machines, but this is labor intense and expensive. Heating and cooking under pressure can destroy nearly 70% aflatoxin. Dry roasting can reduce about 50-70% of aflatoxin and sunlight drying of aflatoxin contaminated feed could reduce the toxin level by more than 70%.

The addition of binding agents can reduce the bioavailability of these compounds in animals, and limit the presence of toxin residues in animal products. In case of aflatoxin B_1 (AFB_1), hydrated sodium calcium aluminosilicates (HSCAS) and phyllosilicates derived from natural zeolites have a high affinity, both *in vitro* and *in vivo*. Zeolites, which are hydrated aluminosilicates of alkaline cations are able to adsorb AFB_1. Bentonites have been shown to be effective for the adsorption of AFB_1. Other clays, such as kaolin, sepiolite and montmorillonite, bind AFB_1 but less effectively than HSCAS and bentonite. Activated charcoal has mixed results against AFB_1.

Although clays are effective against aflatoxins, caution should be exercised to make sure that their inclusion level is not too high and they are free from impurities such as dioxin. When the level of inclusion is very high, which is actually required for them to be effective, there are chances that these compounds can bind minerals and antibiotics like monensin. Some of the binders are not biodegradable and could pose environmental problem.

2.3.2. *Chemical methods*

A variety of chemical agents such as acids, bases (ammonia, caustic soda), oxidants (hydrogen peroxide, ozone, sodium hypochlorite), reducing agents (Bisulphites), chlorinated agents and formaldehyde have been used to degrade mycotoxins in contaminated feeds particularly aflatoxins. However, these techniques are not totally safe, are expensive and not well accepted by consumers.

2.3.3. *Biological / microbiological methods*

The biological decontamination of mycotoxins using yeast *Saccharomyces cerevisiae* and lactic acid bacteria has received much attention. Yeast and lactic acid bacterial cells are known to bind different toxins on the cell wall surface. This will be of immense value in reducing the mycotoxin hazards (Table 2), and effective binding strains of these microbes could eventually be used to minimize aflatoxin exposure and improving overall health in animals.

To tackle the high inclusion levels of clays, cell walls of specific yeasts were studied for their ability to bind aflatoxins. The wealth of data to date has shown that beta-glucans (esterified glucomannans), specific sugars present in the inner cell wall of yeast, can bind aflatoxins. The levels of inclusion of yeast-based binders are much lower than clay-based binders. About 500 gm of glucomannans from yeast cell-wall have the same adsorption capacity as 8 kg of clay. This binder reduces the AFM_1 content of milk by 58% in cows given a diet contaminated with AFB_1 at a concentration of 0.05% of dry mater.

Probiotic strain of *Lactobacillus acidophilus* CU028 has shown to bind aflatoxin. Probiotic fermented milk containing *Lactobacillus casei* and *Lactobacillus rhamnosus* strains alone or in combination with chlorophyllin exhibited protective effect against aflatoxin B_1- induced hepatic damage. Acid treated lactic acid bacteria were able to bind high dosage of aflatoxin in gut conditions.

2.3.4. *Biotransformation*

Dual cultivation of *Aspergillus niger, Mucor racemosus, Alternaria alternata, Rhizopus oryzae* and *Bacillus stearothermophilus* with toxigenic strain of *Aspergillus flavus* results in 70-80% degradation of aflatoxins. Certain microbes are also able to metabolize mycotoxins (*Corynebacterium rubrum*) in contaminated feed or to biotransform them(*Rhizopius, Trichosporon mycotoxinivorans, Rhodotorula rubra, Geotrichum fermentans*). However, these biological processes are generally slow and have a varied efficiency. Ruminants are considered to be relatively resistant to aflatoxins, due to biodegrading and biotransforming ability of rumen microbes compared to monogastric animals. This would be a great asset in biological detoxification of aflatoxins and with the help of genetic engineering techniques, benefits of this can be better realized.

Isolates	Number of aflatoxin B_1 binding strains			
	Percentage of binding			
	<15	15-39	40-59	>60
Yeast				
Saccharomyces cerevisiae	1	8	3	3
Condida krusei	4	5	1	1
Candida parapsilosis				
Trichosporon mucoides	1	-	-	-
Candida catenulanta	1	-	-	-
Bacteria				
Lactobacillus plantarum	-	-	4	1
Lactobacillus fermentum	-	-	-	1
Pediococcus acidilactici	-	-	1	-

Table 2. Aflatoxin binding ability of different strains of yeast and bacteria.

2.4. Dietary manipulations

2.4.1. Hepatotropic nutrients and anti-oxidants

Various nutritional strategies have been employed to alleviate the adverse effects of aflatoxins. Addition of specific amino acids like methionine in excess of their requirement protect the chicks from growth depressing effects of AFB_1, possibly through an increased rate of detoxification by glutathione, a sulfur amino acid metabolite. Supplementation of phenyl alanine has shown to alleviate toxicity of ochratoxin. Addition of vegetable oil (safflower oil, olive oil) to aflatoxin contaminated feed improves the performance of chicks.

Aflatoxins cause toxicity through release of free radicals and lipid peroxidation. Hence, antioxidants could aid in the overall detoxification process in liver and hence may help in alleviation of aflatoxicosis. Butylated hydroxy toluene (BHT) is effective in preventing the adverse effects of AFB_1. Vitamin E and Selenium supplementation also has shown to overcome negative effects of aflatoxin. Of late, there is a growing interest in the use of phytochemicals (curcumin, flavonoids, resveratrol, Allixin, polyphenolics) as antioxidants in increasing the activity of antioxidant enzymes (SOD, catalase, glutathione peroxidase) and neutralizing the free radicals, thus, ameliorating the mycotoxin toxicity.

3. Conclusion

Aflatoxins are common in nature, hence minimizing the contamination is not an a easy task due to the interaction of fungus with environment and feed material. This involves constant attention during the entire process of grain harvest, storage, feed manufacturing and animal production. Most effective methods (physical, chemical, biological, biotechnological) to improve seed production, cultivation, harvest and storage need to be adopted. Use of binders and understanding their mechanism of action is the current concept and research areas in the use of microbes for decontamination and biotransformation of aflatoxins is gaining momentum. Biotechnological intervention in terms of developing transgenic fungal resistant crops and biological control using non-toxigenic, competitive fungal species holds a better promise in managing the problem of aflatoxicosis. Advancement in molecular techniques using fungal oligonucleotide probes with PCR based microarray analysis would help in early forecasting / detection of potential aflatoxin production, suggesting for critical control strategies.

Acknowledgements

The first author wishes to thank and acknowledge the technical and financial support of M/S. Alltech Biotechnology Private Ltd., Bangalore in publishing this chapter. This company is a leading manufacturer and marketer of natural feed supplements including mycotoxin binder. They are present in over 120 countries and operates on ACE Principle- providing solutions to the animal industry which are friendly to animals, consumer and environment

Author details

N.K.S. Gowda[1*], H.V.L.N. Swamy[2] and P. Mahajan[2]

*Address all correspondence to: nksgowda@rediffmail.com

1 National Institute of Animal Nutrition and Physiology, Bangalore, India

2 Alltech Biotechnology Private Ltd., Bangalore , India

References

[1] Bakutis, B., Baliukoniene, V., & Paskevicius, A. (2005). Use of biological method for detoxification of mycotoxins. *Botanica Lithuania.*, 7, 123-129.

[2] Bhatnagar, D., Payne, G. A., Linz, J. E., & Cleveland, T. E. (1995). Molecular biology to eliminate aflatoxin. *Int. News Fats oils Related Mater.*, 6, 262-271.

[3] Byun, J. R., & Yoon, Y. H. (2003). Binding of aflatoxin G_1, G_2 and B_2 by probiotic Lactobacillus spp. *Asian- Australasian Journal of Animal Science.*, 16(11), 1686-1689.

[4] Cleveland, T. E., Cary, J. W., Brown, R. L., Bhatnagar, D., Yu, J., Chang, P. K., Chlan, C. A., & Rajasekaran, K. (1997). Use of biotechnology to eliminate aflatoxin in preharvest crops. *Bulletin of the Institute for Comprehensive Agricultural Sciences., Kinki University*, 5, 75-90.

[5] Cuero, R. G., Duffus, E., Osuji, G., & Pettit, R. (1991). Aflatoxin control in preharvest maize : effects of chitosan and two microbial agents. *Journal of Agricultural Science.*, 117(2), 165-169.

[6] D'Mello, J. P. F., Placinta, C. M., & Macdonald, A. M. C. (1999). Fusarium mycotoxins : a review of global implications for animal health, welfare and productivity. *Animal Feed Science and Technology*, 80, 183-205.

[7] Devegowda, G., Raju, M. V. L. N., & Swamy, H. V. L. N. (1998). Mycotoxins in poultry feed: Methods of control and prevention. *Poult. Punch.*, 6, 73-81.

[8] Dorner, T. W., Cole, R. J., & Wicklow, D. T. (1999). Aflatoxin reduction in corn through field application of competitive fungi. *Journal of Food Protection*, 62(6), 650-656.

[9] Faraj, M. K., Smith, J. E., & Harran, G. (1993). Aflatoxin biodegradation :effects of temperature and microbes. *Mycotoxin Research.*, 97(11), 1388-1392.

[10] Gowda, N. K. S., & Ledoux, D. R. (2008). Use of antioxidants in amelioration of mycotoxin toxicity: a reiview. *Animal Nutrition and Feed Technology.*, 8(1), 1-11.

[11] Gowda, N. K. S., Ledoux, D. R., Rottinghaus, G. E., Bermudez, A. J., & Chen, Y. C. (2008). Efficacy of turmeric (Curcuma longa), containing a known level of curcumin, and a hydrated sodium calcium aluminosilicate to ameliorate the adverse effects of aflatoxin in broiler chicks. *Poultry Science*, 87, 1125-1130.

[12] Gowda, N. K. S., Ledoux, D. R., Rottinghaus, G. E., Bermudez, A. J., & Chen, Y. C. (2009). Antioxidant efficacy of supplemental curcuminoids from turmeric (Curcuma longa L.) powder in broiler chicks fed diet containing aflatoxin B_1. *British Journal of Nutrition*, 102, 1629-1634.

[13] Gowda, N. K. S., Malathi, V., & Suganthi, R. U. (2004). Effect of various chemical and herbal compounds on growth of Aspergillus parasiticus and aflatoxin production. *Animal Feed Science and Technology.*, 116, 281-291.

[14] Gowda, N. K. S., Suganthi, R. U., Malathi, V., & Raghavendra, A. (2006). Efficacy of heat treatment and sun drying of aflatoxin contaminated feed for reducing the harmful biological effects in sheep. *Animal Feed Science and Technology.*, 133, 167-175.

[15] Gowda, N. K. S., Suganthi, R. U., Malathi, V., & Raghavendra, A. (2004). Effect of various chemical and herbal compounds on growth of *Aspergillus parasiticus* and aflatoxin production. *Animal Feed Science and Technology.*, 116, 281-291.

[16] Huwig, A., Freimund, S., Kappeli, O., & Dutler, H. (2001). Mycotoxin detoxification of animal feed by different adsorbents. *Toxicology Letters.*, 122, 1-12.

[17] Joshua, M., Baughman, K., & Vamsi, K. M. (2006). Buffering mitochondrial DNA variation. *Nature Genetics*, 38(11), 1232-33.

[18] Khadem, A.a., Sharifi, S. D., Barati, M., & Borji, M. (2012). Evaluation of effectiveness of yeast, zeolite and active charcoal as aflatoxin absorbents in broiler diets. *Global Veterinaria.*, 8(4), 426-432.

[19] Kumar, M., Verma, V., Nagpal, R., Kumar, A., Behare, P. V., Singh, B., & Aggarwal, P. K. (2012). Anticarcinogenic effect of probiotic fermented milk and chlorophyllin on aflatoxin B_1- induced liver carcinogenesis in rats. *British Journal of Nutrition.*, 107, 1006-1016.

[20] Manafi, M. Counteracting effect of high grade sodium bentonite during aflatoxicosis in broilers. *Journal of Agricultural Science and Technology.*, 14(3), 539-547.

[21] Meronuck, R., & Concibido, V. (1997). Mycotoxins in feed. *Feed Stuffs.*, 7(7), 142-148.

[22] Price, M. S., Yu, J., William, C., Nierman, H., Kim, S., Pritchard, B., Jacobes, C. A., Bhatnagar, D., Cleveland, T. E., & Payne, G. A. (2006). The aflatoxin pathway regulatorAflR induces gene transcription inside and outside of the aflatoxin biosynthetic cluster. *FEMS Microbiology Letters.*, 255(2), 275-279.

[23] Motameny, R., Sadeghi, A. A., Dehghan-Banadaky, M., Chamani, M., & Abolhassani, M. (2012). Effect of some acid treated bacteria on reduction of impure aflatoxin B_1 in ruminant gastrointestinal model. *Journal of American Science.*, 7(12), 213-217.

[24] Munkvold, G. R. (2003). Cultural and genetic approaches to managing mycotoxins in maize. *Annual Review of Phytopathology.*, 41, 99-116.

[25] Murphy, P. A., Hendrich, S., Landgren, C., & Brynat, C. M. (2006). Food mycotoxins: An update. *Journal of Food Science*, 71(5), 51-65.

[26] Murphy, P. A., Hendrich, S., Landgren, C., & Bryant, C. M. (2006). Food mycotoxins: An update. *Journal of Food Science.*, 71(5), R, 51-65.

[27] Oluwafemi, F., & Taiwo, V. O. (2003). Reversal of toxigenic effects of aflatoxin B_1 on cockerels by alcoholic extract of African nutmeg, Monodora myristica. *Journal of the Science of Food and Agriculture.*, 84, 333-340.

[28] Payne, G. A. (1998). Process of contamination by aflatoxin- producing fungi and their impact on crops. In:, *Sinha K.K., Bhatngar, D., (eds). Mycotoxins in Agriculture and Food Safety. New York, Dekker*, 278-306.

[29] Viswanath, P. (2004). Mycotoxin regulations for foods : a global update. *Journal of Food Science and Technology.*, 41(2), 115-123.

[30] Raju, M. V. L. N., & Reddy, M. R. (2000). Prevention and control of mycotoxins in poultry diets. *In: Poultry seminar, 9 th March, 2000, Hyderabad*, 61-81.

[31] Raju, M. V. L. N., & Devegowda, G. (2000). Influence of esterified glucomannan on performance and organ morphology, serum biochemistry and hematology in broilers exposed to individual and combined mycotoxicosis(aflatoxin,ochratoxin and T-2 toxin). *British Poultry Science.*, 41, 640-650.

[32] Shetty, P. H., & Jespersen, L. (2006). Saccharomyces cerevisiae and lactic acid bacteria as protential mycotoxin decontaminating agents. *Trends in Food Sci.Technol.*, 17, 48-55.

[33] Mabbett, T. (1999). The mycotoxin menace. *Feeding Times.*, 4(3), 4-5.

[34] Upadhaya, S. D., Park, M. A., Jong, K., & Ha, . (2010). Mycotoxins and their biotransformation in the rumen: a review. *Asian-Australasian Journal of Animal Science.*, 23(9), 1250-1260.

[35] Yarru, L. P., Settivari, R. S., Gowda, N. K. S., Antoniou, E., Ledoux, D. R., & Rottinghaus, G. E. (2009). Effects of turmeric (Curcuma longa L) on the expression of hepatic genes associated with biotransformation, anti-oxidant and immune systems in broiler chickens fed aflatoxin. *Poultry Science*, 88, 2620-2627.

[36] Yiannikouris, A., & Jouany, J. P. (2002). Mycotoxins in feed and their fate in animals: a review. *Animal Research.*, 51, 81-99.

[37] Magnoli, A. P., Monge, M. P., Miazzo, R. D., Cavaglieri, L. R., Magnoli, C. E., Merkis, C. I., Cristofolini, A. L., Dalcero, A. M., & Chiacchiera, S. M. (2011). Effect of low lev-

els of aflatoxin B1 on performance, biochemical parameters, and aflatoxin B[1] in broiler liver tissues in the presence of monensin and sodium bentonite. *Poultry Science*, 90, 48-58.

[38] Yiannikouris, Alexandros, Gwénaëlle, Andre, Poughon, Laurent , Francüois, Jean , Dussap, Claude-Gilles, Jeminet, Georges, Bertin, Gerard , & Jouan, Jean-Pierre . (2006). Chemical and conformational study of the interactions involved in mycotoxin complexation with alpha-D-glucans. *Biomacromolecules*, 7, 1147-1155.

Section 2

Food and Agriculture

Chapter 7

Occurrence of Aflatoxins in Food

Ayhan Filazi and Ufuk Tansel Sireli

Additional information is available at the end of the chapter

1. Introduction

Mycotoxins are natural contaminants in raw materials, food and feeds [1]. Aflatoxins are toxic metabolites produced by different species of toxigenic fungi, called mycotoxins. The discovery of aflatoxins dates back to the year 1961 following the severe outbreak of turkey "X" disease, in the England, which resulted in the deaths of more than 100.000 turkeys and other farm animals. The cause of the disease was attributed to a feed using thin-layer chromatography (TLC) revealed that a series of fluorescent compounds, later termed aflatoxins, were responsible for the outbreak. The disease was linked to a peanut meal, incorporated in the diet, contaminated with a toxin produced by the filamentous fungus *Aspergillus flavus*. Hence, the name aflatoxins, an acronym, has been formed from the following combination: the first letter, "A" for the genus *Aspergillus*, the next set of three letters, "FLA", for the species *flavus*, and the noun "TOXIN" meaning poison [2].

Aflatoxins (AFs) are difuranocoumarins produced primarily by two species of *Aspergillus* fungus which are especially found in areas with hot, humid climates. *A. flavus* is ubiquitous, favouring the aerial parts of plants (leaves, flowers) and produces B aflatoxins. *A. parasiticus* produces both B and G aflatoxins, is more adapted to a soil environment and has more limited distribution [3]. *A. bombysis, A. ochraceoroseus, A. nomius,* and *A. pseudotamari* are also AFs-producing species, but are encountered less frequently. From the mycological perspective, there are qualitative and quantitative differences in the toxigenic abilities displayed by different strains within each aflatoxigenic species. For example, only about half of *A. flavus* strains produce AFs-producing species more than 10^6 μg kg^{-1} [4].

Among the 18 different types of aflatoxins identified, the major members are aflatoxin B1 (AFB1), B2 (AFB2), G1 (AFG1), G2 (AFG2), M1 (AFM1) and M2 (AFM2). AFB1 is normally predominant in amount in cultures as well as in food products. Pure AFB1 is pale-white to yellow crystalline, odorless solid. AFs are soluble in methanol, chloroform, acetone, acetonitrile. *A. flavus* typically produces AFB1 and AFB2, whereas *A. parasiticus* produce AFG1 and AFG2 as well as AFB1 and AFB2. Four other aflatoxins M1, M2, B2A, G2A which may be produced in minor amounts were subsequently isolated from cultures of *A. flavus* and *A. parasiticus*. A number of closely related compounds namely aflatoxin GM1, parasiticol and aflatoxicol are also produced by *A. flavus*. The order of acute and chronic toxicity is AFB1 > AFG1 > AFB2 > AFG2, reflecting the role played by epoxidation of the 8,9-double bond and also the greater potency associated with the cyclopentenone ring of the B series, when compared with the six-membered lactone ring of the G series. AFM1 and AFM2 are hydroxylated forms of AFB1 and AFB2 [5]. AFM1 and AFM2 are major metabolites of AFB1 and AFB2 in humans and animals and may be present in milk from animals fed on AFB1 and AFB2 contaminated feed [6, 7]. AFM1 may be also present in egg [8], corn [9] and peanut [10, 11].

AFs interact with the basic metabolic pathways of the cell disrupting key enzyme processes including carbohydrate and lipid metabolism and protein synthesis [12]. The health effects of aflatoxins have been reviewed by a number of expert groups. Aflatoxins are among the most potent carcinogenic, teratogenic and mutagenic compounds in nature [13]. The International Agency for Research on Cancer (IARC) has concluded that naturally occurring aflatoxins are carcinogenic to humans (group 1), with a role in etiology of liver cancer, notably among subjects who are carriers of hepatitis B virus surface antigens. In experimental animals there was sufficient evidence for carcinogenicity of naturally occurring mixtures of aflatoxins and of AFB1, AFG1 and AFM1, limited evidence for AFB2 and inadequate evidence for AFG2. The principal tumors were in the liver, although tumors were also found at other sites including the kidney and colon. AFB1 is consistently genotoxic in vitro and in vivo [3].

The Joint Food and Agriculture Organization of the United Nations (FAO)/World Health Organization (WHO) Expert Committee on Food Additives (JECFA) estimated potency values for AFB1 from the epidemiological data. These corresponded to 0.3 cancers/year per 100,000 population per ng aflatoxin/kg body weight (b.w). per day (uncertainty range: 0.05-0.5) in hepatitis B virus antigen positive individuals and 0.01 cancers/year per 100,000 population per ng aflatoxin/kg b.w. per day (uncertainty range: 0.002-0.03) in hepatitis B virus antigen negative individuals. AFM1 has been evaluated separately from AFB1 by the JECFA, because of its potential to be present in milk and milk products of livestock fed on aflatoxins contaminated feed [14]. The JECFA concluded that AFM1 should be presumed to induce liver cancer in rodents by a similar mechanism to AFB1, and that estimates of the potency of AFB1 can be used for determining the risk due to intake of AFM1, including those for populations with a high prevalence of carriers of hepatitis B virus. The carcinogenic potency of AFM1 was estimated to be one-tenth that of AFB1, based on a comparative study in the Fischer rat conducted by Cullen et al. [15].

Humans can be exposed to aflatoxins by the periodic consumption of contaminated food, contributing to an increase in nutritional deficiencies, immunosuppression and hepatocellular carcinoma. AFs have a wide occurrence in different kind of matrices, such as spices, cereals, oils, fruits, vegetables, milk, meat, etc. [5]. About 4.5 billion people, mostly in developing countries, are at risk of chronic exposure to aflatoxins from contaminated food crops [16]. Therefore, in order to avoid the toxicity, the levels of aflatoxins and similar toxic compounds in foodstuffs have to be monitored closely, and to be kept under control continuously. Otherwise, related health effects like acute and chronic intoxications, and even deaths, will still be an issue [17].

In this review, we are presenting a report on the situation of aflatoxin contamination in food and foodstuffs such as oilseeds (peanuts, almonds, pistachios and other tree nuts), cereals (wheat, barley, rice, corn, etc), spices, milk and dairy products and other foods of animal origin (meat, offal, eggs etc) in world. Incidence of contamination will classified as country and type of food and discussed. Also, we are reviewing the scientific literature on aflatoxins in foods and how they affect animal and human health.

2. Occurrence of Aflatoxin in Oilseeds

Oilseed crops are primarily soybeans, sunflower seed, canola, rapeseed, safflower, flaxseed, mustard seed, peanuts and cottonseed, used for the production of cooking oils, protein meals for livestock, and industrial uses. These specific oilseeds are eligible for nonrecourse loans. Other oilseed crops include castor beans and sesame. After extraction of the oil the residue is a valuable source of protein, especially for animal feeding stuffs, as in oil-seed cake or press cake. Oilseeds and their products are mainly consumed as snacks as well as part of the ingredients of certain dishes in human daily diet [18].

According to many reports (Table 1), peanuts are the main susceptible products for aflatoxin contamination [19-21]. Tree nuts such as almonds, walnuts, and pistachios may be contaminated with aflatoxin, though at lower levels than for cottonseed and corn; however, the problem is very significant to producers because: (1) the crop has a high unit value, and (2) much of the crop is sold to European markets that enforce limits significantly lower than those in some countries [22].

For over all sanitary precaution, the European Union has enacted in 1998, very severe aflatoxin tolerance standards of 2 $\mu g\ kg^{-1}$ AFB1 and 4 $\mu g\ kg^{-1}$ total aflatoxins for nuts and cereals for human consumption and this has come into effect from January, 2001. Consumers in the developed world are well aware of the carcinogenic effect of aflatoxins and will thus stay away from a product that has aflatoxin beyond the acceptance level. Exports of agricultural products particularly groundnuts and other oilseeds from developing countries have dropped considerately in recent years resulting in major economic losses to producing countries as a result of this restriction. According to the World Bank estimate, the policy change by the European Union will reduce by 64%, imports of cereals, dried fruits, oil seeds and nuts from nine African countries namely Chad, Egypt, Gambia, Mali, Nigeria, Senegal, South Africa, Sudan and Zimbabwe and this will cost African countries about US $670 million in trade per year. However, the new rule of the EU has been criticized as being too stringent. There is the need for mycotoxin surveillance because of its wide occurrence in contaminated commodities [23].

Aflatoxin contamination of peanut, resulting from invasion by *A. flavus* or *A. parasiticus,* is a major problem in semi-arid tropical regions where plants are primarily rain-fed. Light sandy soils, where peanut is typically cultivated, favor these fungi. While *A. flavus* infection of peanuts does not affect yield, the fungus can produce high levels of aflatoxin in infected nuts, and these toxins can pose serious health risks to humans and animals [24]. The environmental conditions required to induce pre-harvest aflatoxin contamination of peanuts were studied. In the study, peanuts did not become contaminated with aflatoxins in the absence of severe and prolonged drought stress even when the frequency of infection by *A. flavus* and *A. parasiticus* was up to 80%. Also, larger, more mature peanut kernels required considerably more drought stress to become contaminated than did smaller, immature kernels [25]. Peanuts become infected with aflatoxigenic fungus when seed pods come into direct contact with aflatoxigenic fungus in soil. These fungi can invade and produce toxins in peanut kernels before harvest, during drying, and in storage [26].

Dried fruits can be contaminated with aflatoxins. Maximum permitted levels for total aflatoxins in the European Union are 4 μg kg^{-1} in dried fruit intended for direct human consumption. AFB1 was the most common mycotoxin encountered as a natural contaminant in stored samples [25].

Food Type	Country	Contaminated/ Total examined	Aflatoxin	Concentration (ppb)	Method	Reference
Peanut	Argentina	2/50	AFB1 AFG1	435-625 for AFB1 83-625 for AFG1	TLC	[19]
Peanut	Botswana	94/120	Total[a]	12-329	ELISA	[20]
Peanut	China	2/16	AFB1+AFB2	1.96 (mean)	UHPLC	[21]
Peanut	Malaysia	11/13	AFB1	1.47-15.33	ELISA	[27]
Peanut	China	15/65	Total[a]	0.03-28.39	HPLC	[28]
Fresh Peanuts	China	14/35	Total[a]	0.3-7.4	UHPLC-MS/MS	[10]
Musty Peanuts	China	5/5	Total[b]	1.2-1482	UHPLC-MS/MS	[10]
Peanut and products	Trinidad	0/186	AFB1	ND[c]	Charm II	[29]
Peanut and products	Brazil	41/80	Total[a]	43-1099	TLC	[30]
Peanut butter	China	41/50	AFB1	<LOD[d]-68.51	HPLC	[31]

Peanut butter	Sudan	120/120	AFB1	17-170	Fluorometer	[32]
Peanut butters	China	31/33	Total[e]	0.7-96	UHPLC-MS/MS	[10]
Pistachio	Turkey	48/95	Total[a]	0.007-7.72	HPLC	[33]
Pistachio	Iran	3699/10068	AFB1	5.9 ± 41.7 (Mean)	HPLC	[34]
Hazelnut	Egypt	18/20	Total[a]	25-175	TLC	[35]
Walnut	Egypt	15/20	Total[a]	15-25	TLC	[35]
Walnut	China	31/48	Total[a]	0.02-1.20	HPLC	[28]
Hazelnut	Turkey	43/51	Total[a]	<0.625-10	ELISA	[36]
Cacao hazelnut cream	Turkey	38/40	Total[a]	<0.625-10	ELISA	[36]
Unsorted, in-shell almonds	California, USA	10/74	Total[a]	1-107	TLC	[37]
Diced almonds	California, USA	13/27	Total[a]	<LOD-119	TLC	[37]
Sesame	Malaysia	7/8	AFB1	0.54-1.82	ELISA	[27]
Sesame paste	China	37/100	AFB1	<LOD-20.45	HPLC	[31]
Sunflower	Malaysia	6/7	AFB1	1.14-5.33	ELISA	[27]
Nuts	Malaysia	2/7	AFB1	0.66-1.09	ELISA	[27]
Pine nut	China	2/12	Total[a]	0.19-0.25	HPLC	[28]
Dried apricot	Iran	9/30	AFB1	"/0.2	HPLC	[38]
Prune	Iran	2/15	AFB1	"/0.2	HPLC	[38]

a. Total: AFB1+ AFB2+ AFG1+ AFG2

b. Total AFs including AFM1 (ND-64.7 ppb) and AFM2 (ND-3.6 ppb)

c. ND: Not detected

d. LOD: Limit of detection

e. Total AFs including AFM1 (ND-4.2 ppb) and AFM2 (ND-1.8)

Table 1. Aflatoxins in oilseeds.

3. Occurrence of Aflatoxin in Cereals

Cereals and its products are the main foods for human consumption throughout the world. The cereal grains belong to corn, rice, barley, wheat and sorghum are found susceptible to AFs accumulation by aflatoxigenic fungus. The problem of aflatoxins occurring naturally in cereals, especially in rice and corn, has become troublesome because of changing agricultural technology. The aflatoxin problem in cereals is not restricted to any geographic or climatic region. Toxins are produced on cereals, both in the field and in storage; they involve both the grain and the whole plant [39].

Our results showed that rice was significantly more heavily colonized by aflatoxin-producing *Aspergillus* spp. than other cereals, with overall aflatoxin levels being correspondingly higher. But this may be caused by the variations in cultivars used. Additionally, corn is the second susceptible after rice for aflatoxin accumulation by *A. fungus* (Table 2).

Rice and sorghum are the most important staple food crops in many countries. In these countries, the majority of rice is grown in the rainy season. During the rainy season, sun drying of rice, which is practiced by most farmers, may not reduce the moisture content of grains sufficiently to prevent growth of fungi. As a result, rice grains with a moisture content higher than the desired level (>14%) may enter the storage system. The harmful effects of such fungal invasion are discoloration of the grain and/ or husk, loss in viability, loss of quality, and toxin contamination. Sorghum is grown in harsh environments where other crops do not grow well. Improvements in production, availability, storage, utilization, and consumption of this food crop will significantly contribute to the household food security and nutrition of the inhabitants of these areas. Sorghum is typically harvested as early as possible so that fields can quickly be planted with another crop. Sometimes the sorghum harvest coincides with heavy rainfall, hurricanes, and floods, all of which promote infection by mycotoxin-producing fungi [25].

As it is known, the consumption of large amounts of AFs contaminated food by starving people can cause toxic hepatitis (jaundice) and death. Aflatoxin epidemics were reported from India, in 1975 among the Bhils (the largest and most widely distributed tribal group in India), who had consumed corn heavily contaminated with *A. flavus*. The epidemic was characterized by jaundice, rapidly developing ascites, and portal hypertension. Approximately 400 persons were affected by the epidemic [40].

Food Type	Country	Contaminated/ Total examined	Aflatoxin	Concentration (ppb)	Method	Reference
Rice	India	814/1200	AFB1	0.1-308	ELISA	[25]
Rice	India	581/1511	AFB1	"/5	HPTLC	[41]
Rice	Iran	59/71	Total[a]	2.097-10.94	HPLC	[42]
Rice	Canada	99/199	AFB1	<0.002-7.1	HPLC	[43]

Rice	Sweden	57/99	Total[a]	0.1-50.7	HPLC	[44]
Rice	China	23/74	Total[a]	0.15-3.88	HPLC	[28]
Rice	Japan	0/48	Total[a]	ND[b]	HPLC	[45]
Rice	Nigeria	21/21	Total[a]	27.7-371.9	HPLC	[46]
Rice	Turkey	56/100	Total[a]	0.05-21.4	ELISA	[47]
Rice	Tunisia	2/16	Total[a]	2.0-7.5	ELISA	[48]
Rice	Tunisia	0/11	Total[a]	ND[b]	HPLC	[49]
Rice	Malaysia	9/13	AFB1	0.68-3.79	ELISA	[27]
Corn	Tunisia	1/17	Total[a]	0.42	HPLC	[49]
Corn	Malaysia	6/8	AFB1	1.75-8.95	ELISA	[27]
Corn	Brazilian	82/214	AFB1	0.2-129	ELISA	[50]
Corn	China	4/18	Total[a]	5.67 (mean)	UHPLC	[21]
Corn	China	52/84	Total[a]	0.02-1098	HPLC	[28]
Corn	Tunisia	9/21	Total[a]	2.9-12.5	ELISA	[48]
Corn flour	Morocco	16/20	AFB1	0.23-11.2	HPLC	[51]
Wheat	Ethiopia	5/120	AFB1	<1.0-12.3	HPLC	[52]
Wheat	Tunisia	15/51	Total[a]	4.0-12.9	ELISA	[48]
Wheat	Tunisia	10/46	Total[a]	0.15-18.6	HPLC	[49]
Wheat	Malaysia	9/14	AFB1	0.55-5.07	ELISA	[27]
Wheat	India	663/1646	AFB1	"/5	ELISA	[41]
Wheat flour	Morocco	3/17	AFB1	0.03-0.15	HPLC	[51]
Sorghum	Tunisia	36/49	Total[a]	0.4-25.8	HPLC	[49]
Sorghum	Ethiopia	5/82	AFB1	<1.0-25.9	HPLC	[52]
Sorghum	India	1173/1606	AFB1	0.01-263.98	ELISA	[53]
Sorghum	Tunisia	13/17	Total[a]	1.7-67.0	ELISA	[48]
Barley	Tunisia	2/46	Total	0.6-0.8	HPLC	[49]
Barley	Ethiopia	13/115	AFB1	<1.0-11.7	HPLC	[52]
Barley	Tunisia	11/25	Total	3.5-11.5	ELISA	[48]
Oat	Malaysia	5/10	AFB1	0.65-2.85	ELISA	[27]
Cereals	Ethiopia	31/352	AFB1	<LOD-26	HPLC	[52]

a. Total: AFB1+ AFB2+ AFG1+ AFG2

b. ND: Not detected

Table 2. Aflatoxins in cereals.

4. Occurrence of Aflatoxin in Spices

The popularity of hot peppers (*Capsicum annuum* L.), also known as chili peppers, as spices or vegetables and for other uses increases every year. Powdered red pepper is one of the favorite spices in South Asia and is commonly used for flavoring, seasoning, and imparting aroma or coloring to foods. Hot peppers are the principal component of curry and chili powder and can be used to make pepper sauce, red pepper, and paprika [54].

Spices are often contaminated with mycotoxins (Table 3). The climatic conditions prevailing in the tropics are especially favorable for mold contamination and mycotoxin production. Of the different mycotoxins, aflatoxin is the commonest contaminant in spices [55].

Peppers are very susceptible to aflatoxin contamination, which is affected by atmospheric temperature, humidity, insects, and drying and processing conditions. Mold contamination can occur in the field during crop production and during storage when conditions are favorable. Sun drying is a common postharvest practice in some countries, which involves spreading peppers on soil in a single layer. Because of the drying processes are on the soil, some peppers are contaminated with fungus [54, 56].

Substantial aflatoxin contamination of ground red pepper has been reported from Ethiopia, where eight of 60 samples collected from markets, shops and storage facilities were contaminated with AFB1 in concentrations of 250-525 μg kg^{-1} [57]. In Turkey, 24% and 13% of samples of different pepper types contained 7.5-200 [58] and 1.1-97.5 μg kg^{-1} [56] total aflatoxins, respectively. Elshafie et al. [59] were surveyed one hundred and five samples of seven spices (cumin, cinnamon, clove, black pepper, cardamom, ginger, and coriander) for the mycoflora and AFs in the Sultanate of Oman. Twenty fungal species were isolated in which *A. flavus, A. niger. Penicillium, Rhizopus,* and *Syncephalastrum racemosum* were the most dominant. Of the seven spices studied, clove was found to be the least contaminated, while cumin was the most contaminated. None of the 15 selected samples of the spices contaminated by *A. flavus* were found to contain AFs [59]. Cooking revealed that the aflatoxin content of spices did not decrease [60].

Although the mold *A. flavus* grows well on the spices, the production of AFs is lower than in cereals. This indicates that spices are not an ideal substrate for AF formation. It has been shown that essential oils extracted from spices, e.g. cloves, can inhibit mold growth and AF production completely. According to some reports, fungal growth was weak on curcumin, black pepper and white pepper and no AF was detected in black or white pepper after 10 days at 25 °C. According Martins et al. [61], the results of the survey indicate that there is little evidence for significant AF contamination in spices. The majority of samples contained AFs at low levels and others were negative (cardamom, cloves, ginger and mustard). In the meat industry (sausage, dry cured ham, luncheon meat) and in confection of ethnic dishes, AF contamination of spices is probably not relevant as a direct health hazard [61].

Food Type	Country	Contaminated/ Total examined	Aflatoxin	Concentration (ppb)	Method	Reference
Pepper	Turkey	12/90	Total[a]	1.1-97.5	TLC	[56]
Pepper	Turkey	12/49	Total[a]	7.5-200	TLC	[58]
Pepper	Malaysia	4/4	AFB1	0.65-2.1	ELISA	[27]
Ground red pepper	Turkey	33/164	Total[a]	"/5	HPLC	[33]
Ground red pepper	Ethiopia	8/60	AFB1	250-525	ELISA	[57]
Paprika	Spain	8/17	Total[a]	1.8-50.4	HPLC	[62]
Smoked Paprika	Spain	4/4	Total[a]	22.3-83.7	HPLC	[62]
Chilli	Malaysia	8/8	AFB1	0.58-3.5	ELISA	[27]
Chilli	Spain	6/11	Total[a]	1.9-65.7	HPLC	[62]
Cumin	Malaysia	2/3	AFB1	1.89-4.64	ELISA	[27]
Spices	Hungary	23/91	AFB1	0.14-15.7	HPLC	[55]
Spices	Portuqual	34/79	AFB1	1.0-59.0	HPLC	[61]

a. Total: AFB1+ AFB2+ AFG1+ AFG2

Table 3. Aflatoxins in spices.

5. Occurrence of Aflatoxin in Milk and Dairy Products

Most of AFB1 and AFB2 ingested by mammals are eliminated through urine and faeces, however a fraction is biotransformed in the liver and excreted together with milk in the form of AFM1 and AFM2, respectively. AFM1 could be detected in milk 12-24 h after the first AFB1 ingestion, reaching a high level after a few days. The ratio between AFB1 ingested and AFM1 excreted has been estimated to be 1-3% [5].

The system responsible for the biotransformation of AFB1 basically has five mechanisms, represented by reactions of reduction, hydration, epoxidation, hydroxylation and ortho-demethylation. The aflatoxicol is produced by reduction of AFB1 by an NADPH-dependent cytoplasmic enzyme present in the soluble fraction of the liver. The toxicity of aflatoxicol is apparently much smaller than AFB1, but the conversion is reversible and the aflatoxicol can serve as a reservoir toxicity of AFB1 in the intracellular space, it can be converted in this mycotoxin by microsomal dehydrogenase. The aflatoxicol can also be metabolized to AFM1 and AFH1. The hydration process results in a metabolite AFB2a. This compound has the main action the inhibition of enzymes, in the liver and other tissues, causing a reduction in proteic synthesis AFM1 and AFQ1 are results of hydroxylation reaction of AFB1. These compounds have a hydroxyl group, allowing their conjugation with glucuronic acid, sulfate and glutathione, making them very water-soluble substances that can be excreted in the bile, urine and milk. Most of the aflatoxins are excreted between 72 to 96 h after the exposure, with the liver and the kidney retaining the waste for a longer period compare to other tissues [63].

A tolerable daily intake of 0.2 ng kg^{-1} b.w. for AFM1 was calculated by Kuiper-Goodman [64] and this toxin has been categorized by the International Agency for Research on Cancer (IARC) as a class 2B toxin, a possible human carcinogen. In the assessment of cancer risk, the infants are more exposed to the risk because the milk is a major constituent of their diet. It must be also considered that young animals have been found to be more susceptible to AFB1 (and so probably AFM1) than adults. Therefore the presence of AFM1 in milk and milk products is considered to be undesirable [65].

The carcinogenicity of AFM1 may be influenced by the duration and level of exposure. Exposure is most likely to occur through the frequent consumption of milk and milk by-products (infant milk, cheese, butter, yoghurt). Several studies in different countries have reported high or low contamination levels of AFM1 in different categories of milk and dairy samples. These significantly variable AFM1 levels may be due to several influencing factors such as cheese manufacturing procedures and storage, types of cheese, conditions of cheese ripening, analytical methods and finally the geographical and seasonal effects [6].

The concentration of AFM1 is relatively increased in cheese samples because of its affinity to proteins. During cheese making, AFM1 can be decreased in cheese by increasing renneting temperature from 30 to 40°C, decreasing cutting size of curd and increasing press time from 1 to 2 h, which causes more loss of AFM1 in the whey [66]. On a weight basis, however, AFM1 concentration in cheese actually increases. In soft cheese, it becomes 2.5 to 3.3 times higher and in hard cheese, 3.9 to 5.8 times higher than in the milk from which the cheeses were made. Converting milk that may contain aflatoxin into a cheese, such as feta cheese, reduces the exposure of the consumer to this toxin. During pasteurization of milk, about 90% or more of the AFM1 is retained in the milk but during cheese manufacturing, there is a partitioning of AFM1 between the cheese, whey, and brine. During cheese manufacturing, results on the distribution of AFM1 between curd and whey can be variable. This variability has been associated with the type of cheese, the particular cheese-making process applied, the type and degree of milk contamination, and the analytical method employed. Lopez et al. [67] manufactured cheese using artificially AFM1 contaminated milk and found that the greatest proportion of toxin (60%) was in whey, while 40% AFM1 remained in cheese. Some researchers also reported that the greatest proportion of AFM1 was in the curd ranging between 66-80% [68]. About 37% of the AFM1 in milk is lost from the cheese into the whey, and another 30% diffuses from the cheese into brining solution during storage. Thus, the amount that would be ingested in a 30 g serving of cheese made from milk containing 500 ng AFM1/L would be only 35 ng AFM1 compared to 125 ng AFM1 from a 250 g serving of fluid milk. Thus, consumers in a region where there are high aflatoxin levels in milk would be at less health risk if the milk is pasteurized and converted into a cheese such as feta or other white pickled cheese before it is delivered to the consumer [69]. Applebaum et al. [70]

reported that AFM1 concentration in cheese was about four times higher than the cheese milk. The increase in AFM1 concentration in cheese has been explained by the affinity of AFM1 for casein.

The Commission of the European Communities established a limit for AFM1 of 50 ng kg^{-1} for milk and a variable limit for cheese, depending on concentration caused by drying process or processing. Milk containing AFM1 concentrations above the action level must be discarded, causing significant economic loss for the dairy producer. Similar regulations exist in most developed countries.

In this Regulation the Commission stated that "even if AFM1 is regarded as a less dangerous genotoxic carcinogenic substance than AFB1, it is necessary to prevent the presence in milk, and consequently in milk products, intended for human consumption and for young children in particular". The Commission has also set a limit for AFB1 of 5 μg kg^{-1} for supplementary feedstuffs for lactating dairy cattle. However this tolerance level is difficult to observe because the average daily individual intake in a herd should be limited to 40 μg AFB1 per cow, in order to produce milk with less than 50 ng AFM1 per kg [65].

Many factors may affect the formation of aflatoxins in animal feeds. Geographic and climate changes can affect the farm management practices and feed quality. These effects can lead to the wide variations in AFM1 levels in milk (Table 4). The preserved fodder such as silage and hay might have been contaminated by aflatoxin producing fungi and the improper storage led to aflatoxin production. The level of AFM1 in feed in rainy seasons is more than in dry seasons. It can be also probable to use higher amounts of contaminated concentrates in the cold months [71].

Food Type	Country	Contaminated/ Total examined	Aflatoxin	Concentration (ppb)	Method	Reference
Raw Milk[a]	Italy	125/161	AFM1	<0.023	HPLC	[72]
Raw Milk[a]	Greece	40/58	AFM1	0.005-0.055	HPLC	[73]
Raw Milk[a]	North African	35/49	AFM1	0.03-3.13	HPLC	[74]
Raw Milk[a]	Italy	?/310	AFM1	0.002-0.09	HPLC	[75]
Raw Milk[a]	Trinidad	13/212	AFM1	NM[a]	Charm II	[29]
Raw Milk[a]	Slovenia	0/60	AFM1	ND[b]	HPLC	[76]
Raw Milk[a]	Indonesia	65/113	AFM1	5-25	ELISA	[77]
Raw Milk[a]	China	12/12	AFM1	0.16-0.5	ELISA	[78]
Raw Milk[a]	Croatia	NM[a]/61 (one sample exceeded limit EU)	AFM1	0.0006-0.059	ELISA	[79]
Raw Milk[a]	Turkey	43/50	AFM1	<0.03	ELISA	[80]

Raw Milk[a]	Iran	60/60	AFM1	2.0-64.0	HPLC	[81]
Raw Milk[a]	Pakistan	177/232	AFM1	0.002-1.9	ELISA	[82]
Raw Milk[a]	Pakistan	63/120	AFM1	0.004-0.174	HPLC	[83]
Raw Milk[a]	Syria	70/74	AFM1	0.02-0.69	ELISA	[84]
Raw Milk[a]	South Korea	48/100	AFM1	0.002-0.08	HPLC	[85]
Raw Milk[a]	Portugal	25/31	AFM1	<0.005-0.05	HPLC	[86]
Raw Milk[a]	Iran	128/128	AFM1	0.031-0.113	ELISA	[87]
Raw Milk[a]	Iran	117/140	AFM1	<0.01-0.10	ELISA	[88]
Raw Milk[a]	Spain	3/92	AFM1	0.014-0.019	HPLC	[89]
Buffalo raw milk	Pakistan	153/360	AFM1	0.002-0.087	HPLC	[83]
Pasteurized milk[a]	Greece	113/136	AFM1	0.005-0.05	HPLC	[73]
Pasteurized milk[a]	Morrocco	47/54	AFM1	0.001-0.117	HPLC	[51]
Pasteurized milk[a]	Brazil	7/10	AFM1	0.01-0.02	HPLC	[90]
Pasteurized milk[a]	Iran	83/116	AFM1	0.006-0528	ELISA	[91]
Pasteurized milk[a]	Iran	624/624	AFM1	0.045-0.08	ELISA	[92]
Pasteurized milk[a]	Syria	10/10	AFM1	0.008-0.765	ELISA	[84]
Pasteurized milk[a]	Iran	48/48	AFM1	0.01-0.10	ELISA	[88]
Pasteurized milk[a]	Brazil	58/79	AFM1	0.05-0.24	HPLC	[93]
Milk[a] (Raw, pasteurized and powder)	Argentina	18/77	AFM1	0.01-0.03	ELISA	[94]
UHT Milk[a]	Greece	14/17	AFM1	0.005-0.05	HPLC	[73]
UHT Milk[a]	Turkey	75/129	AFM1	Max.0.54	ELISA	[95]
UHT Milk[a]	Turkey	67/100	AFM1	0.01-0.63	ELISA	[96]
UHT Milk[a]	Brazil	40/40	AFM1	0.010-0.5	HPLC	[90]

UHT Milk[a]	Turkey	50/50	AFM1	0.01-0.244	ELISA	[7]
UHT Milk[a]	Iran	116/210	AFM1	0.012-0.249	ELISA	[97]
UHT Milk[a]	Iran	68/109	AFM1	0.006-0.516	ELISA	[91]
UHT Milk[a]	Brazil	53/60	AFM1	0.015-0.5	HPLC	[93]
UHT Milk[a]	Iran	48/48	AFM1	0.01-0.10	ELISA	[88]
UHT Milk[a]	Turkey	14/24	AFM1	<0.01-0.05	HPLC	[98]
UHT-whole milk	Portugal	17/18	AFM1	<0.005-0.059	HPLC	[86]
UHT-semi skimmed milk	Portugal	20/22	AFM1	<0.005-0.061	HPLC	[86]
UHT-skimmed milk	Portugal	23/30	AFM1	<0.005-0.02	HPLC	[86]
UHT-Pasteurized milk	Japan	207/208	AFM1	0.001-0.029	HPLC	[99]
Ewe's milk	Greece	19/27	AFM1	0.005-0.055	HPLC	[73]
Ewe's milk	Greece	27/54	AFM1	<0.005-0.182	ELISA	[100]
Ewe's milk	Syria	13/23	AFM1	0.006-0.634	ELISA	[84]
Goat milk	Greece	12/20	AFM1	0.005-0.05	HPLC	[73]
Goat milk	Syria	7/11	AFM1	0.008-0.054	ELISA	[84]
Milk (ewe, goat and buffalo mix)	Italy	85/102	AFM1	0.05-0.25	ELISA	[101]
Infant milk food, Milk based cereal, weaning food, infant formula and liquid milk	India	76/87	AFM1	0.063-1.012	ELISA	[102]
Milk powder	Brazil	72/75	AFM1	0.01-0.5	HPLC	[90]
Milk powder	China	15/15	AFM1	Max 0.54	ELISA	[78]
Milk powder	Syria	1/8	AFM1	0.012	ELISA	[84]

Milk powder	South Korea	17/24	AFM1	0.083-0.342	HPLC	[103]
Cheese[b]	Iran	66/80	AFM1	0.15-2.41	TLC	[104]
Cheese[b]	China	4/4	AFM1	0.16-0.32	ELISA	[78]
Cheese[b]	Lebanon	75/111	AFM1	0.056-0.315	ELISA	[105]
Cheese[b]	Iran	30/50	AFM1	0.041-0.374	ELISA	[106]
Cheese[b]	Brazil	39/58	AFM1	0.01-0.304	IAC/LC	[107]
Cheese[b]	North African	15/20	AFM1	0.11-0.52	HPLC	[74]
Cheese[b]	Turkey	14/20	AFM1	<0.155	ELISA	[80]
Cheese[b]	Turkey	10/200	AFM1	0.1-0.6	ELISA	[108]
Cheese[b]	Iran	93/116	AFM1	0.052-0.745	ELISA	[109]
Cheese[b]	Turkey	82/100	AFM1	<0.05-0.8	ELISA	[110]
Cheese[b]	Turkey	36/127	AFM1	0.07-0.77	ELISA	[111]
White brined cheese[b]	Turkey	31/50	AFM1	0.1-5.2	Fluorometri	[112]
White brined cheese[b]	Turkey	159/193	AFM1	0.052-0.86	ELISA	[113]
Herby cheese[b]	Turkey	52/60	AFM1	0.16-7.26	Fluorometri	[112]
Cream cheese[b]	Turkey	44/49	AFM1	Max 0.25	ELISA	[36]
Cream cheese[b]	Turkey	8/200	AFM1	0.1-0.7	ELISA	[108]
Cream cheese[b]	Turkey	99/100	AFM1	0.01-4.1	ELISA	[114]
Cream cheese[b]	Iran	68/94	AFM1	58.3-785.4	ELISA	[109]
Kashar cheese[b]	Turkey	47/53	AFM1	"/0.25	ELISA	[36]
Kashar cheese[b]	Turkey	12/200	AFM1	0.12-0.8	ELISA	[108]
Kashar cheese	Turkey	8/28	AFM1	<0.37	ELISA	[80]
Kashar cheese[b]	Turkey	109/132	AFM1	0.05-0.69	ELISA	[96]

Kashar cheese[b]	Turkey	85/100	AFM1	0.05-0.80	ELISA	[110]
Tulum cheese[b]	Turkey	16/20	AFM1	<0.378	ELISA	[80]
Tulum cheese[b]	Turkey	81/100	AFM1	0.05-0.80	ELISA	[110]
Ewe's cheese	Turkey	14/50	AFM1	0.02-2.0	TLC	[6]
Dairy drinks	Brazil	10/12	AFM1	0.01-0.05	IAC/LC	[107]
Milk products	China	66/104	AFM1	Max 0.5	ELISA	[78]
Butter	Turkey	92/92	AFM1	0.01-7.0	ELISA	[114]
Butter	Turkey	25/27	AFM1	Max 0.1	ELISA	[36]
Butter	Turkey	66/80	AFM1	0.01-0.12	ELISA	[115]
Yoghurt	Brazil	49/65	AFM1	0.01-0.529	IAC/LC	[107]
Yoghurt	Italy	73/120	AFM1	<0.032	HPLC	[72]
Yoghurt	Turkey	68/104	AFM1	<0.1	ELISA	[116]
Yoghurt	South Korea	31/60	AFM1	0.017-0.124	HPLC	[103]
Yoghurt	Portugal	2/48	AFM1	0.043-0.045	HPLC	[117]
Fruit yoghurt	Portugal	16/48	AFM1	0.019-0.098	HPLC	[117]
Fruit yogurt	Turkey	7/21	AFM1	<0.1	ELISA	[115]
Strained yoghurt	Turkey	29/52	AFM1	<0.15	ELISA	[116]
Yogurt (whole fat)	Turkey	18/25	AFM1	<0.069	ELISA	[80]
Yoghurt (Semi fat)	Turkey	10/25	AFM1	<0.078	ELISA	[80]
Infant formula	South Korea	18/26	AFM1	0.032-0.132	HPLC	[103]
Infant formula	Iran	116/120	AFM1	0.001-0.014	ELISA	[87]
Dairy dessert	Turkey	26/50	AFM1	<0.08	ELISA	[80]

a.Cow milk, b. Cow cheese

Table 4. Aflatoxins in milk and dairy products.

6. Occurrence of Aflatoxin in Other Foods of Animal Origin

Meat refers to meat flesh, skeletal muscles, connective tissue or fat and others than meat flesh, including brain, heart, kidney, liver, pancreas, spleen, thymus, tongue and tripe that is used as food, excluding the bone and bone marrow and it contains high biological value protein and important micronutrients that are needed for good health throughout life. Residues of aflatoxins and their metabolites could be present in the meat, offal and eggs of animals receiving aflatoxin contaminated feeds (Table 5). In addition to the economic losses, aflatoxin in feeds could pose a risk to human health because of ingestion of aflatoxin containing foods derived from the animals fed the toxin-contaminated diet [118].

Cytochrome P450 enzymes (CYP) (including CYP1A2, CYP3A4 and CYP2A6) in the liver and other tissues convert AFB1 to epoxides (AFB1-8,9-exo-epoxide, and AFB1-8,9-endo-epoxide), and to AFM1, AFP1, AFQ1, and its reduced form aflatoxicol. Of the epoxides, the AFB1-8,9-exo-epoxide can form covalent bonds with DNA and serum albumin resulting in AFB1-N7-guanine and lysine adducts, respectively. Like AFB1, AFM1 can also be activated to form AFM1-8,9-epoxide that binds to DNA resulting in AFM1-N7-guanine adducts. These guanine and lysine adducts have been noted to appear in urine. The metabolites AFP1, AFQ1, and aflatoxicol are thought to be inactive and are excreted as such in urine, or in the form of glucuronyl conjugates from bile in feces [119].

When chicken exposed to AFB1 with contaminated rations, AFB1, AFM1, and aflatoxicol have been detected in liver, kidneys, and thigh muscles. Besides these, AFB2a has also been detected in livers of both broilers and layers on a ration contaminated with a mixture of aflatoxins [120]. In laying hens the effects of exposure to AF are a dose-dependent decrease in egg production and egg quality with increased susceptibility to salmonellosis, candidiasis, and coccidiosis. AFs and some of their metabolites can be carried over from feed to eggs in ratios ranging from 5,000:1 to 66,200:1 and even to 125,000:1, whereas in other trials no measurable residual AFB1 or its metabolites were found in eggs. These contrasting results may be ascribed to the administration of naturally contaminated feeds containing different AF with different levels of toxicity [8].

Wolzak et al. [121] have reported that tissue residues of aflatoxins were highest in kidney, gizzard, and liver (average concentration 3 μg kg^{-1} mass) when broilers were exposed for 4 weeks to a mixture of AFB1 and AFB2. After 7 days of removal of the contaminated feed, aflatoxin residues could not be detected in above tissues. Hussain et al. [120] also indicated that the elimination of AFB1 in chicken increased during longer exposure to AFB1. They fed broiler chicks on rations containing 0, 1.6, 3.2, and 6.4 μg AFB1/kg for 7, 14, or 28 day's age. After 2 to 3 days of exposure, AFB1 could be detected in livers of the birds exposed to 1.6 μg AFB1/kg and higher dietary levels of the toxin. After cessation of toxin feeding, AFB1 residues decreased in livers and muscles of all the chicks, with lower levels at 10 days post-cessation in the chicks exposed to higher toxin levels. They concluded that the residues of AFB1 in tissues increase with increase in dietary concentration of the toxin but decrease with increase in age (or after longer exposure) of broiler chicks. The elimination of AFB1 from tissues was rapid in older birds than in younger birds [120].

Poultry birds fed AF contaminated rations under experimental conditions resulted in the presence of AF residues in their edible tissues like liver and muscles. Residues of AFB1 in liver of broiler and layer birds have been reported to vary from no detection to 3.0 μg kg^{-1} by feeding 250-3310 μg kg^{-1} AFB1 for variable periods [118, 119]. The wide variations in the tissue AF residue concentration suggested that these levels might be influenced by different factors including dietary AF levels, duration of administration, age, type of the birds etc. However, effect of such factors upon concentration of AFB1 residues in poultry meat (liver and muscles) and clearance of AFB1 from the body tissues after withdrawal of dietary AF have not been adequately studied [120].

Dietary contamination of aflatoxins pose a big risk to human health including acute aflatoxicosis, Hepatocellular carcinoma, hepatitis B virus infection, growth impairment in different regions of the World particularly Asian and African countries [122]. European community and many other countries have imposed 2 μg kg^{-1} AFB1 as maximum tolerance level in human food products. Birds fed Afs, following ingestion are rapidly metabolized into nontoxic substances in the body. A rapid decrease in AFB1 residues below the tolerance limits from the muscles and liver within 3 and 7 days of withdrawal of dietary AFB1 and that it may not become a significant human health risk. However, in areas with no regulatory limits on AFB1 levels of poultry feed, the secondary exposure to aflatoxins through consumption of chicken liver and meat derived from the poultry fed AF contaminated feed may pose a risk to consumers health [123].

Food Type	Country	Contaminated/ Total examined	Aflatoxin	Concentration (ppb)	Method	Reference
Chicken Liver	Thailand	248/450	AFB1	Mean 0.6092	HPLC	[118]
Chicken Muscle	Thailand	96/450	AFB1	Mean 0.0451	HPLC	[118]
Meat	Jordan	12/50	Total[a]	0.15-8.32	HPLC	[124]
Fresh Fish	Egypt	10/30	Total[a]	22-70.5	Florometric	[125]
Salted fish	Egypt	12/30	Total[a]	18.5-50	Florometric	[125]
Smoked Fish	Egypt	8/30	Total[a]	32-96	Florometric	[125]
Egg	Jordan	5/40	Total[a]	0.01-6.15	HPLC	[124]
Processed egg	Saudi Arabia	0/25	Total[a]	ND[b]	ELISA	[126]
Unwashed egg	Saudi Arabia	3/25	Total[a]	0.61-1.19	ELISA	[126]

a. Total: AFB1+ AFB2+ AFG1+ AFG2

b. ND: Not detected

Table 5. Aflatoxins in other foods of animal origin.

7. Conclusion

AFs are toxic secondary metabolites produced by *Aspergillus* fungus growing in susceptible agricultural commodities. They can result in major economic losses and can negatively affect animal and human health. This review has sought to summarize the possible AFs contamination in a wide array of agricultural commodities worldwide. AFs contamination can occur both in temperate and tropical regions of the World. Major food commodities affected are cereals, nuts, dried fruit, spices, oil seeds, dried peas and beans and fruit. Regulations for major mycotoxins in commodities and food exist in at least 100 countries, most of which are for aflatoxins, maximum tolerated levels differ greatly among countries [27].

Frequent analytical surveillance program by food control agencies is highly recommended to control the incidence of aflatoxins contamination in food grains to ensure food safety and to protect consumer's health [27]. Some analytical techniques such as thin-layer chromatography (TLC), high performance liquid chromatography (HPLC), two-dimensional thin layer chromatography and enzyme-linked immunosorbent assay (ELISA) have been available for the qualitative and quantitative analysis of AFs. Poor separation, unsatisfied accuracy and low sensitivity limit the application of TLC. Although ELISA is a fast and sensitive method for AFs analysis is liquid chromatography combined with fluorescence detection, which has been extensively studied in various food matrices. However, conventional approach by HPLC in a gradient reversed phase mode typically using columns with 6 μm particles often costs a lot of time to get a complete separation of the target compounds and additionally, in order to improve detection limits of AFB1 and AFG1 a tedious pre- or post- column derivatization must be done [10].

The inability to control and at times even predict AF production makes it a unique challenge to food safety. To avoid aflatoxin problem in food grains, farmers should improve the practice of drying seeds to the required moisture content immediately after harvest. They must also develop proper storage structures by spraying fungicides or some other chemicals to reduce Aspergilli and subsequent toxin accumulation on food grains under storage conditions. Although prevention is the best control strategy, it is not always possible to prevent all mycotoxin contamination. Optimal postharvest storage conditions will minimize consumer exposure to AFs, but decontamination procedures may be needed in some cases. One approach to managing the risks associated with AF contamination is use of an integrated system based on the Hazard Analysis and Critical Control Point (HACCP) approach. This approach involves strategies for prevention, control, good manufacturing practices, and quality control at all stages of production, from the field to the final consumer [25]. Cheap and environmentally sustainable methods that can be applied pre or post-harvest to reduce the contamination of AFs are available. These methods include proper irrigation, choice of genetically resistant crop strains and bio-pesticide management which involves using a non-aflatoxigenic strain of *Aspergillus* that competitively excludes toxic strains. Other methods include sorting and disposal of visibly moldy or damaged seeds, reducing the bioavailability of aflatoxins using clay and chemo-protection [16].

Author details

Ayhan Filazi[1*] and Ufuk Tansel Sireli[2]

*Address all correspondence to: filazi@veterinary.ankara.edu.tr

1 Department of Pharmacology and Toxicology, Faculty of Veterinary Medicine, Ankara University, Turkey

2 Department of Food Hygiene and Technology, Faculty of Veterinary Medicine, Ankara University, Turkey

References

[1] Bosco, F., & Mollea, C. (2012). Mycotoxins in Food. In: Valdez B. (ed), *Food Industrial Processes- Methods and Equipment*, InTech, 169-200, Available from, http://www.intechopen.com/books/food-industrial-processes-methods-and-equipment, (accessed 13 May 2012).

[2] Rustom, I. Y. S. (1997). Aflatoxin in food and feed:occurrence, legistlation and inactivation by physical methods. *Food Chemistry*, 59(1), 57-67.

[3] European Food Safety Authority (EFSA). (2007). Opinion of the scientific panel on contaminants in the food chain on a request from the commission related to the potential increase of consumer health risk by a possible increase of the existing maksimum levels for aflatoxins in almonds, hazelnuts and pistachios and derived products. *The EFSA Journal*, 446, 1-127.

[4] Turner, N. W., Subrahmanyam, S., & Piletsky, S. A. (2009). Analytical methods for determination of mycotoxins: A review. *Analytica Chimica Acta*, 632, 168-180.

[5] Dors, G. C., Caldas, S. S., Feddern, V., Bemvenuti, R. H., Hackbart, H. C. S., Souza, M. M., Oliveira, M. S., Garda-Buffon, J., Primel, E. G., & Badiale-Furlong, E. (2011). Aflatoxins: Contamination, Analysis and Control. In:. Guevara-Gonzalez RG. (Ed), *Aflatoxins- Biochemistry cand Molcecular Biology*, InTech, 415-438, Available from, http://cdn.intechopen.com/pdfs/20401/InTech-Aflatoxins_contamination_analysis_and_control.pdf, (accessed 13 May 2012).

[6] Filazi, A., Ince, S., & Temamogullari, F. (2010). Survey of the occurrence of aflatoxin M1 in cheeses produced by dairy ewe's milk in Urfa city, Turkey. *Veterinary Journal of Ankara University*, 57(3), 197-199.

[7] Gundinc, U., & Filazi, A. (2009). Detection of Aflatoxin M1 Concentrations in UHT Milk Consumed in Turkey markets by ELISA. *Pakistan Journal of Biological Sciences*, 12(8), 653-656.

[8] Zaghini, A., Martelli, G., Roncada, P., Simioli, M., & Rizzi, L. (2005). Mannanoligosaccharides and Aflatoxin B1 in Feed for Laying Hens: Effects on Egg Quality, Aflatoxins B1 and M1 Residues in Eggs, and Aflatoxin B1 Levels in Liver. *Poultry Science*, 84(6), 825-832.

[9] Shotwell, O. L., Goulden, M. L., & Hesseltine, C. W. (1976). Aflatoxin M1. Occurrence in stored and freshly harvested corn. *Journal of Agricultural and Food Chemistry*, 24(3), 683-684.

[10] Huang, B., Han, Z., Cai, Z., Wu, Y., & Ren, Y. (2010). Simultaneous determination of aflatoxins B1, B2, G1, G2, M1 and M2 in peanuts and their derivative products by ultra-high-performance liquid chromatography-tandem mass spectrometry. *Analytica Chimica Acta*, 662(1), 62-68.

[11] Ren, Y., Zhang, Y., Shao, S., Cai, Z., Feng, L., Pan, H., & Wang, Z. (2007). Simultaneous determination of multi-component mycotoxin contaminants in foods and feeds by ultra-performance liquid chromatography tandem mass spectrometry. *Journal of Chromatography A*, 1143, 48-64.

[12] Quist, C. F., Bounous, D. I., Kilburn, J. V., Nettles, V. F., & Wyatt, R. D. (2000). The Effect of dieteray aflatoxin on wild turkey poults. *Journal of Wildlife Diseases*, 36(3), 436-444.

[13] Jackson, L. S., & Al-Taher, F. (2008). Factors Affecting Mycotoxin Production in Fruits. In: Barkai-Golan R, Paster N. (Ed), *Mycotoxins in Fruits and Vegetables*, Academic Press is an imprint of Elsevier, 75-104.

[14] Food and Agriculture Organization of the United Nations/World Health Organization (FAO/WHO). (1998). Evaluation of certain veterinary drug residues in food. *Forty seventh report of the joint FAO/WHO Expert Committee on Food Additives (JECFA)*, World Health Organization Technical Report Series, 876, 1-85.

[15] Cullen, J. M., Ruebner, B. H., Hsieh, L. S., Hyde, D. M., & Hsieh, D. P. (1987). Carcinogenicity of Dietary Aflatoxin M1 in Male Fischer Rats Compared to Aflatoxin B1. *Cancer Research*, 47, 1913-1917.

[16] Shuaib, F. M. B., Ehiri, J., Abdullahi, A., Williams, J. H., & Jolly, P. E. (2010). Reproductive health effects of aflatoxins: A review of the literature. *Reproductive Toxicology*, 29, 262-270.

[17] Becer, U. K., & Filazi, A. (2010). Aflatoxins, Nitrates And Nitrites Analysis In The Commercial Cat And Dog Foods. *Fresenius Environmental Bulletin*, 18(11), 2523-2527.

[18] Princen, L. H. (1983). New oilseed crops on the horizon. *Economic Botany*, 37(4), 478-492.

[19] Fernandez Pinto, V., Patriarca, A., Locani, O., & Vaamonde, G. (2001). Natural co-occurrence of aflatoxin and cyclopiazonic acid in peanuts grown in Argentina. *Food Additives and Contaminants*, 18(11), 1017-1020.

[20] Mphande, F. A., Siame, B. A., & Taylor, J. E. (2004). Fungi, Aflatoxins, and Cyclopiazonic Acid Associated with Peanut Retailing in Botswana. *Journal of Food Protection,* 67(1), 96-102.

[21] Fu, Z., Huang, X., & Mina, S. (2008). Rapid determination of aflatoxins in corn and peanuts. *Journal of Chromatography A,* 1209, 271-274.

[22] Shephard, G. S. (2003). Aflatoxin and Food Safety: Recent African Perspectives. *Journal of Toxicology,* 22(2-3), 267-286.

[23] Negedu, A., Atawodi, S. E., Ameh, J. B., Umoh, V. J., & Tanko, H. Y. (2011). Economic and Health Perspectives of Mycotoxins: A Review. *Continental Journal of Biomedical Sciences,* 5(1), 5-26.

[24] Kumar, K. V. K., Basu, M. S., & Rajendran, T. P. (2008). Mycotoxin research and mycoflora in some commercially important agricultural commodities. *Crop Protection,* 27(6), 891-905.

[25] Reddy, K. R. N., Abbas, H. K., Abel, C. A., Shier, W. T., Oliveira, C. A. F., & Raghavender, C. R. (2009). Mycotoxin contamination of commercially important agricultural commodities. *Toxin Reviews,* 28(2-3), 154-168.

[26] Horn, B. W., Greene, R. L., & Dorner, J. W. (1995). Effect of corn and peanut cultivation on soil populations of Aspergillus flavus and A. parasiticus in southwestern Georgia. *Applied and Environmental Microbiology,* 61(7), 2472-2475.

[27] Reddy, K. R. N., Raghavender, C. R., Salleh, B., Reddy, C. S., & Reddy, B. N. (2011). Potential of aflatoxin B1 production by Aspergillus flavus strains on commercially important food grains. *International Journal of Food Science and Technology,* 161-165.

[28] Wang, J., & Liu, X. M. (2007). Contamination of aflatoxins in different kinds of foods in China. *Biomedical and Envirionmental Sciences,* 20(6), 483-487.

[29] Offiah, N., & Adesiyun, A. (2007). Occurrence of Aflatoxins in Peanuts, Milk, and Animal Feed in Trinidad. *Journal of Food Protection,* 70(3), 771-775.

[30] Freitas, V. P. S., & Brigido, B. M. (1998). Occurrence of aflatoxins B1, B2, G1, and G2 in peanuts and their products marketed in the region of Campinas, Brazil in 1995 and 1996. *Food Additives and Contaminants,* 15(7), 807-811.

[31] Li, F. Q., Li, Y. W., Wang, Y. R., & Luo, X. Y. (2009). Natural occurrence of aflatoxins in Chinese peanut butter and sesame paste. *Journal of Agricultural and Food Chemistry,* 57(9), 3519-3524.

[32] Elshafie, S. Z. B., El Mubarak, A., El -Nagerabi, S. A. F., & Elshafie, A. E. (2011). Aflatoxin B1 Contamination of Traditionally Processed Peanuts Butter for Human Consumption in Sudan. *Mycopathologia,* 171, 435-439.

[33] Set, E., & Erkmen, O. (2010). The aflatoxin contamination of ground red pepper and pistachio nuts sold in Turkey. *Food and Chemical Toxicology,* 48, 2532-2537.

[34] Cheraghali, A. M., Yazdanpanah, H., Doraki, N., Abouhossain, G., Hassibi, M., Ali-abadi, S., Aliakbarpoor, M., Amirahmadi, M., Askarian, A., Fallah, N., Hashemi, T., Jalali, M., Kalantari, N., Khodadadi, E., Maddah, B., Mohit, R., Mohseny, M., Phaghihy, Z., Rahmani, A., Setoodeh, L., Soleimany, E., & Zamanian, F. (2007). Incidence of aflatoxins in Iran pistachio nuts. *Food and Chemical Toxicology*, 45(5), 812-816.

[35] Abdel-Hafez, A. I., & Saber, S. M. (1993). Mycoflora and mycotoxin of hazelnut (Corylus avellana L.) and walnut (Juglans regia L.) seeds in Egypt. *Zentralblatt für Mikrobiologie*, 148(2), 137-147.

[36] Aycicek, H., Aksoy, A., & Saygi, S. (2005). Determination of aflatoxin levels in some dairy and food products which consumed in Ankara, Turkey. *Food Control*, 16, 263-266.

[37] Schade, J. E., McGreevy, K., King, A. D., Mackey, B., & Fuller, G. (1975). Incidence of Aflatoxin in California Almonds. *Applied and Environmental Microbiology*, 29(1), 48-53.

[38] Janati, S. S. F., Beheshti, H. R., Asadi, M., Mihanparast, S., & Feizy, J. (2012). Preliminary survey of Aflatoxins and Ochratoxin A in dried fruits from Iran. *Bulletin of Environmental Contamination and Toxicology*, 88(3), 391-395.

[39] Hesseltine, C. W. (1974). Natural occurrence of mycotoxin in cereals. *Mycopathologia*, 53(1-4), 141-153.

[40] Krishnamachari, K. A. V. R., Bhat, R. V., Nagarajan, V., & Tilak, T. B. G. (1975). Hepatitis due to aflatoxicosis: An outbreak in Western India. *Lancet*, 1, 1061-1063.

[41] Toteja, G. S., Mukheriee, A., Diwakar, S., Singh, P., Saxena, B. N., Sinha, K. K., Sinha, A. K., Kumar, N., Nagaraja, K. V., Baj, G., Krishna Prasad, C. A., Vanchinathan, S., Roy, R., & Sarkar, S. (2006). Aflatoxin B(1) contamination of parboiled rice samples collected from different states of India: A multi-centre study. *Food Additive and Contaminant*, 23(4), 411-414.

[42] Mazaheri, M. (2009). Determination of aflatoxins in imported rice to Iran. *Food and Chemical Toxicology*, 47(8), 2064-2066.

[43] Bansal, J., Pantazopoulos, P., Tam, J., Cavlovic, P., Kwong, K., Turcotte, A. M., Lau, B. P. Y., & Scott, P. M. (2011). Surveys of rice sold in Canada for aflatoxins, ochratoxin A and fumonisins. *Food Additives and Contaminants: Part A*, 28(6), 767-774.

[44] Fredlund, E., Thim, A. M., Gidlund, A., Brostedt, S., Nyberg, M., & Olsen, M. (2009). Moulds and mycotoxins in rice from the Swedish retail market. *Food Additives & Contaminants: Part A*, 26(4), 527-533.

[45] Tanaka, K., Sago, Y., Zheng, Y., Nakagawa, H., & Kushiro, M. (2007). Mycotoxins in rice. *International Journal of Food Microbiology*, 119(1-2), 59-66.

[46] Makun, H. A., Dutton, M. F., Njobeh, P. B., Mwanza, M., & Kabiru, A. Y. (2011). Natural multi-occurrence of mycotoxins in rice from Niger State, Nigeria. *Mycotoxin Research*, 27, 97-104.

[47] Aydin, A., Aksu, H., & Gunsen, U. (2011). Mycotoxin levels and incidence of mould in Turkish rice. *Environmental Monitoring and Assessment*, 178(1-4), 271-280.

[48] Ghali, R., Khlifa, K. H., Ghorbel, H., Maaroufi, K., & Hedili, A. (2008). Incidence of aflatoxins, ochratoxin A and zearalenone in tunisian foods. *Food Control*, 19(9), 921-924.

[49] Ghali, R., Khlifa, K. H., Ghorbel, H., Maaroufi, K., & Hedili, A. (2010). Aflatoxin determination in commonly consumed foods in Tunisia. *Journal of the Science of Food and Agriculture*, 90(14), 2347-2351.

[50] Vargas, E. A., Preis, R. A., Castro, L., & Silva, C. M. G. (2001). Co-occurrence of aflatoxins B1, B2, G1, G2, zearalenone and fumonisin B1 in Brazilian corn. *Food Additives and Contaminants*, 18(11), 981-986.

[51] Zinedine, A., Gonzalez-Osnaya, L., Soriano, J. M., Molto, J. C., Idrissi, L., & Manes, J. (2007). Presence of aflatoxin M1 in pasteurized milk from Morocco. *International Journal of Food Microbiology*, 114, 25-29.

[52] Ayalew, A., Fehrmann, H., Lepschy, J., Beck, R., & Abate, D. (2006). Natural Occurrence of Mycotoxins in Staple Cereals from Ethiopia. *Mycopathologia*, 162(1), 57-63.

[53] Ratnavathi, C. V., Komala, V. V., Kumar, B. S. V., Das, I. K., & Patil, J. V. (2012). Natural occurrence of aflatoxin B1 in sorghum grown in different geographical regions of India. *Journal of the Science of Food and Agricultural*, DOI: 10.1002/jsfa.5646.

[54] Iqbal, Q., Amjad, M., Asi, M. R., & Arino, A. (2011). Assessment of Hot Peppers for Aflatoxin and Mold Proliferation during Storage. *Journal of Food Protection*, 74(5), 830-835.

[55] Fazekas, B., Tar, A., & Kovacs, M. (2005). Aflatoxin and ochratoxin A content of spices in Hungary. *Food Additives and Contaminants*, 22(9), 856-863.

[56] Erdogan, A. (2004). The aflatoxin contamination of some pepper types sold in Turkey. *Chemosphere*, 56, 321-325.

[57] Fufa, H., & Urga, K. (1996). Screening of aflatoxins in Shiro and ground red pepper in Addis Ababa. *Ethiopian Medical Journal*, 34, 243-249.

[58] Demircioglu, S., & Filazi, A. (2010). Detection of aflatoxin levels in red pepper produced in Turkey. *Journal of Veterinary Medical Association*, 81(2), 63-66.

[59] Elshafie, A. E., Al-Rashdi, T. A., Al-Bahry, S. N., & Bakheit, C. S. (2002). Fungi and aflatoxins associated with spices in the Sultanate of Oman. *Mycopathologia*, 155, 155-160.

[60] MacDonald, S., & Castle, L. (1996). A UK retail survey of aflatoxins in herbs and spices and their fate during cooking. *Food Additives and Contaminants*, 13, 121-128.

[61] Martins, M. L., Martins, H. M., & Bernardo, F. (2001). Aflatoxins in spices marketed in Portugal. *Food Additives and Contaminants*, 18(4), 315-319.

[62] Santos, L., Marin, S., Mateo, E. M., Gil-Serna, J., Valle-Algarra, F. M., Patino, B., & Ramos, A. J. (2011). Mycobiota and co-occurrence of mycotoxins in Capsicum powder. *International Journal of Food Microbiology*, 151, 270-276.

[63] Heidtmann-Bemvenuti, R., Mendes, G. L., Scaglioni, P. T., Badiale-Furlong, E., & Souza-Soares, L. A. (2011). Biochemistry and metabolism of mycotoxins: A review. *African Journal of Food Science*, 5(16), 861-869.

[64] Kuiper-Goodman, T. (1990). Uncertainties in the risk assessment of three mycotoxins: aflatoxin, ochratoxin and zearalenone. *Canadian Journal of Physiology and Pharmacology*, 68, 1017-1024.

[65] Prandini, A., Tansini, G., Sigolo, S., Filippi, L., Laporta, M., & Piva, G. (2009). On the occurrence of aflatoxin M1 in milk and dairy products. *Food and Chemical Toxicology*, 47, 984-991.

[66] Mohammadi, H., Alizadeh, M., Bari, M., Khosrowshahi, A., & Tajik, H. (2008). Minimization of aflatoxinM1 content in Iranian white brine cheese. *International Journal of Dairy Technology*, 6(2), 141-145.

[67] Lopez, C., Ramos, L., Ramadan, S., Bulacio, L., & Perez, J. (2001). Distribution of aflatoxin M1 in cheese obtained from milk artificially contaminated. *International Journal of Food Microbiology*, 211-5.

[68] Sengun, I. Y., Yaman, D. B., & Gonul, S. A. (2008). Mycotoxins and mould contamination in cheese: a review. *World Mycotoxin Journal*, 1(3), 291-298.

[69] Motawee, M. M., & McMahon, D. J. (2009). Fate of AflatoxinM1 during Manufacture and Storage of Feta Cheese. *Journal of Food Science*, 74(5), 42-45.

[70] Applebaum, R. S., Brackett, R. E., Wiseman, D. W., & Marth, E. H. (1982). Aflatoxins: toxicity to dairy cattle and occurrence in milk and milk products. *Journal of Food Protection*, 45, 752-777.

[71] Kamkar, A., Khaniki, G. R. J., & Alavi, S. A. (2011). Occurrence of aflatoxin M1 in raw milk produced in Ardebil of Iran. *Iranian Journal of Environmental Health Science and Engineering*, 8(2), 123-128.

[72] Galvano, F., Galofaro, V., Ritieni, A., Bognanno, M., De Angelis, A., & Galvano, G. (2001). Survey of the occurrence of aflatoxin M1 in dairy products marketed in Italy: second year of observation. *Food Additives and Contaminants*, 18(7), 644-646.

[73] Roussi, V., Govaris, A., Varagouli, A., & Botsoglou, N. A. (2002). Occurrence of aflatoxin M1 in raw and market milk commercialized in Greece. *Food Additives and Contaminants*, 19(9), 863-868.

[74] Elgerbi, A. M., Aidoo, K. E., Candlish, A. A. G., & Tester, R. F. (2004). Occurrence of aflatoxin M1 in randomly selected North African milk and cheese samples. *Food Additives and Contaminants*, 21(6), 592-597.

[75] Nachtmann, C., Gallina, S., Rastelli, M., Ferro, G. L., & Decastelli, L. (2007). Regional monitoring plan regarding the presence of aflatoxin M1 in pasteurized and UHT milk in Italy. *Food Control*, 18, 623-629.

[76] Torkar, G. K., & Venust, A. (2008). The presence of yeast, moulds and aflatoxin M1 in raw milk and cheese in Slovenia. *Food Control*, 19, 570-577.

[77] Nuryono, N., Agus, A., Wedhastri, S., Maryudani, Y. B., Setyabudi, F. M. C., Böhm, J., & Razzazi-Fazeli, E. (2009). *Food Control*, 20, 721-724.

[78] Pei, S. C., Zhang, Y. Y., Eremin, S. A., & Lee, W. J. (2009). Detection of aflatoxin M1 in Milk products from China by ELISA using monoclonal antibodies. *Food Control*, 20, 1080-1085.

[79] Bilandzic, N., Vernina, I., & Solomun, B. (2010). Aflatoxin M1 in raw milk in Croatia. *Food Control*, 21, 1279-1281.

[80] Ertas, N., Gonulalan, Z., Yildirim, Y., & Karadal, F. (2011). A survey of concentration of aflatoxin M1 in dairy products marketed in Turkey. *Food Control*, 22, 1956-1959.

[81] Sani, A. M., & Nikpooyan, H. (2012). Determination of aflatoxin M1 in milk by high-performance liquid chromatography in Mashhad (North east of Iran). *Toxicology and Industrial Health*, DOI: 10.1177/0748233711434954.

[82] Sadia, A., Jabbar, M. A., Deng, Y., Hussain, E. A., Riffatd, S., Naveed, S., & Arif, M. (2012). A survey of aflatoxin M1 in milk and sweets of Punjab, Pakistan. *Food Control*, 26, 235-240.

[83] Hussain, I., Anwar, J., Munawar, M. A., & Asi, M. R. (2008). Variation of levels of aflatoxin M1 in raw milk from different localities in the central areas of Punjab, Pakistan. *Food Control*, 19, 1126-1129.

[84] Ghanem, I., & Orfi, M. (2009). Aflatoxin M1 in raw, pasteurized and powdered milk available in the Syrian market. *Food Control*, 20, 603-605.

[85] Lee, J. E., Kwak, B. M., Ahn, J. H., & Jeon, T. H. (2009). Occurrence of aflatoxin M1 in raw milk in South Korea using an immunoaffinity column and liquid chromatography. *Food Control*, 20, 136-138.

[86] Martins, M. L., & Martins, H. M. (2000). Aflatoxin M1 in raw and ultra-high temperature-treated milk commercialized in Portugal. *Food Additives and Contaminants*, 17(10), 871-874.

[87] Oveisi, M. R., Jannat, B., Sadeghi, N., Hajimahmoodi, M., & Nikzad, A. (2007). Presence of aflatoxin M1 in milk and infant milk products in Tehran, Iran. *Food Control*, 18, 1216-1218.

[88] Rahimi, E., Shakerian, A., Jafariyan, M., Ebrahimi, M., & Riahi, M. (2009). Occurrence of aflatoxin M1 in raw, pasteurized and UHT milk commercialized in Esfahan and Shahr-e Kord, Iran. *Food Sec.*, 1, 317-320.

[89] Velasco, M. L. R., Delso, M. M. C., & Escudero, D. O. (2003). ELISA and HPLC determination of the occurrence of aflatoxin M1 in raw cow's milk. *Food Additives and Contaminants*, 20(3), 276-280.

[90] Shundo, L., Navas, S. A., Lamardo, L. C. A., Ruvieri, V., & Sabino, M. (2009). Estimate of aflatoxin M1 exposure in milk and occurrence in Brazil. *Food Control*, 2009, 655-657.

[91] Fallah, A. A. (2010). Assessment of aflatoxin M1 contamination in pasteurized and UHT milk marketed in central part of Iran. *Food and Chemical Toxicology*, 48, 988-991.

[92] Alborzi, S., Pourabbas, B., Rashidi, M., & Astaneh, B. (2006). Aflatoxin M1 contamination in pasteurized milk in Shiraz (south of Iran). *Food Control*, 17, 582-584.

[93] Garrido, N. S., Iha, M. H., Ortolani, M. R. S., & Favaro, R. M. D. (2003). Occurrence of aflatoxins M1 and M2 in milk commercialized in Ribeira˜ o Preto-SP. *Brazil Food Additives and Contaminants*, 20(1), 70-73.

[94] Lopez, C. E., Ramos, L. L., Ramadan, S. S., & Bulacio, L. C. (2003). Presence of aflatoxin M1 in milk for human consumption in Argentina. *Food Control*, 14(1), 31-34.

[95] Unusan, N. (2006). Occurrence of aflatoxin M1 in UHT milk in Turkey. *Food and Chemical Toxicology*, 44, 1897-1900.

[96] Tekinsen, K. K., & Eken, H. S. (2008). Aflatoxin M1 levels in UHT milk and kashar cheese consumed in Turkey. *Food and Chemical Toxicology*, 46, 3287-3289.

[97] Heshmati, A., & Milani, J. M. (2010). Contamination of UHT milk by aflatoxin M1 in Iran. *Food Control*, 21, 19-22.

[98] Gurbay, A., Aydın, S., Girgin, G., Engin, A. B., & Sahin, G. (2006). Assessment of aflatoxin M1 levels in milk in Ankara, Turkey. *Food Control*, 17, 1-4.

[99] Nakajima, M., Tabata, S., Akiyamax, H., Itoh, Y., Tanakak, T., Sunagawa, H., Tyonan, T., Yoshizawaxx, T., & Kumaga, S. (2004). Occurrence of aflatoxin M1 in domestic milk in Japan during the winter season. *Food Additives and Contaminants*, 21(5), 472-478.

[100] Grigoriadou, I. K., Eleftheriadou, A., Mouratidou, A. T., & Katikou, P. (2005). Determination of aflatoxin M1 in ewe's milk samples and the produced curd and Feta cheese. *Food Control*, 16, 257-261.

[101] Anfossi, L., Baggiani, C., Giovannoli, C., D'Arco, G., Passini, C., & Giraudi, G. (2012). Occurrence of aflatoxin M1 in Italian cheese: Results of a survey conducted in 2010 and correlation with manufacturing, production season, milking animals, and maturation of cheese. *Food Control*, 25, 125-130.

[102] Rastogi, S., Dwivedi, P. D., Khanna, S. K., & Das, M. (2004). Detection of aflatoxin M1 contamination in milk and infant milk products from Indian markets by ELISA. *Food Control*, 15, 287-290.

[103] Kim, E. K., Shon, D. H., Ryu, D., Park, J. W., Hwang, H. J., & Kim, Y. B. (2000). Occurrence of aflotoxin M1 in Korean dairy products determined by ELISA and HPLC. *Food Additives and Contaminants*, 17(1), 59-64.

[104] Kamkar, A. (2006). A Study on the occurrence of aflatoxin M1 in Iranian Feta cheese. *Food Control*, 17, 768-775.

[105] Elkak, A., Atat, O., & Abbas, M. (2012). Occurrence of aflatoxin M1 in cheese processed and marketed in Lebanon. *Food Control*, 25, 140-143.

[106] Tavakoli, H. R., Riazipour, M., Kamkar, A., Shaldehi, H. R., & Nejad, A. S. M. (2012). Occurrence of aflatoxin M1 in white cheese samples from Tehran, Iran. *Food Control*, 23, 293-295.

[107] Iha, M. H., Barbosa, C. B., Okada, I. A., & Trucksess, M. W. (2011). Occurrence of aflatoxin M1 in dairy products in Brazil. *Food Control*, 22, 1971-1974.

[108] Yaroglu, T., Oruc, H. H., & Tayar, M. (2005). Aflatoxin M1 levels in cheese samples from some provinces of Turkey. *Food Control*, 16, 883-885.

[109] Fallah, A. A., Jafari, T., Fallah, A., & Rahnama, M. (2009). Determination of aflatoxin M1 levels in Iranian white and cream cheese. *Food and Chemical Toxicology*, 47, 1872-1875.

[110] Sarımehmetoglu, B., Kuplulu, O., & Celik, T. H. (2004). Detection of aflatoxin M1 in cheese samples by ELISA. *Food Control*, 15, 45-49.

[111] Kav, K., Col, R., & Tekinsen, K. K. (2011). Detection of aflatoxin M1 levels by ELISA in white-brined Urfa cheese consumed in Turkey. *Food Control*, 22, 1883-1886.

[112] Tekinsen, K. K., & Tekinsen, O. C. (2005). Aflatoxin M1 in white pickle and Van otlu (herb) cheeses consumed in southeastern Turkey. *Food Control*, 16, 565-568.

[113] Ardic, M., Karakaya, Y., Atasever, M., & Adiguzel, G. (2009). Aflatoxin M1 levels of Turkish white brined cheese. *Food Control*, 20, 196-199.

[114] Tekinsen, K. K., & Gurkan, U. (2008). Aflatoxin M1 levels in butter and cream cheese consumed in Turkey. *Food Control*, 19, 27-30.

[115] Aydemir, Atasever. M., Atasever, M., Oztorun, K., & Urcar, S. (2010). Determination of aflatoxin M1 level in butter samples consumed in Erzurum, Turkey. *The Journal of the Faculty of Veterinary Medicine, University of Kafkas*, 16(Suppl-A), 159-162.

[116] Akkaya, L., Birdane, Y. O., Oguz, H., & Cemek, M. (2006). Occurrence of aflatoxın M1 in yogurt samples from Afyonkarahisar, Turkey. *Bulletin of the Veterinary Institute Pulawy*, 50, 517-519.

[117] Martins, M. L., & Marin, H. (2004). Aflatoxin M1 in yoghurts in Portugal. *International Journal of Food Microbiology*, 91, 315-317.

[118] Bintvihok, A., & Davitiyananda, D. (2002). Aflatoxins and their metabolites residues in chicken tissues from 5 parts (10 provinces) of Thailand. *Thai Journal of Health Research*, 16(1), 37-50.

[119] Yunus, A. W., Razzazi-Fazeli, E., & Bohm, J. (2011). Aflatoxin B1 in Affecting Broiler's Performance, Immunity, and Gastrointestinal Tract: A Review of History and Contemporary Issues. *Toxins*, 3, 566-590.

[120] Hussain, Z., Khan, M. Z., Khan, A., Javed, I., Saleemi, M. K., Mahmood, S., & Asi, M. R. (2010). Residues of aflatoxin B1 in broiler meat: Effect of age and dietary aflatoxin B1 levels. *Food and Chemical Toxicology*, 48, 3304-3307.

[121] Wolzak, A., Pearson, A. M., & Coleman, T. H. (1986). Aflatoxin carryover and clearance from tissues of laying hens. *Food and Chemical Toxicology*, 24, 37-41.

[122] Wild, C. P., & Gong, Y. Y. (2010). Mycotoxins and human disease: a largely ignored global health issue. *Carcinogenesis*, 31(1), 71-82.

[123] Micco, C., Miraglia, M., Onori, R., Brera, C., Mantovani, A., Ioppolo, A., & Stasolla, D. (1988). Long-term administration of low doses of mycotoxins to poultry. 1. Residues of aflatoxin B1 and its metabolites in broilers and laying hens. *Food Additive and Contaminant*, 5, 303-308.

[124] Herzallah, S. M. (2009). Determination of aflatoxins in eggs, milk, meat and meat products using HPLC fluorescent and UV detectors. *Food Chemistry*, 114, 1141-1146.

[125] Hassan, A. A., Hassan, M. A., El Shafei, H. M., El Ahl, R. M. H. S., & El -Dayem, R. H. A. (2011). Detection of Aflatoxigenic Moulds Isolated From Fish and their Products and its Public Health Significance. *Nature and Science*, 9(9), 106-114.

[126] Bahobail, A. A. S., Hassan, S. A., & El -Deeb, B. A. (2012). Microbial quality and content aflatoxins of commercially available eggs in Taif, Saudi Arabia. *African Journal of Microbiology Research*, 6(13), 3337-3342.

Chapter 8

Aflatoxins Importance on Animal Nutrition

Vivian Feddern, Giniani C. Dors, Fernando de C. Tavernari,
Helenice Mazzuco, Anildo Cunha Jr.,
Everton L. Krabbe and Gerson N. Scheuermann

Additional information is available at the end of the chapter

1. Introduction

Mycotoxins are secondary metabolites produced by several fungal species, toxic to humans, animals and plants. Their ingestion, inhalation or dermal absorption may cause different diseases and even death. These compounds have been described for many decades, however in the beginning of the 1960s, they have been chemically characterized due to the discovery of aflatoxins [1]. The word aflatoxin is the combination of 3 other words "a" for *Aspergillus* genus, "fla" for the species *flavus* and toxin, meaning poison [2].

Aflatoxin is the mycotoxin generating the greatest losses and the highest management costs due to its extremely high toxicity on a unit basis, and its long history of stringent regulation. The costs are inversely related to the regulatory level that must be met, and lower concentration allowances will increase the costs of crop management. Several effective ways for the management of mycotoxin contamination in agriculture have been stressed. One strategy to manage mycotoxin contamination and decrease health risks and economic costs is to instruct food producers and handlers on how to minimize mycotoxin contamination, and to encourage the adoption of process-based guidelines such as Good Agricultural Practices (GAPs) before harvest and good manufacturing practices (GMPs) after harvest. These actions would minimize risk throughout the production, handling, and processing chain, and can complement product standards [3]. To control the presence of aflatoxins in foods, many countries established maximum tolerated concentrations through legislation (Van Egmond, 1989a cited [4]).

This chapter focuses on properties of aflatoxins and their occurrence in feeds and animal products as meat, eggs, liver, kidneys and milk. Topics regarding mycotoxins absorbents and legislation in feed ingredients and feeds are also covered.

2. Aflatoxin properties

All mycotoxins are low-molecular-weight natural products (i.e., small molecules) (Bennett, 1987 cited [3]). The hyphal structure of filamentous fungi has evolved to utilize solid substrates efficiently by growing over surfaces and penetrating into solid matrices. Moulds are able to secrete enzymes to break down complex macromolecular compounds and utilize them for growth and metabolism. They can absorb low molecular weight nutrients, produce and secrete secondary metabolites, which are also relatively low molecular weight compounds but not associated with the process of growth and primary metabolism (Bushell, 1989 cited by [4]).

Some aflatoxin derivatives are products of animal metabolism following ingestion of the mould metabolites, which are divided into the B and G groups based on their blue or green fluorescence under UV light when absorbed to solid substrates. *A. parasiticus* is the most toxigenic species, the majority of strains producing both B and G toxins (Van Egmond, 1989a cited by [4]).

Aflatoxins are crystalline substances, freely soluble in moderately polar solvents such as chloroform, methanol, dimethyl sulfoxide; they dissolve in water to the extent of 10-20 mg L^{-1}. Some important physical and chemical properties of aflatoxins are given in Table 1 [5].

				UV absorption max (ε(L mol⁻¹ cm⁻¹)), methanol	
Aflatoxin	*Molecular formula*	*Molecular weight*	Melting Point	*265 nm*	*360-362 nm*
B_1	$C_{17}H_{12}O_6$	312	268-269	12,400	21,800
B_2	$C_{17}H_{14}O_6$	314	286-289	12,100	24,000
G_1	$C_{17}H_{12}O_7$	328	244-246	9,600	17,700
G_2	$C_{17}H_{14}O_7$	330	237-240	8,200	17,100

Table 1. Physical and chemical properties of aflatoxins.

Due to the important structural diversity (Figure 1) of mycotoxins and the variations in their metabolism, it is difficult to edit general rules. Thus, each toxin and respective metabolites have to be investigated as a particular case. The chemical and physical properties of aflatoxins are described as follows [6]:

i. *Description*: Colorless to pale-yellow crystals. Intensely fluorescent in ultraviolet light, emitting blue (aflatoxins B_1 and B_2) or green (aflatoxin G_1) and green–blue (aflatoxin G_2) fluorescence, from which the designations B and G were derived, or blue–violet fluorescence (aflatoxin M_1)

ii. *Melting-points*: see Table 1.

iii. *Absorptionspectroscopy*: see Table 1.

iv. *Solubility*: Very slightly soluble in water (10–30 μg mL^{-1}); insoluble in non-polar solvents; freely soluble in moderately polar organic solvents (e.g., chloroform and methanol) and especially in dimethyl sulfoxide

v. *Stability*: Unstable to ultraviolet light in the presence of oxygen, to pH extremes (< 3, > 10) and to oxidizing agents

vi. *Reactivity*: The lactone ring is susceptible to alkaline hydrolysis. Aflatoxins are also degraded by reaction with ammonia or sodium hypochlorite.

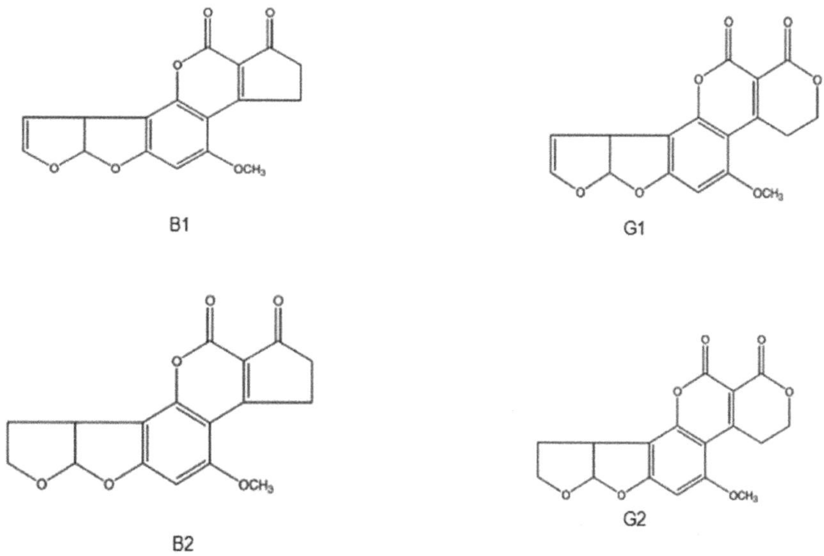

Figure 1. Chemical structures of aflatoxins [7].

Aflatoxins are quite stable in many foods and are fairly resistant to degradation. The effectiveness of some processes in reducing concentrations of aflatoxins in food can be affected by many factors, such as the presence of protein, pH, temperature and length of treatment. Commercial processing of raw commodities using cleaning regimes including the removal of broken particles, milling and sorting can reduce aflatoxin concentration considerably [5].

Naturally occurring aflatoxins (as a group) and other 107 agents were evaluated as carcinogenic to humans (Group 1). There is sufficient evidence in humans for the carcinogenicity of aflatoxins, being liver cancer (hepatocellular carcinoma) the main effect. Aflatoxin M_1, the metabolite of aflatoxin B_1 found in milk of lactating mammals was classified in Group 2B as possibly carcinogenic to humans [8]. Carcinogenicity of naturally occurring mixtures of aflatoxins B_1, G_1 and M_1 is also demonstrated in experimental animals. The intake of these toxins over a long period of time in very low concentrations may be highly dangerous. These compounds can enter the food chain, mainly, by ingestion through the diet of humans and animals (Miraglia et al., 1996 cited by [11]).

There is strong evidence that the carcinogenicity of aflatoxins operates by a genotoxic mechanism of action that involves metabolic activation to a genotoxic epoxide metabolite, formation of DNA adducts, and modification of the TP53 gene. In humans, hepatocellular carcinomas from areas of high exposure to aflatoxins, up to 50% of tumors have been shown to harbor a specific point mutation in the TP53 tumor suppressor gene [9]. Table 2 shows aflatoxin main producing species and toxic effects, pointed by the International Agency for Research on Cancer (IARC).

Aflatoxins	*Main producing species*	*CAS No.*	*Toxic effect*
B_1	*A. flavus, A. parasiticus, A. nomius, A. bombycis*	1162-65-8	Hepatotoxic, genotoxic, carcinogenic, immunomodulation
B_2	*A. flavus, A. parasiticus, A. nomius, A. bombycis*	7220-81-7	Limited evidence for carcinogenicity
G_1	*A. parasiticus, A. nomius, A. bombycis*	1165-39-5	Sufficient evidence of carcinogenicity
G_2	*A. parasiticus, A. nomius, A. bombycis*	7241-98-7	Inadequate evidence for carcinogenicity

Table 2. Main producing species and effects of aflatoxins [6,10].

AFB_1 is the most potent carcinogenic substance naturally produced by *Aspergillus* species. Indeed, AFB_1 is classified by IARC as Group 1 carcinogen [10]. This compound is certainly acutely toxic to humans, is probably responsible for liver necrosis following chronic exposure, and may be involved in the epidemiology of human liver cancer in some parts of the world perhaps synergistically with hepatitis B virus (Van Egmond, 1989a cited by [4]).

After ingestion, aflatoxin B is metabolized by enzymes to generate a reactive 8,9-epoxide metabolite that can be bound to DNA as well as to serum albumin forming aflatoxin-N-7 guanine and lysine adducts, respectively. Covalent binding to DNA is considered to be a critical step in aflatoxin hepatocarcinogenesis [11].

Determination of these metabolites was solved by developing enzyme linked immunosorbent assay (ELISA) methods (Vidyasagar et al., 1997 and Nayak, et al., 2001 cited by [11]). The biosynthesis of aflatoxins is induced by sugars. The induction is associated with the transcriptional activation of the pathway genes and the pathway regulatory gene, aflR. The regulation of aflatoxin biosynthesis had been examined by manipulating the transcription of aflR. Studies concerning this topic showed that constitutive overexpression of the pathway transcriptional regulatory gene aflR led to higher transcript accumulation of pathway genes and increased aflatoxin production (Flaherty and Payne, 1997 cited by [11])s

Aflatoxins are metabolized in ruminants by the liver and are excreted in the bile. AFB_1 increases the apparent protein requirement of cattle. When significant quantities are consumed, the metabolite M_1 appears in milk within 12 hours. Research suggests M_1 is not as carcinogenic or mutagenic as B_1, but it does appear to be as toxic as its parent compound [12]. When an animal ingests food contaminated with an AFB_1, from 0.5 to 5% of the toxin ingested is biotransformed in the liver into AFM_1 (Hussein and Brasel, 2001 cited by [13].

On a worldwide basis about 35% of *A. flavus* strains produce aflatoxins and only the B group. These molds occur in warmer parts of the world and aflatoxins may be produced in a wide range of tropical and subtropical food commodities (Van Egmond, 1989a cited by [4]). The presence of fungi does not necessarily imply the presence of toxins. The fungi species can produce aflatoxins on commodities in the field under stress conditions or in storage when high moisture and warm temperature propitiate their growth (Schuster et al. 1993 cited by [11]). The minimum moisture content of foods that allows the growth of *A. flavus* is around 85% relative humidity (0.85 water activity), and temperature of 25-30 ºC. In cereals with high starch conten as rice, maize, sorghum, wheat, barley, the moisture content in the grain is 18-18.5%. The subsequent drying does not affect the existing level of aflatoxin because it resists drying and roasting temperatures (Pitted, 1998; Sabino, 1996; Wilson and Payne, 1994 cited by [14]). Because of the weather, aflatoxins are commonly found in South America, Africa, Asia and Australia [15].

3. Aflatoxins occurrence in feeds

Time of harvest has been shown to be important in influencing the occurrence and levels of aflatoxin because *Aspergillus* does not compete well with other molds when corn presents more than 20% moisture. Harvesting corn when moisture content is above 20% followed by rapid drying to at least 14% moisture content within 24 to 48 hours of harvest can inhibit *Aspergillus* growth and toxin production. Contaminated grains and their byproducts are the most common sources of aflatoxin. Corn silage may also be a source of aflatoxins, because the ensiling process does not destroy toxins already present in silage [12].

On the farm, more than one mold or toxin may be present in the contaminated feed, which often makes definitive diagnosis of aflatoxicosis difficult. The prognosis of aflatoxicosis depends upon the severity of liver damage. Once overt symptoms are noticed the prognosis is poor. Treatment should be directed at the severely affected animals in the herd and further poisoning prevented. Aflatoxicosis is typically a herd rather than an individual cow problem. If aflatoxicosis is suspected, feed should be analyzed immediately. If aflatoxins are present, the source should be eliminated immediately. Levels of protein in feed and vitamins A, D, E, K and B should be increased as the toxin binds vitamins and affects protein synthesis. Good management practices to alleviate stress are essential to reduce the risk of secondary infections which must receive immediate attention and treatment [12].

Importantly, it has been demonstrated that simple measures can significantly reduce the risk of mycotoxin exposure on farm. Storage of grain at appropriate moisture content (below 130 g

kg^{-1}), inspection of grain regularly for temperature, insects and wet spots will limit the possibility of fungal development in feeds and feedstuffs as discussed before. The risk of feed contamination will be reduced in animal units with rapid turnover of feed because there will be less time for fungal growth and toxin production [17]. Aflatoxin is just one of many mycotoxins that can adversely affect animal health and productivity. Care regarding animal feed must be extended not only to the nutritional and economic value, but also to food quality [13].

Decades of animal studies have demonstrated that chronic exposure to aflatoxins in animals can also cause growth inhibition and immune suppression [18]. Nursing animals may be affected, and AFM_1 may be excreted in the milk of dairy cattle and other dairy animals. This in turn poses potential health risks to both animals and humans that consume that milk. Chronic aflatoxin exposure in animals can result in impaired reproductive efficiency, reduced feed conversion efficiency, increased mortality rates, reduced weight gain, anemia, and jaundice. In the case of laying hens, aflatoxicosis causes an enlarged fatty liver and lowered egg production [19].

Sex and age of animals have also an influence on AFB_1 susceptibility. For instance, males are more susceptible than females and young animals of all species are more susceptible than mature animals to the effects of aflatoxin [12,16]. Feed refusal, reduced growth rate and decreased feed efficiency are the predominant signs of chronic aflatoxin poisoning. In addition, listlessness, weight loss, rough hair coat and mild diarrhea may occur. Anemia along with bruises and subcutaneous hemorrhage are also symptoms of aflatoxicosis. The disease may also impair reproductive efficiency, including abnormal estrous cycles (too short and too long) and abortions. Other symptoms include impaired immune system response, increased susceptibility to disease, and rectal prolapse [12].

A study identified and quantified aflatoxins (AFB_1, AFB_2, AFG_1 and AFG_2) from poultry feed and their recovery, together with their metabolites (AFM_1, AFM_2, AFP_1 and aflatoxicol) in litter. Hens were divided in 3 groups and fed with 2 AFB_1 concentrations: 30 and 500 ppb, besides the control group. Feed samples of the 3 groups presented significant difference with AFB_2 and AFG_2, whereas in litter samples, there were significant differences for AFG_2 in the 500 ppb group. Poultry litter had traces of AFM_1, AFM_2, AFP_1 (can be considered as a demethylated AFB_1) and aflatoxicol with no significant differences among treatments [20].

The presence of molds in foodstuffs causes the appearance of flavors and odors that reduce palatability and affect feed consumption by animals as well as reduce the nutritional value of foods. Mycotoxins, in turn, affect the digestion and metabolism of nutrients in animal production, resulting in nutritional and physiological disorders, besides a negative effect on the immune system [21].

It was reported main effects caused by aflatoxins during swine growth and termination phases. When feed was contaminated with 10-100 ppb, productivity losses without noticeable clinical signs were observed. When this level was 200-400 ppb, reduced growth and feed efficiency occurred. At 400-800 ppb of aflatoxins in feed, there were liver diseases (friable or yellow-tan liver). After 800-1200 ppb of aflatoxins administration in feed, reduction of food intake and growth was observed. Finally, at 1200-2000 ppb, jaundice, coagulopathy, anorexia and even mortality may happen. Not only swine is affected by aflatoxins but all species, being the main clinical signs and lesions reported as decreased weight gain, digestive disorders, liver disease, anorexia, ataxia, tremors and death [22].

A total of 480 poultry feed samples from Rio de Janeiro state were collected monthly during one year and analyzed, being the main fungal species found *P. citrinum* (35% of the samples) followed by *A. flavus* (25%) which is the main aflatoxin producer microorganism. AFB_1 levels ranged from 1.2 to 17.5 ppb. There were no significant differences ($P<0.001$) between all months tested except February and March when the highest and lowest AFB_1 production was found [23]. In Pakistan, a total of 216 samples of poultry feed ingredients were assayed, being found maximum 191.65 ppb for AFB_1, 86.85 ppb for AFB_2, 89.80 ppb for AFG_2 and 167.82 ppb for AFG_1. Minimum aflatoxins were produced in the winter season. The temperature varies from 10 to 45 ºC in this country, favorable to *Aspergillus* growth [24].

Recently [25], a survey reported the association of mycotoxins with hematological and biochemical profiles in broilers. The authors performed meta-analysis using data from 98 articles, totaling 37,371 broilers. Some conclusions of this review were that mycotoxins reduced ($P<0.05$) the hematocrit (−5%), hemoglobin (−15%), leukocytes (−25%), heterophils (−2%), lymphocytes (−2%), uric acid (−31%), creatine kinase (−27%), creatinine (−23%), triglycerides (−39%), albumin (−17%), globulin (−1%), total cholesterol (−14%), calcium (−5%), and inorganic phosphorus (−12%). Mycotoxins also altered ($P<0.05$) the concentrations of alkaline phosphatase, alanine aminotransferase and aspartate aminotransferase. The total protein concentration in blood was 18% lower ($P< 0.05$) in broilers challenged by aflatoxins compared with that of the unchallenged ones. The inclusion of antimycotoxin additives in diets with aflatoxins altered ($P<0.05$) some variables (uric acid, creatinine, alkaline phosphatase, alanine aminotransferase, aspartate aminotransferase, and γ-glutamyl transferase) in relation to the group that received diets with the mycotoxin and without the additive.

Another recent study assumes that aflatoxins could compromise the macrophages functions; in particular, co-exposure to AFB_1, AFB_2, AFM_1 and AFM_2 may exert interactions which can significantly affect immunoreactivity [26].

4. Aflatoxins occurrence in animal products

When focusing on how mycotoxins play a role in food safety, attention should be limited to mycotoxins that are known to be transferred from feed to food of animal origin, as this food represents a significant route of exposure for humans [27]. Apart from their toxicological effects in affected animals, the carry-over through animal derived products, such as meat, milk and eggs into the human food chains is an important aspect of mycotoxin contamination. FAO has estimated that up to 25% of the world's food crops and a higher percentage of the world's animal feedstuffs are significantly contaminated by mycotoxins.

Aflatoxin or ochratoxin residues in meat are uncommon and rarely found [28]. However, it's more common in organs especially liver. This organ may have its lipid content increased over three fold when 20 ppm aflatoxin is incorporated in broiler feed [29].

The problem in the egg production is that the long-term or short-term hen's exposure, via dietary sources, to low concentrations of certain mycotoxins causes contamination of eggs. This is the case of aflatoxins, which have a high impact in both, human and animal health, causing significant losses in the egg industry, considering the deleterious effect on egg production and quality.

In laboratory studies it was proved that aflatoxin can decrease egg production and increase liver fat (fatty liver syndrome). This classical study established the typical symptoms associated with acute or chronic aflatoxicosis, observed until today in field conditions [30].

A distinctive sequence of events during acute aflatoxicosis in laying hens (30 weeks-old) in a four week experiment with increasing aflatoxin doses in the diet of 0; 1.25; 2.5; 5.0 and 10.0 $\mu g\ g^{-1}$ [31]. Results indicated that egg production was decreased by about 70% from the control value at $10 \mu g\ g^{-1}$ concentration in the diet and the liver size was increased significantly by 5 and 10 $\mu g\ g^{-1}$ dietary concentrations of aflatoxin and the liver lipid increasing dramatically by a smaller dose of 2.5 $\mu g\ g^{-1}$. Table 3 shows the dramatic effect of aflatoxin in the liver function [31]. The obtained data suggest that plasma and yolk lipids respond to the inhibition of lipid synthesis and transport from the liver during aflatoxicosis induced by the dietary treatments. The liver malfunction results in an increase in its fat content and a decrease in the levels of plasma lipids.

Dose ($\mu g\ g^{-1}$)	*Liver lipids (%)*	*Plasma lipids (g 100 g^{-1})*
0.0	21.2±1.1	2.6±0.3
1.25	24.4±1.8	2.6±0.4
2.5	32.7±1.8	2.1±0.2
5.0	35.6±4.9	1.9±0.1
10.0	46.5±4.7	1.7±0.2

Adapted from ref. [31]. Values are means ± Standard error of the mean

Table 3. Response of liver lipid and plasma lipid during aflatoxicosis in laying hens.

Zaghini et al. [32] supported the previous finding showing the effects of AFB_1 on egg quality and chemical parameters. In the study, 44 weeks-old laying hens were submitted to a diet containing 2.5 ppm of AFB_1 and by the end of the second and third weeks of the trial, changes were observed like decreased egg weight (from 73.76 g to 72.5 g, week 0-4th, respectively) and reduced shell weight, as indicated by the decline in the percent shell of eggs laid by the hens fed the AFB_1 contaminated diet from 10.49% to 10.19%. In the same study, aflatoxin also influenced color parameters, which were probably related to interference of AFB_1 with lipid metabolism and pigmentary substances deposition in yolk. Additionally, all livers collected from the hens administered the mycotoxin group were positive for AFB_1.

Astonishingly, as little as 0.2 mg kg^{-1} (or 0.2 ppm) of the metabolite AFB_1 has been documented to reduce egg production and egg mass in laying hens from 22 to 40 weeks of age

[33]. The most important economic effect of poultry ingesting AF-contaminated feed would be the increase of the mortality index [34] and in addition, aflatoxins intake can decrease productivity due to hepatic [35]; immunological [36] and renal damages [37].

In a detailed study focusing on the effects of aflatoxin chronic intoxication in renal function of laying hens (13 weeks-old), aflatoxins were evident at 17 weeks of the intoxication period. Final concentrations were 0, 0.46, 0.98 and 1.53 mg of aflatoxins per kg of feed, respectively and birds ingested the contaminated diets during 17 and 42 weeks. Body weight of intoxicated hens, showed a tendency to decrease, being significant in 1.0 and 1.5 mg kg^{-1} of feed concentrations in both times of the intoxication period. Evidence of tubular damage in kidney was found as a result of a decreased concentration of Ca^{++} and PO_4^{+3} in plasma or even a decreased Ca^{++} absorption from the gut. Also, microscopic lesions of glomerular and tubular structures like inflammatory and degenerative processes of the renal structures in hens kidneys were found. Additionally, the authors pointed out that the renal lesions occurred more frequently in larger doses of AF and over a long period of exposure to the toxin (42 weeks intoxication period) [38].

Other authors concluded that aflatoxins may have direct or indirect effect or both, on functionality of the gastrointestinal tract. Results indicated that specific activity of the intestinal maltase and disaccharidase increased quadratically, by feeding up to 1.2 mg kg^{-1} aflatoxins and declined at 2.5 mg kg^{-1} concentration in the study and the intestinal crypt depth (but not villus length) increased linearly with increasing the level of aflatoxins in the experimental diets [39].

Hens were fed three levels of aflatoxin that might approximate contamination under field conditions [40]. Pure AFB_1 was prepared and mixed in the diet as follows: 0.1 ppm for 10 days; 0.2 ppm for 12 days and 0.4 ppm for 15 days. Results confirmed that AFB_1 fed to hens was transmitted into eggs in measurable amounts at all levels and was found in both, albumen and yolks. The average amounts of aflatoxin distributed between albumen and yolk were 2.2 and 3.6 ppb, respectively. Even at the concentration of 0.1 ppm of AFB_1 in the layer diet, the transmission into eggs occurred as an average of 0.23 ppb.

Mainly aflatoxins and ochratoxin A may be found as residues at significant levels in muscles and muscle foods when contaminated feed is distributed to farm animals. Meat contamination may also result from toxigenic mold development during ripening and ageing. In muscles, only low levels are found, often below detection limits of the methods used, even after exposure of the animals to high doses of AFB_1. In ruminants, many studies evaluated aflatoxin transfer into the milk of lactating cows. However, as for other species, residues can be found in liver and kidney that are edible parts of these animals [41].

It was reported [41] that processing conditions during ageing of hams may allow aflatoxin synthesis. Thus, is important to conduct research evaluating the production of AFB_1 during meat processing and ageing. Studies show that frequency of processed meat contamination with AFB_1 was low and the toxin level within meat was usually <10 ng g^{-1} (ppb). It is not clear whether AFB_1 was produced during meat processing or was present before at the residual level in muscles. The contamination of spices and additives added during meat processing may also represent a source of mycotoxin. Besides, spice addition may lead to a secondary contamination of the final product with aflatoxins.

Table 4 shows residues of one or more aflatoxins in different hen tissues.

Animal species	*Dose or exposure time*	*Tissues*	*Residues (µg kg⁻¹)*	*Metabolites*	*Reference*
Laying hens	2.5 ppm AFB_1 in feed for 4 weeks	Muscle	0.08±0.03		[32]
		Eggs	0.24±0.07 and 0.25±0.09		
		Liver	4.13±1.95		
		Eggs	<0.5 and <0.01		
Layer breeder hens	$OTA+AFB_1$ mg kg^{-1}			$OTA+AFB_1$	[42]
	0 + 0	Liver/Muscle/Kidney	ND + ND		
	0 + 5	Muscle	ND + 0.03		
		Kidney	ND + 0.25		
		Liver	ND + 1.44		
	3 + 5	Muscle	0.34 + 0.02		
		Kidney	2.80 + 0.27		
		Liver	1.98 + 0.26		
	5 + 5	Muscle	0.51 + 0.02		
		Kidney	2.81 + 0.27		
		Liver	2.21 + 0.11		

Table 4. Residues of aflatoxin in animal products combined or not with other mycotoxin.

Feeding one mycotoxin alone (OTA) resulted in significantly higher residue levels in liver, kidney and breast muscles of hens than their counterpart birds kept on a diet concurrently contaminated with both OTA and AFB_1 [42]. In this study, a total of 72 White Leghorn layer breeder hens at 45 weeks of age were submitted to diets containing different combinations of these mycotoxins (some are shown in Table 4, focusing mainly on AFB_1 residues). In liver of hens fed OTA alone at 5 mg kg^{-1} (ppm) feed, residues level was 22.54±1.48 (mean±SD) ppb, as compared to significantly lower residual concentration of 2.21±0.42, in the same levels of OTA when administered in combination with AFB_1. Residues of OTA were significantly higher in liver than in kidneys of the hens fed OTA alone, in all experimental groups. However, feeding OTA in combination with AFB_1 resulted in higher deposition of OTA in kidneys than in livers. Residues of AFB_1 were significantly higher in liver and breast muscles of the birds kept on AFB_1 contaminated feed compared with those fed OTA and AFB_1 concurrently. When the maximum dosage (5 ppm) was administered, residues of OTA and AFB_1 were also the maximum in the liver, i.e. 22.54±1.48 and 1.44±0.21 ppb, respectively, while a minimum concentration of residues of both mycotoxins was found in the breast muscles of the laying hens. Residues of AFB_1 in the eggs appeared at day 5 of toxin feeding and disappeared at day 6 of withdrawal of AFB_1 contaminated diet. As in case of tissues, residues of OTA and AFB_1 are significantly lower in eggs obtained from hens fed both toxins in combination, compared to those fed each mycotoxin alone, possibly due to their protein binding potentials.

Another study shows that concentration of AFB_1 residues in liver and muscles increased with toxin ingestion time and were at its highest levels (6.97 ppb in liver and 3.27 ppb in muscle) on the last day (7th) of feeding AFB_1 contaminated ration. Broiler chicks of 7, 14 and 28 days of age fed same level of AFB_1 showed lower tissue residues in older birds compared with younger ones. Birds given 1600 and 3200 ppb AFB_1 for 7 days at 28 days of age had no detectable AFB_1 levels in liver and muscles after 3 and 8 days of withdrawal of contaminated feed. A rapid decrease in AFB_1 residues below the tolerance limits from muscles and liver within 3 and 7 days of withdrawal of dietary AFB_1 in this study confirm the rapid metabolism of aflatoxins in the body of chicken and that it may not become a significant human health risk. However, in areas with no regulatory limits on AFB_1 levels of poultry feed, the secondary exposure to aflatoxins through consumption of chicken liver and meat derived from the poultry fed contaminated feed may pose a risk to consumers health [43].

Product	*Aflatoxin*	*Positive/total of samples*	*Range (ppb)*	*Method*
Corn	AFs	76 / 246	2-906	TLC or ELISA
Corn	AFB_2	33 / 292	1-17	TLC
Feed	AFB_1	14 / 96	11-287	TLC
Eggs	AFB_1	2 / 210	2-5	TLC
Eggs	AFM_1	0 / 210	-	TLC
Swine liver	AFB_1	1 / 43	27	TLC
Chicken liver	AFB_1	3 / 6	1.2-3.2	TLC/HPLC

Table 5. Mycotoxin levels in vegetable and animal products. Adapted from ref. [45]. See this review to obtain the references of original publication.

The impact of subchronic exposure of AFB_1 on the tissue residues of enrofloxacin and its metabolite ciprofloxacin was examined in broilers. Broiler chickens given either normal or AFB_1 (750 ppb diet) supplemented diets for 6 weeks received enrofloxacin (10 mg kg^{-1} day^{-1}, p.o.) for 4 days and thereafter, residue levels were determined at 1, 5 and 10 days after the last treatment. In AFB_1-unexposed broiler chickens, enrofloxacin was detected in all the tissues. After 24 h of treatment cessation, concentrations of enrofloxacin were up to 0.85 $\mu g\ g^{-1}$ in the following order: liver>skin+fat>muscle>kidney. The parent drug was not found in any of the tissues except liver 10 days after the last dose of enrofloxacin. Ciprofloxacin was not detectable in any tissue. In AFB_1-exposed broiler chickens, higher concentrations of enrofloxacin and ciprofloxacin were found in different tissues, compared with tissues of control broiler chickens. After 24 h of the last dose of enrofloxacin, concentrations up to 4.53 $\mu g\ g^{-1}$ were found of the parent drug in the order skin+fat>liver>kidney>muscle. The parent drug persisted in all the tissues except muscle for 10 days. Ciprofloxacin was detected in muscle and skin plus fat 24 h after termination of enrofloxacin administration and it persisted only in muscle for 10 days. The metabolite was not detectable in kidney [44].

A review carried out in Brazil [45] showed high variability among the results (Table 5). For instance, corn contamination with aflatoxins reached 906 ppb, above those levels allowed by legislation (20 ppb). This fact indicates the need for quality control in the reception of this ingredient in the feed mill with the use of rapid tests for mycotoxins. Regarding products of animal origin, major problems were not observed in eggs and tissues of swine and poultry (Table 5). However, among chicken liver samples 50% tested positive but with relatively low levels. Anyway, attention should be paid with liver consumption when there are evidences of corn contamination. From Table 5 data, it can be noted that contaminated feed samples achieved up to 287 ppb AFB_1, above values allowed by legislation.

A survey with hens fed AFB_1 via moldy rice powder feed showed residues in eggs and tissues (kidneys, liver, muscle, blood, and ova) [46]. Hens were fed for 7 days with a contaminated diet (8 μg g^{-1}) followed by additional 7 days on an aflatoxin-free diet. Eggs were collected over the entire 14-day period. The study showed that aflatoxicol (R0), a carcinogenic metabolite of AFB_1, was found in all samples but blood (Table 6). Levels of R0 and AFB_1 were approximately the same in eggs, ova, kidneys, and liver. In eggs, the levels of R0 and AFB_1 (0.02 to 0.2 ng g^{-1}) increased steadily for 4 or 5 days until reaching a plateau and then decreased after B_1 withdrawal at the same rate as they increased. After 7 days of withdrawal, only trace amounts of R0 (0.01 ng g^{-1}) remained in eggs. All samples from hens sacrificed immediately before aflatoxin withdrawal contained R0 or R0+AFB_1. R0 was the only aflatoxin detected in muscle. Seven days after aflatoxin withdrawal, B_1 (0.08 ng g^{-1}) was found in one of nine livers and R0 (0.01-0.04 ng g^{-1}) in eight of nine muscles analyzed, but no aflatoxins were found in any other tissues. Interestingly, the transfer of aflatoxins into eggs is right after administration, since B_1 (0.03 ng g^{-1}) and R0 (0.02 ng g^{-1}) residues were found in eggs laid 1 day after contaminated feed was administered. This indicates that toxins penetrate the egg through eggwhite since yolk was already formed before this period. Aflatoxin apparently can enter the egg at any stage of its development. This is because it takes 7 to 8 days for each oocyte to develop into a mature ovum (yolk) and 24 hours for the egg oviposition.

Tissue	*R0*	*AFB_1*	*AFM_1*
Ova	0.25	0.24	ND
Kidneys	0.10	0.25	0.05
Liver	0.20	0.46	ND
Muscle	0.08	ND	ND
Blood	ND	0.05	0.10

Table 6. AFB_1 and its metabolites aflatoxicol (R0) and AFM_1 (ng g^{-1}) after contaminated diet ingestion with AFB_1 (8 μg g^{-1}). Values in ppb; ND = Not detected [46].

5. Legislation in feed and feed ingredients

In the last decades, only aflatoxins and, to a lesser extent, ochratoxin A were regulated in foods from animal origin. For other toxins, the risk management was based on the control of the contamination of food from vegetal origin intended for both human and animal consumption. Nowadays, other mycotoxins are included. Regulatory values or recommendations are mainly built on available knowledge on toxicity and potential carryover of these molecules in animal. Therefore, by limiting animal exposure through feed ingestion, one can guarantee against the presence of residues of mycotoxins in animal-derived products. However, accidental high levels of contamination may lead to a sporadic contamination of products coming from exposed animals [41].

Tolerance levels of mycotoxins in foods are needed to ensure product quality and consumer health. The limits differ among countries, i.e., depending on the product and the country there are different tolerance levels for each mycotoxin, but it is certain that their presence in foods has been widely researched and new standards were required over the years, in the last decade. Table 7 shows an average of mycotoxin variation depending on the type of food, required as maximum standard in different countries.

*Mycotoxins**	*Feed[1]*	*Corn[2]*	*Soybean[3]*
Aflatoxin B_1, ppb	1.5 - 50	1 – 50	30 - 50
Aflatoxin B_1, B_2, G_1 and G_2, ppb	0 - 75	5 – 50	20 - 50
Deoxynivalenol, ppb	5 - 1500	-	-
Toxin HT2, ppb	25 - 100	-	-
Ochratoxin, ppb	5 - 300	50 – 300	-
Zearalenone, ppb	-	0.5 – 200	-

[1] Relative to feed and concentrates for all categories and phases of animal
[2] Corn and byproducts
[3] Soybean and byproducts
* Source: Adapted from Resolution RDC Nº7 [47] and EUR-LEX [48].

Table 7. Variation among different countries regarding maximum tolerance limits of mycotoxins.

In Brazil, the most recent resolution on mycotoxins in food is the RDC 07/2011 [47] which establishes maximum tolerated levels for aflatoxins (AFB_1+AFB_2+AFG_1+AFG_2 and AFM_1), ochratoxin A, deoxynivalenol, fumonisins (FB_1 + FB_2), patulin and zearalenone, admissible in ready-to-eat foods and raw materials. To adapt to the new standard required in 2011, the producers of 14 food categories should meet the requirements until 2016. Table 8 shows standard values set for corn, which is the main ingredient added to feed in the country.

Brazil, like different countries, also follows the recommendation to keep mycotoxin levels as low as possible. For that, better practices and technologies in the production, handling, storage, processing and packaging should be accomplished in order to prevent that contaminated food is sold or consumed.

Mycotoxins	*Corn and byproducts**
Aflatoxins B_1, B_2, G_1 and G_2, ppb	20
Deoxynivalenol, ppb	3000
Fumonisin B_1+B_2, ppb	5000
Zearalenone, ppb	400

Table 8. Maximum tolerated levels for mycotoxins according to Resolution RDC 07/2011 [47].*The maximum tolerated levels refer to results obtained by methodologies that comply with the performance criteria established by Codex Alimentarius.

6. Aflatoxin binders and strategies to reduce toxicity to farm animals

Adsorbents are necessary and important and may have great impact on improving animal production and health, providing greater security to consumers of animal products, due to the reduction and/or removal of mycotoxins in these products.

Considering that aflatoxins were the first discovered mycotoxins, there are many data available searching for binders and other methods to reduce toxicity in animals. However, due to methodologies used for evaluation, there is certain degree of variation in results.

The most common additives used in animal diets are aluminosilicates, produced synthetically or extracted from clay mines. There are also other alternatives to reduce aflatoxin toxicity, as presented.

a. *Clay derived sorbents:*

This type of binder is basically composed by single or blended type of clay. The most common clay is hydrated sodium calcium aluminosilicate (HSCAS). However, there are other sort of clays which can be used as toxin binders, like sodium or calcium bentonites and zeolites. Not often, any particular varying sort of clay can be used as well. It has to be considered that such materials can be synthesized industrially or obtained from mines around the world. In the case of natural sources (mines) it has to be considered that each source may present specific particularities in terms of composition, which can impact the binding capacity, and even clays obtained from the same place, can vary from batch to batch, that has to be well controlled throughout quality control.

There is much information available in the literature comparing those different clays [49]. These authors for example, compared zeolite, bentonite and HSCAS for AFB_1 binding capacity using *in vitro* method, simulating gastrointestinal fluids. This method seems to be the most frequent technology adopted for such assays, including a double condition of pH (3.0 and 7.0).

Those researchers find that zeolite and bentonite aflatoxin binding capacity varied according to the pH used for the assay, and both clays were less effective than HSCAS (Figure 2).

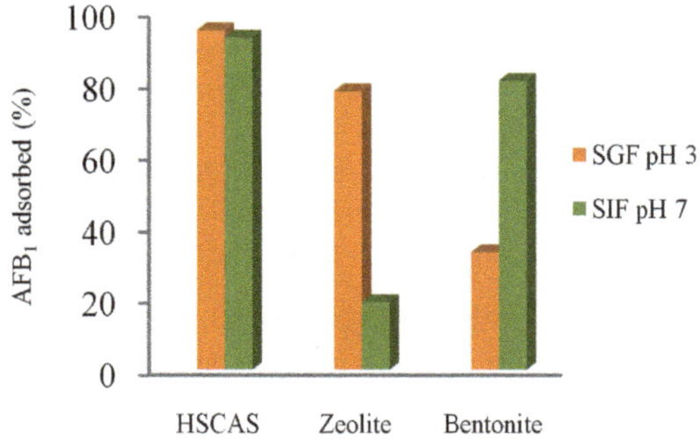

Figure 2. Adsorption percentage of AFB_1 (8 μg/ml) to sorbents (0.5% w/v) in simulated gastrointestinal fluid at pH 3 and pH 7 [49].

It is important to establish the correct inclusion rate to animal diets in order to optimize the binding response. In Figure 3, it can be observed how those three types of clays perform under the same pH (7.0) when increasing doses are included.

It has been shown that montmorillonite (0.5%) added to the diet containing 5 ppm of aflatoxin has proven its effectiveness in preventing the effects of aflatoxicosis in broilers [50].

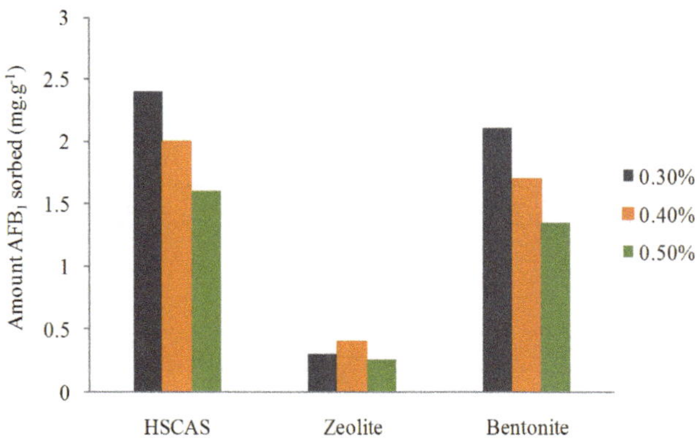

Figure 3. Amount of aflatoxin B_1 adsorbed on sorbents at different concentrations of the adsorbents in simulated intestinal fluid at pH 7 [50].

Based on the data presented in Figures 2 and 3, it is clear that assays condition (especially pH) and toxin:sorbent dosing rate are extremely important. These conditions should be considered when product performance reports are compared. However, when evaluating [51] nine different toxin binders (4 activated charcoals, 3 sodium bentonites, 1 calcium bentonite and 1 esterified glucomannan) all products presented adsorption above 95% of AFB_1, regardless of the pH used (3.0, 7.0, 10.0 and the original pH of each product).

Other methods can be used to evaluate toxin adsorbents, as *in vivo* trials. In this case, the most frequent inconsistency when comparing research reports is related to the source used to obtain aflatoxin (synthetic crystalline vs natural aflatoxin obtained by fermentation) and also the aflatoxin level used in the specific assay. When comparing analytical reports with field accepted levels of aflatoxins, there is much difference. One reason for that is the difference between the experimental conditions (well controlled) where animals are not submitted to stress situation in comparison to the real farm condition.

Another additional evaluation that should be performed is the presence of aflatoxin in specific organs, like liver. Low level of AFB_1 (50 ppb) on broiler performance was studied on biochemical parameters and aflatoxin presence in liver tissue, when monensin and sodium bentonite were added to the feed, from 18 up to 46 days of age. The authors concluded that monensin and AFB_1 compete for adsorption sites on sodium bentonites, indicating a non-selective adsorption capacity of this particular binder. The researchers comment as well that different substances, such as coccidiostats, vitamins, minerals, aminoacids or other dietary components, could affect the ability of the adsorbent to bind low levels of aflatoxin. In addition, significant levels of AFB_1 in livers indicate that this determination is important not only for diagnosis of aflatoxicosis in broilers, but also for quality control of avian products [52].

b. *Organic sorbents:*

The most well-known natural toxin binders are yeast based products. Glucans are yeast cell wall constituents. Those compounds have been submitted to esterification process generating a new additive with toxin binding capacity, called esterified glucomannan (EGM). Efficacy of EGM was tested against mycotoxins naturally present in broiler feed [53], being 0.05% EGM efficient to counteract the adverse effects of mycotoxins (Table 9).

*Mycotoxin**	*EGM (%)*	*LW (g)*	*FI (g)*	*FCR (g g⁻¹)*
----	----	1,391.2 ᵇ	3,017.6 ᵇ	2.17 ᵇ
----	0.05	1,441.4 ᶜ	2,994.0 ᵇ	2.07 ᵃ
+++	----	1,258.8 ᵃ	2,803.4 ᵃ	2.22 ᶜ
+++	0.05	1,381.0 ᵇ	2,952.6 ᵇ	2.15 ᵇ
SEM		7.25	20.01	0.015

* Aflatoxin 168 ppb, ochratoxin 8.4 ppb, zearalenone 54 ppb and T2-Toxin 32 ppb [53].

Table 9. Efficacy of esterified glucomannans (EGM) on broiler live weight (LW), feed intake (FI) and feed conversion ratio (FCR) fed with a mycotoxin contaminated diet, from one up to 35 days.

c. *Other strategies:*

The use of mechanisms that improve animal health and physiology can be helpful. One example is the use of probiotics which have been used to ameliorate mycotoxicosis. Activated charcoal has been used to prevent animal intoxication by several compounds, including mycotoxins. Plant extracts with specific mode of action, like liver protection, have been used as well to reduce the toxicity of some mycotoxins, specially aflatoxin.

Milk thistle (*Silybum marianum*), which is a medicinal herb found in Pakistan, has been used to treat liver diseases. This herb was tested in poultry feed contaminated with AFB_1 (80 ppb for the first week and 520 ppb from the second until the fifth) at a dose of 1%. The results indicated that milk thistle is effective as hepatoprotectant and growth promoter in the presence of AFB_1 in the feed [54]. Protection against the negative effects of aflatoxin on performance of broiler chickens was observed when broilers were fed during 4 weeks with 0.05% ETE (ethanolic turmeric extract, *Curcuma longa*) plus 3 ppm aflatoxins [55].

Another report concluded that dietary citric acid supplementation can be used as an additive to degrade aflatoxins in the ration as well as to promote growth performance in young broiler chickens. Results showed that aflatoxins in the diet, at a concentration of 39 ppb were almost degraded (92%) by the acidification procedure (up to 50 g kg^{-1}) [56].

Other alternatives to degrade aflatoxins have been tested, like the use of microorganism. Bacteria (*Nocardia corynebacteroides*, NC) showed ability to degrade AFB_1 [57]. In a trial performed with broiler fed AFB_1 (800-1,200 ppb) NC was safe to the birds and showed protection to the animal, indicating that it can be used as a tool to detoxify feed contaminated with AFB_1 at high levels [58].

Humic acid, generated during matter decomposition, has binding capacity for many molecules. The use of oxihumate was evaluated as AFB_1 binder, *in vitro* and *in vivo* [35]. Oxihumate showed a high *in vitro* affinity for AFB_1. *In vivo* trial showed that oxihumate decreased adverse effects caused by AFB_1 on broiler body weight and also protective effect against liver damage, stomach and heart hyperplasia, acting positively preserving standard blood parameters. Enzyme degradation of aflatoxin has been tested as well. Data suggest that lactoperoxidase can be used to hydrolyze aflatoxin [59].

d. *Other aspects:*

There are many contradictory data available in the main scientific journals. One and probably the main reason for that is the way the trials have been performed, differing in terms of toxin levels, and environmental condition of the trials. Under a real field condition, the challenges the animals suffer are far stronger then under experimental situation. Toxin binder, especially clays, may affect the cation binding capacity of feeds and consequently influencing water intake and feed consumption. Also, the effects of none, medium (1 g kg^{-1} feed) and high (2.5 g kg^{-1} feed) inclusion levels of HSCAS was evaluated in broiler mycotoxin free diets [60]. The data suggest that increasing HSCAS to diets may modify performance, internal organ weights, gastrointestinal and biochemical parameters. However, other authors [61] did not see effect as the consequence of toxin binders (EGM) presence on broiler body weight and feed efficiency.

Detoxifying agents	Mycotoxin[3]								
	AFB_1	CPA	DAS	DON	Fusaric acid	NIV	OTA	T2 Toxin	ZEA
HSCAS	+	-					-	-	
Clinoptilolite	+/-	-							
Modified nanomontmorilonite	+								
Mg K aluminosilicate	+/-								
Sodium bentonite	+								
Ca montmorillonite	+								
Synthetic crystalline aluminosilicate	+/-								
Acidic phyllosilicate		-							
Zeolite	+		-			-		-	
Diatomaceous earth							+		
Charcoal							-		
Superactivated charcoal	+/-							+/-	
BHT[1]	+								
Cell wall *Saccharomyces cerevisiae*	+								
Yeast glucomannans	+								
Esterified glucomannans	+			+/-	+		+	-	+
Xylanase				-					
Live yeast culture residue	+								
Nocardia corynebacteroides	+/-								
Eubacterium			+	+				+	
Yeast *Trichosporon mycotoxinivorans*							+		-
Saccharomyces cerevisiae	+								
Ammonia	+								
Calcium propionate	+/-								
PVPP[2]	+/-			-					

[1]BHT = butylhydroxytoluene; [2]PVPP = Polyvinylpolypyrrolidone; [3]Mycotoxins AFB_1 = aflatoxin B_1, CPA = cyclopiazonic acid; DAS = diacetoxyscirpenol, DON = deoxynivalenol; NIV = nivalenol, OTA = ochratoxin A, ZEA = zearalenone.
Cells highlighted in dark gray indicate that the product has shown positive effects in counteracting deleterious effects of mycotoxins, while light gray color depicts that the product was not effective.
+: positive effect of the mycotoxin-detoxifying agent;
- : negative effect of the mycotoxin-detoxifying agent;
+/-: positive effect of the mycotoxin-detoxifying agent on some parameters, no effect on other parameters.

Table 10. Mycotoxin detoxifying agents tested *in vivo* in poultry. Adapted from ref. [62].

Additionally, different regions across the world have been dealing with mycotoxin subject in different ways. In US for instance, no toxin binders are officially registered, as a consequence of the control quality assumed for feedstuffs. In EU, since 2009, toxin sequestrants have been considered as a sort of feed additive, and a scientific group of specialists, namely European Food Safety Authority [62] have been working on re-evaluation of the efficacy and biological effects of detoxifying agents in animals. In Tables 10 and 11, a summary of the outcome of that technical group is presented.

Detoxifying agents	AFB_1	DON-NIV	Fumonisin	OTA	T2 Toxin	ZEA
HSCAS	+	-				
Montmorilonite		-				-
Sodium bentonite	+	-				
Calcium bentonite	+					
Zeolite	+					+
Sepiolite	+					
Palygorskite	+					
Ammonium carbonate		-				
Charcoal			+/-			
Polyvinylpolypyrrolidone		-				
Yeast glucomannans	+	+/-			-	+
Apple pommace		+				
Alfafa						+
Content of large intestine of hens		+				
Eubacterium		+				
Combination of Eubacterium BBSH 797 with dried yeast and clays		-				

Cells highlighted in dark gray indicate that the product has shown positive effects in counteracting deleterious effects of mycotoxins, while light gray color depicts that the product was not effective.
+: positive effect of the mycotoxin-detoxifying agent; -: negative effect of the mycotoxin-detoxifying agent;
+/-: positive effect of the mycotoxin-detoxifying agent on some parameters, no effect on other parameters.

Table 11. Mycotoxin detoxifying agents tested *in vivo* in pigs. Adapted from ref. [62].

In other regions, like South America, due to the climate and grain production conditions, mycotoxin has been a significant challenge along the past decades. This particular situation has been forcing the development of research groups which are involved with commercial sequestrants evaluation. As a consequence, the maximum acceptable aflatoxin limit has been established for different raw materials and feed, as well the specificity and inclusion levels of toxin binders in animal feed. However there are many different criteria for toxin adsorbents registration in different countries, some demanding an extend documentation about product efficacy and others are less restrictive.

7. Conclusions

There is a risk of contamination of meat products, eggs and milk with mycotoxins, although literature shows great variability due to the ingredients contamination which are included in feed. Ingredients should be analyzed before their entrance into the silo.

Mycotoxin contamination of ingredients used in feeds for broilers and laying hens is a reality in Brazil, since there are deficiencies in storage, handling and harvesting. Effects on animal production may vary depending on dose and time of administration and/or combination among mycotoxins which are not fully elucidated yet.

In general, mycotoxin residues tend to decrease rapidly after removal of the contaminated diet, which allows "to clean" broilers when feeding a diet free of mycotoxins few days before slaughter. This management option is not viable for laying hens and dairy cattle, since eggs and milk are generated continuously.

Apart from their toxicological effects in affected animals, the carry-over through animal derived products, such as eggs into the human food chains is an important aspect of aflatoxin contamination. Aflatoxins have a high impact in both, human and animal health, causing significant losses in the egg industry, considering the deleterious effect on egg production and quality. There is scarce literature in Brazil regarding egg contamination by aflatoxins.

After mycotoxin contamination of raw materials and feeds, effects can be minimized by using adsorbents which inhibit intestinal absorption of mycotoxins and can thereby prevent their deleterious effects in poultry production.

Author details

Vivian Feddern[1*], Giniani C. Dors[2], Fernando de C. Tavernari[1], Helenice Mazzuco[1], Anildo Cunha Jr.[1], Everton L. Krabbe[1] and Gerson N. Scheuermann[1]

*Address all correspondence to: vivian.feddern@embrapa.br

1 Embrapa Suínos e Aves, Concórdia, SC, Brazil

2 Instituto Federal Catarinense (IFC) – Campus Concórdia, Concórdia, SC, Brazil

References

[1] Ramos Girona, A. J., Sillué, S. M., & Almenar, V. S. (2011). Micotoxinas. Introducción histórica. Ramos Girona AJ (ed.). *Micotoxinas y micotoxicosis*, Madrid, AMV Ediciones, 1-18.

[2] Bakırdere, S., Bora, S., Bakırdere, E., Aydın, F., Arslan, Y., Komesli, O., et al. (2012). Aflatoxin species: their health effects and determination methods in different foodstuffs. *Central European Journal of Chemistry*, 10(3), 675-85.

[3] Atanda, S. A., Aina, J. A., Agoda, S. A., & U. OE, P. PO. (2012). Mycotoxin Management in Agriculture: a Review. *Journal of Animal Science Advances*, 2(Suppl. 3.1), 250-60.

[4] Moss, M. O. (1996). Mycotoxins. *Mycological Research*, 100(5), 513-23.

[5] European Mycotoxins Awareness Network. (2012). The Aflatoxins. Available from, http://services.leatherheadfood.com/eman/Factsheet.aspx?ID=6, accessed 08 oct. 2012.

[6] IARC Monographs on the Evaluation of Carcinogenic Risks to Humans. (2002).

[7] Kussak, A., Andersson, B., & Andersson, K. (1995). Determination of aflatoxins in airborne dust from feed factories by automated immunoaffinity column clean-up and liquid chromatography. *Journal of Chromatography A*, 708(1), 55-60.

[8] IARC Monographs on the Evaluation of Carcinogenic Risks to Humans. (2012). *Agents Classified by the IARC Monographs*, 1-105, Available from, http://monographs.iarc.fr/ENG/Classification/, accessed 28 June 2012.

[9] IARC Monographs on the Evaluation of Carcinogenic Risks to Humans. (2012). *A Review of Human Carcinogens: Chemical Agents and Related Occupations*, 100F, Available from, http://monographs.iarc.fr/ENG/Monographs/vol100F/mono100F-23.pdf, accessed 28 June 2012.

[10] IARC Monographs on the Evaluation of Carcinogenic Risks to Humans. (1993). Some Naturally Occurring Substances: Food Items and Constituents. *Heterocyclic Aromatic Amines and Mycotoxins*, 56.

[11] Papp, E., -Otta, K. H., Záray, G., & Mincsovics, E. (2002). Liquid chromatographic determination of aflatoxins. *Microchemical Journal*, 73(1-2), 39-46.

[12] Cassel, E. K. , Campbell, B., Draper, M., & Epperson, B. (2012). Aflatoxins hazards in grain/Aflatoxicosis and livestock. South Dakota State University (SDSU). Cooperative Extension Service / College of Agriculture & Biological Sciences / USDA. *Text adapted with permission from University of Maryland Cooperative Extension Service Fact Sheet* [444 & 445], FS 907: PDF by CES. October 2001, http://igrow.org/up/resources/02-2033-2012.pdf, accessed 08 oct. 2012.

[13] Gonçalez, E., Pinto, M. M., Manginelli, S., & Felicio, J. D. (2004). Dairy cows poisoned with cottonseed meal naturally contaminated with aflatoxins. *Ciência Rural*, 34(1), 171-4.

[14] Maia, P. P., Bastos, M. E. P., & de Siqueira, . (2007). Aflatoxins in pet foods- A Review. *Revista da FZVA*, 14(1), 235-257.

[15] Akande, K., Abubakar, M., Adegbola, T., & Bogoro, S. (2006). Nutritional and health implications of mycotoxins in animal feeds: A review. *Pakistan Journal of Nutrition*, 5(5), 398-403.

[16] Arias, S. L., Mary, V. S., Theumer, M. G., & Rubinstein, H. R. (2011). Micotoxicosis en animales de laboratorio. *In: Ramos Girona AJ (ed.). Micotoxinas y micotoxicosis*, Madrid, AMV Ediciones, 299-328.

[17] Bryden, W. L. (2012). Mycotoxin contamination of the feed supply chain: Implications for animal productivity and feed security. *Animal Feed Science and Technology*, 173(1-2), 134-58.

[18] Khlangwiset, P., Shephard, G. S., & Wu, F. (2011). Aflatoxins and growth impairment: A review. *Critical Reviews in Toxicology*, 41(9), 740-55.

[19] Wu, F., Narrod, C., Tiongco, M., & Liu, Y. (2011). The health economics of aflatoxin: global burden of disease. *Working Paper 4*, February Available from, http://www.ifpri.org/sites/default/files/publications/aflacontrol_wp04.pdf, accessed 04 May 2012.

[20] Cortés, G., Carvajal, M., Méndez-Ramírez, I., Ávila-González, E., Galván, C., Castillo-Urueta, P., et al. (2010). Identification and quantification of aflatoxins and aflatoxicol from poultry feed and their recovery in poultry litter. *Poultry Science*, 89, 993-1001.

[21] Bünzen, S., & Haese, D. (2006). Controle de micotoxinas na alimentação de aves e suínos. *Revista Eletrônica Nutritime*, 3(1), 299-304.

[22] Mallmann, C. A., Tyska, D., Mallmann, A. O., Mallmann, B. A., & Dilkin, P. (2008). Interação das micotoxinas e seus efeitos nos suínos. *In: Simpósio Brasil Sul de Suinocultura (1.: Chapecó, SC). Anais do I Simpósio Brasil Sul de Suinocultura*, Concórdia: Embrapa Suínos e Aves, 82-85.

[23] Oliveira, G., Ribeiro, J., Fraga, M., Cavaglieri, L., Direito, G., Keller, K., et al. (2006). Mycobiota in poultry feeds and natural occurrence of aflatoxins, fumonisins and zearalenone in the Rio de Janeiro State, Brazil. *Mycopathologia*, 162(5), 355-62.

[24] Alam, S., Shah, H., Khan, H., & Magan, N. (2012). The Effect of Substrate, Season, and Agroecological Zone on Mycoflora and Aflatoxin Contamination of Poultry Feed from Khyber Pakhtunkhwa, Pakistan. *Mycopathologia*, 174(4), 341-9.

[25] Andretta, I., Kipper, M., Lehnen, C. R., & Lovatto, P. A. (2012). Meta-analysis of the relationship of mycotoxins with biochemical and hematological parameters in broilers. *Poultry Science*, 91(2), 376-82.

[26] Bianco, G., Russo, R., Marzocco, S., Velotto, S., Autore, G., & Severino, L. (2012). Modulation of macrophage activity by aflatoxins B_1 and B_2 and their metabolites aflatoxins M_1 and M_2. *Toxicon,* 59(6), 644-50.

[27] FAO and IFIF. (2010). Good practices of the feed industry implementing the Codex Alimentarius Code of Practice in Good Animal Feeding. FAO Animal production and Health Manual n.9. Rome. Section 1. Health Hazards associated with animal feed.

[28] Biswas, A. K., Kondaiah, N., Anjaneyulu, A. S. R., & Mandal, P. K. (2010). Food Safety Concerns of Pesticides, Veterinary Drug Residues and Mycotoxins in Meat and Meat Products. *Asian Journal of Animal Sciences,* 4, 46-55.

[29] Wyatt, R. D., Briggs, D. M., & Hamilton, P. B. (1973). The Effect of Dietary Aflatoxin on Mature Broiler Breeder Males. *Poultry Science,* 52(3), 1119-23.

[30] Hamilton, P. B., & Gralich, J. D. (1971). Aflatoxin as possible cause of fatty liver syndrome in laying hens. *Poultry Science,* 50, 800-4.

[31] Huff, W. E., Wyatt, R. D., & Hamilton, P. B. (1975). Effects of dietary aflatoxin on certain egg yolk parameters. *Poultry Science,* 54, 2014-8.

[32] Zaghini, A., Martelli, G., Roncada, P., Simioli, M., & Rizzi, L. (2005). Mannanoligosaccharides and aflatoxin B_1 in feed for laying hens: effects on egg quality, aflatoxins B_1 and M_1 residues in eggs, and aflatoxin B_1 levels in liver. *Poultry Science,* 84(6), 825-32.

[33] Azzam, A. H., & Gabal, M. A. (1998). Aflatoxin and immunity in layer hens. *Avian Pathology,* 27, 570-7.

[34] Jand, S. K., Kaur, P., & Sharma, N. S. (2005). Mycoses and mycotoxicosis in poultry: A review. *Indian Journal of Animal Science,* 75, 465-76.

[35] Jansen Van Rensburg, C., Van Rensburg, C. E. J., Van Ryssen, J. B. J., Casey, N. H., & Rottinghaus, G. E. (2006). In vitro and in vivo assessment of humic acid as an aflatoxin binder in broiler chickens. *Poultry Science,* 85, 1576-83.

[36] Tessari, E. N., Oliveira, C. A., Cardoso, A. L., Ledoux, D. R., & Rottinghaus, G. E. (2006). Effects of aflatoxin B_1 and fumonisin B_1 on body weight, antibody titres and histology of broiler chicks. *British Poultry Science,* 47, 357-64.

[37] Glahn, R. P., Van Campen, D., & Dousa, T. P. (1994). Aflatoxin B_1 reduces Na(+)-P(i) co-transport in proximal renal epithelium: Studies in opossum kidney (OK) cells. *Toxicology,* 92, 91-100.

[38] Martinez-de-Anda, A., Valdivia, A. G., Jaramillo-Juarez, F., Reys, J. L., Ortiz, R., Quezada, T., et al. (2010). Effects of aflatoxin chronic intoxication in renal function of laying hens. *Poultry Science,* 89, 1622-8.

[39] Applegate, T. J., Schatzmayer, G., Pricket, K., Trache, C., & Jiang, Z. (2009). Effect of aflatoxin culture on intestinal function and nutrient loss in laying hens. *Poultry Science,* 88, 1235-41.

[40] Jacobson, W. C., & Wiseman, H. G. (1974). The transmission of aflatoxin B_1 into eggs. *Poultry Science*, 53, 1743-5.

[41] Bailly, J. D., & Guerre, P. (2009). Mycotoxins in meat and processed meat products. *In: Toldrá F (ed.). Safety of Meat and Processed Meat, Food Microbiology and Food Safety,* New York, Springer, 83-124, Available from, http://www.springerlink.com/content/m0562t7598299h5j/, accessed 16 April 2012.

[42] Hassan, Z. U., Khan, M. Z., Khan, A., Javed, I., & Hussain, Z. (2012). Effects of individual and combined administration of ochratoxin A and aflatoxin B_1 in tissues and eggs of White Leghorn breeder hens. *Journal of the Science of Food and Agriculture*, 92, 1540-4.

[43] Hussain, Z., Khan, M. Z., Khan, A., Javed, I., Saleemi, M. K., Mahmood, S., & Asi, M. R. (2010). Residues of aflatoxin B_1 in broiler meat: Effect of age and dietary aflatoxin B_1 levels. *Food and Chemical Toxicology*, 48(12), 3304-7.

[44] Kalpana, S., Aggarwal, M., Srinivasa Rao, G., & Malik, J. K. (2012). Effects of aflatoxin B_1 on tissue residues of enrofloxacin and its metabolite ciprofloxacin in broiler chickens. *Environmental Toxicology and Pharmacology*, 33, 121-6.

[45] Rodriguez-Amaya, D. B., & Sabino, M. (2002). Mycotoxin research in Brazil: the last decade in review. *Brazilian Journal of Microbiology*, 33, 1-11.

[46] Trucksess, M. W., Stoloff, L., & Young, K. (1983). Aflatoxicol and aflatoxins B_1 and M_1 in egg and tissues of laying hens consuming aflatoxin-contaminated feed. *Poultry Science*, 62, 2176-82.

[47] Resolution RDC Nº 7. (2011). Technical Regulation on maximum tolerated levels for mycotoxins in food. from February 18, http://www.brasilsus.com.br/legislacoes/anvisa/107378-7.html , accessed 28 June 2012.

[48] EUR LEX. (2012). http://eur-lex.europa.eu/pt/index.htm, accessed 25 June 2012.

[49] Thieu, N. Q., & Pettersson, H. (2008). In vitro evaluation of the capacity of zeolite and bentonite to adsorb aflatoxin B_1 in simulated gastrointestinal fluids. *Mycotoxin Research*, 24, 124-9.

[50] Batina, P. N., Lopes, S. A., Santurio, J. M., de Souza, C., & Martins, D. B. (2005). The effects of the addition of sodic montmorilonite on the feeding diet on the biochemical profile of broiler chicken intoxicated by aflatoxin. *Ciência Rural*, 35(4), 826-31.

[51] Diaz, D. E., Hagler, W. M., Jr Hopkins, B. A., & Whitlow, L. W. (2003). Aflatoxin Binders I: In vitro binding assay for aflatoxin B_1 by several potential sequestering agents. *Mycopathologia*, 156(3), 223-6.

[52] Magnoli, C. E., Astoreca, A. L., da Rocha Rosa, C. A., & Dalcero, A. M. (2011). Micotoxicosis en aves. *Ramos Girona AJ (ed.). Micotoxinas y micotoxicosis*, Madrid, AMV Ediciones, 221-239.

[53] Aravind, K. L., Patil, V. S., Devegowda, G., Umakantha, B., & Ganpule, S. P. (2003). Efficacy of esterified glucomannan to counteract mycotoxicosis in naturally contaminated feed on performance and serum biochemical and hematological parameters in broilers. *Poultry Science*, 82(4), 571-6.

[54] Muhammad, D., Chand, N., Khan, S., Sultan, A., Mushtaq, M., & Rafiullah, . (2012). Hepatoprotective role of milk thistle (*Silybum marianum*) in meat type chicken fed aflatoxin B_1 contaminated feed. *Pakistan Veterinary Journal*, 32(3), 443-6.

[55] Rangsaz, N., & Ahangaran, M. G. (2011). Evaluation of turmeric extract on performance indices impressed by induced aflatoxicosis in broiler chickens. *Toxicology and Industrial Health*, 27(10), 956-60.

[56] Salgado-Tránsito, L., Arjona-Román, J. L., Moreno-Martínez, E., & Méndez-Albores, A. (2011). Effect of citric acid supplemented diets on aflatoxin degradation, growth performance and serum parameters in broiler chickens. *Archivos de Medicina Veterinaria*, 43(3), 215-22.

[57] Ciegler, A., Lillehoj, E. B., Peterson, R. E., & Hall, H. L. (1966). Microbial detoxification of mycotoxins. *Applied Microbiology*, 14, 934-9.

[58] Tejada-Castañeda, Z. I., Avila-Gonzalez, E., Casaubon-Huguenin, M. T., Cervantes-Olivares, R. A., Vásquez-Peláez, C., Hernández-Baumgarten, E. M., et al. (2008). Biodetoxification of aflatoxin-contaminated chick feed. *Poultry Science*, 87(8), 1569-76.

[59] Doyle, M. P., & Marth, E. H. (1978). Degradation of aflatoxin by lactoperoxidase. *Zeitschrift für Lebensmitteluntersuchung und Forschung A*, 166(5), 271-3.

[60] Basalan, M., Gungor, T., Aydogan, I., Hismiogullari, S. E., Erat, S., & Erdem, E. (2006). Effects of Feeding Mycotoxin Binder (HSCAS) on Gastrointestinal Environment and Metabolism in Broilers at Later Ages. *Archiva Zootechnica*, 9, 5-9.

[61] Girish, C. K., & Devegowda, G. (2004, 8-10 September). Efficacy of modified glucomannan (Mycosorb) and HSCAS to alleviate the individual and combined toxicity of aflatoxin and T-2 toxin in broiler chickens. Istanbul, Tukey. *XXII'nd World Poultry Congress*.

[62] EFSA, European Food Safety Authority (2009). Review of mycotoxin-detoxifying agents used as feed additives: mode of action, efficacy and feed/food safety: Scientific report submitted to EFSA by AFSSA, CODA-CERVA, INRA Clermont-Ferrand, INRA Toulouse, IRTA, ISPA; 192p. Grant number: CFP/EFSA/FEEDAP/2009/01. Available from http://www.efsa.europa.eu/en/scdocs/doc/22e.pdf

Chapter 9

Occurrence of Aflatoxin M_1 in Raw and Pasteurized Goat Milk in Thailand

Suthep Ruangwises, Piyawat Saipan and
Nongluck Ruangwises

Additional information is available at the end of the chapter

1. Introduction

Aflatoxins are a group of structurally related mycotoxins produced by certain species of the genus *Aspergillus,* particularly *A. flavus, A. parasiticus* and *A. nomius,* which can grow on a variety of food and feed commodities [1]. Aflatoxin production is influenced by several factors: for example, temperature and humidity [2]. It has been shown that aflatoxin B_1 (AFB_1) is the most potent hepatocarcinogen of this group of mycotoxins. Aflatoxin M_1 (AFM_1) is a hydroxylated metabolite of AFB_1 produced by the hepatic microsomal cytochrome P450, and is secreted in the milk of mammals that have consumed AFB_1-contaminated foods. AFM_1 is also a hepatocarcinogen and is classified in Group 1 as carcinogenic to humans by the International Agency for Research on Cancer [3]. In terms of food safety and public health concerns, exposure to AFM_1 through milk products is considered to be a serious problem.

According to worldwide regulations for mycotoxins in food and feed compiled by the Food and Agriculture Organization of the United Nations, 60 countries have already established regulatory limits for AFM_1 in raw milk and milk products. The report also indicates that the limits vary from ND (not detectable) to 15 μg/L [4]. The values of 0.05 μg/L and 0.5 μg/L are the two most prevalent regulatory limits for AFM_1 in milk products, enforced in 34 and 22 countries, respectively. The maximum permitted level for AFM_1 established by the European Community is 0.025 μg/kg for infant formulae and follow-on formulae, including infant milk and follow-on milk, while the limit for raw milk and heat-treated milk is 0.05 μg/kg [5]. The U.S. regulatory standard for AFM_1 is 0.5 μg/L [4]. There are still several countries, including Thailand, that have not yet established regulatory limits for AFM_1 in dairy products.

The law that regulates the quality of milk products in Thailand is the Notification of the Ministry of Public Health No. 265, which regulates only cow milk products. However, the law does not specify the regulatory standards for AFM_1 but states that "*...milk products may be contaminated with aflatoxins at a level that is not harmful to human health*" [6]. The only guideline that regulates the quality of raw goat milk is the Thai Agricultural Standard TAS 6006-2008 of the National Bureau of Agricultural Commodity and Food Standards, Ministry of Agriculture and Cooperatives [7]. Like Notification No. 265 for cow milk products, the TAS 6006-2008 guideline does not specify the recommended limit for AFM_1 in goat milk.

In Thailand, the number of dairy goats is approximately 5% that of dairy cows [8–10]. Goat milk is consumed by only a small percentage of the country's population, particularly Thai people who have an allergy to cow milk. Goat milk has been shown to form finer and softer curds than cow milk following acidification under conditions similar to those in the stomach, thus making it more readily digested [11]. It has been reported that micellar caseins of human and goat milk were 96% hydrolyzed by pepsin and trypsin in *in vitro* studies, while the hydrolytic rate of cow milk was 76–90% [12]. With the knowledge that goat milk is more easily digested, some Thai adults prefer goat milk products. As a result, the number of dairy goats in Thailand has been gradually increasing in recent years. In 2009, the number of dairy goats in Thailand was 20,830; the numbers increased to 22,630 and 33,363 in 2010 and 2011, respectively [8–10].

Thailand is administratively divided into four regions: central, north, northeast and south. The central region was selected for this study, since this region has the highest number of dairy goats and the highest rate of goat milk production, accounting for approximately 60% of the national total [8–10]. There are no internationally published reports regarding the quality and levels of AFM_1 in goat milk produced in Thailand.

The purpose of this study was to investigate whether the concentrations of AFM_1 in raw and pasteurized goat milk produced in Thailand are within the acceptable level for consumption.

2. Materials and methods

2.1. Chemicals

AFM_1 reference standard (from *Aspergillus flavus*) was purchased from Sigma-Aldrich (St. Louis MO, USA). $AflaM_1^{TM}$ immunoaffinity columns were obtained from Vicam (Nixa MO, USA). Solvents (HPLC grade) – acetonitrile, methanol, and water – were purchased from Merck (Darmstadt, Germany).

2.2. Milk sample collection and sample preparation

Raw goat milk samples were collected from private farms, while pasteurized goat milk samples were purchased from supermarkets in the central region of Thailand. In Thailand, commercial pasteurized milk is produced by heat treatment, either at 63 °C for 30 min or at 72 °C for at least 15 s [6]. All milk samples were collected over three years: January–February of the years 2009–2011. Both types of milk samples were frozen at –20 °C until analysis (within one month from the collection date for raw milk, or 2 months from the manufacturing date for pasteurized milk). A total of 90 milk samples were collected and analyzed in this study.

2.3. Extraction and determination of aflatoxin M_1

The extraction procedure was performed using the manufacturer's recommendations, as previously described by Ruangwises et al. [13]. Briefly, 50 ml of raw milk or pasteurized milk sample was pipetted into a 50-ml plastic centrifuge tube. Milk samples were defatted by centrifugation at 3,500 *g* for 20 min at 4°C. Fat was separated; the resulting skimmed milk was then transferred into a 50-ml plastic syringe with a Luer tip which was attached to an immunoaffinity column. The skimmed milk was allowed to flow into the column by gravity at a flow rate of approximately 1 ml/min. After the skimmed milk had run through, 20 ml of HPLC water was used to wash the column. AFM_1 was eluted from the column with 1.25 ml of acetonitrile:methanol (3:2) and 1.25 ml of HPLC water. The eluate (a total volume of 2.5 ml) was filtered through a nylon syringe filter for HPLC with pore size 0.45 μm (Whatman, UK). AFM_1 in the final solution was determined using HPLC. Each milk sample was extracted and analyzed for AFM_1 in duplicate.

2.4. Instrument

A complete liquid chromatographic system (ProStar; Varian, Palo Alto CA, USA) consisted of a HPLC pump (model 240), an auto injector (model 410), a column oven (model 510), and a fluorescence detector (model 363). The HPLC conditions for analysis of AFM_1 were as follows: column, Spherisorb ODS-2 (Waters, Milford MA, USA); column temperature, 40 °C; mobile phase, water:methanol:acetonitrile (57:23:20); flow rate, 1 ml/min; and detector, fluorescence spectrophotometer (excitation 360 nm; emission 440 nm).

2.5. Determination of limit of quantification

The Q2B procedure of the U.S. Food and Drug Administration [14] was used for determination of the limit of quantification (LOQ) for AFM_1. Milk samples (50 ml) were fortified with standard AFM_1 at four concentrations of 0.025, 0.050, 0.125 and 0.250 μg/L, while blank samples were not fortified with standard AFM_1. Concentrations of AFM_1 in AFM_1-fortified milk samples and blank samples were quantified as described above in Section 2.3 using $AflaM_1^{TM}$ immunoaffinity columns. All samples were analyzed for AFM_1 in duplicate.

Individual linear regression lines were obtained from least-square regression analyses of the residual peak areas versus the four concentrations of fortified AFM_1 (0.025, 0.050, 0.125 and 0.250 μg/ml). The residual peak areas were peak areas of AFM_1-fortified samples minus the peak area of blank sample. A total of 12 regression lines (six regression lines each for intraday and interday analyses) were obtained by least-square linear regression. The LOQ of the method was calculated using the equation $LOQ = 10\ \sigma/S$, where σ is the standard deviation of *y*-intercepts and S is the average slope of the 12 linear regression analyses [14].

2.6. Statistical analysis

A randomized block experiment was used to evaluate the differences in AFM_1 concentrations in the two types of milk samples and among the three collection years. Duncan's multiple comparison test was applied to obtain significance levels between the raw milk and pasteurized milk, and among each year of individual milk products ($P < 0.05$). SPSS Statistics version 17.0 for Windows was used for statistical analysis.

3. Results and discussion

Table 1 shows the results of analysis and a regression line obtained from least-square analysis of Sample A, of which the slope and y-intercept were used for the calculation of LOQ. Twelve regression lines (six lines each for intraday and interday analyses) were performed in this study; slopes and y-intercepts of all 12 analyses are presented in Table 2. The calculation for LOQ was based on the equation LOQ = 10 σ/S, where σ and S are the standard deviation of y-intercepts and the average slope of the 12 regression lines, respectively. In this study, the standard deviation of y-intercepts was 173.69 mV × L/μg and the average slope was 180,518 mV. The calculated LOQ was (10 * 173.69)/180,518 = 0.01 μg/L. The accuracy of the method, expressed as % recovery, ranged from 88.8% to 94.1%, with an average value of 90.8%. The precision of the method, expressed as %RSD (percent relative standard deviation), ranged from 1.1% to 7.5%. Table 3 summarizes the accuracy and precision of determination of AFM_1 in goat milk samples fortified with AFM_1 at four concentrations, with intraday and interday analyses. HPLC chromatograms of standard AFM_1 (10 μg/L), a goat milk sample contaminated with AFM_1 (0.05 μg/L), and an uncontaminated goat milk sample are presented in Figure 1. The retention time for AFM_1 under the conditions in this study was approximately 6.8 min.

Table 4 shows the incidence and concentrations of AFM_1 in raw and pasteurized goat milk samples. The incidence of AFM_1 in raw goat milk collected in 2009, 2010 and 2011 was 46.7% (7/15), 66.7% (10/15) and 60.0% (9/15), respectively, while the incidence in pasteurized milk was 53.3% (8/15), 46.7% (7/15) and 53.3% (8/15), respectively. The total incidence of positive samples with respect to 90 samples analyzed in this study was 54.4% (49/90). Of the 49 positive samples, only 7 samples (14.3%) were contaminated with AFM_1 above the EU standard of 0.05 μg/L. The three-year average concentrations of AFM_1 found in the raw and pasteurized milk samples were 0.043 and 0.040 μg/L, respectively. The maximum concentration found in this study was 0.086 μg/L, which was far below the U.S. regulatory limit of 0.5 μg/L. In this study, statistical analysis showed that there were no significant differences in AFM_1 concentrations among the raw and pasteurized milk samples and across the two types of milk samples collected over a three-year period.

When compared to cow milk, goat milk has a lower percentage of positive samples and lower AFM_1 concentrations. Ghanem and Orfi [15] reported that the average concentration of AFM_1 in raw goat milk (0.019 μg/L, n = 11), collected from markets in Syria between April 2005 and April 2006, was less than that in raw cow milk (0.143 μg/L, n = 74); the percentage of positive samples of goat milk (7 samples, 63.6%) was also less than that of cow milk (70 samples, 94.6%). Hussain et al. [16] found that 6 (20%) of 30 raw goat milk samples were contaminated with AFM_1 at an average concentration of 0.002 μg/L, while 15 (37.5%) of 40 raw cow milk samples were contaminated with an average AFM_1 level of 0.014 μg/L. Rahimi et al. [17] reported that the incidence of AFM_1 in raw goat and cow milk samples collected from Ahvaz in Khuzestan province, Iran, between November 2007 and December 2008, was 31.7% (19/60) and 78.7% (59/75), respectively. Concentrations of AFM_1 in raw milk samples of both species were 0.0301 and 0.0601 μg/L, respectively.

AFM_1 added (μg/L)	Peak area[1] (mV)	Residual peak area[2] (mV)
0	6,410	-
0.025	11,126.5	4,716.5
0.050	16,144.5	9,734.5
0.125	29,251	22,841
0.250	52,773	46,363

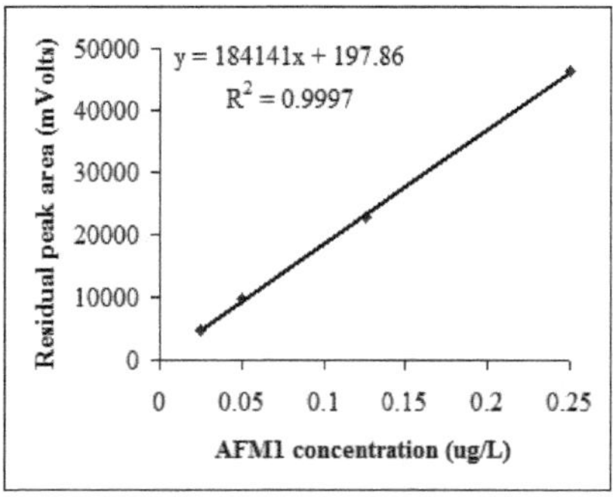

slope = 184,141; *y*-intercept = 197.86

[1] Average value of two determinations

[2] Residual peak area = peak area of AFM_1-fortified sample – peak area of blank sample

Table 1. Linear regression analysis of AFM_1-fortified sample A for the determination of LOQ

Sample	Slope (mV × L/µg)	*y*-intercept (mV)
Intraday (*n* = 6)		
A	184,141	197.86
B	180,733	293.38
C	183,706	141.26
D	179,857	549.02
E	180,039	207.84
F	181,224	109.74
Interday (*n* = 6)		
G	181,454	127.39
H	175,861	432.76
I	185,285	223.45
J	179,462	442.02
K	175,904	339.74
L	178,545	639.60
Overall (n = 12)		
Mean	**180,518 (S)**	**308.67**
SD	**2,955.5**	**173.69 (σ)**

Table 2. Slopes and *y*-intercepts of 12 regression lines used for determination of LOQ for AFM_1

AFM_1 added (µg/L)	**Intraday (*n* = 6)**			**Interday (*n* = 6)**		
	Found[a] (µg/L)	**%RSD[b]**	**Recovery (%)**	**Found[a] (µg/L)**	**%RSD[b]**	**Recovery (%)**
0.025	0.023 ± 0.001	4.3	92.1	0.024 ± 0.002	7.5	94.1
0.050	0.046 ± 0.001	2.2	91.9	0.046 ± 0.002	3.5	91.4
0.125	0.112 ± 0.003	2.7	89.3	0.111 ± 0.004	3.9	88.8
0.250	0.225 ± 0.002	1.1	89.8	0.222 ± 0.005	2.1	89.0

[a] Values are mean ± SD

[b] % RSD = percent relative standard deviation.

Table 3. Accuracy and precision of determination of AFM_1 in goat milk

Year	Samples analyzed	Positive[1] (%)	AFM_1 concentration (ng/ml)[2]		AFM_1 incidence[3]	
			Mean	Range	0.010–0.050 µg/L	> 0.05 µg/L
Raw milk						
2009	15	7 (46.7)	0.042 ± 0.012	0.022–0.061	6 (85.7)	1 (12.5)
2010	15	10 (66.7)	0.049 ± 0.018	0.025–0.086	8 (80.0)	2 (20.0)
2011	15	9 (60.0)	0.036 ± 0.015	0.018–0.066	8 (88.9)	1 (11.1)
Total	45	26 (57.8)	0.043 ± 0.017	0.018–0.086	22 (84.6)	4 (15.4)
Pasteurized milk						
2009	15	8 (53.3)	0.039 ± 0.017	0.015–0.075	7 (87.5)	1 (12.5)
2010	15	7 (46.7)	0.045 ± 0.015	0.022–0.061	6 (85.7)	1 (14.3)
2011	15	8 (53.3)	0.035 ± 0.019	0.014–0.073	7 (87.5)	1 (12.5)
Total	45	23 (51.1)	0.040 ± 0.016	0.014–0.073	20 (87.0)	3 (13.0)
Overall	**90**	**49 (54.4)**	**0.041 ± 0.016**	**0.014–0.086**	**42 (85.7)**	**7 (14.3)**

[1]Numbers in parentheses are percentages for each year

[2]Means and ranges of AFM_1 concentrations in the positive samples

[3]AFM_1 incidence of the positive samples

Numbers in parentheses are percentages with respect to the positive samples

Table 4. Incidence and concentrations of AFM_1 in raw and pasteurized goat milk samples collected within the central region of Thailand

High incidence and concentrations of AFM_1 in cow milk have also been found in Thailand. Ruangwises and Ruangwises [18] reported that all of 240 raw cow milk samples collected from 80 milk tanks at a milk collecting center in the central region of Thailand were found to be contaminated with AFM_1 at an average concentration of 0.070 µg/L. For pasteurized milk samples, our previous studies showed that AFM_1 was found in 349 (83.1%) of 420 pasteurized milk samples, collected from 40 provinces in all four regions of Thailand from May 2006 to January 2008, with AFM_1 concentrations ranging between 0.012 and 0.114 µg/L [13,19].

Table 5 shows the incidence and concentrations of AFM_1 in raw and pasteurized goat milk from various countries. For raw goat milk, Assem et al. [20] found that all of the three raw milk samples collected from markets in Lebanon between March–July 2010 contained AFM_1 less than the LOQ of 0.005 ng/ml. Ozdemir [21] found that the mean concentration of AFM_1 in 93 positive samples out of 110 raw milk samples collected from the city of Kilis, Turkey, from March–April 2006 was 0.019 µg/L. For pasteurized milk, Oliveira and Ferraz [22] determined the concentrations of AFM_1 in 12 pasteurized goat milk samples collected from the state of Sao Paulo, Brazil, and found that 7 samples (58.3%) were contaminated with an average concentration of 0.034 µg/L.

The levels of AFM_1 in goat milk are influenced by both feeding practices and the types of feedstuffs. Virdis et al. [23] determined the concentrations of AFM_1 in goat milk collected from two groups of farms with different feeding practices – extensive and intensive farms – in Sardinia, Italy, between the years 2003 and 2004. In extensive farms, goats were principally fed on grass and naturally growing bushes which were often present in marginal areas, supplemented with low levels of concentrates consisting of broad bean (*Vicia faba*) and garden pea (*Pisum sativum*). In intensive farms, goats were mainly fed silo maize, maize grains, and alfalfa (*Medicago sativa*). The incidence of AFM_1 in goat milk samples from extensive and intensive farms was 11.2% (9/80) and 71.4% (20/28), respectively. Concentrations of AFM_1 found in positive samples from both farms were 0.009 and 0.0177 ng/ml, respectively.

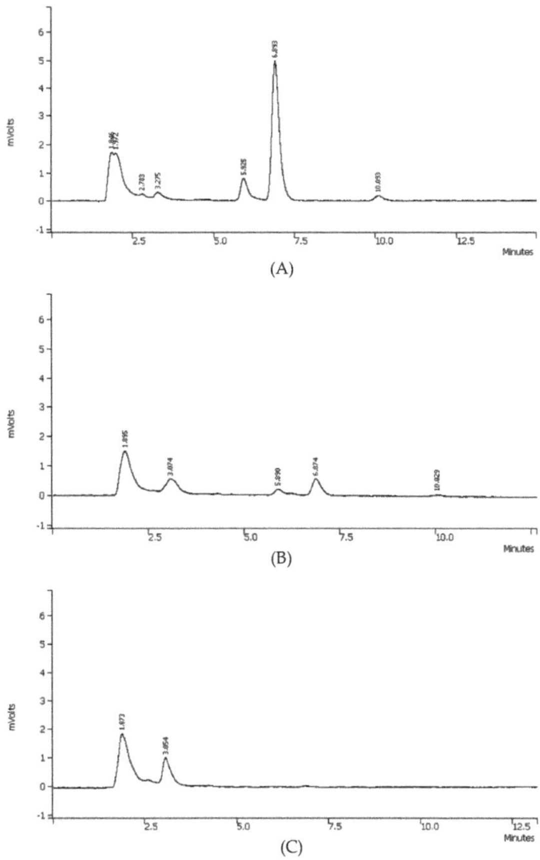

Figure 1. HPLC chromatograms of AFM_1 with a retention time of approximately 6.8 min: (A) standard 10 μg/L AFM_1, (B) goat sample contaminated with 0.05 μg/L AFM_1, and (C) uncontaminated goat milk sample

Country	Year	Samples analyzed	Positive (%)	Concentration (µg/L)[1]	Reference
Raw milk					
Lebanon	Mar–July 2010	3	0 (0)	< 0.005	Assem et al. (2011)
Iran	Nov 2007 – Dec 2008	60	19 (31.7)	0.0301 ± 0.0183	Rahimi et al. (2010)
Pakistan	Jan–Dec 2007	30	6 (20)	0.002 ± 0.005	Hussain et al. (2010)
Turkey	Mar–Apr 2006	110	93 (84.5)	0.019 (0.005–0.117)[2]	Ozdemir (2007)
Syria	Apr 2005 – Apr 2006	11	7 (63.6)	0.019 ± 0.0138 (0.008–0.054)	Ghanem and Orfi (2009)
Thailand	Jan 2008 – Feb 2011	45	26 (57.8)	0.036 ± 0.015 (0.011–0.064)	Present study
Pasteurized milk					
Brazil	Oct 2004 – May 2005	12	7 (58.3)	0.072 ± 0.048	Oliveira and Ferraz (2007)
Thailand	Jan 2008 – Feb 2011	45	23 (51.1)	0.034 ± 0.014 (0.010–0.058)	Present study

[1] Concentrations of AFM_1 in positive samples

[2] Values in parentheses are ranges

Table 5. Incidence and concentrations of AFM_1 in raw and pasteurized goat milk in various countries

The observation that the incidence and concentrations of AFM_1 in goat milk are relatively lower than those in cow milk can be explained in terms of the feeding procedure and the carry-over rate of AFB_1 in feedstuffs to AFM_1 in the milk. Cows are generally fed with several major AFB_1-contaminated feedstuffs: corn, cotton seed, and concentrated feed. Unlike cows, goats are fed with fresh grass but not corn or cotton seed; the main AFB_1-contaminated feedstuffs fed to goats are concentrate feedstuffs. Motawee et al. [24] explained the different feeding patterns of cows and goats in Egypt. Cows are generally kept in enclosed areas and fed with a large proportion of AFB_1-contaminated feedstuffs, with a short period of time for grazing on pasture; while goats are allowed to graze on pasture in the morning and are brought back into the enclosed areas for concentrate feedstuffs in the evening. Hussain et al. [16] explained that goats in Pakistan are mainly fed by grazing on pasture. AFB_1-contaminated feedstuffs – corn, cotton seed, and concentrate feed – are not used to feed goats. In Thailand, the feeding procedures for cows and goats are similar to those in Egypt and Pakistan [25].

The carry-over rate of AFB_1 in feedstuffs to AFM_1 in milk is relatively lower in goats than in cows. The carry-over rates in cows have been reported to vary from 0.3% to 6.2%, with a mean value of 1.81% (n = 42) [26]. In Thailand, Ruangwises and Mhosatanun [27] determined the carry-over rates during the early lactation period (the first 4 weeks of lactation) in nine cows fed with feedstuffs naturally contaminated with AFB_1. The carry-over rates ranged between 1.96% and 3.12%, with an average value of 2.02%. For goats, Smith et al. [28] reported an average carry-over rate of 0.55% in three goats which were fed with feedstuffs containing 100 ppb AFB_1. Mazzette et al. [29] found an average carry-over rate of 0.26% in three goats within 72 h after receiving a single oral dose of 0.8 mg of AFB_1.

This study showed that 49 samples (54.4%) of the 90 goat milk samples collected within the central region of Thailand in January–February of the years 2009–2011 were contaminated with AFM_1 equal to or more than the LOQ of 0.01 μg/L. Concentrations of AFM_1 were not significantly different among the raw and pasteurized milk samples and across the two types of milk samples collected over three years. Of the 49 positive samples, 7 samples (14.3%) had AFM_1 greater than the EU regulatory limit of 0.05 μg/L. All 90 goat milk samples contained AFM_1 below the U.S. regulatory limit of 0.5 μg/L. This study presents the first internationally published report on the contamination of AFM_1 in raw and pasteurized goat milk produced in Thailand. The present study and our three previous reports on the occurrence of AFM_1 in cow milk products [13,18,19] suggest that regulatory standards be adopted for AFM_1 to ensure the quality of raw milk and milk products in Thailand.

Acknowledgments

This study was financially supported in part by The 40th Anniversary of Khon Kaen University Fund. The authors thank Mrs. Chailai Kuwattananukul and Sunan Rangseekansong for their technical assistance, and Mr. Christopher Salisbury, Chiang Mai University, Thailand, for reviewing the manuscript.

Author details

Suthep Ruangwises[1*], Piyawat Saipan[2] and Nongluck Ruangwises[3]

*Address all correspondence to: suthep.r@chula.ac.th

1 Department of Veterinary Public Health, Faculty of Veterinary Science, Chulalongkorn University, Bangkok, Thailand

2 Department of Veterinary Public Health, Faculty of Veterinary Medicine, Khon Kaen University, Khon Kaen, Thailand

3 Department of Pharmaceutical Chemistry, Faculty of Pharmacy, Mahidol University, Bangkok, Thailand

References

[1] JECFA (Joint FAO/WHO Expert Committee on Food Additives). (2001). Safety Evaluation of Certain Mycotoxins in Food: Aflatoxin M_1. WHO Food Additives Series. (47), Geneva, Switzerland: World Health Organization.

[2] Jay, J. M., & Loessner, Golden. D. A. (2005). Modern Food Microbiology (7th edition). New York: Springer Science and Business Media.

[3] IARC (International Agency for Research on Cancer). (2002). IARC Monographs on the Evaluation of Carcinogenic Risks to Humans,. 82, Some Traditional Herbal Medicines, Some Mycotoxins, Naphthalene and Styrene. Lyon, France: World Health Organization (WHO), IARC Press.

[4] FAO (Food and Agriculture Organization of the United Nations). (2004). Worldwide Regulations for Mycotoxins in Food and Feed in 2003. *Food and Nutrition Paper,* 81. Rome: FAO.

[5] European Commission. Commission Regulation (EC) No. (2006). Setting Maximum Levels for Certain Contaminants in Foodstuffs. *Official Journal of the European Union,* 364-5, L 364/24.

[6] Ministry of Public Health. (2003). Notification of the Ministry of Public Health. B.E. 2545): Cow's Milk., *The Royal Thai Government Gazette* [265], , 120 Special Issue 4D (January 2003) (in Thai).

[7] Ministry of Agriculture and Cooperatives. (2008). Thai Agricultural Standard (TAS) 606-2008 Raw Goat Milk. The Royal Thai Government Gazette, 139(139D), August 2008) (in Thai).

[8] DLD (Department of Livestock Development). (2009). Livestock Infrastructure Information in Thailand:. Bangkok: Department of Livestock Development, Ministry of Agriculture and Cooperatives; 2009 (in Thai).

[9] DLD. (2010). Livestock Infrastructure Information in Thailand:. Bangkok: Department of Livestock Development, Ministry of Agriculture and Cooperatives; 2010 (in Thai).

[10] DLD. (2011). Livestock Infrastructure Information in Thailand:. Bangkok: Department of Livestock Development, Ministry of Agriculture and Cooperatives; 2011 (in Thai).

[11] Park, Y. W. (2007). Rheological Characteristics of Goat and Sheep Milk. *Small Ruminant Research,* 68, 73-87.

[12] Jasinska, B. (1995). The Comparison of Pepsin and Trypsin Action on Goat, Cow, Mare and Human Caseins. *Roczniki Akademii Medycznej w Białymstoku,* 40, 486-493.

[13] Ruangwises, N., Saipan, P., & Ruangwises, S. (2011). Estimated Daily Intake of Aflatoxin M_1 in Thailand. In: Guevara-González RG (ed.). Aflatoxins- Biochemistry and

Molecular Biology. Rijeka: InTech; Available from:, http://www.intechopen.com/articles/show/title/estimated-daily-intake-of-aflatoxin-m1 -in-thailand.

[14] U.S. FDA (United States Food and Drug Administration). (1996). Guidance for Industry- Q2B Validation of Analytical Procedures: Methodology. Rockville, MD: Department of Health and Human Services, Center for Drug Evaluation and Research;, Available from:, http://www.fda.gov/downloads/Drugs/GuidanceComplianceRegulatoryInformation/Guidances/ucm073384.pdf.

[15] Ghanem, I., & Orfi, M. (2009). Aflatoxin M1 in Raw, Pasteurized and Powdered Milk Available in the Syrian Market. *Food Control*, 20, 603-605.

[16] Hussain, I., Anwar, J., Asi, M. R., & Munawar, Kashif. M. (2010). Aflatoxin M_1 Contamination in Milk from Five Dairy Species in Pakistan. *Food Control*, 21, 122-124.

[17] Rahimi, E., Bonyadian, M., Rafei, M., & Kazemeini, H. R. (2010). Occurrence of Aflatoxin M_1 in Raw Milk of Five Dairy Species in Ahvaz, Iran. *Food Chemistry and Toxicology*, 48, 129-131.

[18] Ruangwises, N., & Ruangwises, S. (2010). Aflatoxin M_1 Contamination in Raw Milk within the Central Region of Thailand. *Bulletin of Environmental Contamination and Toxicology*, 85, 195-198.

[19] Ruangwises, S., & Ruangwises, N. (2009). Occurrence of Aflatoxin M_1 in Pasteurized Milk of the School Milk Project in Thailand. *Journal of Food Protection*, 72, 1761-1763.

[20] Assem, E., Mohamad, A., & Oula, E. A. (2011). A Survey on the Occurrence of Aflatoxin M1 in Raw and Processed Milk Samples Marketed in Lebanon. *Food Control*, 22, 1856-1858.

[21] Ozdemir, M. (2007). Determination of Aflatoxin M_1 Levels in Goat Milk Consumed in Kilis Province. *Ankara Universitesi Veteriner Fakultesi Dergisi*, 54, 99-103.

[22] Oliveira, C. A. F., & Ferraz, J. C. O. (2007). Occurrence of Aflatoxin M_1 in Pasteurized, UHT Milk and Milk Powder from Goat Origin. *Food Control*, 18, 375-378.

[23] Virdis, S., Corgiolu, G., Scarano, C., Pilo, A. L., & De Santis, E. P. L. (2008). Occurrence of Aflatoxin M_1 in Tank Bulk Goat Milk and Ripened Goat Cheese. *Food Control*, 19, 44-49.

[24] Motawee, Bauer. J., & Mc Mahon, D. J. (2009). Survey of Aflatoxin M_1 in Cow, Goat, Buffalo and Camel Milks in Ismailia-Egypt. *Bulletin of Environmental Contamination and Toxicology*, 83, 766-769.

[25] DLD. (2010). Transition to Organic Goat Milk Production System. Bangkok: Department of Livestock Development, Ministry of Agriculture and Cooperatives.

[26] Pettersson, H. (2004). Controlling Mycotoxins in Animal Feed. In: Magan N, Olsen M (eds.). Mycotoxins in Food: Detection and Control. Cambridge, England: Woodhead Publishing;, 262-304.

[27] Ruangwises, S., & Mhosatanun, B. (1996). Relation of aflatoxin B_1 in dairy cow feed and aflatoxin M_1 in milk. Final Research Report. Chulalongkorn University (Bangkok); (in Thai).

[28] Smith, Phillips. T. D., Ellis, J. A., Harvey, R. B., Kubena, L. F., Thompson, J., & Newton, G. (1994). Dietary Hydrated Sodium Calcium Aluminosilicate Reduction of Aflatoxin M_1 Residue in Dairy Goat Milk and Effects on Milk Production and Components. *Journal of Animal Science*, 72, 677-682.

[29] Mazzette, A., Decandia, M., Acciaro, M., Fenu, A., Francesconi, A. H. D., & Battacone, G. (2009). Excretion of Aflatoxin M_1 in Milk of Goats Fed Diet Contaminated by Aflatoxin B_1. *Italian Journal of Animal Science*, 8, 631-633.

Chapter 10

Aflatoxin in Fish Flour from the Amazon Region

Ariane M. Kluczkovski and
Augusto Kluczkovski Junior

Additional information is available at the end of the chapter

1. Introduction

The Amazon region is well-known for biodiversity and nutritious food. The sustainable use of wildlife is considered strategically as an action for the conservation of the natural tropical environments and maintaining biodiversity [1]. The fish trade market, for example, comprises different countries and consumers with different intention of use, and requires a process chain of frozen fish as support. On the other hand, the Fish waste management has been of the problems with the greatest inpect on teh environment. Most of the waste is discarded or used in other fish products. Treated fish waste has found many applications among which the most important are animal feed, biodiesel/biogas, dietic products (chitosan), natural pigments (after extraction), food-packaging applications (chitosan), cosmetics (collagen), enzyme isolation, soil fertilizer and moisture maintenance in foods (hydrolysates)[2]. The muscle of some fish species with low fat, for example, can be useful in the flour production. In Brazil, the fish industries waste provides environmental pollution due to the inadequate disposable ways, in most of the cases. About 50% of the biomass produced by the industries is discarded along the process. Thus, there is an increasing interest for other ways of profitability of those wastes, since a high amount of fish protein has been lost [3]. With the increasing world population, it became necessary to search for alternative foods, to increase the demand and supply. These sources of food should be nutritious, have good sensory characteristics and be low cost, to achieve much of the population [4]. The alternative that has grown tremendously in the market is to concentrate the protein of raw materials. The protein concentrate which has a high nutritional value and has a low cost of raw material used, aims to provide a product with the human element constructor, no fat, avoiding the intake of saturated fats cause high cholesterol, obesity and other consequences negative health [3]. Thus, a more directed waste recovery of slaughtered animals can be used in the form of direct consumption by humans, or indirectly by means of the feeding [4]. This protein concentrate product could be obtained by other sources of animals such as fish [5] and soy or nuts [6, 7]. The fish flour is an alternative protein source for the natives of the Amazon region and came from the Indian culture. Beyond that, the fish flour is used as animal feed in some regions. The flour of the acari-bodó (*Liposarcus pardalis*) fish, for example, is called "piracui" and it is considered the "classic" fish flour. In the Tupi language, "piracuí" means fish dry fire, pounded in a mortar, grind him to powder, sifted, put into and kept in a smokehouse. The idea was born of conserving food for all the Indian tribes of the Amazon. At the time of low water (drought) had plenty of food (hunting and fishing). And at the time of the great waters (floods), food was difficult because the fish moved around. Thus was born the idea of storing food: game meat (boiling them with herbs that will retain for several days). The native long dominate technique resulting in dehydration of fish product known as piracuí. Only one type of fish derived from fish muscle, dried and shredded, which represents a major source of protein, average of 70% protein of optimum digestibility in the diet of some population especially the poor [8].The production involves an artisanal drying process, with the raw material of the fish waste or the whole meat from the fish. The flour is regularly sailed in a bulk in local markets of the Amazon region at the Amazon environmental conditions with temperature above 25°C and relative humidity (RH) above 70%. In most of the markets it is sailed between other products. The illustration of the fish flour presentation is presented in Figure 1 (a and b).

(a) Fish Flour (Piracuí) sold in the market

(b) Fish Flour (Piracuí)

Figure 1. Fish Flour (Piracuí) presentation

It is consumed as ingredient in the local cuisine or as a protein source. Some authors reported the final product proximate composition of flour and protein concentrate as described in Table 01.

The protein levels around 78% from piracuí seems to be higher than other protein concentrate obtained from other fish ranging from 57.4 to 77.8g% [10]. The fish flour is sailed in common markets and there is no color or granulometry standard, since each artisanal production region has different procedures of process. It can be visible in the product the presence of bones and collagen fibers such as showed in Figure 2.

	Proximate Composition of fish products			
	Peixoto Castro[9] [a]		Murueta [10] [b]	Romanelli&Schmidt [11][c]
Calories[d]	356.8[e]	350.5[f]	3888.4-5015.9	-
Moisture [g]	7.3	11.8	69.9-82.5	2.4-3.8
Protein[g]	76.4	75.5	57.4-77.8	36.8-63.4
Lipids[g]	4.7	4.7	0.6-16.5	22.2-52.5
Ash[g]	9.4	6.5	8.1-20.2	2.3-12.4

[a]Samples of piracui from *Liposacus pardalis*; [b] Protein concentrate from nine different fish species (range); [c] Samples of Viscera Flour from *Caiman yacare* ; [d] Expressed in kcal; [e] Piracuí done by grilled fish; [f] Piracuí done by cooked fish; [g] expressed in g%.

Table 1. Proximate composition of fish and fish products samples according different authors.

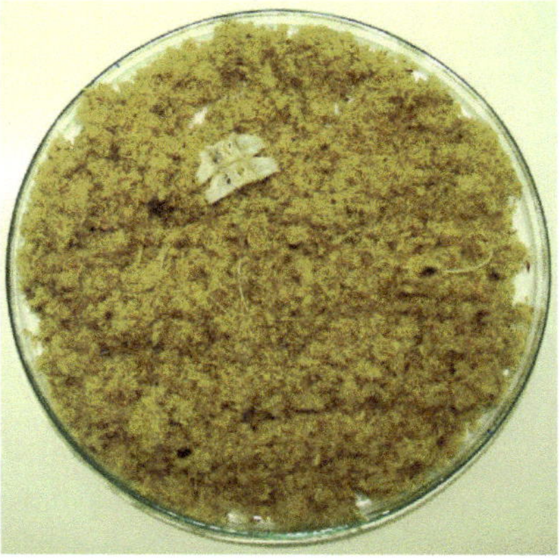

Figure 2. Small bones in Fish Flour (Piracuí)

The production of piracuí takes some stages and the flowchart is described in Figure 3. The fish or fish waste are washed and, the fish is eviscerated. They are cooked in an oven (100°C) and Sodium Chloride 2% is added. Then, a stage of drying is applied with temperatures of 60 to 80°C for 50 to 60 min. The material is cooled in room temperature and packaged in polyethylene bags and stored at room temperature.

The low water activity (*aw*) and moisture content (*mc*) levels in the product can increase the stability and shelf life, because the flour does not require refrigeration or low temperatures of storage, and can be kept in the environmental conditions. This is an advantage of the fish flour for some Amazon communities, because they are geographically far from the power energy supplies to keep poultry food. On the other hand, some environmental conditions from the Amazon region, such as high temperature (>30°C) and RH >80% associated to the poor safety conditions of the process can favor the contamination, especially by fungi that can be toxigenic, such as the mycotoxin producers [12]. The aflatoxin is one of those metabolic produced by some fungi strains with carcinogenic action to human beings and their level in food supply must be studied [13].

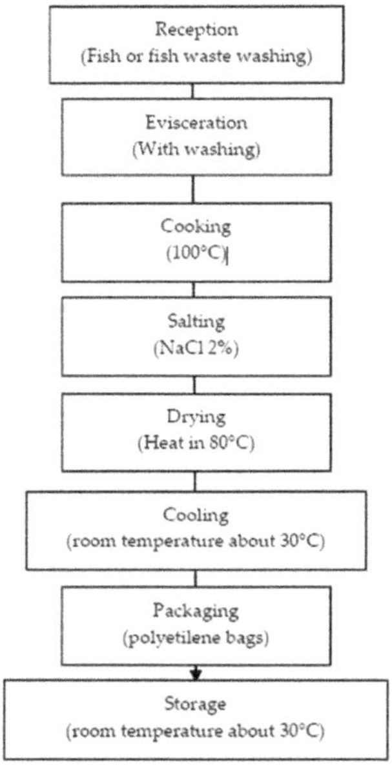

Figure 3. Flowchart of general Fish Flour Process

They have been reported, not only in nuts and vegetable products, but also in animal feed and meat products. Some aflatoxigenic moulds have been isolated from salted fish samples such as *Candida* spp., *Rhodotorulla* spp. and *Aspergillus* spp. [14]. Concerning the possibility of aflatoxigenic moulds in animal feed and to prevent contamination in the Amazon region consumers diet, a work was carried out in order to evaluate the presence of aflatoxin in fish flour samples from the Amazon Region a work was carried out concerning the evaluation of water activity (*aw*), moisture content (*mc*), aflatoxigenic fungi strains and total aflatoxin.

2. Material and methods

The total of 30 (thirty) samples (500g each) of fish flour from Brazil were collected from public markets at the Amazon region. The samples were sailed in a bulk.

The methods of analysis were:

a. *Aflatoxigenic moulds:* the samples were prepared and examined according to the technique recommended by APHA [15]. The identification of isolated mould and yeast genera was carried out according to Pitt & Hocking [16]. For the evaluation of aflatoxigenic strains, we used the method of incubation of strains from coconut agar (5-7 d, 26-28° C) [17]. To the *Aspergillus* spp. strain, only that identified as *A. flavus* were tested. After incubation the colonies were observed in UV light. The fluorescence indicative of the presence of aflatoxins was observed at the reverse of the plate.

b. Total aflatoxin (B_1+B_2+G_1+G_2): the samples were analyzed by HPLC [18]. The Limit of Quantification (LOQ) was 0.95 μg/kg. Five points were used to build an analytical curve, in order to obtain the correlation coefficient (R) values for LOD and LOQ. Each point corresponded to a mean of five injections of each extract. The recoveries for each aflatoxin (B_1, B_2, G_1 and G_2) were: 91.0; 75.0; 95.0 and 92.0%, respectively.

Sample preparation: the samples were visually inspected in order to identify the presence of bones. The samples were finely ground in a mill (particle size <100 μm) and homogenized;

Chemicals: aflatoxin standards and trifluoroacetic acid (TFA) were purchased by Sigma-Aldrich while acetonitrile, methanol (HPLC grade) and n-hexane were purchased by Nuclear;

Instrumentation: The HPLC operating conditions were as follows: Colum type and size: C18 Supelco; 25cm x 4.6 mm id; 5 micron particle size; Temperature: room temperature 25C; Mobile phase: deionized water: acetonitrile:methanol:water (8:27:65) and the flow rate was fixed at 1.0 ml min^{-1}; membrane filter and degassed in an ultrasonic bath for 25 min prior to use;

Standards preparation: the aflatoxin B_1, B_2, G_1 and G_2 standards (1.0 mg of each aflatoxin) in capped amber bottles) were used to the working solutions were prepared according to the AOAC [19] procedure by injecting 1 ml of acetonitrile into each vial to dissolve the aflatoxins.

Extraction and clean-up: 20 g of sample was extracted with 80 mL acetonitrile:water (9:1) mixture for 30 min by shaking under high speed and then filtered using a N°. 04 Whatman filter paper. A 1 mL portion of the filtrate was loaded on a multifunctional column and passed through at a flow rate of 2 mL/min. Then 1mL of acetonitrile:water (9:1) was applied to the column for 5 times. The filtrates were combined and evaporated to dryness under nitrogen and the residue was used for the derivatisation.

Derivatization: a 100 μl of the TFA solution and 300 μl of n-hexane were added to the residue from the sample extracted or to the aflatoxin work standards, vortexed for 30 s and kept in the dark for 15 minutes in room temperature. Nine hundred microlitres of acetonitrile:water (9:1) was added to the vial and vortexed for 30 s. The mixture was left to stand to allow the two layers to be separated. Twenty microlitres of the derivatized product (bottom layer) was injected into the HPLC column.

3. *Water activity* (*aw*): was determined in triplicate in an Aqualab series 3TE instrument (Decagon, USA) at 25±0.1°C;

(d)Moisture Content (mc): the *mc* levels were determined by the gravimetric method [19];

3. Results and discussion

3.1. Aflatoxigenic moulds

All the samples (100%) presented fungi growth. The *Aspergillus spp.* was identified in 85% of the samples and 15 isolated were obtained and tested concerning aflatoxin production as showed in Table 02. According to other authors, *Aspergillus* spp. was the most frequent strain reported in feed and fish products. Hassan et al. [14] also found levels of the presence of *Aspergillus* spp. (66.6%). The *Penicillium* spp. was found in 43% of the samples and 90% of the isolated tested for aflatoxin production was negative.

Fungi Strains	Incidence in the samples (%)	Number of isolated tested[a]	Toxigenic Strains	
			Positive	Negative
Aspergillus spp.	85	15	85%	15%
Penicillium spp.	43	10	10%	90%

[a] Strains identified as *Aspergillus flavus*

Table 2. Fungi and aflatoxin production in fish four

In our work, from the *Aspergillus* spp strains identified as *A. flavus*, 85% presented aflatoxin production as showed in Figure 4. Alinezhad et al. [20] reported *A. flavus* (60.66%) isolated from feed ingredients as well as pellet feed. Among 37 *A. flavus* isolates, 19 (51.35%) were able to produce AFB_1 in the range of 10.2 to 612.8 μg/g fungal dry weight. The aflatoxigenic behavior with fluorescence was showed in Figure 4.

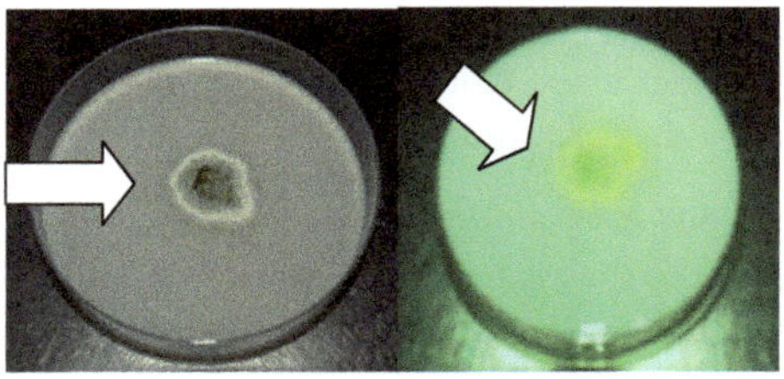

Figure 4. Aflatoxigenic behavior with fluorescence production from *A. flavus* from fish flour

The presence of aflatoxigenic fungi strains can be explained by the environment contamination, since the product was disposable in room temperatures with no regards of hygienic standards. The Brazilian regulation does not require the fungi analysis in fish or fish products [21]. In the process of fish flour temperatures of > 60-80° C for 60 min are applied with the binomial time-temperature acting on the microbiological control. The fish flour is rich in protein and nutrients to be spoiled by aflatoxigenic fungi strains. Adding NaCl 2%, during the process seems to not affect efficiently as a preservative factor to avoid fungi strains. In cured fish, for example, slight inhibition of mycelial growth and/or sporulation was recorded when isolates were cultured in basal medium containing 5% sodium chloride. On the other hand, the extent of inhibition increased with increasing salt concentrations, at 25% level, all the species had their growth completely inhibited [22].

3.2. Aw, Mc and total aflatoxin

The samples presented the following results with mean (range) described in Table 03. The *(a) aw*: 0.65 (0.64-0.70); *(b) mc*: 15.5 (10.0-20.8) % and *(c)* total aflatoxin (B_1+B_2+G_1+G_2): 10.5 (1.5-18.0) μg/kg. The aflatoxin was found in 20% of the samples under the LOQ. All the positive samples were under the limit of the Brazilian regulation for animal feed of 50 μg/kg [23]. The 05 (five) positive samples for aflatoxin belong to the group of samples with *A. flavus* isolated in the fungi test, and identified as aflatoxin producers. This fact, confirms the association between the presence of aflatoxigenic strains and the aflatoxin production in fish flour samples.

Number of Samples	Aw	Mc %	Total Aflatoxin μg/kg[a]	
	Mean (range)	Mean (range)	Positive samples	Mean (range)
30	0.65 (0.64-0.70)	15.5 (10.0-20.8)	05 (20%)	10.5 (1.5-18.0)

[a]Total aflatoxin= B_1+B_2+G_1+G_2

Table 3. *Aw, Mc* and Total Aflatoxin in fish flour from the Amazon region

The aflatoxin production in the fish flour could be affected by the levels of *aw* and *mc*. Those parameters have shown to allow the toxigenic fungi strains into the aflatoxin production, as showed in other dry food, such as nuts [24]. In previous work [25], the *aw* levels ranged from 0.1-0.90 and the microbiological stability of piracuí was showed at *aw* < 0.6 if *mc* will be below 10g%. The levels of our findings of *mc* were higher than 10%, so these levels must be concerned, because in *aw* below 0.6, there was reported shortly halophilic bacteria growth. Our results, concerning the *mc* levels were below 18.6%, reported by Santos & Freitas [26].

4. Conclusions

Despite the levels of aflatoxin in the samples below the limits of the Brazilian regulation for animal feed, the mycotoxin must be avoided. The studies in this matter are necessary, especially in areas such as the Amazon communities where the consumption of fish or fish products occurs 6 d/week, with 6.1 g/capital/d [27]. This data confirms that "piracuí" has an economic relevance [28]. Concerning the levels of aflatoxin found in the samples, it is necessary a work of good manufacture practices and safety storages conditions. The results of *aw* and *mc* this research provide data for the study of materials that can be used as packaging for storage of piracuí due the product to be frequently traded in the Amazon region. Concerning the significance of the fish flour for the Amazon region consumers, other studies are necessary to evaluate other toxicological aspects and the risk analysis.

Author details

Ariane M. Kluczkovski* and Augusto Kluczkovski Junior

Federal University of Amazonas, Faculty of Pharmaceutical Sciences, Manaus, Am, Brazil

References

[1] Rodrigues, E. C., Bressan M.C., Vicente Neto, J., Vieira, J. O. , faria, P. B., Ferrão, S. P. B., Andrade, P. L. Quality and chemistry composition of commercial cuts of alligator swanpland meat (Cayman yacare). Ciênc. Agrotec. 2007, 3(2), 448-455.

[2] Arvanitoyannis, I.S. & Kassaveti, A. Fish industry waste: treatments, environmental impacts, current and potential uses. International Journal of Food Science and Technology 2008, 43, 726–745.

[3] Camilo, A. G.; Fonseca, G. G.; Cavenaghi, A.D., Azambuja, S.P.H. Obtenção de concentrado proteico a partir de carne mecanicamente separada de pescado pintado. UFGD. MSc. 2011.

[4] Romanelli, P. F & Schmidt, J. Study of the utilization of the pantanal alligator's viscera for meat four. Ciênc. e Tecn. de Alimentos, 2003, 23: 131 - 139.

[5] Souza, J. F., Bitencourt, N. N., Gomes, C. S.; Oliveira, J. K., Santos, R. M., Reis, I. A. O., Nunes, M. L. Nunes, Narain, N. Desenvolvimento e caracterização físico-química e sensorial de nuggets formulados com concentrado protéico de pescado – Marine Beef, Scientia Plena 2010, 6(3), 2-4.

[6] Meza, B. E., Verdini R. A., Rubiolo A.C. Effect of freezing on the viscoelastic behaviour of whey protein concentrate suspensions. Food Hydrocolloids 2010, 24, 414–423.

[7] Gloria, M. & Regitano-d'Arce, M. A .B. Concentrado e Isolado protéico de Torta de Castanha do Pará:Obtenção e Caracterização Química e Funcional. Cienc. Tecnol. Aliment 2000, 20 (2), 240-245.

[8] Peixoto, C. Produção e estabilidade durante a estocagem de concentrado proteico de peixe (piracui) de acari-bodo, Pterygoplichtys multiradiatus, hancok (1828) e aruana Osteoglossum bicirrhosum (Vandelli, 1829). MSc. INPA/UFAM; 110p; 1999.

[9] Marueta, J. H. C., Del Toro, M. de L. A. N. & Carreno, F. G. Concentrates of fish protein from by catch species produced by various drying processes. Food Chemistry 2007, 100 705-711.

[10] Pacheco, A. M. & Scussel, V. M., 2006. Effects of processing and factory storage on aflatoxin contamination of in-shell Brazil nuts. Proceedings of the 9th International Working Conference on Stored Product Protection, 159-164.

[11] IARC. International Agency of Research on Cancer. Evaluation of carcinogenic risks to humans: some naturally occuring substances: aromatic amines and mycotoxins. Lyon.. IARC Monographs, 56. p. 245-39, 1997.

[12]] Hassan, A. A.; Hassan, M. A., Howayda, M. E., El Ahl, R. M. H. Abd El-Dayem, R.H.. Detection of aflatoxigenic moulds isolated from fish and their products and its public health significance. Nature and Science 2001, 9 (9), 106- 114.

[13] APHA. American public health association. 2001. Compendium of met. for the mic. exam. of foods . 4th Ed. American Public Health Association, Washington, DC.

[14] Pitt, J.I. and Hocking, A.D. 1999. Fungi and Food Spoilage. 2 ed. Aspen Publishers, Maryland, USA. 593p.

[15] Davis, N., Iyer, S., Diener, U. Improved method of screening for aflatoxin with a coconut agar medium. App. and Env. Microbiology 1987, 53:7, 1593-1595.

[16] Khayoon, W. S., Saad, B., Yan, C. B., Hashim, N.H., Ali, A. S. M., Salleh, M. I., Salleh, B. Determination of aflatoxins in animal feeds by HPLC with multifunctional column clean-up. Food Chemistry 2010, 118, 882-886.

[17] AOAC. Association of Official Analytical Chemists. 2005. Official methods of analysis of AOAC international (17 th ed.). Gaithersburg, MD, USA:AOAC International.

[18] Alinezahd, S.; Tolouee M.; Kamalzadeh A.; Motalebi A. A., Nazeri M., Yasemi M., Shams-Ghahfarokhi M., Tolouei R., Razzaghi-Abyaneh M. Mycobiota and aflatoxin B1 contamination of rainbow trout (Oncorhinchus mykiss) feed with emphasis to Aspergillus section Flavi. Iranian Journal of fisheries 2011, 10(3), 363-374.

[19] Brazil. Ministerio da Saude, ANVISA. RDC 12, 02/01/2001.

[20] Diyaolu, S. A. & Adebajo. L. O. Effects of sodium chloride and relative humidity on growth and sporulation of moulds isolated from cured fish. Food Nahrung 1994, 38(3), 311–317.

[21] BRASIL. Ministério da Agricultura. Portaria n. 07, de 09 de novembro de 1988. Estabelece os padrões mínimos das diversas matérias primas empregadas na alimentação animal. Diário Oficial da União, Poder Executivo, Brasília, DF, 09 de novembro de 1988.

[22] Arrus, K., Blank, G., Clear, R., Holley, R.A. Aflatoxin production by Aspergillus flavus in Brazil nuts. J. of Stored ducts Research 2005, 41, 513-527.

[23] Lourenço, L. F. H., Santos, D. C., Ribeiro, S.C.A., Almeida H., Araujo, E.A.F .Study of adsorption isoth.erm and microbiological quality of fish meal type "piracuí" of Acari-Bodo (Liposarcus pardalis, Castelnau, 1855). Procedia - Food Science 2011, 1, 455-462.

[24] Santos, J. R. C. & Freitas, J. A. Characteristics and quality of a piracuí fish derived product Rev. ciênc. agrár. 2004, 41, 25-46.

[25] Cerdeira, R.G.P. Fish consumption and others food itens by the riverine population of the Lago grande de Monte Alegre, PA-Brazil, Acta Amazonica 1997, 27(3):213-228.

[26] Almeida, J.C. Economic evaluation of protein concentrate production by Amazon fish. MSc. INPA, 165 p. 2009.

Section 3

Chemico-Biological Interactions and Human Health

Review of the Biological and Health Effects of Aflatoxins on Body Organs and Body Systems

Godfrey S. Bbosa, David Kitya, A. Lubega,
Jasper Ogwal-Okeng , William W. Anokbonggo and
David B. Kyegombe

Additional information is available at the end of the chapter

1. Introduction

Aflatoxins are a group of naturally occurring carcinogens that are known to contaminate different human and animal food stuffs. Aflatoxins are poisonous by-products from soil-borne fungus *Aspergillus,* which is responsible for the decomposition of plant materials [1-9]. The occurrence of aflatoxins foods and food products vary with geographic location, agricultural and agronomic practices. The susceptibility of food product to fungal attack occurs during pre-harvest, transportation, storage, and processing of the foods [1, 2, 4, 6, 9, 10]. The problem of aflatoxin contamination of the food products is a common problem in tropical and subtropical regions of the world especially in the developing countries such as the sub-Saharan countries with poor practices and where the environmental conditions of warm temperatures and humidity favors the growth fungi [1, 2, 4, 6, 9, 10]. The various food products contaminated with aflatoxins include cereals like maize, sorghum, pearl millet, rice and wheat; oilseeds such as groundnut, soybean, sunflower and cotton; spices like chillies, black pepper, coriander, turmeric and zinger; tree nuts such as almonds, pistachio, walnuts and coconut; and milk and milk products [11]. The aflatoxins were initially isolated and identified as the causative agent in Turkey X disease that caused necrosis of the liver in 1960 and over 100,000 turkeys died in England and USA and the death was attributed to the consumption of a mould-contaminated peanut meal [2, 6, 9, 12, 13]. Very high concentrations of aflatoxins are most often found in nutritive seeds such as maize, nuts and cereal grains in Africa and rice in China and Southeast Asia [2, 6, 9, 12-14].

Difuranocoumarins	Type of aflatoxin	*Aspergillus specie(s)*
Difurocoumarocyclopentenone series	Aflatoxin B_1 (AFB_1)	*A. flavus, A. arachidicola, A. bombycis, A. minisclerotigenes, A. nomius, A. ochraceoroseus, A. parasiticus, A. pseudotamarii, A. rambellii, Emericella venezuelensis*
	Aflatoxin B_2 (AFB_2)	*A. arachidicola, A. flavus, A. minisclerotigenes, A. nomius, A. parasiticus*
	Aflatoxin B_{2a} (AFB_{2a})	*A. flavus*
	Aflatoxin M_1 (AFM_1)	*A. flavus, A. parasiticus*; metabolite of aflatoxin B_1 in humans and animals and comes from a mother's milk
	Aflatoxin M_2 (AFM_2)	Metabolite of aflatoxin B_2 in milk of cattle fed on contaminated foods
	Aflatoxin M_{2A} (AFM_{2A})	Metabolite of AFM_2
	Aflatoxicol (AFL)	*A. flavus*, metabolite of AFB_1
	Aflatoxicol M_1	Metabolite of AFM_1
Difurocoumarolactone series	Aflatoxin G_1 (AFG_1)	*A. arachidicola, A. flavus, A. minisclerotigenes, A. nomius, A. Parasiticus*
	Aflatoxin G_2 (AFG_2)	*A. arachidicola, A. flavus, A. minisclerotigenes, A. nomius, A. parasiticus*
	Aflatoxin G_{2A} (AFG_{2A})	Metabolite of AFG_2
	Aflatoxin GM_1 (AFG_1)	*A. flavus*
	Aflatoxin GM_2 ($AFGM_2$)	Metabolite of AFG_2
	$AFGM_{2A}$	Metabolite of $AFGM_2$
	Aflatoxin B_3 (AFB_3)	***Aspergillus*** species not defined
	Parasiticol (P)	*A. flavus*
	Aflatrem	*A. flavus, A. minisclerotigenes*
	Aspertoxin	*A. flavus*
	Aflatoxin Q_1 (AFQ_1)	Major metabolite of AFB_1 in in vitro liver preparations of other higher vertebrates

Table 1. Summary of the major aflatoxins produced by the *Aspergillus* species of Moulds

Aflatoxins are a group of approximately 20 related fungal metabolites produced primarily by the fungi *Aspergillus flavus* and *A. parasiticus* [15-18]. Aflatoxins belongs to a group of difuranocoumarins that are classified into two broad groups according to their chemical structure and they include the difurocoumarocyclopentenone series (AFB_1, AFB_2, AFB_{2A}, AFM_1, AFM_2, AFM_{2A} and aflatoxicol) and the difurocoumarolactone series (AFG_1, AFG_2, AFG_{2A}, $AFGM_1$, $AFGM_2$, $AFGM_{2A}$ and AFB_3) [15-19], (Table 1 and figure 1).

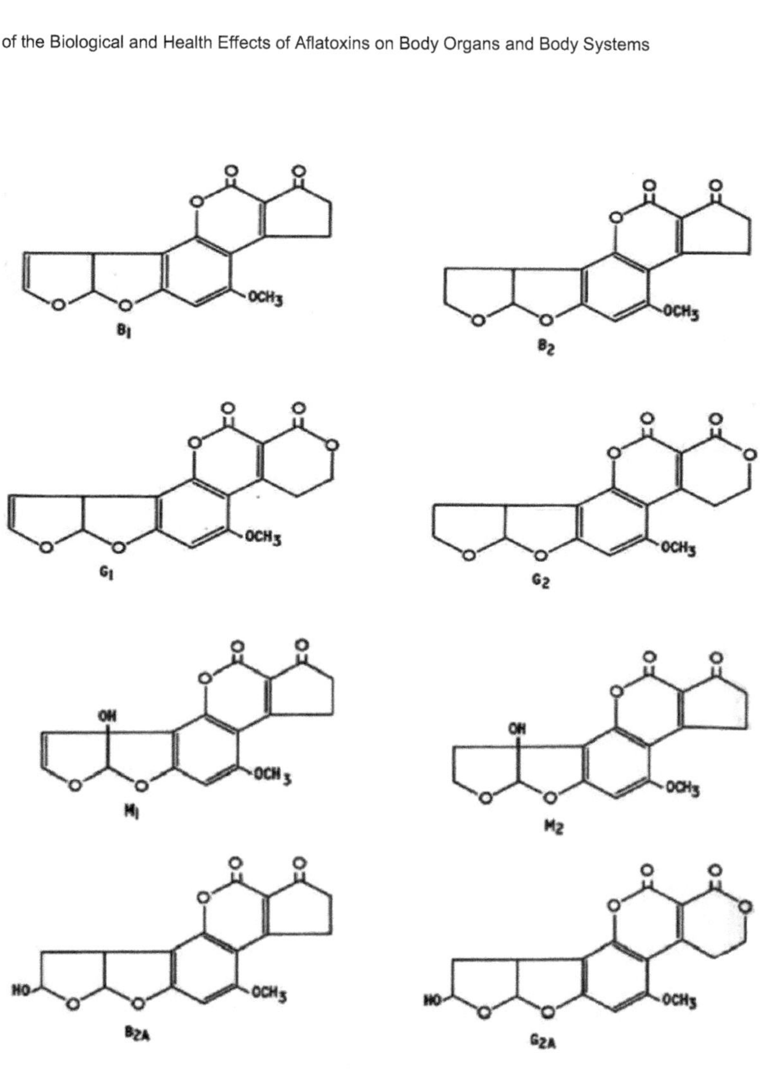

Figure 1. Structures of the major aflatoxins B_1, B_2, G_1, G_2, M_1, M_2, B_{2A} and G_{2A} (Adopted from Reddy, 2012)[16]

The four major naturally known aflatoxins produced by the *Aspergillus* species of mold include AFB_1, AFB_2, AFG_1 and AFG_2 where the "B" and "G" refer to the blue and green fluorescent colors produced under UV light on thin layer chromatography plates, while the subscript numbers 1 and 2 indicate major and minor compounds, respectively. Whereas the B designation of aflatoxins B_1 and B_2 result from the exhibition of blue fluorescence under UV-light, while the G designation refers to the yellow-green fluorescence of the relevant structures under UV-light [2, 6, 9, 12, 13]. The metabolic products of aflatoxins, M_1 and M_2 were first isolated from milk of lactating animals fed on Moldy grains contaminated with aflatoxin hence, the M designation [2, 4]. These toxins have closely similar structures (Figure 1) and form a unique group of highly oxygenated, naturally occurring heterocyclic compounds. Aflatoxins B_2 and G_2 were established as the dihydroxy derivatives of B_1 and G_1, respectively. Whereas, aflatoxin M_1 is 4-hydroxy aflatoxin B_1 and aflatoxin M_2 is 4-dihydroxy aflatoxin B_2. Of the four major aflatoxins (B_1, B_2, G_1 and G_2), G_2 occurs in high quantities though less toxic while AFB_1 is the most toxic of all the aflatoxin. The World Health Organization (WHO) classifies AFB_1 as a class 1 carcinogen [4, 6, 9, 18]. The aflatoxins display potency of toxicity, carcinogenicity, mutagenicity in the order of AFB_1> AFG_1> AFB_2> AFG_2 [15-19]. The extent of toxicity depends on the organ affected especially the liver. The lethal toxicity of aflatoxin B_1 varies in different animals from extremely susceptible (Sheep, Rat, Dog) to resistant species (Monkey, Chicken, Mouse). However, there are no toxicity in humans though epidemiological data from studies in Africa, South Africa, South East Asia and India implicate aflatoxins in the incidence of liver cancer especially the hepatobiliary carcinoma and death of children due to malnutrition, kwashiorkor and marasmus [20, 21]. Aflatoxins have been associated with various diseases like aflatoxicosis and other health problems in humans, livestock and domestic animals globally.

2. Absorption, distribution, metabolism, excretion and mechanisms of action of aflatoxins

Aflatoxins are highly liposoluble compounds and are readily absorbed from the site of exposure usually through the gastrointestinal tract and respiratory tract into blood stream [22, 23]. Human and animals get exposed to aflatoxins by two major routes (a) direct ingestion of aflatoxin-contaminated foods or ingestion of aflatoxins carried over from feed into milk and milk products like cheese and powdered milk as well as other animal tissues mainly as AFM_1 [22](b) by inhalation of dust particles of aflatoxins especially AFB_1 in contaminated foods in industries and factories [24]. After entering the body, the aflatoxins are absorbed across the cell membranes where they reach the blood circulation. They are distributed in blood to different tissues and to the liver, the main organ of metabolism of xenobiotics. Aflatoxins are mainly metabolized by the liver to a reactive epoxide intermediate or hydroxylated to become the less harmful aflatoxin M1 [25, 26]. In humans and susceptible animal species, aflatoxins especially AFB_1 are metabolized by cytochrome P450 (CYP450) microsomal enzymes to aflatoxin-8,9-epoxide, a reactive form that binds to DNA and to albumin in the blood serum, forming adducts and hence causing DNA damage [25, 26]. Various CYP450 enzymes isoforms occur in the liver and they metabolize aflatoxin into a reactive oxygen species (aflatoxin-8,9-epoxide), which may then bind to proteins and cause acute toxicity (aflatoxicosis) or to DNA and induce liver cancer [25, 26]. The predominant human CYP450 isoforms involved in human metabolism of AFB_1 are CYP3A4 and CYP1A2. Both enzymes catalyze the biotransformation of AFB_1 to the highly reactive *exo*-8,9-epoxide of AFB1[27]. CYP 1A2 is also capable of catalyzing the epoxidation of AFB_1 to yield a high proportion of *endo*-epoxide and hydroxylation of AFB1 to form aflatoxin M_1 (AFM_1), which is a poor substrate for epoxidation [27] and less potent than AFB_1 [28]. This is generally considered as the major detoxification metabolic pathway for aflatoxins. The CYP3A4 is the major CYP450 enzyme responsible for activation of AFB_1 into the epoxide form and also form

AFQ_1, a less toxic detoxification metabolite. The CYP3A5 metabolizes AFB1 mainly to the *exo*-epoxide and some AFQ_1 [29]. However, polymorphism studies with CYP3A5 have indicated that, this enzyme isoform is not expressed by most people especially in Africans [28]. Studies in Gambian children showed that aflatoxin cross the placenta and transported to the fetus and the new born where they can cause detrimental effects [28]. The CYP3A7 is a major CYP450 enzyme isoform in human fetal liver and metabolizes AFB_1 to the 8, 9- epoxide that may cause fetal defects to the developing fetus [30].

The epoxidation of AFB_1 to the exo-8, 9-epoxide is a critical step in the genotoxic pathway of this carcinogen. The binding of AFB1 to DNA and DNA adduction by AFB_1 exo-8,9 epoxide has been reported to cause a functional changes of DNA conformation [31].The epoxide is highly unstable and binds with high affinity to guanine bases in DNA to form afltoxin-N7-guanine [32]. The aflatoxin-N7-guanine has been shown to be capable of forming guanine (purine) to thymine (pyrimidine) transversion mutations in DNA and hence affecting the p53 suppressor gene in the cell cycle [33, 34]. The p53 gene is important in preventing cell cycle progression when there are DNA mutations, or signaling apoptosis. The mutations have been reported to affect some base pair locations more than others especially in the third base of codon 249 of the p53 gene in the region corresponding to the DNA binding domain of the corresponding protein [13, 34]and this appears to be more susceptible to aflatoxin-mediated mutations than nearby bases [35]. AFB1 induces the transversion of base G to base T in the third position of codon 249 and similar mutations have been observed in hepatocellular carcinoma (HCC) in high AFB_1 contaminated food in regions in East Asia and Africa [34, 36, 37].

Epoxide hydrolase and glutathione-S-transferase (GST) are both involved in hepatic detoxification of activated AFB_1, but the GST-catalyzed conjugation of glutathione to AFB_1-8,9-epoxides is thought to play the most important role in preventing epoxide binding to target macromolecules like DNA and various cell proteins [38]. Glutathione pathway is reported to play a vital role in the detoxification of AFB_1 [39, 40]. The AFB_1 8,9 *exo* and *endo*epoxides are conjugated by glutathione to form AFB-mercapturate and the reaction is catalyzed by glutathione S-transferase (GST) [39, 40]. The glutathione-aflatoxin conjugate is transported from the cells with an ATP-dependent multidrug-resistance protein through an accelerated process [39]. Despite a preference for conjugating the more mutagenic AFB_1 exo-epoxide isomer, the relatively low capacity for GST-catalyzed detoxification of bio-activated AFB_1 in lung may be an important factor in the susceptibility of the lung to AFB_1 toxicity [4, 8, 41].The *exo* and *endo* epoxide can also be converted non-enzymatically to AFB_1-8,9-dihydrodiol which in turn can slowly undergo a base-catalysed ring opening reaction to a dialdehyde phenolate ion [27]. AFB_1 dialdehyde can form Schiff bases with lysine residues in serum albumin forming aflatoxin-albumin complex [42]. Also the aflatoxin dialdehyde are reduced to a dialcohol in a NADPH-dependent catalyzed reaction by aflatoxin aldehyde reductase (AFAR) [43]. However the guanine alkylation by aflatoxin B_1 produces *exo*-8,9-epoxide which is the reactive form and a carcinogen to the liver and the reaction is more than 2000 times more efficient in DNA than in aqueous solution [44], (Figure 2).

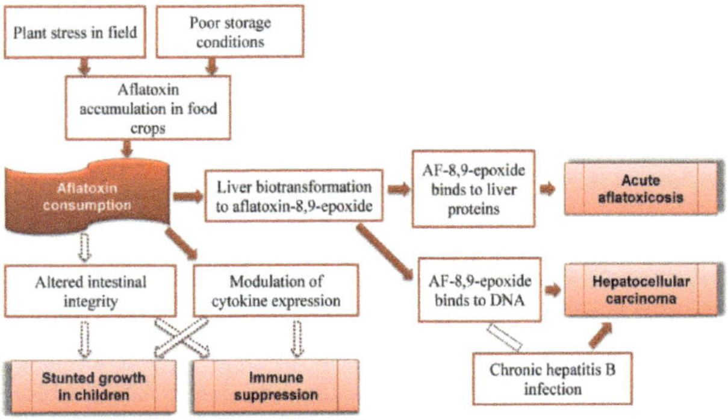

Figure 2. Aflatoxin disease pathways in humans (Adopted from Wu, 2010; Wu, 2011)[10, 26]

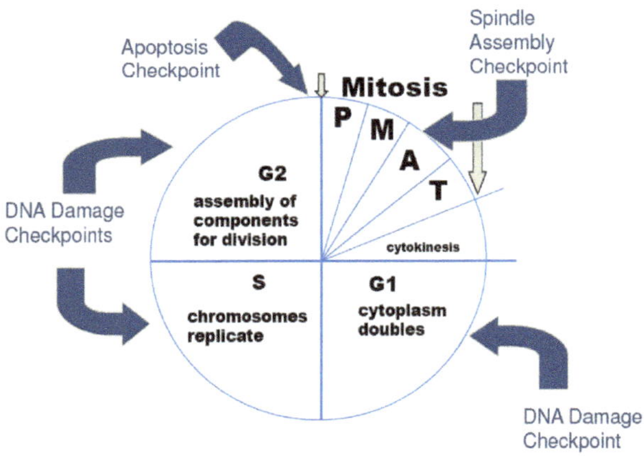

Figure 3. Various check points that can be damaged by binding of aflatoxins and AF-8,9-epoxide causing the deregulation of the cell cycle; P –prophase, M-Metaphase, A- Anaphase, T- Telophase, S- Synthetic DNA phase, G1 and G2 – Gaps (growth phase) [47-49]

2.1. Effect of aflatoxins on mitochondrial DNA

The reactive aflatoxin-8,9-epoxide preferentially binds to mitochondrial DNA (mitDNA) during hepatocarcinogenesis as compared to nuclear DNA that hinder ATP production and FAD/NAD-linked enzymatic functions and this causes the disruption of mitochondrial functions in the various parts of the body that require production of energy in the form of ATP [45]. Aflatoxin damage to mitochondria can lead to mitochondrial diseases and may be responsible for aging mechanisms [45]. It is reported that certain mitochondrial diseases result from the ability of the nucleus to detect energetic deficits in its area. The nucleus attempts to compensate for the ATP shortages by triggering the replication of any nearby mitochondria but unfortunately, the response promotes replication of the very mitochondria that are causing the local energy deficit hence aggravating the problem [46]. The AFB_1 also binds to DNA and cause structural DNA alterations that lead to gene mutations as well as changes in the length of the telomeres and the check points in the cell cycle [47-49]. The binding of AFB_1 to DNA at the guanine base in liver cells corrupt the genetic code that regulates cell growth, thereby leading to formation of tumors ([45-49]. The damage to mitDNA is caused by adduction and mutations of mitochondrial membranes leading to increased cell death (apoptosis) as well as disruption of energy production (production of ATP) [46, 49, 50]. The reactive aflatoxin-8, 9-epoxide can affect the mitotic (M) phase, growth process (G1 and G2 phase) and DNA synthesis (S phase) in the cell cycle by disrupting the various check points that regulate the cell cycle development and proliferation leading to deregulation of the cell and hence cancer development [47-49], (Figure 3).

However in resistant rodents, their mitDNA is protected from aflatoxins from DNA adducts that effect mitochondrial transcription and translation [46-49]. The mycotoxin alters energy-linked functions of ADP phosphorylation and FAD- and NAD-linked oxidizing substrates and α-ketoglutarate-succinate cytochrome reductases [46-49].

2.2. Effect of aflatoxins on mitochondrial structure

AFB causes ultrastuctural changes in mitochondria [46-49]and also induces mitochondrial directed apoptosis thus reducing their function [20, 29, 48-51]. Also the aflatoxins may affect the telomere length and the various check point in the cell cycle causing further damage to the regulatory processes of the cell cycle [51]. Also the extent of aflatoxin binding to DNA and its damage, the level of different proteins changes from cell cycle and apoptotic pathways such as c-Myc, p53, pRb, Ras, protein kinase A (PKA), protein kinase C (PKC), Bcl-2, NF-kB, CDK, cyclins and CKI contribute to the life or death decision making process that may contribute to the deregulation of the cell proliferation leading to cancer development [34, 48, 49](Figure 3).

2.3. Role of glutathione in detoxification of aflatoxins and their metabolites

However like in hepatic detoxification of aflatoxins and other chemicals, GSH act as antioxidant and has many functions in membrane maintenance and stability as well as in reducing oxidative stress factors and the high reactive oxygen species (ROS) produced from the process of lipid peroxidation [38-41, 46, 52-56]. The increased depletion of GSH leads to abnormally high levels of ROS found in cells affected by aflatoxin due to uncoupling of metabolic processes resulting from the lack of GSH for GSH-peroxidase catalysis of O_2 to H_2O_2 leading to lipid peroxidation and compromised cell membranes. Its reduction further enhances the damage to critical cellular components (DNA, lipids, proteins) by the 8,9 epoxides. However the most serious adverse effects of the AFB1-8,9-epoxide metabolite is that it reacts with amino acids in DNA and forms an adduct [38-41, 46, 52-55]. The adduct are fairly resistant to DNA repair processes and this causes gene mutation that leads to liver cancers especially the hepatocellular carcinomas [38-41, 46, 52-55].

2.4. The role of cytoplasmic reductase in detoxification of AFB_1

Also in the hepatocytes, AFB_1 are converted to other different classes of metabolites by cytoplasmic reductase such as aflatoxicol and by microsomal mixed-function oxidase system to form AFM_1, $AGFQ_1$, AFP_1 and AFB_1 -epoxide (the most toxic and carcinogenic derivative) and these metabolites may be deposited in various body tissues as well as in edible animal products [38-41, 46, 52-55]. These metabolites other than the AFB_1 are less toxic and are conjugated with other molecules that enhance their rapid elimination from the body [22]. The metabolite AFQ_1 has very little cancer-causing potential and they are usually excreted in urine with little effect on the body.

2.5. Effect of aflatoxins on protein synthesis

The aflatoxin binds and interferes with enzymes and substrates that are needed in the initiation, transcription and translation processes involved in protein synthesis. They interacts of with purines and purine nucleosides and impair the process of protein synthesis by forming adducts with DNA, RNA and proteins [57]. Aflatoxin also inhibits RNA synthesis by interacting with the DNA-dependent RNA polymerase activity and thus causes degranulation of endoplasmic reticulum. Also the reduction in protein content in body tissues like in skeletal muscle, heart, liver and kidney could be due to increased liver and kidney necrosis [58]. AFB_1 is a potent mutagenic, carcinogenic, teratogenic, and immunosuppressive and all these may interfere with normal process of protein synthesis as well as inhibition of several metabolic systems thus causing damages to various organs especially the liver, kidney and heart [59, 60].

2.6. Role of aflatoxins in cancer

Aflatoxins especially AFB_1, AFG_1 and AFM_1 are the most toxic, naturally occurring carcinogens known with AFB_1 the most hepatocarcinogenic compound, causing various cancers of the liver and other body organs in humans and animals [4, 14, 45, 61]. Aflatoxin's cancer-causing potential is due to its ability to produce altered forms of DNA adducts. The primary disease associated with aflatoxin intake is hepatocellular carcinoma (HCC, or liver cancer). This disease is the third-leading cause of cancer death globally [4, 45, 61], with about 550,000–600,000 new cases each year. The incidence of liver cancer has been consistently higher in men than in women with a sex ratio ranging from 2 to 3 in most countries [9]. Eighty-three percent of these cancer deaths occur in East Asia and sub-Saharan Africa [62-64]. Hepatocellular carcinoma (HCC) is one of the most common cancers worldwide with extremely poor prognosis. The majority of cases occur in south-east Asia and sub-Saharan Africa where the major risk factors of chronic infection with hepatitis B and C viruses (HBV and HCV) as well as dietary exposure to aflatoxins are a problem [9, 25, 61, 65]. Aflatoxin B_1, the most commonly occurring and potent of the aflatoxins is associated with a specific AGG to AGT amino acid transversion mutation at codon 249 of the p53 gene in human HCC, providing mechanistic support to a causal link between exposure and disease [25, 26, 66, 67]. Liver cancer has an increasing incidence that parallels the rise in chronic hepatitis B (HBV) and hepatitis C (HCV) infection [25, 67, 68]. Chronic infection with hepatitis B virus (HBV) or hepatitis C virus (HCV) can progress to advanced liver disease, including cirrhosis and hepatocellular carcinoma (HCC), a form of primary liver cancer [25, 61, 67, 68]. HCC is the third leading cause of cancer-related mortality worldwide [69]. The data show that individuals positive for the hepatitis B virus and exposed to aflatoxin in the diet are about 60 times of risk for developing hepato-biliary carcinoma or liver cancer [26, 66, 67] especially in poor developing countries worldwide [67]. Reports have shown that a number of interactions exist between HBV and aflatoxins in development of hepatocellular carcinoma in humans. They may include the fixation of AFB_1-induced mutations in the presence of liver regeneration and hyperplasia induced by chronic HBV infection, the predisposition of HBV-infected hepatocytes to aflatoxin induced DNA damage, an increase in susceptibility to chronic HBV infection in aflatoxin exposed individuals and oxidative stress exacerbated by co-exposure to aflatoxins and chronic hepatitis infection [61](Figure 4).

In humans, epidemiological studies in Africa, Southeast Asia, USA and other countries of the west where there is a high incidence of hepatocellular carcinoma, have revealed an association between cancer incidence and the aflatoxin content of the diet [5, 6, 70]. Aflatoxin B_1 (AFB_1) is a major risk factor in the pathogenesis of liver cancer in Asia and sub-Saharan Africa [71]. Aflatoxin B_1 is a potent liver carcinogen in a variety of experimental animals. It causes liver tumours in mice, rats, fish, marmosets, tree shrews and monkeys following administration by various routes. Types of cancers described in research animals include hepatocellular carcinoma (rats) colon and kidney (rats), cholangiocellular cancer (hamsters), lung adenomas (mice), and osteogenic sarcoma, adenocarcinoma of the gall bladder and carcinoma of the pancreas (monkeys) [5, 6, 12, 70].

3. Health effects of aflatoxins on human and animals (Aflatoxicosis)

Aflatoxicosis is a condition caused by aflatoxins in both humans and animals. It occurs in two general forms (1) the acute primary aflatoxicosisis produced when moderate to high levels of aflatoxins are consumed. Specific acute episodes of disease may include hemorrhage, acute liver damage, edema, alteration in digestion, absorption and/or metabolism of nutrients, and possibly death [5, 6, 12, 69, 70]. Acute dietary exposure to AFB_1 has been implicated in epidemics of acute hepatic injury [13, 72]. Evidence of acute aflatoxicosis in humans has been reported worldwide especially in the third world countries like Taiwan, Uganda, India, Kenya and many others [7]. (2) The chronic primary aflatoxicosis results from ingestion of low to moderate levels of aflatoxins (USAID, 2012). The effects are usually subclinical and difficult to recognize. Some of the common symptoms are impaired food conversion and slower rates of growth with or without the production of an overt aflatoxin syndrome [9]. The chronic forms of aflatoxicosis include (1) teratogenic effects associated with congenital malformations (2) mutagenic effects where aflatoxins cause changes (mutations) in the genetic code, altering DNA and these changes can be chromosomal breaks, rearrangement of chromosome pieces, gain or loss of entire chromosomes, or changes within a gene (3) the carcinogenic effect in which the carcinogenic mechanisms have been identified such as the genotoxic effect where the electrophilic carcinogens alter genes through interaction with DNA and thus becoming a potential for DNA damage and the genotoxic carcinogens that are sometimes effective after a single exposure, can act in a cumulative manner, or act with other genotoxic carcinogens which affect the same organs [50, 60]. Chronic effects of aflatoxin has been reported to impair the normal body immune function by either by reducing phagocytic activity or reduce T cell number and function as observed immunological suppression in animal model. Aflatoxins have also been reported to interfere with nutrition in a dose response relationship between exposure to aflatoxin and rate of growth in infants and children [4, 9, 20, 50, 60]. Aflatoxins also causes nutrient modification like vitamin A or D in animal models and thus making them unavailable for the normal body physiology and hence leads to nutritional deficiencies [7, 20].

The contamination of foods and feeds with aflatoxin can cause serious consequences in human and animal health. It is estimated that more than 5 billion people in developing countries worldwide are at risk of chronic aflatoxin exposure due to consumption of aflatoxin-contaminated foods and of these more than 4 billion people develop aflatoxin related liver cancer especially the hepatocellular carcinoma [64, 69, 73, 74]. Aflatoxin exposure is mainly a problem in poor and developing countries with poor regulatory authorities in food processing and storage as well as with high levels of malnutrition. Aflatoxins have also been linked with kwashiorkor and marasmus in most of the sub-Saharan countries in children [20]. Many people in these countries experience chronic aflatoxicosis associated with long-term exposure to low to moderate levels of aflatoxin in the food supply chain. AFB_1, AFB_2 and AFM have been detected in liver, gall bladder, spleen, heart, muscle and kidney [75]. Aflatoxin B_1 exposure results in both steatosis and accumulation of fat and necrosis or cell death of liver cells. The amount of aflatoxins consumed contributes to the mutagenic, carcinogenic, teratogenic, and immunosuppressive health effects in the body. The adverse effect of aflatoxins in humans ranges from acute hepatic toxicity to chronic disease such as liver cancer, haemorrhages, oedema, and even immediate death. Prolonged consumption of aflatoxins has also been reported to cause impaired immune function and malnutrition and stunted growth in children and a number of disabilities and death [7, 76, 77]. Human studies have reported that aflatoxins cause an increase in circulating alpha tumor necrosing factor, suggesting that these mycotoxins are also immunotoxic in humans. Due to the aflatoxin body immunosuppressant, it has been associated with HIV and tuberculosis [66, 67](Figure 2). Aflatoxins also pose a threat to developing fetuses and they are transferred from mother to infant in breast milk. Aflatoxins have been reported to be associated with a Reye-like Syndrome in Thailand, New Zealand, Czechoslovakia, the United States, Malaysia, Venezuela, and Europe [4, 50, 78].

All species of animals are susceptible to aflatoxicosis and the susceptibility of individual animals to aflatoxicosis varies considerably depending on dose, duration of exposure, species, age, sex and nutrition. AFB_1, AFB_2 and AFM have been detected in liver, gall bladder, spleen, heart, muscle and kidney of growing swine when protein and protein-free portions of the diet were separately fed [75]. Chronic exposure of aflatoxins to animals causes immunosuppression and also interferes with protein metabolism and multiple micronutrients that are critical to health due to adduct formation. These adduct are responsible for mutations, cancer, immunosuppression, lung injury and birth defects [46]. In animals, the aflatoxins cause liver damage, decreased milk production, reduced reproductively and suppressed immunity in animals consuming low dietary concentrations. The aflatoxicosis syndrome in animals may also be characterized by vomiting, abdominal pain, pulmonary oedema, convulsions, coma, and death with cerebral edema and fatty involvement of the liver, kidneys, and heart. In dairy and beef cattle, the signs of acute toxicosis include anorexia, depression, dramatic drop in milk production, weight loss, lethargy, gastrointestinal dysfunctions such as ascitis, icterus, tenesmus, abdominal pain, bloody diarrhoea, decreased feed intake and efficiency; weight loss, jaundice, abortion, hepatoencephalopathy, blindness, walking in circles, ear twitching, frothy mouth, photosensitization, bleeding and death [4, 6, 22, 79]. In poultry, beside inappetance, weight loss, decreased egg production, leg and bone problems, poor pigmentation, fatty liver, kidney dysfunction, bruising and death, suppression to natural immunity and susceptibility to parasitic, bacterial and viral infections can occur [6, 22], (Figure 4).

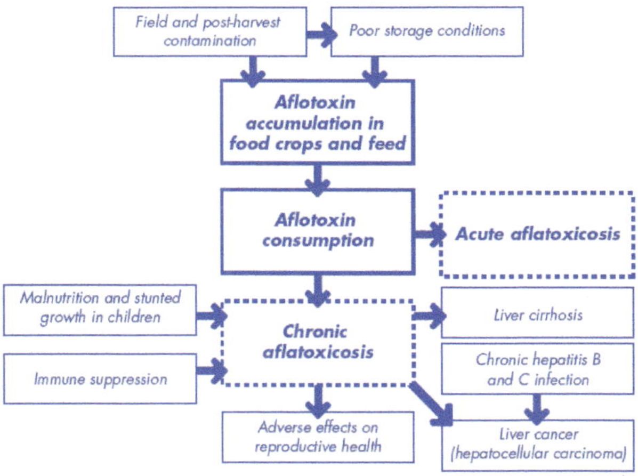

Figure 4. Aflatoxin disease pathways in humans (Adopted from Wu, 2010; USAID, 2012; WHO, 2011; Wu and Tritscher, 2011) [7, 26, 80]

4. Biological effect of aflatoxins on the body organs and body systems

Aflatoxins have been reported to affect the various body organs like the liver, kidneys, lungs, brain, testes and many endocrine and exocrine organs, the heart, skeletal muscles and the different body systems.

4.1. Role of aflatoxins in hepatic injury and other body organs and tissues

Aflatoxins have been reported to cause liver cirrhosis as well as liver cancers [4, 6, 7, 26, 80]. Hepatic injury can be acute or chronic form caused by a variety of toxic agents like aflatoxins, chemicals and drugs, trauma and infectious agents [2, 4, 6, 7, 26, 61, 76, 80, 81]. The reduced level of total protein is indicative of the toxic effect of AFB_1 to the liver due to the failure in synthesis of the proteins and kidney in which aflatoxins are known to impair protein biosynthesis by forming adducts with DNA, RNA and proteins, inhibits RNA synthesis, DNA-dependent RNA polymerase activity and causes degranulation of endoplasmic reticulum [58-60]. Acute hepatic injury due to aflatoxin causes a rise in serum enzymes including aspartate aminotransferase, lactate dehydrogenase, glutamate dehyrogenase, gamma-glutamyltransferase and alkaline phosphatase and bilirubin that reflect liver damage as well as other biochemical changes such as proteinura, ketonuria, glycosuria and hematuria [4, 5, 40]. The other frequently used liver enzymes are the alkaline phosphatase (ALP) and Gamma-glutamyltransferase and gamma-glutamyltranspeptidase (GGT and GGTP) that indicate obstruction to the biliary system, either within the liver or in the larger bile channels outside the liver [9, 45, 61]. The presence of jaundice and neurological disorders due to brain damage leading to hepatic encephalopathy are associated with liver failure. Chronic liver failure leads to accumulation of metabolites in circulation such as ammonia and fatty acids that eventually lead to brain damage and hence hepatic encephalopathy [40, 82]. The liver failure makes it unable to detoxify ammonia, the product of protein and amino acid metabolism leading to hyperammonemia that may cross the blood brain barrier leading to increased synthesis of glutamate neurotransmitters henceleading to cytotoxicity of the brain cells and hence the hepatic encephalopathy [82-84]. AFB_1 has been reported to cause pallor discoloration of liver and enlargement of liver and kidneys, congestion of liver parenchyma, cytoplasmic vaculation or fatty change of hepatocytes, necrosis of hepatocytes and newly formed bile ducts, mononuclear and heterophilic cell infiltration are reported in aflatoxin fed broiler chicks [85]. It is also reported that there is a decrease in protein content in skeletal muscle, heart, liver and kidney in aflatoxin-fed animals due to the AFB_1's potent mutagenic, carcinogenic, teratogenic, immunosuppressive and its ability to inhibits several metabolic systems such as protein synthesis thus leading to liver, kidney and heart damage [58-60]. In chicken, the activity of serum or plasma enzymes like the sorbitol dehydrogenase, glutamic dehydrogenase, lactate dehydrogenase, alkaline phosphatase, acid phosphatase, aspartate aminotransferase and alanine aminotransferase were reported to be increased in aflatoxicated chickens [22].

4.2. Effect of aflatoxins on the central nervous system

In the brain or central nervous system, the neurons have a high metabolic rate but little capacity for anaerobic metabolism and subsequently, inadequate oxygen flow to the brain kills the neuronal brain cells within minutes. Some compounds damage neurons or neurotoxic and thus inhibit their function. Mycotoxins especially aflatoxins and its metabolites and other products such as the reactive oxygen species (ROS) like the AFB-8,9-epoxides may interfere with the normal functioning of the nerve cells by forming DNA adducts, protein adducts, oxidative stress factors, mitochondrial directed apoptosis of the nerve cells as well as inhibiting their synthesis of protein, RNA and DNA [40, 44, 47, 50, 52, 54]. Aflatoxins also cause abnormalities in mitochondrial DNA, structure and function, including defective oxidative phosphorylation in the brain cells [29, 49, 50, 54]. The oxidative stress may result in damage to critical cellular macromolecules such as DNA, lipids and proteins. Cellular fatty acids are readily oxidized by ROS to produce lipid peroxyl radicals which can subsequently propagate into MDA that may interact with cellular DNA to cause DNA-MDA adduct that may affect energy production in the brain [29, 49, 50, 54]. The role of ROS has been postulated in the development of aging and chronic degenerative diseases, inflammatory diseases and brain cancers [52]. Aflatoxins may also deplete the myelin sheath of the nerves, an important substance that covers the nerves and hence become exposed to insults. Mycotoxins especially aflatoxins have been reported to be toxic to various aspects of brain chemistry and their function [4, 50, 82]. AFB_1 also alters the levels of various biogenic amines (neurotransmitters) and their precursors in rat and mouse brains. Acute AFB_1 treatment in experimental animals has been reported to cause a decrease in regional brain acetylcholinesterase enzymes that may affect the cognitive functions as well as memory and learning of the individual while chronic exposure increases adenohypophyseal acetylcholinesterase [24]. Aflatoxin causes a decrease in dopamine, serotonin and alterations in the levels of the precursor's tyrosine and tryptophan [86-88]. Deficiencies in these neurotransmitter lead to neurological symptoms such as neurocognitive decline and alteration of sleep cycle and symptoms of brain damage like dullness, restlessness, muscle tremor, convulsions, loss of memory, epilepsy, idiocy, loss of muscle coordination, and abnormal sensations [89, 90]. AFB_1 has also been reported to increase the central and peripheral nervous system Na^+/K^+-ATPase, β-glucuronidase and β-galactosidase while inhibiting the Mg^{2+}-ATPse in experimental animals and this also is important in the normal functioning of the glutamate neurotransmitter and their NMDA receptors [24, 53, 91-93]. The liver failure makes it unable to detoxify ammonia, the product of protein and amino acid metabolism leading to hyperammonemia that may cross the blood brain barrier leading to increased synthesis of glutamate neurotransmitters hence leading to cytotoxicity of the brain cells and hence the hepatic encephalopathy [82-84]. Toxic encephalopathy was originally described in children with Reye's syndrome associated with consumption of Aflatoxin B_1 and/or salicylates [78] and subsequently in cases of aflatoxicosis in canines and Chinese children were reported [94]. Aflatoxins also have been linked to Reye's syndrome that is characterized by symptoms of encephalopathy and fatty degeneration of the viscera. It is a pediatric disease characterized by cerebral edema and neuronal degeneration. Toxic encephalopathy due to aflatoxins involves multiple symptoms like loss of balance, recent memory decline, headaches, lightheadedness, spaciness/disorientation, insomnia, loss of coordination [4, 18, 50, 82]. Aflatoxins have been reported to be associated with a Reye-like Syndrome in Thailand, New Zealand, Czechoslovakia, the United States, Malaysia, Venezuela and Europe [4, 9, 24, 50, 78]. Aflatoxins especially AFB_1 have been reported to cause tumors in both the central and peripheral nervous system and several nonepithelial neurogenic tumors like the schwannomas, gliomas, meningiomas and granular cell tumors have been reported [24].

4.3. Effect of aflatoxins on the gastrointestinal tract (GIT)

The gastrointestinal tract (GIT) is the main route of entry of aflatoxins as a result of consumption of aflatoxin-contaminated foods especially AFB_1. It is also the main route of excretion aflatoxin metabolites from the bile. The aflatoxins, metabolites and AF-8,9-epoxides have been reported to cause intestinal tumors especially the human colon cancers like colon carcinomas and similar results have been reported in experimental animals [24]. Aflatoxins have also been reported to cause serious acute effects on the GIT [95]. Aflatoxins have been implicated as potential factors in the increased incidence of human gastrointestinal and hepatic neoplasms in Africa, Philippines and China [22]. Aflatoxins have been reported to cause digestive system effects such as diarrhea, vomiting, intestinal hemorrhage, and liver necrosis and fibrosis [89]. Aflatoxins have been reported also to damage the integrity of the pancreas. In domestic animals, aflatoxins cause changes in the GIT physiology especially decreased rumen motility and function in cows [24]. In birds, aflatoxins interfere with intestinal morphology, sialic acid production and apparent digestible energy [96].

4.4. Effect of aflatoxins on the respiratory system

Aflatoxins have reported to have serious acute effects on the respiratory systems [95].The respiratory tract is the only organ system with vital functional elements in constant and direct contact with the environment [97]. Many people working in food industries as their occupational setting get exposed to aflatoxins especially AFB_1 when they inhale aflatoxin-contaminated dusts like during grain shelling and processing and have been reported to have a higher incidences of upper respiratory tract and lung cancers [24, 95]. In experimental animals, AFB_1 was reported to induce 100% pulmonary adenomas. In the respiratory tract, aflatoxins may also be converted to active metabolites like in the nasal mucosa [23]. It is also reported that the intranasal administration of AFB_1 lead to formation of tissue-bound metabolites in subtentacular cells, bowman's glands and in neuronal cells in the olfactory mucosa but there is no evidence that AFB_1 may induce tumours in olfactory bulbs [98]. Epoxide hydrolase and glutathione-S-transferase (GST) are both involved in hepatic detoxification of activated AFB_1 but the GST-catalyzed conjugation of glutathione to AFB_1-8,9-epoxides is thought to play more important role in preventing epoxide binding to target macromolecules [23, 89, 99]. However, the low capacity for GST-catalyzed detoxification of bio-activated AFB_1 in lung may be an important factor in the susceptibility of the lung to AFB_1 toxicity ([41]. Nose-only inhalation exposure of rats to AFB_1 aerosols suppressed alveolar macrophage (AM). Intratracheal administration of AFB_1 also suppressed the release of tumor necrosis factor-alpha from AMs and impaired systemic innate and acquired immune defenses as well as suppression of peritoneal macrophage phagocytosis and the primary splenic antibody response thus leading to suppression of respiratory tract defenses system [99].

4.5. Effect of aflatoxins on the cardiovascular system, blood and blood cells

Aflatoxins have reported to have serious acute effects on the cardiovascular systems including vascular fragility and hemorrhaging in tissues [58, 89, 95] as well as heart damage and teratogenic effects [59, 60]. It is reported that there is a decrease in protein content of the muscles of these tissues and organs as well as inhibition of their metabolic processes attributable by the aflatoxin consumption of contaminated foods [59, 60].

4.6. Effect of aflatoxins on the blood and blood cells

The aflatoxins and its metabolites as well as the generated reactive oxygen species(ROS) has been reported to have a deleterious effects on the bone and blood cells as well as induction of cancers on the hemopoietic system in bone marrow and lymphoid organs where blood, blood cells and blood components are produced [52]. The blood system can be damaged by agents that affect blood cell production (bone marrow), the components of blood (platelets, red blood cells, and white blood cells), or the oxygen-carrying capacity of red blood cells or impair blood clotting and their poor growth rates. Oxidative damage by the AFB_1 on human lymphocytes has been reported [100] and significant declines in both the proportion of peripheral blood lymphocytes and in the percentages of ANAE-positive peripheral blood lymphocytes (T-lymphocytes) in a dose dependent manner has been observed [101]. Aflatoxins have been linked to anemia in pregnancy [7, 102] and alterations in erythrocytes during induced chronic aflatoxicosis in rabbit also have been reported [103, 104]. Aflatoxin causes hematopietic suppression and anemia, decrease in total erythrocytes, packed-cell volume and hemoglobin [16] as well as toxicity to red blood cells [103]. Aflatoxin is known to produce hemolytic anemia by decreasing the circulating mature erythrocytes [104]and consequently the spleen appear congested because of an unusually high concentration of inorganic iron and debris from the circulation [103, 104]. In birds, AFB_1 is reported to causes hematological changes [105]. Aflatoxicosis has been reported to cause lymphocytopenia and monocytopenia and increased percentage of neutrophil counts [106]. In cattle, aflatoxins are reported to cause blood coagulation defects that may involve impairment of prothrombin, factors VII and X and possibly factor IX and similar effects are reported in dogs [5]. Generally aflatoxins have been reported to depress growth and alter many aspects of humoral and cellular immunity and thus affecting the hematological parameters [101, 107].

4.7. Effect of aflatoxins on the urinary system

The kidney is susceptible to many toxic agents due to the high amount of blood it receives and about 20-25% of blood that flows in at rest coupled with the large amounts of circulating toxicants that reach the kidneys [89]. The kidneys also have high oxygen and nutrient requirements because of their workload and therefore filters one-third of the blood reaching them and reabsorb 98-99% of the salt and water. Different parts of the nephrone are exposed to aflatoxins especially the AFB_1 and its metabolites leading to nephrotoxicity before it is excreted in the urine [24, 58]. The aflatoxin induced reduction in protein content has been reported to be due to increased necrosis of the kidney [58-60, 90]. AFB_1 has been reported to cause kidney tumors in experimental animals and a mixture of AFB and AFG was observed to cause renal and hepatic tumors in 80% of hamsters [24]. There were also renal lesions with features of megalocytosis in the proximal renal tubules. In Africa, birds exposed to AFB_1 were reported to develop fatty and hemorrhagic kidney syndrome, thickening of the glomerular basement membrane, abnormal development of glomerular epithelial cells and degenerative changes in renal tubular cells, congestion and parenchyma hemorrhage [24, 85]. In other animals, there was a reduction in the glomerular filtration rate, glucose reabsorption and tubular transport of electrolytes and organic anions, reduced activities of renal glutamate-oxaloacetate and pyruvate transaminases and alkaline phosphatase in rats attributed to by the aflatoxins and their metabolites as well as the generated ROS. There was induced aggregation and loss of chromatin, mitochondrial degeneration and loss of microvilli induced by AFB1 in cultured kidney cell lines [24, 85].

4.8. Effect of aflatoxins on the endocrine system

Aflatoxin especially AFB has been reported to interfere with the functioning of the various endocrine gland by disrupting the enzymes and their substrates that are responsible for the synthesis of the various hormones. Aflatoxins and their metabolites as well as the generated ROS have been reported to cause various cancers in different endocrine glands like pituitary gland, granulosa cell tumors of the ovary and adenomas and adenocarcinomas of the adrenal gland, kidneys, thyroid gland, ovaries, testes, thyroid gland, parathyroid glands and endocrine pancreas [4, 90, 108]. The plasma testosterone and luteinizing hormone (LH) concentrations have been reported to reduce in aflatoxin-fed birds [90]. In laboratory animals, aflatoxin causes delayed maturation of both males and females [4, 22, 90, 109]. Aflatoxicosis in white leghorn males chicken decreased feed consumption, body weight, testes weight and semen volume (Sharlin et al., 1980) and decreased plasma testosterone values [22].

4.9. Effect of aflatoxins on the reproductive system

In humans exposed to chronic aflatoxin-contaminated foods, it has been reported that higher concentrations of aflatoxins occur in the semen of infertile men [3]. It is also associated with low birth weight, a risk factor for jaundice in infants as well as presence of AFM in maternal breast milk where it can cause deleterious effect in the newborns [102]. In Nigeria, about 37% of the infertile men had aflatoxin in their blood and semen hence contributing to the incidence of infertility in Nigerians [110]. Experimental results indicate that certain agents like aflatoxins can interfere with the reproductive capabilities of sexes, causing sterility, infertility, and abnormal sperm, low sperm count, and/or affect hormone activity in animals. Aflatoxins have been reported to disrupt the reproductive system in both male and female animals after ingestion of aflatoxin-contaminated foods. Aflatoxins also cause pathological alterations in the form of coagulative necrosis especially in the growing and mature follicles and decrease in number and size of graffian and growing follicles with increased number of atretic follicles and small areas of degenerative changes in experimental animals [111]. AFB_1 has been reported to have a deleterious effect on the reproductive capacity of laboratory and domestic female animals where they cause reductions in ovarian and uterine sizes, increases fetal resorption, implantation loss and intra-uterine death in the aflatoxin exposed female rats [111]. They also cause a reduction in the primary spermatocytes and spermatids [112] and affect the morphology of the sperm cells produced [113]. Stillbirths were reported in the 15th to the 18th days of pregnancy in rats [108]. The levels of plasma testosterone, plasma 5a-DHT and absolute and relative testes weights were reported in experimental animals of aflatoxin-treated males remained low in all age groups and a delay in the onset of sexual maturation during aflatoxicosis [114]. In cows, aflatoxins affected the reproductive system by causing abortion, the birth of weak, deformed calves, reduced fertility due to reduced vitamin A levels [109]. The teratogenic effects of AFB_1 were described as enlarged eye sockets and enlarged liver of embryos [60]. In poultry, AFB1 cause a reduction in semen volume, testes weight, spermatocrit and plasma testosterone as well as a reduction in egg output [24].

5. Effect of aflatoxins on the immune system

Chronic consumption of aflatoxin-contaminated foods has been reported to cause immunosuppression in both humans and animals worldwide [7, 89]. In human, aflatoxins affect both the cellular and humoral immune responses where they alter immunological parameters in participants with high AFB_1 levels resulting in impairments in cellular immunity hence decreasing the host resistance to infections [115-117]. Aflatoxin exposure has been shown to cause immune suppression, particularly in cell-mediated responses [115-117]. Chronic exposures of the individual to aflatoxins depress the phagocytic efficiency of the phagocytes and the delayed hypersensitivity reactions in birds [24]. Aflatoxins also deplete the cell populations of the thymus; reduce the bone marrow and the red and white blood cells count, macrophage numbers and the phagocytic activity of the cells [24]. It also depresses the T-cell-dependent functions of splenic lymphocytes in mice. The natural killer cell function of the peripheral blood lymphocytes are also affected by aflatoxins especially AFB1 [24]. A reduction in the leukocyte immunophenotypes in peripheral blood, $CD4^+$ T cell proliferative response, $CD4^+$ T and $CD8^+$ T cell cytokine profiles and monocyte phagocytic activity were reported. Children in developing countries appear to be naturally exposed to aflatoxin through their diet at levels that compromise the immune system. In general, the proportion of childhood growth stunting is directly correlated with the proportion of the population living below the national poverty line and is inversely correlated with gross domestic product per capita [7, 45]. As is the case with liver cancer, childhood stunting is prominent in regions such as Southeast Asia and Sub-Saharan Africa, where aflatoxin exposure through consuming contaminated food is common [7, 45]. It has been reported that the immunosuppression and nutritional effects of chronic aflatoxin exposure may be linked to the high prevalence of HIV in Southern Africa [7, 74, 118, 119]. The CD4 proteins that have been weakened by aflatoxin exposure have been reported to correlate positively with HIV infection [116]. Also high aflatoxin levels have been reported to increase risk of developing tuberculosis in HIV positive individuals. Persons who are exposed to aflatoxin and are HIV positive have decreased plasma vitamin A and vitamin E in the blood, although there was no interaction detected between aflatoxin and HIV infection [120]. HIV infection is likely to increase aflatoxin exposure by two possible routes: (1) HIV infection decreases the levels of antioxidant nutrients that promote the detoxification of aflatoxin, or (2) the high degree of co-infection of HIV-infected people with hepatitis B also increases the biological exposure to aflatoxin [7, 118, 119]. Aflatoxin induce immunosuppression and increases susceptibility of toxicated birds and animals to bacterial, viral and parasitic infections [58]. It also affects the lymphoid follicles of caecum thus depleting the lymphocytes that may contribute to the observed immunosuppression [117]. Aflatoxin decreases the concentrations of immunoglobulins IgM, IgG and IgA in birds as well as decrease complement activity in chickens [22, 121]. The low dose of AFB_1 slightly decrease both mRNA and protein levels of lymphocytic IL-2, IFNγ and it preferentially affects macrophage functions as well as IL-1α, IL-6 and TNF production by these cells [121, 122]. Aflatoxin suppression of the immune system therefore subjects the individual to high risk of susceptible to infectious diseases like parasitic, bacterial and viral infections [123].

6. Conclusion

Chronic consumption of aflatoxin-contaminated foods is a common problem in both humans and animals worldwide especially in poor developing nations of south East Asia and sub-Saharan Africa where there is poor food harvesting, processing and storage of food and food products thus allowing the growth of mold on them. Aflatoxins, their metabolites, the aflatoxin-8,9-epoxide and the generated ROS causes deleterious effects on the various body organs and body systems including the development of cancers especially the liver cancer mainly due to AFB_1 exposure. Aflatoxins are also responsible for the suppression of both the humoral and cell-mediated immunity and thus making individuals susceptible to infectious diseases. Aflatoxins also responsible for the malabsorption of various nutrients thus leading to nutritional deficiencies, impaired immune function, malnutrition and stunted growth and hence the development of kwashiorkor and marasmus in infants. Aflatoxins also can affect almost all the different body systems and hence the health of the affected individuals especially in poor developing nations of south East Asia and sub-saharan Africa where there is poor food harvesting, processing and storage thus allowing the growth of mold on them.

Author details

Godfrey S. Bbosa[1*], David Kitya[2], A. Lubega[1], Jasper Ogwal-Okeng [1],
William W. Anokbonggo[1] and David B. Kyegombe[3]

*Address all correspondence to: godfossa@yahoo.com

1 Department of Pharmacology and Therapeutics, Makerere University College of Health sciences, Kampala, Uganda

2 Department of Surgery, Mbarara University of Science & Technology Medical School, Mbarara, Uganda

3 Department of Pharmacology and Toxicology, Kampala International University, School of Health Sciences, Ishaka Campus, Busyenyi, Uganda

References

[1] Bankole, S. A., & Adebanjo, A. (2003). Review of mycotoxins in food in West Africa: current situation and possibilities of controlling it. *African Journal of Biotechnology*, 2(9), 254-263.

[2] Bennett, J. W., & Klich, M. (2003). Mycotoxins. *Clinical Microbiology Reviews*, 16(3), 497-516.

[3] Gupta, R. C. (2011). Aflatoxins. *Ochratoxins and Citrinins. Reproductive and Developmental Toxicology*, 55, 753-761.

[4] INCHEM Principles of evaluating chemical effects on the aged population: International Programme on chemical Safety- Environmental Health Criteria 144 World Health Orgnization, Geneva,(1993). http://www.inchem.org/documents/ehc/ehc/ehc144.htm (Accessed on 19th June 2012) (1993).

[5] Aflatoxins, N. L. M. (2002). National Library of Medicine. Hazardous Substance Data Base. *Toxnet (National Data Network).*

[6] Thrasher, J. D. (2012). Aflatoxicosis in animals. *Aflatoxins and Health*, www.alphaboostjuice.com/AFLATOXICOSIS_IN_ANIMALS.pdf.

[7] USAID. (2012). Aflatoxin: A Synthesis of the Research in Health, Agriculture and Trade. *Feed the Future: The Office of Regional Economic Integration USAID East Africa Regional Mission Nairobi, Kenya*, www.eastafrica.usaid.gov/...esearch_in_Health_Agriculture_and_Trade/pdf, 10-15.

[8] Whitlow, L.W.W., Hagler, M. J., & Diaz, D. E. (2010). mycotoxinsin in feeds. *Feedstuffs*, http://fdsmagissues.feedstuffs.com/fds/Reference_2010 13_MycotoxinsinFeeds.pdf, 74.

[9] WHO. (2000). Hazardous Chemicals in Humans and Environmental Health: International Programme on Chemical safety, Geneva, Switzerland. *World Health Organisation*, http://whqlibdoc.who.int/hq/2000/WHO_PCS_00.1.pdf, 7-9.

[10] Wu, F., et al. (2011). *The Health economics of aflatoxins: Global burden of disease Aflacontrol Working Paper 4 FebruaryInternational Food Policy Research Institute. 2033 K Street, NW Washington, DC 20006-1002 USA*, 1-16.

[11] Lopez, C., et al. (2002). Aflatoxin B1 in human serum: Aflatoxin B1 content in patients with hepatic diseases. *Medicina (Buenos Aires)*, 313-316.

[12] Otsuki, T., Wilson, J. S., & Sewadeh, M. (2002). A Race to the Top? *A Case Study of Food Safety Standards and African Exports. Development Research Group (DECRG), World Bank, 1818 H Street NW, Washington DC 20433 USA. 1424_wps 2563.pdf.*

[13] Sudakin, D. L. (2003). Dietary aflatoxin exposure and chemoprevention of cancer: A clinical review. *Journal of Toxicology and Clinical Toxicology*, 41, 195-204.

[14] Kitya, D., Bbosa, G. S., & Mulogo, E. (2009). Aflatoxin levels in common foods of South Western Uganda: a risk factor to hepatocellular carcinoma. European Journal of Cancer Care. 10.1111/j.1365-2354.2009.01087.x , 1-6.

[15] Cortés, G., et al. (2010). Identification and quantification of aflatoxins and aflatoxicol from poultry feed and their recovery poultry litter . *Poultry Science*, 89(5), 993-1001.

[16] Reddy, S. V., & Waliyar, F. (2012). Properties of aflatoxin and its producing fungi. *Aflatoxins*, http://www.icrisat.org/aflatoxin/aflatoxin.asp, (Accessed on 8th June 2012).

[17] Smith, J. E., & Sivewright-Henderson, R. Mycotoxins and animal foods. *CRC Press*, 978-0-84934-904-1, 614.

[18] Thrasher, J., , D., & Crawley, S. L. (2009). The Biontaminants and Complexity of Damp Indoor Spacs: More than Meets the Eyes. *Toxicology and Industrial Health*, http://drthrasher.org/page63.html, (Accessed on 10th June 2012), 25, 583-616.

[19] Salhab, A. S., et al. (1977). Aflatoxicol M1: A new metabolite of aflatoxicol. *Xenobiotica*, http://toxicology.usu.edu/endnote/Aflatoxicol-new-metabolic.pdf, 401-408.

[20] Peraica, M., et al., Toxic effects of mycotoxins in humans. (1999). *Bulletin of the World Health Organization.*, http://whqlibdoc.who.int/bulletin/1999/9 bulletin_1999_77%289%29_754-766.pdf, 754-766.

[21] Thomas, A. E., et al. (2005). Toxicity of aflatoxins from selected consumables in Lagos (Nigeria). *Electronic Journal of Environmental, Agricultural and Food Chemistry*, 4(6), 1111-1116.

[22] Agag, B. I. (2004). Mycotoxins in foods and feeds : Aflatoxins. *Association of Universal Bullettin of Environmental Research*, 7(1), 173-191.

[23] Larsson, P., & Tjalve, H. (2000). Intranasal instillation of Aflatoxin B1 in rats: Bioactivation in the nasal mucosa and neuronal transport to the olfactory bulb. *Toxicological Science*, 55, 383-391.

[24] Coulombe, R. A., & Jr , . (1994). Nonhepatic disposition and effects of aflatoxin B1. *The Toxicology of Aflatoxins: Human Health, Veterinary and Agricultural significance.*, http://toxicology.usu.edu/endnote/Nonhepatic-disposition.pdf., 89-101.

[25] Wild, C. P., & Montesano, R. (2009). A model of interaction: Aflatoxins and hepatitis viruses in liver cancer aetiology and prevention. *Cancer Letters*, 286, 22-28.

[26] Wu, F., & Khlangwiset, P. (2010). Health economic impacts and cost-effectiveness of aflatoxin reduction strategies in Africa: Case studies in biocontrol and postharvest interventions. *Food Additives & Contaminants*, 27, 496-509.

[27] Guengerich, F. P., et al. (1998). Activation and detoxication of aflatoxin B1. *Mutatation Research*, 402, 121-128.

[28] Wild, C. P., & Turner, P. C. (2002). The toxicology of aflatoxins as a basis for public health decisions. *Mutagenesis*, 17, 471-481.

[29] Wang, H., et al. (1998). Structure-function relationships of human liver cytochromes 450A: aflatoxin B1 metabolism as a probe. *Biochemistry*, 37, 12536-12545.

[30] Kitada, M., et al. (1998). Mutagenic activation of aflatoxin B1 by 450HFLa in human fetal livers. *Mutation Research*, 227, 53-58.

[31] Raney, V. M., Harris, T. M., & Stone, M. P. (1993). DNA conformation mediates aflatoxin B1-DNA binding and the formation of guanine N7 adducts by aflatoxin B1 8,9-exo-epoxide. *Chemical Research in Toxicology*, 6(1), 64-68.

[32] Guengerich, F.P., Forging the links between metabolism and carcinogenesis.Mutation Research, (2001). , 195-209.

[33] Bailey, E. A., et al. (1996). Mutational properties of the primary aflatoxin B1-DNA adduct. *Proceedings of the National Academy of Sciences, USA.*, 93, 1535-1539.

[34] Li, D., et al. (1993). Aberrations of 53 gene in human hepatocellular carcinoma from China. *Carcinogenesis*, 14, 169-173.

[35] Aguilar, F., Hussain, S. P., & Cerutti, P. (1993). Aflatoxin B1 induces the transversion of G-->T in codon 249 of the 53 tumor suppressor gene in human hepatocytes. *PNAS*, 90(18), 8586-90.

[36] Gerbes, A. L., & Caselmann, W. H. (1993). Point mutations of the 53 gene, human hepatocellular carcinoma and aflatoxins. *Journal of Hepatology*, 19, 312-315.

[37] Mace, K., et al. (1997). Aflatoxin B1-induced DNA adduct formation and 53 mutations in CYP450-expressing human liver cell lines. *Carcinogenesis*, 18, 1291-1297.

[38] Sherratt, P. J., & Hayes, J. D. (2001). Glutathione S-transferase. *Enzyme systems that m,etabolise drugs and other xenobiotics*, 9, 320-351.

[39] Farombi, E. O., & Nwaokeafor, I. A. (2005). Anti-oxidant mechanisms of kolaviron: studies on serum lipoprotein oxidation, metal chelation and oxidative membrane damage in rats. *Clininical and Experimental Pharmacology and Physiology*, 32, 667-674.

[40] Johnson, W. W., et al. (1997). Conjugation of highly reactive aflatoxin B1 exo-8,9-epoxide catalyzed by rat and human glutathione transferases: estimation of kinetic parameters. *Biochemistry*, 36, 3056-3060.

[41] Stewart, R. K., Serabjit-Singh, C. J., & Massey, T. E. (1996). Glutathione S-transferase-catalyzed conjugation of bioactivated aflatoxin B1 in rabbit lung and liver. *Toxicology and Applied Pharmacology*, 140, 499-507.

[42] Sabbioni, G., & , C. P. (1991). Wild, Identification of an aflatoxin G1-serum albumin adduct and its relevance to the measurement of human exposure to aflatoxins. *Carcinogenesis*, 12, 97-103.

[43] Knight, L. P., et al. (1999). cDNA cloning, expression and activity of a second human aflatoxin B1 metabolizing member of the aldo-keto reductase superfamily, AKR7A3. *Carcinogenesis*, 20, 1215-1223.

[44] Brown, K. L., et al. (2009). Inherent Stereospecificity in the Reaction of Aflatoxin B1 8,9-Epoxide with Deoxyguanosine and Efficiency of DNA Catalysis. *Chemical Research in Toxicology*, 22(5), 913-917.

[45] WHO. (2008). World Health Statistics. *World Health Organisation, Geneva*, Retrieved from, http://www.who.int/whosis/whostat/EN_WHS08_Full.pdf.

[46] Wallace, D. C. (1997). Mitochondrial DNA in aging and disease. *Scientific American*, 40-47.

[47] Ezekiel, C. N., et al. (2011). Studies on Dietary Aflatoxin-induced Genotoxicity using two In vivo bioassays. *Archives of Applied Science Research*, 3(2), 97-106.

[48] Jacotot, E., Ferri, K. F., & Kroemer, G. (2000). Apoptosis and cell cycle: distinct checkpoints with overlapping upstream control. *Pathological Biology (Paris)*, 48(3), 271-279.

[49] Vermeulen, K., Berneman, Z. N., & Bockstaele, D. R. V. (2003). Cell cycle and apoptosis. *Cell proliferation*, 36, 165-175.

[50] Thrasher, J. D., & Crawley, S. L. (2012). Neurotoxicity of Mycotoxins. http://www.drthrasher.org/page189.html, (Accessed on 10th June 2012).

[51] Hornsby, P. J. (2007). Senescence: As an Anticancer Mechanism. *Journal of Clinical Oncology*, 25(14), 1852-1857.

[52] Halliwell, B. (2007). Oxidative stress and cancer: have we moved forward? *Biochemistry Journal*, 401, 1-11.

[53] Schubert, D., & Piasecki, D. (2001). Oxidative Glutamate Toxicity Can Be a Component of the Excitotoxicity Cascade. *The Journal of Neuroscience*, 21(9), 7455-7462.

[54] Verma, R. J. (2005). Aflatoxin Cause DNA Damage. *International Journal of Human Genetics*, 4(4), 231-236.

[55] Zhang, X., et al. (2005). Expression of cytochrome 450 and other biotrnasormation genes in fetal and adult human nasal mucosa. *Drug Metabolism and Disposition*, 33, 1423-1428.

[56] Liu, J., et al. (1999). Effect of salvia miltiorrhiza on aflatoxin B1-induced oxidative stress in cultured rat hepatocytes. *Free Radical Research*, 31, 559-568.

[57] Clifford, J. I., & Rees, K. R. (1967). The Interaction of Afiatoxins with Purines and Purine Nucleosides. *Biochemistry Journal*, 103, 467-471.

[58] Sharma, V., et al. (2011). Ameliorative Effects of Curcuma Longa and Curcumin on Aflatoxin B1 Induced Serological and Biochemical Changes In Kidney of Male Mice. *Asian Journal of Biochemical and Pharmaceutical Research*, 1(2), 338-351.

[59] Mohammed, A. M., & Metwally, N. S. (2009). Antiaflatoxicogenic activities of some aqeous plant extracts against AFB1 induced Renal and Cardiac damage. *Journal of Pharmacology and Toxicology*, 4(1), 1-16.

[60] Wangikar, P. B., et al. (2005). Teratogenic effects in rabbits of simultaneous exposure to ochratoxin A and aflatoxin B1 with special reference to microscopic effects. *Toxicology*, 215, 37-47.

[61] Beckingham, I. J. (2001). ABC of liver, pancreas and gall bladder. *British Medical Journal*, http//www.portal-en.tbzmed.ac.ir/CmsModules/Teacher/Download.aspx?/pdf, 1-49.

[62] Kirk, G. D., Bah, E., & Montesano, R. (2006). Molecular epidemiology of human liver cancer: Insights into etiology, pathogenesis and prevention from The Gambia. *Carcinogenesis*, 27, 2070-2082.

[63] Parkin, D. M., et al. (2002). Global Cancer Statistics. *A Cancer Journal for Clinicians*, 55, 74-108.

[64] Strosnider, H., et al. (2006). Workgroup Report: Public Health Strategies for Reducing Aflatoxin Exposure in Developing Countries. *Environmental Health Perspectives*, 114, 1989-1903.

[65] Deng, Z. L., & Ma, Y. (1998). Aflatoxin sufferer and 53 gene mutation in hepatocellular carcinoma. *World Journal of Gastroenterology*, 4(28).

[66] Groopman, J. D., Kensler, T. W., & Wild, C. P. (2008). Protective Interventions to Prevent Aflatoxin-Induced Carcinogenesis in Developing Countries. *Annual Review of Public Health*, 29, 187-203.

[67] Liu, Y., & Wu, F. (2010). Global Burden of Aflatoxin-Induced Hepatocellular Carcinoma: A Risk Assessment. *Environmental Health Perspectives*, 118, 818-824.

[68] Henry, S. H., Bosch, X. F., & Bower, J. C. (2002). Mycotoxins and food safety. *Advances in Experimental Medicine and Biology*, 504(4), 229-233.

[69] Liu, Y., et al. (2012). Population attributable risk of aflatoxin-related liver cancer: Systematic review and meta-analysis. *European Journal of Cancer*.

[70] IARC. (1972). *Monographs on the evaluation of the carcinogenic risk of chemicals to man.Geneva: World Health Organization, International Agency for Research on Cancer*, Present Multivolume work, S7.

[71] Scholl, P. F., et al. (2006). Quantitative Analysis and Chronic Dosimetry of the Aflatoxin B1 Plasma Albumin Adduct Lys-AFB1 in Rats by Isotope Dilution Mass Spectrometry. *Chemical Research in Toxicology*, 19(1), 44-49.

[72] Farombi, E. O. (2006). Aflatoxin contamination of foods in developing countries: Implications for hepatocellular carcinoma and chemopreventive strategies: Review. *African Journal of Biotechnology*, 5(1), 001-014.

[73] Shephard, G. S. (2008). Risk assessment of aflatoxins in food in Africa. *Food Additives & Contaminants: Part A: Chemistry, Analysis, Control, Exposure & Risk Assessment*, 25(10), 1246-1256.

[74] Williams, J. H., et al. (2004). Human aflatoxicosis in developing countries: a review of toxicology, exposure, potential health consequences, and interventions. *American Journal Clinical Nutrition*, 80, 1106-1122.

[75] Murthy, T. R. K., et al. (1975). Aflatoxin B1, B2 and M were detected in liver, gall bladder, spleen, heart, muscle and kidney of growing swine when protein and protein-free portions of the diet were separately fed. *Journal of Animal Science*, 41(5), 1339-1347.

[76] Barrett, J. R. (2005). Liver Cancer and Aflatoxin: New Information from the Kenyan Outbreak. *Environmental Health Perspectives*, 113(12), A 837-A838.

[77] Gong, Y., Hounsa, A., & Egal, S. (2004). Postweaning exposure to aflatoxin results in impaired child growth: a longitudinal study in Benin, West Africa. *Environmental Health Perspective*, 112, 1334-1338.

[78] Dvorakova, I., et al. (1977). Aflatoxin and encephalopathy with fatty degeneration of viscera (Reye). *Annals of Nutrition Aliment*, 31, 977-989.

[79] Fapohunda, S. O., et al. (2007). Enzyme-related aflatoxin production in vital organs of rats fed with Aspergillus species- inoculated rat chow. *Journal of Biology and Environmental Science*, 1(1), 1-41.

[80] Wu, F., & Tritscher, A. (2011). Aflatoxins a global public health problem: Aflatoxins-health impact, Jan 2011. *World Health Organization*, http://www.agriskmanagementforum.org/farmd/sites/agriskmanagementforum.org/files/WHO-Aflatoxin-public health issue.pdf, 1-18.

[81] Bommakanti, A. S., & Waliyar, F. (2012). Importance of aflatoxis in human and livestock health. *Aflatoxin*, http://www.icrisat.org/aflatoxin/health.asp, (Accessed on 8th June 2012).

[82] Butterworth, R. F. (2000). Complications of cirrhosis. III. Hepatic encephalopathy. *Journal of Hepatology*, 32(1), 171-180.

[83] Bémeur, C., Desjardins, P., & Butterworth, R. F. (2010). Role of Nutrition in the Management of Hepatic Encephalopathy in End-Stage Liver Failure: Review Article. *Journal of Nutrition and Metabolism*, doi:10.1155/2010/489823:, 1-12.

[84] Cauli, O. (2010). Brain Aquaporin 4 in Hyperammonemia. *Medicina Universitaria*, 12(46), 47-53.

[85] Hussain, Z., Khan, M. Z., & Z.u, Hassan. (2008). Production of aflatoxins from Aspergillus flavus and Acute aflatoxicosis in young broiler chicks. *Pakistan Journal of Agricutlural Sciences*, 45(1), 95-102.

[86] Columre, R. A., & Sharma, R. P. (1985). Effect of repeated exposure of aflatoxin B1 on brain biogenic amines and metabolites in the rat. *Toxicology and Applied Pharmacology*, 80, 496-501.

[87] Jayasekra, S., et al. (1989). Alteration of biogenic amines in mouse brain regions by alkylating agents. I. Effects of aflatoxin B1 on brain monoamines concentrations and activities of metablozing enzymes. *Archives Environmental Contamination and Toxicology*, 18, 396-403.

[88] Weekley, L. B., et al. (1989). Differential changes in rat brain tryptophan, serotonin and tyrosine levels following acute aflatoxins B1 treatment. *Toxicology Letter*, 47, 173-177.

[89] Harriet, A. M. (2003). Is indoor mold contamination a threat to health? *Journal of Environmental Health*, http://130.88.242.202/medicine/Aspergillus/articlesoverflow/12971049.pdf, 62(2), 0022-0892.

[90] Lakkawar, A. W., Chattopadhyay, S. K., & Johri, T. S. (2004). Experimental aflatoxin B1 toxicosis in young rabbits- A clinical and patho-anatomical study. *Slovenian Veterinary Research*, 41, 73-81.

[91] Arundine, A., & Tymianski, M. (2004). Molecular mechanisms of glutamate-dependent neurodegeneration in ischemia and traumatic brain injury. *Cellular and Molecular Life Sciences*, 61, 657-668.

[92] Facci, L., Leon, A., & Skaper, S. D. (1990). Excitatory amino acid neurotoxicity in cultured retinal neurons: involvement of N-methyl-D-aspartate (NMDA) and non-NMDA receptors and effect of ganglioside GM1. *Journal of Neuroscience Research*, 27(2), 202-210.

[93] Ferreira, I. L., Duarte, C. B., & Carvalho, A. P. (1999). Ca2+ influx through glutamate receptor-associated channels in retina cells correlates with neuronal cell death. *European Journal of Pharmacology*, 302(1-3), 153-162.

[94] Lye, M. S., et al. (1995). An outbreak of acute hepatic encephalopathy due to severe aflatoxicosis in Malaysia. *American Journal of Tropical Medicine and Hygiene*, 53, 68-72.

[95] Gursoy, N., et al. (2008). Changes in spontaneous contractions of rat ileum by aflatoxin in vitro. *Food Chemistry and Toxicology*, 46(6), 2124-2127.

[96] Applegate, T. J., et al. (2008). Effect of aflatoxin culture on intestinal function and nutrient loss in laying hens. *Poultry Science*, 88(6), 1235-1241.

[97] Kussak, A., Andersson, B., & Andersson, K. (1995). Determination of aflatoxins in airborne dust from feed factories by automated immunoaffinity column clean-up and liquid chromatography. *Journal of Chromatography A.*, 708(1), 55-60.

[98] Mézes, M. (2008). Mycotoxins and other contaminants in rabbit feeds. 9th World Rabbit Congress- June 10-13, 2008- Verona- Italy. *Nutrition and Digestive Physiology*,

http://world-rabbit-science.com/WRSA-Proceedings/Congress-2008-Verona/Papers/N2-Mezes.pdf, 491-505.

[99] Jakab, G. J., et al. (1994). Respiratory aflatoxicosis: suppression of pulmonary and systemic host defenses in rats and mice. *Toxicology and Applied Pharmacology*, 125(2), 198-205.

[100] Amstad, P., et al. (1984). Evidence for membrane-mediated chromosomal damage by aflatoxin B1 in human lymphocytes. *Carcinogenesis*, 5, 719-723.

[101] Tuzcu, M., et al. (2010). Effects of Aflatoxin on the Proportions of Peripheral Blood Leukocytes and Alpha-Naphtyl Acetate Esterase (ANAE) Positive Lymphocytes in the Mouse. *Kafkas Univ Vet Fak Derg*, 16(2), 337-341.

[102] Shuaib, F. M. B., et al. (2010). Association between Anemia and Aflatoxin B1 Biomarker Levels among Pregnant Women in Kumasi, Ghana. *American Journal of Tropical Medicine and Hygiene*, 83(5), 1077-1083.

[103] Verma, R. J., & Raval, P. J. (1991). Cytotoxicity of aflatoxin on red blood corpuscles. *Bulletin of Environmental Contamination and Toxicology*, 47(3), 428-432.

[104] Verma, R. J., & Raval, P. J. (1992). Alterations in erythrocytes during induced chronic aflatoxicosis in rabbits. Bulletin of Environmental Contamination and Toxicology, . , 49(6), 861-865.

[105] Dietert, R. R., et al. (1983). Hematological Toxicology following Embryonic Exposure to Aflatoxin-B1. *Experimental Biology and Medicine*, 173(4), 481-485.

[106] D°onmez, N., et al. (2012). Research ArticleEffects of Aflatoxin on Some Haematological Parameters and Protective Effectiveness of Esterified Glucomannan in Merino Rams. *The Scientific World Journal*, 10.1100/2012/342468:, 1-4.

[107] Marin, D. E., et al. (2002). Changes in performance, blood parameters, humoral and cellular immune responses in weanling piglets exposed to low doses of aflatoxin1. *Journal Animal Science*, 80, 1250-1257.

[108] Goerttler, K., et al. (1980). Effects of Aflatoxin B1 on Pregnant Inbred Sprague-Dawley Rats and Their F1 Generation. *A Contribution to Transplacental Carcinogenesis 1, 2, 3 JNCI Journal of the Natational Cancer Institute*, 64(6), 1349-1354.

[109] AAFRD,. (2003). Moldy Feed and Reproductive Failure in Cows. *Alberta Agricultural, Food and Rural Development (AAFRD).*, http://www1.agric.gov.ab.ca/$department/deptdocs.nsf/all/agdex849/$file/666-5 .pdf?

[110] Uriah, N., Ibeh, I. N., & Oluwafemi, F. (2001). A Study on the Impact of Aflatoxin on Human Reproduction. *Laboratory Report. African Journal of Reproductive Health / La Revue Africaine de la Santé Reproductive*, 5(1), 106-110.

[111] El -Azab, S. M., et al. (2009). Study of aflatoxin B1 as a risk factor that impair the reproductive performance in females- Egypt. *The Internet Journal of Toxicology*, http://

www.ispub.com:80/journal/the-internet-journal-of-toxicology/1 study-of-aflatoxin-b1-as-a-risk-factor-that-impair-the-reproductive-performance-in-females-egypt.html, (Accessed on 10th June 2012), 6(1).

[112] Hasanzadeh, S., & Rezazadeh, L. (2012). Effects of aflatoxin B1 on the growth processes of spermatogenic cell series in adult male rats. *Comparative Clinical Pathology*, http://rd.springer.com/article/10.1007/s00580-012-1445-2.

[113] Fapohunda, S. O., et al. (2008). Aflatoxin-mediated Sperm and Blood Cell Abnormalities in Mice Fed with Contaminated Corn. *Mycobiology*, 36(4), 255-259.

[114] Clarke, R. N., Doerr, J. A., & Ottinger, M. A. (1987). Age-Related Changes in Testicular Development and Reproductive Endocrinology Associated with Aflatoxicosis in the Male Chicken. *Biology of Reproduction*, 36, 117-124.

[115] Jiang, Y., et al. (2005). Aflatoxin B1 albumin adduct levels and cellular immune status. *Ghanaians International Immunology*, 17(6), 807-814.

[116] Jiang, Y., et al. (2008). Aflatoxin-related immune dysfunction in health and in human immunodeficiency virus disease. *Clinical Developmental Immunology*, 1-12.

[117] Sahoo, P. K., Chattopadhyay, S. K., & Sikdar, A. (1996). Immunosuppressive effects of induced aflatoxicosis in rabbits. *Journal of Applied Animal Research*, 9, 17-26.

[118] Williams, J. H., et al. (2005). Connecting the Dots: Logical and Statistical Connections between Aflatoxin Exposure and HIV/AIDS. *Peanut Collaborative Research Support Program.*

[119] Williams, J. H., et al. (2010). HIV and hepatocellular and esophageal carcinomas related to consumption of mycotoxin-prone foods in Sub-Saharan Africa. *American Society for Nutrition*, 92, 154-160.

[120] Ayodele, F. O. (2007). Association between exposure to aflatoxin and status of HIV-infected adults in Ghana. *University of Alabama at Birmingham.*

[121] Giambrone, J. J., et al. (1978). Effect of aflatoxin on the humoral and cell-mediated immune systems of the chicken. *American Journal of Veterinary Research*, 39(2), 305-308.

[122] Dugyala, R. R., & Sharma, R. P. (1996). The effect of aflatoxin B1 on cytokine mrna and corresponding protein levels in peritoneal macrophages and splenic lymphocytes. *International Journal of Immunopharmacology*, 18(10), 599-608.

[123] Fernández, A., et al. (2000). Effect of aflatoxin on performance, hematology, and clinical immunology in lambs. *Canadian Journal of Veterinary Research*, 64(1), 53-58.

Chapter 12

The Significance of Glutathione Conjugation in Aflatoxin Metabolism

Tahereh Ziglari and Abdolamir Allameh

Additional information is available at the end of the chapter

1. Introduction

We all are exposed through air, food, drinks and skin contacts to harmful compounds throughout the period of our lifetime, including, a variety of pharmaceuticals and food-derived carcinogen metabolite (e.g. N-acetoxy-PhIP), [52], plant toxins (such as *glycoalkaloids* in nightshades[1], *cyanogenic glucosides* [2], or *pyrrolizidine alkaloids* in some herbs and herbal teas), xenobiotics[3] producing during early human pregnancy, fungal and bacterial toxins such as aflatoxins[4]; and cyanotoxin[5]; as well as free radicals and hydroperoxides. Many of these compounds are lipophilic and the organism can get rid of them only through metabolism.

Biotransformation has been conveniently categorized into three distinct phases, which act in a tightly integrated manner. Phases I and II enzymes catalyze the conversion of a lipophilic, non-polar xenobiotic into a more water-soluble and therefore less toxic metabolite, which can then be more easily excreted from the body. Phase I biotransformation seems to be enzymes that catalyzes oxidation, reduction or hydrolyze reactions, it usually converts substrates to more polar forms by introducing or unmasking a functional group (e.g., —OH, —NH2, or —SH). Phase I consist primarily of microsomal enzymes, which are found abundantly in the liver, gastrointestinal tract, lung and kidney, consisting of families and subfamilies of enzymes that are classified based on their amino acid sequence identities or similarities. [84]. Many of the enzymes like monooxygenases are found in the endoplasmic

1 - Plants like potatoes, tomatoes, peppers, egg plant, tobacco, some spices.

2- *Like bitter almond, cassava root, sorghum root, lima bean, fruit seed, etc.*

3 - Chemical compounds foreign to the human organism without nutritional value

4 - A group of mycotoxins of which aflatoxin B1 is the most potent hepatocarcinogen

5 - A toxin producing by cyanobacteria of which microcystin-LR is predominant

reticulum membrane, but others such as the dehydrogenases for example alcohol dehydrogenases and peroxidases located in the cytoplasm, while still others such as monoamine oxidase are localized in mitochondria. Monooxygenases are also known as mixed function oxidases because in a typical reaction, one molecule of oxygen is consumed (reduced) per substrate molecule: one oxygen atom appearing in the product and the other in a molecule of water. The reaction scope of monooxygenases includes heteroatom oxidation, aromatic and aliphatic hydroxylation, epoxidation, and Baeyer-Villiger oxidation. There are two major types of microsomal monooxygenase, both of which require NADPH as an external reductant: the cytochrome P450 (CYP) system and flavin-containing monooxygenases. The mechanism of CYP is a complex cascade of individual steps involving the interaction of protein redox partners and consumption of reducing equivalents, usually in the form of NADPH. The iron heme containing enzyme, CYP, consists of two enzymes: NADPH–cytochrome P450 reductase and CYP. It is involved in the oxidative metabolism of many endogenous substances such as steroids and bile acids, as well as the detoxication of a wide variety of xenobiotics. It can oxidize AFB_1 to several products. Only one of these, the 8,9-exo-epoxide, appears to be mutagenic and the others are detoxification products. P4503A4, which can both activate and detoxicate AFB1, is found in the liver and the small intestine. [33], [52]. Flavincontaining monooxygenases catalyze an NADPH- and an oxygen-requiring oxidation of substances (primarily xenobiotics) bearing functional groups containing nitrogen, sulfur, or phosphorus. The properties of the CYPs electron transport systems have also been reported [77].

In detoxification pathway, a series of enzyme-catalyzed processes with broad specificities convert the toxic substances into less toxic metabolites by chemical reactions within the body. Although biotransformation reactions take place within cytoplasm and mitochondria but they mostly happen within endoplasmic reticulum (E.R). Cell types also differ in their biotransforming potential for example cells located near the major points of xenobiotic entry into the body such as liver, lung, and intestine possess greater concentrations of biotransforming enzymes than others [52].

Phase II conjugation reactions which generally act follow phase I activation consists of reactions in which metabolites containing appropriate functional groups are conjugated with substances such as glucuronate, glutamate, sulfate, reduced glutathione or uridine diphosphate (UDP)-glucuronic acid to finally discharge them through urine or bile. In general, conjugation dramatically improves solubility, which then promotes rapid excretion. Among the several types of conjugation reactions which are present in the body, including glucuronidation, sulfation, and glutathione and amino acid conjugation, glutathione which is catalyzed by glutathione S-transferases, is the major phase II reaction in many species [52]. With the exception of acetylation, methylation and fatty acid conjugation, the strategy of phase II biotransformation is to convert a xenobiotic to a more hydrophilic form via the attachment of a chemical moiety which is ionizable at physiological pH. This metabolic transformation also results in reduced affinity of the compound for its cellular target. [67], [23].

In animals, elimination of the soluble compounds from cells and excretion of biotransformed molecules from the body referred to as phase III. It has been suggested that the phase III of detoxification system to be called antiporter activity. Antiporter activity is an important factor in the first pass metabolism of pharmaceuticals and other xenobiotics. The antiporter is an energy-dependent efflux pump, which pumps xenobiotics out of a cell, thereby decreasing the intracellular concentration of xenobiotics. In eukaryotic organisms, they are actively excreted or compartmentalized in the vacuole by ATP-dependent GS-X pumps [42], [27]. Indeed, as the glutathionylated moiety is hydrophilic, the conjugate cannot usually simply re-diffuse back into the cell [77]. Antiporter activity in the intestine appears to be co-regulated with intestinal phase I CYP3A4 enzyme. This observation suggests the antiporter may support and promote detoxification. Possibly, its function of pumping non-metabolized xenobiotics out of the cell and back into the intestinal lumen, may allow more opportunities for phase I activity to metabolize the xenobiotic before it is taken into circulation. Although, most literature on detoxification refers to liver enzymes, as the liver is the site of the majority of detoxification activity for both endogenous and exogenous compounds, however, the first contact the body with the majority of xenobiotics take places in the gastrointestinal tract. Intestinal mucosa possesses enzyme systems capable of various types of biotransformation of xenobiotics [52]. Among the detoxification pathways, glutathione conjugation pathway is the prominent route of AFB1 inactivation in liver of mammalians. Depending on the availability of cellular GSH and the activation of glutathione S-transferase subclasses, detoxification of AFB1 is facilitated [24].

2. Glutathione

Glutathione is a ubiquitous thiol-containing isotripeptide (γ-glu-cys-gly, FW 307.3), consisting of glycine, glutamic acid and cysteine molecules which was first discovered by Sir Fredrick Gowland Hopkins in 1920s, synthesized de novo in mammalian cells (Figure 1). This water soluble antioxidant compound is an unusual peptide in that the peptide bond between the glutamate residue and the cysteine residue is formed with the γ-carboxylate group of the former rather than the α-carboxylate group. Today along with β-carotene, ascorbic acid (vitamin C), α-tocopherol (vitamin E) and flavonoids *etc.*, GSH[6] is commonly referred to as an antioxidant [17], which neutralizes free radicals due to the high electron-donating capacity of its sulfydryl (-SH) group, [13], and prevents damage to important cellular components, implicates in the cellular defense against xenobiotics. Glutathione status is a highly sensitive indicator of cell functionality and viability. Its levels in human tissues normally range from 0.1 to 10 mM, being most focused in liver (up to 10 mM) and in the spleen, kidney, lens, erythrocytes and leukocytes and its emptying be joined to a variety of diseases. Under normal conditions, glutathione is predominantly present in its reduced form, with only a small proportion present in its fully oxidized state [20].

Moreover, the GSH/GSSG[7] pair with their high reduction potential participates in maintaining other cellular thiol in a reduced state. Finally, GSH tends to a substrate or cofactor in some of

6 - Glutathione, reduced form

7 - Glutathione, oxidized state

GSH linked enzymes. There are a number of GSH linked enzymes that are involved in cellular protection against toxic substances. The glyoxalase I and II which are responsible for catalyzing the conversion of methylglyoxal (a by-product in glycolysis) to lactic acid are among these enzymes [76]. Glutathione reductase (GR) which catalyzes the reduction of GSSG using NADPH as a reductant is also a glutathione-linked enzyme involved in cell protection. GR is important to keep the high cellular reductive potential. Selenium dependent glutathione peroxidase are other GSH-linked enzymes that catalyze the reduction of peroxides using GSH as the reducing agent [7]. Finally, last but not the least, glutathione transferases are also GSH dependent enzymes with many properties among which catalyzing the conjugation of GSH to various electrophilic compounds is one of the most investigated function [25].

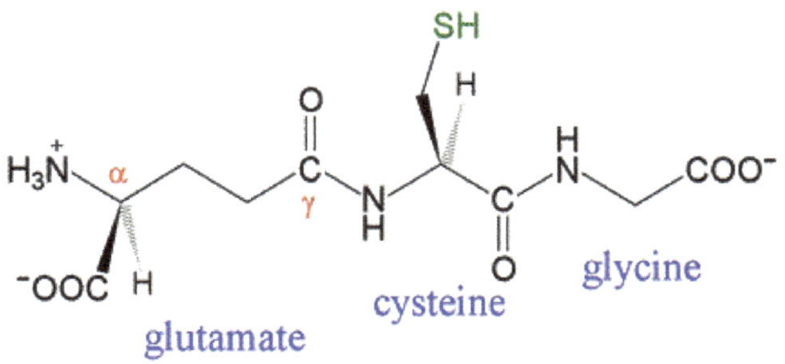

Figure 1. Structure of reduced glutathione; glutamate is linked in an isopeptide bond (*via* its γ-carboxyl group) to cysteine, which in turn forms a peptide linkage with glycine

3. Glutathione S-transferase

Glutathione *S*-transferases (GST, EC 2.5.1.18), which first discovered as enzymes in 1961 [12], are abundant proteins encoded by a highly divergent, ancient gene family. These major cellular detoxification enzymes present mostly in liver and kidney as well as intestine. In spite of 40 years of research the exact function of this protein is more complex than ever, but it has been found that these intracellular dimeric proteins, play a major role in the intracellular transport of endogenous compounds, metabolizes various electrophilic xenobiotics, ligand transport and thus protects cells against toxic effects [31], [87], [85]. GST catalyzes the conjugation of glutathione on the sulfur atom of cysteine to various electrophiles and catalyses the conjugation of various electrophiles with GSH, detoxifying both exogenously and endogenously derived toxic compounds [13].

3.1. Classification and structure

The superfamily of the glutathione transferases are divided into at least four major families of proteins, namely cytosolic or soluble GSTs, mitochondrial GSTs, microsomal GSTs and bacterial fosfomycin-resistance proteins [39], [6], [69]. The cytosolic GSTs (cGSTs) have been subgrouped into numerous divergent classes on the basis of their chemical, physical and structural properties [39], [70]. The mitochondrial GSTs, also known as kappa class GSTs, are soluble enzymes that have been characterized in eukaryotes [65]. The third GST family comprises membrane-bound transferases called membrane-associated proteins involved in ecosanoid and glutathione metabolism, but these bear no similarity to soluble GSTs [44]. Representatives of all three families are also present in prokaryotes but the fourth family is found exclusively in bacteria [4]. The mammalian soluble GSTs are so far divided into eight classes based on their amino acid sequences including: Alpha (α), Kappa (κ), Mu (μ), Omega (ω), Pi (π), Sigma (σ), Theta (θ) and Zeta (ζ), [11], [78], [64], [40]. GSTs are named using a letter corresponding to their class membership and Arabic numerals after the subunit composition (*e.g.* GST A1-1 is a homodimeric alpha class GST consisting of two subunit 1).

3.2. Presence of GST in cells

3.2.1. Microbial GST

For a long time, GST enzymes from microbial sources were neglected and were not systematically studied. One of the reasons for this was the poor activity of microbial GSTs with CDNB[8] as a model substrate for GST activity, which led to the conclusion that these enzymes are rare in unicellular organisms [81], [77]. The first evidence for the presence of GSTs in bacteria was reported more 30 years ago by Takashi Shishido who showed the presence of GST activity in a strain of *Escherichia coli* [71]. Since then, GSTs have been found to be broadly distributed in aerobic prokaryotes, but not in anaerobic bacteria [59]. The absence of the enzyme in these microorganisms is consistent with the lack of GSH [28]. Bacterial glutathione transferases are part of a superfamily of enzymes that play a key role in cellular detoxification. Bacterial GSTs are implicated in a variety of distinct processes such as the biodegradation of xenobiotics, protection against chemical and oxidative stresses and antimicrobial drug resistance. In addition to their role in detoxification, bacterial GSTs are also involved in a variety of distinct metabolic processes such as the biotransformation of dichloromethane, the degradation of lignin and atrazine, and the reductive dechlorination of pentachlorophenol [4], [51].

3.2.2. GSTs of fungi and yeasts

Until recently, relatively little was known about the presence and role of GST in fungi. However, expression of GST has been reported in some fungal species such as *Issatchenkia orientalis,* [73], *Phanerochaete chrysosporium, Yarrowia lipolytica, Mucor circinelloides* [70] *Schizosaccharomyces pombe, Aspergillus nidulans, Aspergillus parasiticus, Aspergillus flavus, Aspergillus fumigates* [Burns et al., 2005] *Saccharomyces cerevisiae,* and *Cunninghamella elegans,* [70] [48] *etc.* However, the role of the enzyme in fungi, particularly toxigenic strains, is not well understood [2]. Although it has been shown that GST has a significant role in detoxifi-

8 - 1-chloro-2,4-dinitrobenzene

cation of aflatoxin and there is a possibility that this enzyme catalyses the conjugation of GSH to AFB_1- epoxide to excrete its derivatives from the body, in 1988 and for the first time Saxena et al. reported that the relation of cytosolic GSH S-transferases from *A..flavus* to aflatoxin synthesis. In truth, they showed that in contrast to other cells that GST has a critical function to break down the aflatoxin, in aflatoxigenic *Aspergillus* spp., there is positive correlation between the GST activity and aflatoxin production [68], factors influencing aflatoxin formation such as growth period, medium etc., always enhanced GST activity in the toxigenic strain. Since the non-toxigenic strain produces no aflatoxin, these factors have little effect on its GST activity. Our experience with GSH-conjugation system using inducers/ inhibitors of aflatoxin metabolism in fungi also show a positive correlation of aflatoxin synthesis and GST activity in *Aspergillus* species [2], [88].

3.2.3. Plant GSTs

Plant GSTs are a family of multifunctional enzymes involved in the intracellular detoxification of xenobiotics and toxic compounds produced endogenously [54], [26]. Most of the enzymes are stress-inducible and play a role in the protection of plants from adverse effects of stresses. However, the activities of different GSTs have been detected and characterized in many plants, including maize, wheat, tobacco, soybean, barley, chickpea, peanut, sorghum, and sugarcane [20], [21], [22], [75].

3.2.4. Mammalian GST

The isoenzymes of glutathione transferase have been most widely studied in rat liver. Six basic transferases in rat liver liver have been characterized. In rabbit, GST catalyzes the conjugation of activated AFB_1 with glutathione. In an experiment to assess the abilities of lung and liver GSTs to conjugate AFB_1-8, 9-epoxide, it has been shown that alpha-class and mu-class GSTs are of similar importance in catalyzing the reaction in the lung. The human glutathione S-transferase, possess both enzymatic and non-enzymatic functions and are involved in many important cellular processes, such as, phase II metabolism, stress response, cell proliferation, apoptosis, oncogenesis, tumor progression and drug resistance. The nonenzymatic functions of GSTs involve their interactions with cellular proteins, such as, Jun N-terminal kinase,(JNK), tumor necrosis factor receptor-associated factor-2 (TRAF2), apoptosis-signal-regulating kinase 1 (ASK), serine/threonine kinases (PKA, PKC), and tissue transglutaminase 2 (TGM2), during which, either the interacting protein partner undergoes functional alteration or the GST protein itself is post-translationally modified and/or functionally altered [53], [74].

3.3. Different functions of GST

3.3.1. The metabolic function of GSTs

GSTs have been reported to involve in steroid metabolism by catalyzing the isomerization of Δ^5-androstene-3, 17-dione to, Δ^4-androstene-3, 17-dione, and biosynthesis of prostaglandins. GST M2-2 is a prostaglandin E synthase in the brain cortex [8] and rat GST A1-1 and GST

A3-3 catalyze the reduction of PGH_2 to PGF_2. The isomerization reaction of PGF_2 to PGD_2 is also catalyzed by sigma class of GST. PGD_2, PGE_2 and PGF_2 act as hormones that bind to G-protein coupled receptors which regulate other hormones and neurotransmittors. Prostaglandin D_2 and E_2 are unstable and will easily be converted to prostaglandin J_2 and A_2, respectively and their derivatives inhibit NFκB, [66] a family of transcription factors that regulate the transcription of genes important for inflammatory processes. There are interesting speculations that GSTs might block other anti-inflammatory pathways by catalyzing the conjugation of GST to PGJ_2 [38], [25]

3.3.2. The ligandin function of GSTs

Because of exhibiting a ligand binding function, glutathione tranferases, have been known as ligandin, a function, which involves the noncovalent binding of nonsubstrate hydrophobic ligands such as heme, bilirubin, various steroids, and conceivably some lipophilic anticancer drugs as well. Although GSTs are generally viewed as playing a protective role in foreign compound metabolism, they can also catalyze reactions that lead to toxification. Examples include the GST dependent metabolism of 1,2-dibromoethane and related haloalkanes and probably also metabolism of the 6-thiopurine prodrug azathioprin [60], [5]. Similarly, the cytotoxicity of the polypeptide antibiotic neocarzinostatin is greatly enhanced by thiols such as GSH, although in this case there is no apparent requirement for GST catalysis [25], [18]. [82].

3.3.3. The regulatory function of GSTs

In addition to above functions, GSTs also are responsible for interacting the proteins and enzymes. For example GST P1-1 interacts with c-Jun N-terminal kinase 1 (JNK1) suppressing the basal kinase activity. GST P1-1 also has a role in protection and cell survival after exposure to H2O2 but not against UV-induced apoptosis [1]. Whereas, mouse GST M1-1 protects cells against both UV-and H2O2-induced cell death and binds to apoptosis signal-regulating kinase 1 (ASK1), inhibits its kinase activity [16]. Moreover, mouse GST A4-4 has also been proposed to interact with JNK and prevent cells from 4-hydroxynonenal induced apoptosis [15], [25].

3.3.4. The detoxification function of GSTs

As enzymes, GSTs are involved in many different detoxification reactions. They are commonly referred to as phase II enzymes. They catalyze the conjugation of GSH to a wide variety of endogenous and exogenous electrophilic toxic compounds. The GSH conjugates are excreted as mercapturic acids by the phase III metabolic pathway [41]. GST P1-1, GST M1-1 and GST A1-1 have been shown to catalyze the inactivation process of α, β unsaturated carbonyls like acrolein, (a cytotoxic compound present in tobacco smoke), propenals, (generated by oxidative damage to DNA) and alkenals, (formed by oxidative damage to lipids) [25], [70].

3.4. GST and aflatoxin

3.4.1. Introduction

Study on GSTs of *Aspergillus flavus* stems from its ability to synthesize the aflatoxin. Aflatoxins are one of the major causes of liver cancer in certain regions of Africa and Asia [83], [61]. These secondary metabolites which primarily produced by some *Aspergillus* spp. are ubiquitous, and under favorable conditions can grow on a wide variety of agricultural commodities. Aflatoxins are major concern with to public health and the most important toxicological interest in aflatoxins has concentrated on aflatoxin B1, largely due to its acute toxicity and carcinogenicity in humans and animals. [3], [62], [88]. Genetic studies on aflatoxin biosynthesis in *A. flavus* and *A. parasiticus* has been led to the cloning of 25 clustered genes within a 70 kb DNA region responsible for the enzymatic conversions in the aflatoxin biosynthetic pathway [86].

3.4.2. Primary metabolism of aflatoxin B_1

Once inside the body and for toxicity to occur, AFB_1 undergoes enzymatic conversion to electrophilic endo and exo stereoisomers of AFB_1-8,9-epoxide by the action of mixed function mono-oxygenase enzyme systems, CYPs are an intensively studied family of enzymes with currently approximately 4,000 known members. They have been found in almost all branches of the "tree of life", ranging from microorganisms over plants to mammalians. CYP enzymes are classified into families identified by a number (e.g., 1, 2, 3, and 4), subfamilies identified by a letter (e.g., 2A, 2B, 2D, and 2E), and specific members identified by another number (e.g., CYP2E1 and CYP2A6) [47], [19].

In human, five CYP gene families, namely; CYP1, CYP2, CYP3, CYP4 and CYP7 are believed to play crucial roles in hepatic as well as extra-hepatic metabolism and elimination of xenobiotics [50], [58], [84]. This superfamily of hemoproteins aids in the oxidation of various substrates such as steroids, eicosanoids, pharmaceuticals, pesticides, pollutants, and carcinogens [57]. As mentioned earlier, they bioactivate AFB_1 to an electrophilic, highly reactive and unstable metabolite known as aflatoxin-8,9-epoxide, which binds to guanine residues in nucleic acids, leading to irreversible damage in DNA and causing hepatocarcinoma in humans, primates, and ducks [32], [84]. However, only AFB_1 exo-epoxide (AFBO), binds appreciably to DNA (Figure 2). The AFBO is highly unstable, and it reacts with cellular nucleophiles and can induce mutations by alkylating DNA, principally at the N^7 position of guanine forming the 8,9-dihydro- 8-(N^7-guanyl)-9-hydroxy-AFB_1. In addition, AFBO can bind to proteins and other critical cellular nucleophiles [43], [63]. Initial studies reported that concentrations of AFB_1 which are likely to be achieved in the liver following ingestion of "real-world" concentrations of AFB_1 are bioactivated to AFBO primarily by CYP1A2, whereas much higher concentrations are catalyzed by CYP3A4 [30], [46], [79]. A recent study demonstrated a dominant contribution of CYP3A4 homologues in AFBO production. AFB1 metabolism studies in human liver microsomal preparations indicate a predominant role for CYP3A4 and that its expression level was an important determinant of the AFB1 disposition in human liver [45]. Specific CYP3A4 inhibitors like troleandomycin have been shown to inhibit AFBO production [29], while inducers of CYP3A4 activity such as 3-methylcholanthrene and rifampicin, increase AFB_1 metabolism in cultured human hepatocytes [49].

CYP1A homologues also metabolize AFB_1 to produce the detoxified metabolite AFM_1, whereas CYP3A enzymes[9], produce another detoxified metabolite, aflatoxin Q_1 (AFQ_1), the major metabolite of AFB_1 (Figure 2). [33]. Although both CYP1A and CYP3A isoforms oxidize AFB_1, there are conflicting reports on their relative importance [63].

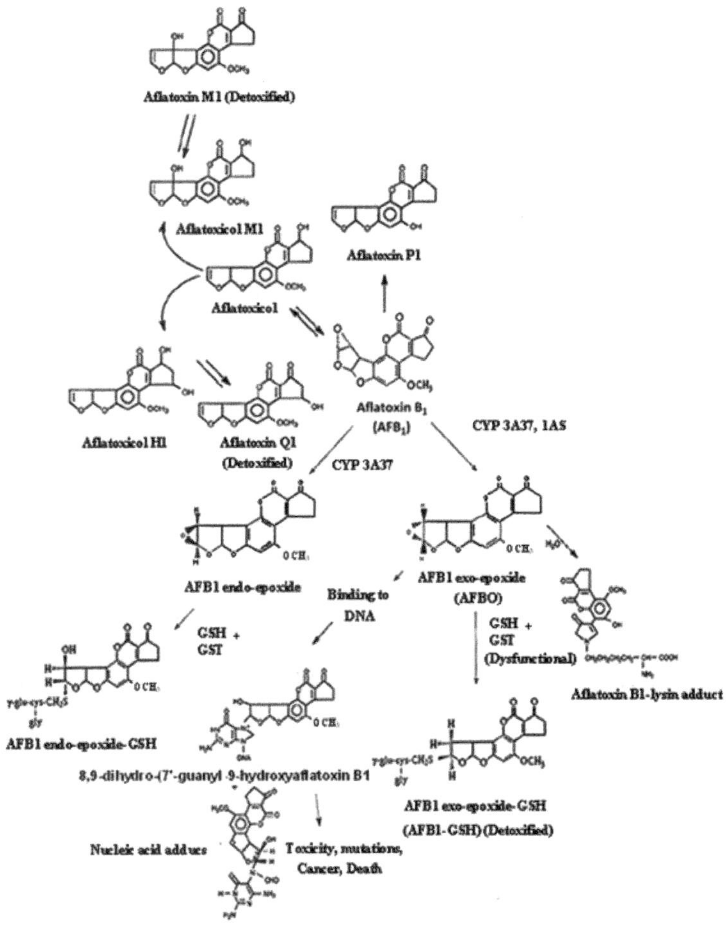

Figure 2. Bioactivation of AFB_1 to exo and endo-epoxides and subsequent GST-catalyzed conjugation with GSH.

9 - P450 III AY and in the fetal liver P450 III A6

CYPs may also catalyze demethylation to aflatoxin P (AFP_1) of the parent AFB_1 molecule, resulting in products less toxic than AFB_1. Other major metabolites in the human include aflatoxicol (AFL), AFLH1, AFB2á and AFB_1-2, 2-dihydrodiol [80].

3.4.3. Secondary metabolism of aflatoxin B_1

Oxidative metabolism of AFB_1 by cytochrome P450 results in the formation of several products such as AFB_1-epoxide which serve as substrates for phase II detoxification enzymes. Phase II enzymes such as GSTP1 and GSTA1, found in several mammalian species and non-tumorous liver tissues [14] are the first step in the mercapturic acid pathway, which leads to the excretion of the xenobiotics. Because conjugation of the electrophilic AFB_1-8,9-epoxide with GSH is an alternative fate to binding to nucleophilic centers in cellular macromolecules, GSTs play a critical role in the protection of tissues from the deleterious effects of bioactivated AFB_1, and tissues vary considerably in both GST concentration and distribution of specific GST isoforms. Two stereoisomers of AFB_1-8,9-epoxide were identified: AFB_1 exo-epoxide and AFB_1 endo-epoxide, and their corresponding GSH conjugates; AFB_1 exo-epoxide-GSH and AFB_1 endo-epoxide-GSH. It has been reported that only the exo-epoxide effectively interacts with DNA and was at least 500-fold more potent as a mutagen than the endo stereoisomer. [43], [72].

Throughout the animal kingdom, significant variations exist in the susceptibility of different species to AFB1. Man and rats are sensitive to AFB1 but mice can tolerate this mycotoxin. [35]. In man and rat as well as many mammalian species, AFB_1-8,9-epoxide is efficiently conjugated with reduced glutathione. Little is known about the identity of the GST which is responsible for detoxifying activated AFB_1. To date, the catalytic conjugation of AFB_1-8,9-epoxide has only been reported using rat and mouse GST as enzyme source and the ability of GST in other species to catalyze this reaction has not been described. In the investigation on hepatic rat GST responsible for catalyzing the conjugation of AFB_1-8,9-epoxide with GSH, it has been shown that the alpha class but not mu-class of GST possess greatest ability to metabolize activated AFB_1. Although the rat pi-class GST cannot catalyze this reaction it might be expected that the theta-class enzyme GST is active towards AFB_1-8,9-epoxide. By contrast with the rat, the mouse exhibits high constitutive levels of GST activity towards AFB_1-8,9-epoxide and alpha-class GST in Swiss-Webster mice possess high activity towards AFB_1-8,9-epoxide and can protect against DNA-binding by AFB_1 metabolites. Neither the murine mu-class nor pi-class GST can detoxify activated AFB_1 and all the activity towards this substrate is contributed by the alpha-class GST. It can be concluded that in the mouse the theta-class enzymes do not play a major role in the detoxification of activated AFB_1. Hamster liver contains significant levels of AFB_1-GSH-conjugating activity but the GST involved have not been characterized. In human liver, GST does not appear to play as important a role in providing protection against AFB_1 as the rodent GST. The *in vitro* studies have suggested that in comparison with rodents, relatively little AFB1-GSH conjugate is produced by human liver, but insufficient data exist to be certain that this reaction is not of physiological importance in man, particularly as an aflatoxin mercapturate has been detected in the urine of marmoset monkeys treated with AFB_1. The ability of human alpha-class

GST to detoxify activated AFB_1 has not been examined systematically. Three separate alpha-class isoenzymes, which represent the dimeric combinations of two distinct subunits (B1 & B2) have been described in human liver. Furthermore, it is not known whether man possesses inducible GST and if so, whether these might be involved in AFB_1 metabolism [34]. Nevertheless, it has been shown that in the humans, the GST with the highest activity toward AFB_1 exo-epoxide is the polymorphic hGSTM1-1 which is absent in about 50% of individuals in most human populations. This suggests that AFB_1-epoxide individuals lacking the beneficial effects of hGSTM1-1 may be at elevated risk. Indeed some reports suggest that the GSTM1 genetic polymorphism may affect AFB_1 detoxification in human liver. In contrast to the liver, the lung is composed of many different cell types and expression of GSTs in different human lung cell types is heterogenous. Thus certain cell types with low levels of GSTs or lacking specific GST isoforms may be at higher risk of AFB_1 toxicity [72]. GSTP was also demonstrated to significantly increase in early hepatocarcinogenesis and hepatocellular carcinoma compared to their adjacent normal tissues. Loss of GSTP1 has been suggested to increase the risk of DNA damage and mutation. Moreover, up-expression of GSTA was suggested to protect liver cells against oxidative stress via an extracellular signal-regulated kinases (ERKs) and p38 kinase (p38K)-related pathway, as well as through the inhibition of H2O2-induced apoptosis to inhibit reactive oxygen species (ROS)- induced lipid peroxidation. It was suggested that inactivated or down-regulated GSTP1 and GSTA1 genes could increase genomic damage when individuals were exposed to carcinogens. [14]. GSTs have also been shown to exhibit GSH-dependent peroxidase activity and thus may be involved in resistance to oxidative stress. Cytosolic GSTs have been identified in almost all organisms, with mammalian GSTs the most clearly characterized [Burns et al. 2005].

Besides the formation of GSH conjugates, glucuronide and sulfate conjugates of AFB_1 have also been described in a variety of species including rat, mouse, monkey and trout. The ability to form these alternative secondary metabolites may be of considerable physiological importance in species, like the trout, that are unable to produce AFB_1-GSH conjugates. Before AFB_1 can form glucuronide and sulfate conjugates it requires to be hydroxylated. The primary metabolites AFM_1, AFP_1, and AFQ_1 can readily form glucuronide or sulfate conjugates. Whilst such conjugation reactions may aid excretion of aflatoxin, their toxicological value is unclear as such hydroxylated metabolites are not particularly harmful because they are not subject to 8,9-epoxidation. However, it has been proposed that AFB_1 is itself capable of forming glucuronide and sulfate conjugate; these reactions might entail a molecular rearrangement possibly involving the addition of water to the keto group in the cyclopentone ring, that result in the introduction of a hydroxyl group into the AFB_1 structure. This proposal is of particular interest as it enables the direct detoxification of AFB_1 through reactions that may not involve cytochrome P450. These workers have also proposed that amines, thiols and alcohols might also be conjugated to AFB_1 via the keto group in the cyclopentone ring [34].

Alternatively, the AFB_1-epoxide can hydrolyse spontaneously to AFB_1-dihydrodiol. This is not a true detoxification process as the dihydrodiol product can rearrange at neutral pH values to form a dialdehydic phenolate ion. This AFB_1-dialdehyde can undergo Schiff-base formation with primary amine groups in proteins and is therefore likely to be cytotoxic.

Recently, a novel AFB_1-aldehyde reductase (AFB_1-AR) purified from ethoxyquin (EQ)-treated rat liver has been shown to metabolize the dialdehyde form of AFB_1-dihydrodiol to an AFB_1-dialcohol and its relative importance in AFB_1 detoxification may be considerable [35]. The toxicity of AFB_1 is selective towards certain species. In contrast with the mouse and hamster, the rat, guinea pig and man are susceptible to the hepatotoxic effects of AFB_1 [34]. The toxicity of the mycotoxin is based on a balance between the rate of primary activation of AFB_1 and the rate of detoxification of primary metabolites or repair of cellular damage, determined by the relative activity of enzymes responsible for these reactions; the differential toxicity of AFB_1, between species is thought to be due mainly to different levels of activity of xenobiotic-metabolizing enzymes. In this regard, the livers of mice which are resistant to the hepatoxic effects of AFB_1 contain high concentrations of a Yc-type GST subunit [55] that has considerable GSH conjugating activity towards AFB_1-epoxide [34], [37], [10], [9]. By contrast, the Fischer rat, an inbred strain that is five times more susceptible to AFB_1- induced liver cancer than the Wistar rat [34], possesses 20-fold less hepatic AFB_1-GSH-conjugating activity than the mouse. Fischer rats can, however, be protected against AFB_1 by treatment with the antioxidant EQ. It has shown that following EQ-treatment the livers of Fischer rats express a GST subunit that is immunochemically related to the mouse Yc subunit [35]. Moreover, this inducible polypeptide (Yc2, subunit 10) has high activity towards AFB_1-epoxide [35]. Thus, the Yc2 subunit is thought to confer protection against AFB_1, and its induction by EQ is likely to be one of the key mechanisms for the protective action of this anti-carcinogen [56].

The transport of foreign compounds out of cells can be achieved by at least two distinct families of efflux pump, both of which may provide protection against AFB_1 by helping eliminate the mycotoxin from target cells. The best characterized of these two pumps is P-glycoprotein, the product of the *mdr* 1 gene which has been studied extensively because of its involvement in acquired resistance to anticancer drugs. The other pump is the glutathione S-conjugate carrier which is responsible for the transport of endogenous compounds such as oxidized glutathione and leukotriene C4 as well as glutathione conjugates of foreign compounds an example of which might be S-(2,4-dinitrophenyl)glutathione. Both pump systems are ATP-dependent and are inhibited by vanadate but differ in that P-glycoprotein appears to have specificity towards hydrophobic compounds whereas the glutathione S-conjugate carrier is as its name suggests specific for leukotrienes and drug-glutathione conjugates. Although it is not known whether P-glycoprotein is able to transport AFB_1 the broad specificity of this efflux pump and its activity towards hydrophobic drugs suggests that this is likely. It also appears highly probable that the glutathione S-conjugate carrier is responsible for the transport of AFB_1 conjugated with GSH. Both P-glycoprotein and the glutathione S-conjugate carrier are expressed in the liver which is compatible with the hypothesis that these pumps could be involved in the efflux of AFB_1 and its metabolites. The involvement of P-glycoprotein in AFB_1 transport is supported by the fact that aflatoxin has been shown to induce mRNA encoding this protein in mouse liver. [36].

Relatively little is known about the enzymes responsible for the removal of AFB_1 that is bound covalently to DNA in mammalian cells. Exposure of cells to AFB_1 results in the formation of three major adducts. Of these, trans-2,3-dihydro-2-(N^7-guanyl)-3-hydroxy AFB_1

(AFB_1- N^7 G) is the most abundant. It is chemically unstable and is lost spontaneously from DNA *in vitro* to yield apurinic sites. The other two adducts, 2,3-dihydro-2-(N-formyl-2,3,6-triamino-4-oxopyrimidine-N-yl)-3-hydroxy AFB_1 and 8,9-dihydro-8-(2-amino-6-formamide-4-oxo-3,4-dihydropyrimid-5-yl formamido)-9-hydroxy AFB_1 (AFB_1-FAPY and AFB_1III respectively) are not spontaneously but appear to be removed catalytically by DNA repair enzymes. The loss of AFB_1-DNA adducts *in vivo* is biphasic and this occurs through two distinct mechanisms. Following exposure to AFB_1, all adduct species are removed rapidly until less than 1000 adducts per cell remain. Once this point is reached the AFB_1-FAPY and AFB_1 III adducts are no longer removed and only AFB1-N^7 G is lost but at a much slower rate from the cell [36].

3.4.4. Conclusion and future directions

Evidences presented in this review article clearly show that glutathione conjugation to aflatoxin metabolites which has been detected in aflatoxin-producing fungi as well as liver tissues of mammalians play a crucial role in reducing the interaction of aflatoxins with cellular macromolecules. However further studies is needed to answer the main questions about the contribution of glutathione conjugation system in removing aflatoxin in different cellular systems. The future direction of this topic is to find out experimental-based answers to the following questions:

1. What is the relationship between the rate of aflatoxin metabolism and the level of aflatoxin-GSH conjugate formation?
2. Which classes of glutathione S-transferases in each cellular system is directly responsible for involvement of aflatoxin-GSH conjugate formation
3. What is the relationship between the efficiency of glutathione conjugation system and toxic action of aflatoxins in different cell systems.

Acknowledgements

It is the time to express the deepest gratitude to Mr. Hamed Foroozesh for whole-hearted support and for his devoted care. Thanks for his vast knowledge and skill in many areas and his assistance.

Author details

Tahereh Ziglari[1*] and Abdolamir Allameh[2]

*Address all correspondence to: Tz_572@usc.edu

1 Herman Ostrow School of Dentistry, University of Southern California, Los Angeles, California, USA

2 Department of Biochemistry, Faculty of Medical Sciences, Tarbiat Modares University, Tehran, I.R. Iran

References

[1] Adler, V., Yin, Z., Fuchs, S. Y., Benezra, M., Rosario, L., Tew, K. D., Pincus, M. R., Sardana, M., Henderson, C. J., Wolf, C. R., Davis, R. J., & Ronai, Z. (1999). Regulation of JNK signaling by GSTp. *EMBO J.*, 18, 1321-34.

[2] Allameh, A., Razzaghi-Abyaneh, M., Shams, M., Rezaee, M. B., & Jaimand, K. (2002). Effect of neem leaf extract on production of aflatoxins and activities of fatty acid synthetase, isocitrate dehydrogenase, and glutathione S-transferase in Aspergillus parasiticus. *Mycopathologia*, 154, 79-84.

[3] Allameh, A. A., Ziglari, T., & Rasooli, I. (2011). Phytoinhibition of growth and aflatoxin biosynthesis in toxigenic fungi. *Chapter 15 in the book: aflatoxin, detection, measurement and control. InTech publication,*, 978-9-53307-711-6.

[4] Allocati, N., Federici, l., Masulli, M., & Dillio, C. (2009). Glutathione transferases in bacteria. *The FEBS Journal*, 276, 58-75.

[5] Anders, M. W., Lash, L., Dekant, W., Elfarra, A. A., & Dohn, D. R. (1988). Biosynthesis and biotransformation of glutathione S-conjugates to toxic metabolites. *Crit. Rev. Toxicol*, 18(4), 311-341.

[6] Armstrong, R. N. (2000). Mechanistic diversity in a metalloenzyme superfamily. *Biochemistry*, 39, 13625-13632.

[7] Arthur, J. R. (2000). The glutathione peroxidases. *Cell. Mol. Life Sci.*, 57, 1825-35.

[8] Beuckmann, C. T., Fujimori, K., Urade, Y., & Hayaishi, O. (2000). Identification of mu-class glutathione transferases M2-2 and M3-3 as cytosolic prostaglandin E synthases in the human brain. *Neurochem. Res.*, 25(5), 733-8.

[9] Beutler, T. M., & Eaton, D. L. (1992a). Complementary DNA cloning, messenger RNA expression and induction of alpha-class glutathione S-transferases in mouse tissues. *Cancer Res*, 52, 314-318.

[10] Beutler, T. M., Slone, D., & Eaton, D. L. (1992b). Comparison of the aflatoxin B1-8,9-epoxide conjugating activities of two bacterially expressed alpha class glutathione S-transferase isozymes from mouse and rat. *Biochem. Biophys. Res. Commun*, 188(2), 597-603.

[11] Board, P. G., Coggan, M., Chelvanayagam, G., Easteal, S., Jermiin, L. S., Schulte, G. K., Danley, D. E., Hoth, L. R., Griffor, M. C., Kamath, A. V., Rosner, M. H., Chrunyk,

B. A., Perregaux, D. E., Gabel, C. A., Geoghegan, K. F., & Pandit, J. (2000). Identification, Characterization, and Crystal Structure of the Omega Class Glutathione Transferases. *Journal of Biological Chemistry*, 275(32), 24798-24806.

[12] Booth, J., Boyland, E., & Sims, P. (1961). An enzyme from rat liver catalyzing conjugations with glutathione. *Biochem. J*, 79, 516-24.

[13] Cancado, G. M. A., De Rosa Jr, V. E., Fernandez, J. H., Maron, L. G., Jorge, R. A., & Menossi, M. (2005). Glutathione S-transferase and aluminum toxicity in maize. *Functional Plant Biology*, 32, 1045-1055.

[14] Chen, Y.l., Tseng, H. S., Kuo, W. H., Yang, S. F., Chen, D. R., & Tsai, H. T. (2010). Glutathione S-Transferase P1 (GSTP1) gene polymorphism increases age-related susceptibility to hepatocellular carcinoma. *BMC Medical Genetics*, 11, 46.

[15] Cheng, J. Z., Singhal, S. S., Sharma, A., Saini, M., Yang, Y., Awasthi, S., Zimniak, P., & Awasthi, Y. C. (2001). Two distinct 4-hydroxynonenal metabolizing glutathione S-transferase isozymes are differentially expressed in human tissues. *Biochem. Biophys. Res. Commun.*, 282, 1268-74.

[16] Cho, S. G., Lee, Y. H., Park, H. S., Ryoo, K., Kang, K. W., Park, J., Eom, S. J., Kim, M. J., Chang, T. S., Choi, S. Y., Shim, J., Kim, Y., Dong, M. S., Lee, M. J., Kim, S. G., Ichijo, H., & Choi, E. J. (2001). Glutathione S-transferase mu modulates the stress-activated signals by suppressing apoptosis signal-regulating kinase 1. *J. Biol. Chem.*, 276(16), 12749-55.

[17] Cotgreave, I. A., & Gerdes, R. G. (1998). Recent trends in glutathione biochemistry-glutathione-protein interactions: a molecular link between oxidative stress and cell proliferation? *Biochem Biophys Res Commun*, 242, 1-9.

[18] De Graff, W. G., Russo, A., & Mitchell, J. B. (1985). Glutathione depletion greatly reduces neocarzinostatin cytotoxicity in Chinese hamster V79 cells. *J. Biol. Chem*, 260, 8312-5.

[19] Diaz, G.j., Murcia, H. W., & Cepeda, S. M. (2010). Cytochrome P450 enzymes involved in the metabolism of aflatoxin B1 in chickens and quail. *Poultry Science association Inc*, 89, 2461-2469.

[20] Dixon, D. P., Cummins, L., Cole, D. J., & Edwards, R. (1998). Glutathione-mediated detoxification systems in plants. *Current opinion in Plant Biology*, 258-266.

[21] Dixon, D. P., Lapthorn, A., & Edwards, R. (2002). Plant glutathione transferases. *Genome Biol*, 3(3), 1-10.

[22] Dixon, D. P., Mc Ewen, A. G., Lapthorn, A. J., & Edwards, R. (2003). Forced evolution of a herbicide detoxifying glutathione transferase. *J. Biol. Chem.*, 278(26), 23930-23935.

[23] Dusinska, M., Staruchova, M., Horska, A., Smolkova, B., Collins, A., Bonassi, S., & Volkovova, K. (2012). Are glutathione S-transferases involved in DNA damage signalling? Interactions with DNA damage and repair revealed from molecular epi-

demiology studies. *Mutation Research/Fundamental and Molecular Mechanisms of Mutagenesis. In press.*

[24] Eaton, D. L., & Gallagher, E. P. (1994). Mechanisms of aflatoxin carcinogenesis. *Annu. Rev. Pharmacol. Toxicol.*, 34, 135-172.

[25] Edalat, M. (2002). Multiple functions of glutathione transferase. *Dissertation for the Degree of Doctor of Philosophy in Biochemistry presented at Uppsala University.*

[26] Edwards, R., Dixon, D. P., & Walbot, V. (2000). Plant glutathione Stransferase: enzymes with multiple functions in sickness and in health. Trends Plant Sci. , 5(5), 193-198.

[27] Ernst, R., Klemm, R., Schmitt, L., Kuchler, K., & Helmut, Sies. L. P. (2005). Yeast ATP-Binding Cassette Transporters: Cellular Cleaning Pumps. *Methods in Enzymology*, 400, 460-484.

[28] Fahey, R. C. (2001). Novel thiols of prokaryotes. *Annual Review of Microbiology*, 55, 333-356.

[29] Gallagher, E. P., Wienkers, L. C., Stapleton, P. L., Kunze, K. L., & Eaton, D. L. (1994). Role of human microsomal and human complementary DNA-expressed cytochromes P4501A2 and P4503A4 in the bioactivation of aflatoxin B1. *Cancer Research*, 54, 101-108.

[30] Gallagher, E. P., Kunze, K. L., Stapleton, P. L., & Eaton, D. L. (1996). The kinetics of aflatoxin B1 oxidation by human cDNA-expressed and human liver microsomal cytochromes P450 1A2 and 3A4. *Toxicology and Applied Pharmacology*, 141, 595-606.

[31] George, S. G. (1994). Enzymology and molecular biology of phase II xenobiotic conjugating enzymes in fish. *Aquatic toxicology, molecular biochemical and cellular perspectives (Eds.: D.C. Malins and G.K.Ostrander). Lewis Publishers, Ann Arbor.*, 27-86.

[32] Guengerich, F. P., Gillam, E. M., & Shimada, T. (1996a). New applications of bacterial systems to problems in toxicology. *Critical Reviews in Toxicology*, 26, 551-583.

[33] Guengerich, F. P., Johnson, W. W., Ueng, Y. F., Yamazaki, H., & Shimada, T. (1996b). Involvement of Cytochrome 450Glutathione S-Transferase, and Epoxide Hydrolase in the Metabolism of Aflatoxin B1 and Relevance to Risk of Human Liver Cancer. *Environment Health Perspectives*, 104, 557-562.

[34] Hayes, J. D., Judah, D. J., Mclellan, L. I., & Neal, G. E. (1991a). Contribution of the glutathione S-transferases to the mechanisms of resistance to aflatoxin B_1. *Pharmacol. Ther*, 50, 443-472.

[35] Hayes, J. D., Judah, D. J., Mc Lellan, L. I., Kerr, L. A., Peacock, S. D., & Neal, G. E. (1991b). Ethoxyquin-induced resistance to aflatoxin B_1 in the rat is associated with the expression of a novel alpha-class glutathione S-transferase subunit, Yc2, which possesses high catalytic activity for aflatoxin B_1-8,9-epoxide. *Biochem. J*, 279, 385-398.

[36] Hayes, J. D., Judah, D. J., Mc Lellan, L. I., & Neal, G. E. (1991c). Contribution of the glutathione S-transferases to the mechanisms of persistence to aflatoxin B1. *Pharmac. Ther*, 50, 443-472.

[37] Hayes, J. D., Judah, D. J., Neal, G. E., & Nguyen, T. (1992). Molecular cloning and heterologous expression of a cDNA encoding a mouse glutathione S-transferase Yc subunit possessing high catalytic activity for aflatoxin B1-8,9-epoxide. *Biochem. J.*, 285, 173-180.

[38] Hayes, J. D., & Strange, R. C. (2000). Glutathione S-transferase polymorphisms and their biological consequences. *Pharmacology*, 61(3), 154-66.

[39] Hayes, J. D., Flanagan, J. U., & Jowsey, I. R. (2005). Glutathione transferases. *Annu Rev Pharmacol Toxicol*, 45, 51-88.

[40] Hu, B., Deng, L., Wen, C., Yang, X., Pei, P., Xie, Y., & Luo, S. (2012). Cloning, identification and functional characterization of a pi-class glutathione S-transferase from the freshwater mussel Cristaria plicata. *Fish & Shellfish Immunology*, 32(1), 51-60.

[41] Ishikawa, T. (1992). ATP dependent glutathione S-conjugate transport pump. *Trends Biochem. Sci*, 17, 463-8.

[42] Ishikawa, T., Li, Z. S., Lu, Y. P., & Rea, P. A. (1997). The GS-X pump in plant, yeast, and animal cells: Structure, function, and gene expression. *Bioscience Reports*, 17, 189-207.

[43] Iyer, R., Coles, B., Raney, K. D., Thier, R., Guengerich, F. P., & Harris, T. M. (1994). DNA adduction by the potent carcinogen aflatoxin B1: mechanistic studies. *Journal of the American Chemical Society*, 116, 1603-1609.

[44] Jakobsson, P. J., Morgenstern, R., Mancini, J., Ford-Hutchinson, A., & Persson, B. (1999). Common structural features of MAPEG- a widespread superfamily of membrane-associated proteins with highly divergent functions in eicosanoid and glutathione metabolism. *Protein Sci*, 8, 689-692.

[45] Kamdem, L. K., Meineke, I., Godtel-Armbrust, U., Brockmoller, J., & Wojnowski, L. (2006). Dominant contribution of P450 3A4 to the hepatic carcinogenic activation of aflatoxin B1. *Chemical Research in Toxicology*, 19, 577-586.

[46] Kelly, J. D., Eaton, D. L., Guengerich, F. P., & Coulombe Jr, R. A. (1997). Aflatoxin B1 activation in human lung. *Toxicology and Applied Pharmacology*, 144, 88-95.

[47] Klaassen, C. D. (2001). Casarett and Doull`s Toxicology: The Basic Science of Poisons. *6th ed. McGraw-Hill, New York, N.Y.*

[48] Krajewski, M. P., Kanawati, B., Fekete, A., Kowalski, N., Schmitt-Kopplin, P., & Grill, E. (2012). Analysis of Arabidopsis glutathione-transferases in yeast. *Phytochemistry In Press.*

[49] Langouet, S., Coles, B., Morel, F., Becquemont, L., Beaune, P., Guengerich, F. P., Ketterer, B., & Guillouzo, A. (1995). Inhibition of CYP1A2 and CYP3A4 by oltipraz re-

sults in reduction of aflatoxin B1 metabolism in human hepatocytes in primary culture. *Cancer Research*, 55, 5574-5579.

[50] Lewis, D. E. (2003). P450 structures and oxidative metabolism of xenobiotics. *Pharmacogenomics*, 4, 387-395.

[51] Li, C., Su, X., Li, Y., Li, T., Sun, C., Zhou, T., & Liu, H. (2012). Two classes of glutathione S-transferase genes with different response profiles to bacterial challenge in Venerupis philippinarum. *Fish & Shellfish Immunology*, 32(1), 219-222.

[52] Liska, D. J. (1998). The detoxification enzyme system. *Alternative Medicine Review*, 3(3), 187-198.

[53] Lo, H. W., & Ali-Osman, F. (2007). Genetic polymorphism and function of glutathione S-transferases in tumor drug resistance. *Current Opinion in Pharmacology*, 7, 367-374.

[54] Mannervik, B., & Danielson, U. H. (1988). Glutathione transferases structure and catalytic activity. *Crit. Rev. Biochem.*, 23(2), 283-337.

[55] Mc Lellan, L. I., Kerr, L. A., Cronshaw, A. D., & Hayes, J. D. (1991). Regulation of mouse glutathione S-transferases by chemoprotectors. Molecular evidence for the existence of three distinct alpha-class glutathione S-transferase subunits, Ya1, Ya2, and Ya3, in mouse liver. *Biochem. J.*, 276, 461-469.

[56] Mc Lellan, L., Judah, D. J., Neal, G. E., & Hayes, J. D. (1994). Regulation of aflatoxin B1-metabolizing aldehyde reductase and glutathione S-transferase by chemoprotectors. *Biochem. J*, 300, 117-124.

[57] Parikh, A., Gillam, E. M., & Guengerich, F. P. (1997). Drug metabolism by Escherichia coli expressing human cytochromes P450. *Nature Biotechnology*, 15, 784-788.

[58] Pascussi, J. M., Gerba-Chaloin, I. S., Drocourt, L., Maurel, P., & Vilarem, M. J. (2003). The expression of CYP2B6, CYP2C9 and CYP3A4 genes: a tangle of networks of nuclear and steroid receptors. *Biochim. Biophys. Acta*, 1619, 243-253.

[59] Piccolomini, R., Aceto, A., Allocati, N., Faraone, A., & Di Ilio, C. (1991). Purification of a GSH-affinity binding protein from Bacteroides fragilis devoid of glutathione transferase activity. *FEMS Microbiol Lett*, 82, 101-106.

[60] Pickett, C. B. (1989). Glutathione S-transferases. Gene structure, regulation, and biological function. *Annu. Rev. Biochem*, 58, 743-764.

[61] Razzaghi-Abyaneh, M., Allameh, A., & Shams, M. (2000). Screening of aflatoxin- producing mould isolates based on fluorescence production on a specific medium under ultraviolet light. *Acta Medica Iranica.*, 38, 67-73.

[62] Razzaghi-Abyaneh, M., Shams-Ghahfarokhi, M., Allameh, A., Kazeroon-Shiri, M., Ranjbar-Bahadori, S., Mirzahoseini, H., & Rezaee, M. B. (2006a). A survey on distribution of Aspergillus section Flavi in corn field soils in Iran: population patterns

based on aflatoxins, cyclopiazonic acid and sclerotia production. *Mycopathologia*, 161, 183-192.

[63] Rawal, S., Kim, J. E., & Jr , R. C. (2010). Aflatoxin B1 in poultry: Toxicology, metabolism and prevention. *Research in Veterinary Science*, 89, 325-331.

[64] Revathy, K. S., Umasuthan, N., Lee, Y., Choi, C. Y., Whang, I., & Lee, J. (2012). First molluscan theta-class glutathione S-transferase: Identification, cloning, characterization and transcriptional analysis post immune challenges. *Biochemistry and Molecular Biology*, 162(1-3), 10-23.

[65] Robinson, A., Huttley, G. A., Booth, H. S., & Board, P. G. (2004). Modelling and bioinformatics studies of the human kappa-class glutathione transferase predict a novel third glutathione transferase family with similarity to prokaryotic 2-hydroxychromene-2-carboxylate isomerases. *Biochem J.*, 379, 541-552.

[66] Rossi, A., Kapahi, P., Natoli, G., Takahashi, T., Chen, Y., Karin, M., & Santoro, M. G. (2000). Anti-inflammatory cyclopentenone prostaglandins are direct inhibitors of IkappaB kinase. *Nature*, 403, 103-8.

[67] Sacco, J. C. (2006). Phase II biotransformation of xenobiotics in polar bear (Ursus maritimus) and channel catfish (Ictalurus punctatus). *A dissertation presented to the graduate school of the University of Florida in partial fulfillment of the requirements for the degree of doctor of philosophy. Waxman.*

[68] Saxena, M., Mukerji, K. G., & Raj, H. G. (1988). Positive correlation exists between glutathione S-transferase activity and aflatoxin formation in Aspergillus flavus. *Biochem J*, 245, 567-70.

[69] Scarcella, S., Lamenza, P., Virkel, G., & Solana, H. (2012). Expression differential of microsomal and cytosolic glutathione S-transferases in Fasciola hepatica resistant at triclabendazole. *Molecular and Biochemical Parasitology*, 181(1), 37-39.

[70] Sheehan, D., Meade, G., Foley, V. M., & Dowd, CA. (2001). Structure, function and evolution of glutathione transferases: implications for classification of non-mammalian members of an ancient enzyme superfamily. *Biochemical Journal*, 360, 1-16.

[71] Shishido, T. (1981). Glutathione S-transferase from Escherichia coli. *Agric Biol Chem*, 45, 2951-2953.

[72] Stewart, R. K., Serabjit-Singh, C. J., & Massey, T. E. (1996). Glutathione S-Transferase-Catalyzed Conjugation of Bioactivated Aflatoxin B1 in Rabbit Lung and Liver. *Toxicology and Applied Pharmacology*, 140, 499-507.

[73] Tamaki, H., Yamamoto, K., & Kumagai, H. (1999). Expression of two glutathione S-transferase genes in the yeast Issatchenkia orientalis is induced by o-dinitrobenzene during cell growth arrest. *J. Bacteriol.*, 181, 2958-2962.

[74] Tew, K. D., Manevich, Y., Grek, C., Xiong, Y., Uys, J., & Townsend, D. M. The role of glutathione S-transferase P in signaling pathways and S-glutathionylation in cancer. *Free Radical Biology and Medicine*, 51(2), 229-31.

[75] Thom, R., Cummins, I., Dixon, D. P., Edwards, R., Cole, D. J., & Lapthorn, A. J. (2002). Structure of a Tau class glutathione S-transferase from wheat active in herbicide detoxification. Structure of a Tau class glutathione S-transferase from wheat active in herbicide detoxification. *Biochemistry*, 41(22), 7008-7020.

[76] Thornalley, PJ. (1993). The glyoxalase system in health and disease. *Mol Aspects Med*, 14, 287-371.

[77] Todorova, T. (2007). Glutathione S-transferases and oxidative stress in Saccharomyces cerevisiae. *Thesis. A cotutelle project between the University Louis Pasteur, Strasbourg and the Sofia University St. Kliment Ohridski. France.*

[78] Umasuthan, N., Revathy, K. S., Lee, Y., Whang, I., Choi, C. Y., & Lee, J. (2012). A novel molluscan sigma-like glutathione S-transferase from Manila clam, Ruditapes philippinarum: Cloning, characterization and transcriptional profiling. *Toxicology & Pharmacology*, 155(4), 539-550.

[79] Van Vleet, T. R., Mace, K., & Coulombe Jr, R. A. (2002). Comparative aflatoxin B1 activation and cytotoxicity in human bronchial cells expressing cytochromes P450 1A2 and 3A4. *Cancer Research*, 62, 105-112.

[80] Verma, R. J. (2004). Aflatoxin cause DNA damage. *Int j Hum Genet.*, 4(4), 231-236.

[81] Vuilleumier, S. (1997). Bacterial glutathione S-transferases: What are they good for? *Journal of Bacteriology*, 179, 1431-1441.

[82] Waxman, D. J. (1990). Glutathione S-transferase: Role in Alkylating agent resistance and possible target for modulation chemotherapy. *A review. Cancer*, 50, 6449-6454.

[83] World Health Organization International Agency for Research on Cancer (IARC),. (1993). IARC Monographs on the Evaluation of Carcinogenic Risks to Humans.

[84] Xu, C., Yong-Tao, C., & Kong, A. T. (2005). Induction of Phase I, II and III Drug Metabolism/Transport by Xenobiotics. *Arch Pharm Res*, 28(3), 249-268.

[85] Yamuna, A., Saravana, Bhavan. P., & Geraldine, P. (2012). Glutathione S-transferase and metallothionein levels in the freshwater prawn Macrobrachium melcolmsonii exposed to mercury. *J.Environ.Biol*, 33, 133-137.

[86] Yu, J., Bhatnagar, D., & Cleveland, T. E. (2004). Completed sequence of aflatoxin pathway gene cluster in Aspergillus parasiticus. *FEBS Lett*, 564, 126-30.

[87] Yu, S. J. (1996). Insect glutathione S-transferases- Review article. *Zoological Studies*, 35, 9-19.

[88] Ziglari, T., Allameh, A., Razzaghi-Abyaneh, M., Khosravi, A. R., & Yadegari, M. H. (2008). Comparison of glutathione S-transferase activity and concentration in aflatoxin-producing and their non-toxigenic counterpart isolates. *Mycopathologia*, 166, 219-226.

Chapter 13

Synergistic Interaction Between Aflatoxin and Hepatitis B Virus in Hepatocarcinogenesis

Michael C. Kew

Additional information is available at the end of the chapter

1. Introduction

Based on the number of new cases of cancer in humans each year, hepatocellular carcinoma (HCC) is the sixth most common cancer worldwide [20], the fifth in males and the seventh in females [37]. In the most recently reported year, 748,000 new cases of the tumour were recorded, constituting 9.2% of all new cancers [20]. Furthermore, the number of new cases of the tumour continues to increase year by year. Not only is HCC common, it also carries an especially grave prognosis, ranking third in annual cancer mortality rates. In the year mentioned, the total death rate from the tumour was 695,900. Of the patients who died, 93% did so within 12 months of the onset of symptoms. This 12 month fatality ratio in HCC (0.93 - 0.96) is the highest of any human tumour.

HCC does not have a uniform geographical distribution. Rather, of all the new cases of the cancer recorded during recent years, approximately 84% occurred in resource-constrained (developing) countries [20], particularly in sub-Saharan Africa and the Asia Pacific region. In these regions the dominant cause of HCC is chronic hepatitis B virus (HBV) infection. This infection is almost invariably acquired very early in life, either as a result of perinatal transmission of the virus or of horizontal transmission in infancy or early childhood, times at which the infection very often becomes chronic [42]. The tumour resulting from the HBV infection frequently occurs at a young or relatively young age, and it carries a particularly grave prognosis.

In addition to chronic HBV infection, the other major cause of HCC in these high-risk regions is dietary exposure to aflatoxins, the toxic secondary metabolites of the fungi, *Aspergillus flavus* and *Aspergillus parasiticus.* These viral and fungal risk factors are largely responsible for the striking geographical variation in incidence of HCC. Both aflatoxin exposure and chronic HBV infection are more common in rural than in urban dwellers in resource-constrained regions [42, 63]. In these regions, the association between aflatoxin exposure and the development of HCC is closest in sub-Saharan Africa [51].

Aflatoxins are structurally-related difuranocoumarin derivatives, some of which are mutagenic and carcinogenic in humans and animals [89, 87]. These toxins are widely distributed in nature. Because atmospheric humidity and moisture content of plants are important factors in determining growth of, and toxin production by, these moulds, contamination of crops occurs mainly in tropical and sub-tropical climates with high humidity and temperature. These conditions exist in sub-Saharan Africa, the Asia Pacific region, and parts of South America. Contamination is particularly likely to occur in subsistence farming communities in regions with these climates and where regulations to control exposure to the fungi are either non-existent or unenforceable in practice.

In these regions, the moulds contaminate a variety of staple foods, especially maize and groundnuts [89, 34, 87]. Because most rural dwellers can afford only limited food variation, these staples make up a significant portion of their diets. Contamination of crops with aflatoxins occurs either during their growth or as a result of their storage under conditions that promote fungal growth and toxin production [31, 32, 33]. Exposure begins *in utero* as a result of trans-placental transmission of the toxins [86] and in the postnatal period as a result of breast-feeding [85], and continues throughout life. Exposure increases with increasing age - for example, in Malaysia evidence of exposure was more common in the population aged 31 to 50 years than that aged 18 to 30 years [46].

Approximately 4.5 billion of the world's population are believed to be exposed to aflatoxins [88]. Between 25,200 new cases of HCC each year (or 4.6% of all cases of the tumour worldwide) and 155,000 new cases each year (or 28.2% of all cases of the tumour world wide) may be attributed to this exposure [51]. It has been estimated that aflatoxins play a causative role in at least 4.6% and at most 28.2% of all cases of HCC worldwide [51]. These large ranges stem from the considerable uncertainty and variability in data on cancer potency factors, HBV prevalence, aflatoxin exposure, and other risk factors [51].

Although the parent aflatoxin molecule is harmless, it is converted by members of the cytochrome 450 superfamily into electrophilic intermediates that are mutagenic and carcinogenic [90, 87, 40, 71]. Of the four naturally occurring aflatoxins, aflatoxin B_1 (AFB_1), B_2, G_1 and G_2, toxigenic strains of *A. flavus* typically produce only aflatoxins B_1 and B_2, whereas most strains of *A. parasiticus* produce all of the aflatoxins [18]. AFB_1 is the most potent experimental hepatocarcinogen known to man -- no animal model exposed to the toxin thus far has failed to develop HCC. AFM_1, the hydroxylation product of AFB_1, is found in milk and milk products when animals intended for dairy production consume aflatoxin-contaminated feed [70]. In rodents exposed to AFB_1 in equivalent doses to those occurring in humans, levels of aflatoxin adduct in the serum have correlated with levels of hepatic DNA damage and with development of HCC [83].

AFB_1 is the aflatoxin most often found in contaminated human foodstuffs [78], and exposure to AFB_1 is causally related to the development of HCC in humans [34]. The correlation between the degree of exposure to AFB_1 and the incidence of HCC is direct. It has been estimated that by reducing dietary AFB_1 levels to below detectable limits in Asia and sub-Saharan Africa, between 72,800 and 98,800 new cases of HCC could be prevented each year [49].

The major human cytochrome P450 (CYP) enzymes involved in aflatoxin metabolism are CYP3A4, 3A5, 3A7 and 1A2, and the predominant site of metabolism is the liver [87, 39]. AFB_1 is metabolized to an AFB_1-8,9-*exo*-epoxide and, to a lesser extent, an AFB_1-8,9-*endo*-epoxide. The *exo*-epoxide binds to DNA to form the predominant promutagenic 8,9-dihydro-8-(N^7-guanyl)-9-hydroxy AFB_1 (AFB_1-N^7-Gua) adduct. AFB_1-N^7-Gua can result in two secondary forms, an apurinic site and a more stable ring-opened AFB_1-formamidopyrimidine (AFB_1-FABY) adduct, which is far more persistent *in vivo*. This adduct causes G to T transversion mutations [28, 4, 87], the most prevalent of which are targeted to the site of the original adduct. AFB_1-FABY exists as a mixture of two rotameric forms. In *Escherichia coli* AFB_1-FABY induced a six-fold higher G to T mutation frequency than AFB_1-N^7-Gua, with mutations also occurring adjacent to the site of adduct formation [82]. AFB_1-FABY also resulted in blocked replication. Subsequent studies showed that the form of AFB_1-FABY normally present in double-stranded DNA is mutagenic, whereas the dominant species in single-stranded DNA blocks replication [5].

Chronic liver injury and regenerative hyperplasia are critical to the development of HCC [30]. AFB_1-induced DNA adducts may therefore be fixed as mutations consequent to an HBV-related increase in cell proliferation and hyperplasia. Inflammation and oxidative stress associated with chronic active hepatitis and aflatoxin exposure may also result directly in DNA damage and mutations [52].

The 'DNA damage checkpoint response' acts as an anti-tumour mechanism against genotoxic agents. By playing a central role in co-ordinating DNA repair and cell cycle progression, 'DNA damage checkpoint response' proteins play a key role in preventing mutations [66]. Genotoxic doses of AFB_1 induce an incomplete and inefficient 'DNA damage checkpoint response', which may contribute to the carcinogenic properties of the toxin [27].

AFB_1 has a geographical distribution similar to that of chronic HBV infection, colonizing a variety of foodstuffs in the same Far Eastern and sub-Saharan African countries. Accordingly, a synergistic interaction between the hepatocarcinogenic effects of HBV and AFB_1 would offer a plausible explanation for the very high incidence of HCC, and perhaps also the young age of the patients, in these regions.

2. Evidence for a synergistic hepatocarcinogenic interaction between aflatoxins and hepatitis B virus

Although a study in Guanxi, China published in the mid-1980s showed that HCC occurring in individuals infected with HBV who lived in villages with a "high" consumption of aflatoxins had a mortality rate that was 10 times higher than that in individuals living in villages with a "low" consumption [93], other early studies of the consequences of exposure to aflatoxins did not include data on the HBV status of the populations studied. All of these studies showed a statistically significant increase in incidence of HCC in those individuals who were exposed to the fungal toxin [31, 53, 32]. But, given the high frequency of chronic HBV infection in this region at that time, the probability exists that at least some, and possibly a significant number, of the subjects included in these early studies were also chronically infected with HBV and that the virus, rather than the exposure to aflatoxins, could have caused the malignant transformation or, at least, contributed to it. In two studies, one earlier and the other later, the roles of AFB_1 and HBV in explaining the varying frequencies of HCC in different areas of Swaziland in southern Africa [61] and in Guanxi Province in China [93] were assessed. Both analyses concluded that with simultaneous exposure to the two potential carcinogenic agents, AFB_1 exposure was the more important determinant of geographical variation in the incidence of HCC than was HBV infection, at least in those regions. However, no attempt was made in either study to evaluate a possible interactive hepatocarcinogenic effect between the two risk factors.

The first published evidence consistent with synergism between AFB_1 and HBV in the genesis of HCC was provided by experiments in which transgenic mice over-expressing the large envelope polypeptide of HBV were fed $AFB_{1.}$ These mice produced more rapid and extensive hepatocyte dysplasia than did their unexposed littermates, and HCCs developed [67]. Shortly thereafter, further experimental evidence for a positive interaction between AFB_1 and another member of the *Hepadnaviridae* family, the woodchuck hepatitis virus, in the development of HCC was presented [3]. Woodchucks infected with woodchuck hepatitis virus and exposed to AFB_1 developed, after six to 26 months of exposure, a high incidence of preneoplastic foci of altered hepatocytes followed by hepatocellular adenomas and HCCs. Moreover, woodchucks infected with woodchuck hepatitis virus had earlier been shown to have enhanced activation of the biologically inactive AFB_1 to AFB_1-8,9-epoxide [17]. The development of liver tumours was also reported in ducks infected with duck hepatitis virus and exposed to AFB_1 [14] and in tree shrews (*Tupaia belangeri chinensis*) infected with HBV and exposed to AFB_1 [48].

Following the introduction of methods to measure aflatoxin metabolites and aflatoxin-DNA adducts in urine and aflatoxin-albumin adducts in serum, biomarkers that were a far more accurate and reliable indicator of AFB_1 exposure than the hitherto used food sampling and dietary questionnaires, five large cohort studies were undertaken in Shanghai and Qidong county, China and in Taiwan to assess the carcinogenic effects of AFB_1 and HBV alone and in combination. In four of the studies an hepatocarcinogenic effect of AFB_1 alone was shown, with increased odds ratios ranging from 1.9 to 32.0 with a mean ratio of 13.7 [65, 64, 54, 59] (Table 1). These studies proved that exposure to AFB_1 alone could cause malignant transformation of hepatocytes in humans. The fifth study failed to show an increased odds ratio of AFB_1 exposure alone [79].

As expected, these studies (including the one that did not show an increased odds ratio for the development of HCC for AFB_1 alone [79]) confirmed an hepatocarcinogenic effect of HBV alone - odds ratios ranged from 3.3 to 17.4, with a mean ratio of 10.0 (Table 1).

	HBV alone	AFB_1 alone	HBV and AFB_1
	RR (95% CL)*	RR (95% CL)	RR (5% CL)
[65]	4.8 (1.2 -19.7)	1.9 (0.5 - 7.5)	60.1 (6.4 - 561.8)
[64]	7.3 (2.2 - 24.4)	3.4 (1.1 - 10.0)	59.4 (15.6 - 212)
[80]	17.4 (3.6 – 143.4) 425.4)	0.3 (0 - 3.6)	70.0 (11.5 –
[54]	17.0 (2.8 - 103.9)	17.4 (3.4 - 90.3)	67.6 (12.2 - 373.2)
[59]	3.3 (1.3 - 8.3)	32.0 (4.0 - 255.8)	40.7 (12.7- 130.9)

Table 1. Findings in five studies comparing the risk of HBV infection alone, dietary exposure to AFB1 alone, and the two risk factors together in the genesis of HCC. * Relative risk (95% confidence limits).

A synergistic interaction between AFB_1 exposure and chronic HBV infection in causing HCC was evident in each of the five studies - odds ratios ranged from 40.7 to 70.0 with a mean of 59.6 (Table 1). In three of these studies there was a striking multiplicative effect, and in the other two a sub-multiplicative effect between exposure to AFB_1 alone and exposure to AFB_1 in the presence of chronic HBV infection in inducing HCC, in comparison with each carcinogen alone. The study which did not show an increased odds ratio with AFB_1 alone had the highest odds ratios for both HBV infection alone and for co-existing AFB_1 exposure and HBV infection [79]. The finding in this study that exposure to AFB_1 alone did not increase the risk of HCC development [79] could conceivably be the source of the erroneous view held by some hepatologists and oncologists that AFB_1 alone does not cause HCC and is important only as a co-carcinogen with HBV.

In other investigations, also in countries with high rates of contamination of foodstuffs by AFB_1, only individuals chronically infected with HBV were studied and the influence of AFB_1 exposure in further increasing their risk of HCC development was analysed. In Qidong county, China, over a 10-year prospective follow-up period, the risk of HCC in male carriers of the virus was shown to be increased 3-fold (95% confidence limits 1.2, 8.7) in those with detectable urinary levels of AFB_1 metabolites in comparison with those without these metabolites [72]. This result was later confirmed in a longer follow-up of the same cohort of HBV carriers, when the risk of HCC was increased 3.5-fold (95% confidence limits 1.5, 8.1) [57]. A dose-response relationship between urinary AFB_1 metabolites and the risk of HCC was shown in HBV carriers in Taiwan [94]. Comparing high and low urinary levels of the aflatoxin metabolite, AFM_1, a multivariate-adjusted odds ratio of 6.0 (95% confidence limits1.2, 29) was calculated. The risk was greater (odds ratio 10.0: 95% confidence limits 1.6 ; 60.9) when both AFM_1 and AFB_1-N^7- gua metabolites were tested for and detected in the urine. In another study performed in chronic carriers of HBV in the same country, a statistically significant relationship was noted between detectable levels of AFB_1 adducts in serum and the risk of HCC, with an age-adjusted odds ratio of 2.0 (95% confidence limits 1.1, 3.7) [73]. A recent meta-analysis has shown that the population attributable risk of developing HCC in individuals exposed to dietary aflatoxins is 17%, with the risk being 21% in those individuals also chronically infected with HBV [49]. Individuals infected with HBV alone have a population attributable risk of 8.8% [49]. If the one study in the meta-analysis that contributed most to heterogeneity in the analysis was excluded, the summarised odds ratio of HCC (with 95% confidence limits) was 73 (36 to 148.3) for the combined effects of AFB_1 and HBV, 11.3 (6.75 to 18.9) for HBV alone, and 6.37 (3.74 to 10.86) for AFB_1 alone [49]. The effect of a synergistic interaction between AFB_1 and HBV on the age of onset of HCC was specifically addressed in a study of Taiwanese patients. HBV-infected patients in whom tumour tissue was shown by histochemical staining to be positive for AFB_1-N^7-gua adducts were on average 10 years younger than those with adduct-negative tumours [10].

Although they have had limitations, various animal models with natural hepatitis viral infections have been used to examine the interaction between hepadnaviruses and AFB_1 [84]. In woodchucks and tree shrews, animal species with hepadnaviral-induced liver pathology similar to that observed in HBV-infected humans, the administration of AFB_1 resulted in a higher incidence of liver tumours than in infected animals not receiving AFB_1 [92, 3]. Moreover, HBsAg transgenic mice over-expressing the large envelope protein of HBsAg in the liver developed HCC when exposed to aflatoxin, whereas their littermates not exposed to carcinogens did not [67, 47].

In those human populations in which an interaction between the fungal toxin and HBV has been described, the infection is predominantly acquired in infancy or early childhood. During the early years of HBV infection, a state of immune tolerance towards the virus exists and little if any cellular damage occurs. With loss of this tolerance, the ongoing infection results in recurring cell damage. Exposure to AFB_1 in contaminated foodstuffs also occurs in young children [81].

Nevertheless, it is likely, certainly in China and Taiwan, where perinatal transmission of HBV is the predominant mode of infection, and also probably in Africa, where slightly later horizontal infection is the major route of infection, that the HBV carrier state is established, not before exposure to, but before heavy exposure to the toxin.

3. Possible mechanisms of interaction between AFB_1 and HBV in hepatocarcinogenesis

A number of possible mechanisms for the interaction between HBV and AFB_1 in causing HCC have been suggested. One is that HBV infection directly or indirectly sensitizes hepatocytes to the carcinogenic effects of AFB_1. One way in which this may be accomplished is that the specific cytochrome P450s that metabolize AFB_1 to AFB_1-8,9-epoxide may be induced either by chronic hepatitis caused by HBV infection or by the presence of the virus itself. Induction of these phase I enzymes has been described in HBV transgenic mice [21, 6], where this effect appeared to result from hepatocyte injury induced by the virus rather than the presence of the virus *per se* [6]. The observation that Gambian and Taiwanese children and adolescents chronically infected with HBV have higher concentrations of AFB_1 adducts than uninfected individuals [2, 76, 11] is consistent with this mechanism. But studies in adults in China, Taiwan and The Gambia have either failed to show a significant difference in serum AFB_1 adduct levels between HBsAg-positive and -negative subjects [23, 79, 12] or showed only a marginally significant difference [74]. Moreover, a study in woodchucks with chronic woodchuck hepatitis virus infection did not show enhanced activation of AFB_1 [75, 44].

The generated aflatoxin-8,9-epoxide has been shown to bind to proteins, causing acute toxicity, or to DNA inducing changes that over time increase the risk of malignant transformation [26]. DNA damage can also increase the chance of integration of the viral DNA into the host genome [16]. This effect could be exerted directly by AFB_1 or indirectly by oxidative stress induced by chronic viral hepatitis.

A guanine to thymine transversion at the third base of codon 249 of the p53 tumour suppressor gene (arginine to serine substitution; 249^{ser}, R249S) is present in between 40 and 66% of HCC patients in regions with heavy dietary exposure to AFB_1 [28, 4, 45, 30]. The mutation is also detectable in circulating cell-free DNA from the plasma of HCC patients and healthy subjects from these regions [77]. The exact timing of the development of the 249^{ser} mutation remains uncertain, although it has been shown to be an early event. The mutation abrogates the normal functions of p53, including those in cell cycle control, DNA repair, and apoptosis, thereby contributing to the multistep process of hepatocarcinogenesis. This mutation is extremely uncommon in tumors other than HCC [58].

A specific and close association between this inactivating mutation, the presence of AFB_1 biomarkers, and the development of HCC was recognised in epidemiological studies in regions with high or low AFB_1 exposure rates [4, 60, 15, 19, 45, 30, 62, 22], and evidence that the mutation induced chromosomal instability was found [62]. Arising from the observation of the co-existence of the p53 mutation and AFB_1 exposure, the presence of the 249^{ser} mutation was believed to be a primary genetic event in hepatocarcinogenesis. It occurs early in the series of events leading to AFB_1-associated HCC, and may thus provide an early biomarker of exposure to the fungal toxin and AFB_1-induced hepatocarcinogenesis [36].

But the findings have been inconsistent with support for an aetiological association being provided by some but not all studies. In an investigation of Taiwanese patients with HCC the mutation was present in 36.3% of HBV-infected patients with HCC, compared with 11.7% of those without HBV markers [79]. In a second analysis in Taiwan, all of the 249^{ser} mutations occurred in patients positive for HBsAg, giving an odds ratio of 10.0 (95% confidence limits 1.6; 17.5) [54]. In a study in The Gambia patients positive for HBsAg alone had an increased relative risk of 10, those with 249^{ser} mutation alone of13, and those with both an estimated risk of 399 [45]. Other studies, however, showed a similar, but not a statistically significant trend [68, 19], and in yet other analyses from a variety of countries no association could be found [listed in reference: 69]. Furthermore, in a meta-analysis of 49 published studies using a method that takes into account both within-study and study-to-study variability, little evidence for HBV-AFB_1 interaction in modulating the 249^{ser} mutation was found [69]. In addition, the absence of the 249^{ser} mutation from the serum of patients from countries with a low incidence of HBV-induced HCC to date suggests that chronic HBV infection alone is insufficient to result in the development of the 249^{ser} mutation [82].

Another suggested possibility is that the activity of phase II detoxification enzymes (glutathione S transferase (GST) and epoxide hydrolase(EPHX)) may play a role in the genesis of HCC induced jointly by AFB_1 and HBV [94; 73; 56]. A multiplicative interaction in the genesis of HCC in West African and Chinese patients was demonstrated between HBV infection and mutations of EPHX [56]: patients with chronic HBV infection but with normal EPHX alleles were at a 15-fold increase in risk, and those with both HBV infection and at least one EPHX mutant were at a 77-fold increased risk. In further studies in these patients a positive interaction between HBV and AFB_1 seemed to depend on the presence of a polymorphism of the GST M1, GST T1, and EPHX genes that are normally responsible for converting the carcinogenic AFB_1 -8,9-epoxide to non-reactive metabolites [56, 8, 94, 73]. But again no consistent pattern has emerged. In one analysis in Taiwan the risk of HCC formation was greater in HBV carriers who had the GST M1 null genotype compared with the non-null genotype [94], in a second study the risk appeared to depend on the presence of a GST T1 null genotype [73], and in a third the risk was considerably greater in those with null genotypes of both GST M1 and GST T1 [8].

Another possible mechanism for a carcinogenic interaction between AFB_1 and HBV is that increased hepatocyte necrosis and proliferation cause by chronic HBV infection increases the likelihood of both AFB_1 mutations, including 249^{ser}, and the subsequent clonal expansion of cells containing these mutations [13]. Chronic necroinflammatory hepatic disease, including that resulting from HBV infection, results in the generation of oxygen and nitrogen reactive species [50, 35]. Both of the latter are mutagenic, but, in addition, increased oxidative stress has been shown to induce 249^{ser} mutations [29].

The HBV x gene is frequently included in sequences of the virus that are integrated into cellular DNA [43]. AFB_1-DNA adducts are normally repaired by the nucleotide excision repair pathway. The HBV X protein interferes with the nuclear excision repair pathway [38; 43] and might, by this means, favour persistence of existing mutations or impaired DNA. DNA repair is also compromised by the rapid cell turnover rate in chronic hepatitis. In the presence of dietary exposure to AFB_1, the HB X protein may contribute to the uncontrolled cell proliferation in other ways. The transcription of $p21^{waf1/cip1,}$ which induces cell cycle arrest at the G_1-S checkpoint, is activated by HB X protein in a dose-dependent manner in the presence of functional p53. This transcription is, however, repressed by HB X protein when p53 is not functional or is functional at a low level [1]. The expression of HB X protein also correlates with an increase in the overall frequency of DNA mutations in transgenic mice and a 2-fold increase the incidence of the 249^{ser} mutation in transgenic mice exposed to AFB_1 [55].

Altered methylation of genes may play a role in hepatocarcinogenesis [43]. For example, the methylation status of the human *ras* association domain gene (RASSF1A) and the P16 gene has been incriminated in the pathogenesis of HCC [95]. No association was found between methylation status and P53 status {95]. A statistically significant association was, however, found between RASSF1A methylation status and the level of AFB_1-DNA adducts in HCC tissues [95].

An understanding of the mechanisms responsible for the heightened risk of malignant transformation in patients chronically infected with HBV and exposed to AFB_1 is far from complete, and there is clearly a need for further research to be undertaken into the pathogenetic mechanisms involved in this interaction between the two common hepatocarcinogens in resource-constrained geographical regions.

Author details

Michael C. Kew[1*]

Address all correspondence to: michael.kew@uct.ac.za

1 Department of Medicine, Faculty of Health Sciences, University of Cape Town, Department of Medicine, Faculty of Health Sciences, University of the Witwatersrand, Johannesburg, South Africa

References

[1] Ahn, J. I., Jung, E. Y., Kwun, H. J., et al. (2002). Dual effects of hepatitis B virus X protein on the regulation of cell cycle depending on the status of cellular p53. *J Gen Virol*, 83, 2765-2772.

[2] Allen, S. J., Wild, C. P., Wheeler, J. G., et al. (1992). Aflatoxin exposure and hepatitis B infection in rural Gambian children. *Trans Roy Soc Trop Med Hyg*, 86, 426-430.

[3] Bannasch, P., Khoshkhou, N. I., Hacker, H. J., et al. (1995). Synergistic hepatocarcinogenic effect of hepadnaviral infection and dietary aflatoxin B_1 in woodchucks. *Cancer Res*, 55, 3318-3330.

[4] Bressac, B., Kew, M. C., Wands, J. R., et al. (1991). Selective G to T mutations of p53 gene in hepatocellular carcinoma from southern Africa. *Nature Lond*, 350, 429-430.

[5] Brown, K. L., Deng, J. Z., Iver, R. S., et al. (2006). Unravelling the aflatoxin-FABY conundrum: structural basis for differential replicative processing of isomeric forms of the formaminopyrimidine DNA adduct of aflatoxin B_1. *J Am Chem Soc*, 128, 15188-15199.

[6] Chemin, I., Ohgaki, H., Chisari, F. V., et al. (1999). Altered expression of hepatic carcinogen metabolizing enzymes with liver injury in HBV transgenic mice lineages expressing various amounts of hepatitis B surface antigen. *Liver* , 19, 81-87.

[7] Chen, C. J., Liang, K. Y., Chang, A. S., et al. (1991). Effect of hepatitis B virus, alcohol drinking, cigarette smoking and familial tendency on hepatocellular carcinoma. *Hepatology*, 13, 398-406.

[8] Chen, C. J., Wang, L. Y., Lu, S. N., et al. (1996). Elevated aflatoxin exposure and increased risk of hepatocellular carcinoma. *Hepatology*, 24, 38-42.

[9] Chen, C. J., Yu, M. W., Liaw, Y. F., et al. (1996). Chronic hepatitis B carriers with null genotypes of glutathione S-tranferase M1 and T1 polymorphisms who are exposed to aflatoxin are at increased risk of hepatocellular carcinoma. *Am J Hum Genet*, 59, 128-134.

[10] Chen, C. J., Zhang, Y. J., Lu, S. N., et al. (1992). Aflatoxin B_1 adducts in smeared liver tissue of hepatocellular carcinoma patients. *Hepatology*, 16, 1150-1155.

[11] Chen, S. Y., Chen, C. J., Chou, S. R., et al. (2001). Association of afltoxin B_1-albumin adduct levels with hepatitis B surface antigen status among adolescents in Taiwan. *Cancer Epidemiol Biomarkers Prev*, 10, 1223-1226.

[12] Chen, S. Y., Chen, C. J., Tsai, W. Y., et al. (2000). Associations of plasma aflatoxin B_1-albumin adduct level and genetic polymorphisms of glutathione-S- transferase MI and T1. *Nutr Cancer*, 38, 179-185.

[13] Chisari, F. V., Klopchin, K., Moriyama, T., et al. (1989). Molecular pathogenesis of hepatocellular carcinoma in hepatitis B transgenic mice. *Cell* , 59, 1145-1156.

[14] Cova, L., Wild, C. P., Mehrotta, V., et al. (1990). Contribution of aflatoxin B1 and hepatitis B virus infection in the induction of liver tumors in ducks. *Cancer Res*, 50, 2156-2163.

[15] Coursaget, P., Depril, N., Chabaud, M., et al. (1993). High prevalence of mutations at codon 249 of the p53 gene in hepatocellular carcinoma in Senegal. *Br J Cancer*, 67, 1395-1397.

[16] Dandri, M., Burda, M. R., Burkle, A., et al. (2002). Increase in de novo HBV DNA integrations in response to oxidative DNA damage or inhibition of poly(ADP-ribosyl)ation. *Hepatology*, 35, 217-223.

[17] De Flora, S., Hietane, E., Bartsch, H., et al. (1989). Enhanced metabolic activation of chemical hepatocarcinogenesis in woodchucks infected with woodchuck hepatitis virus. *Carcinogenesis*, 10, 1099-1106.

[18] Dorner, J. W. (2004). Biological control of crops. *J Toxicol Toxin Rev*, 23, 445-450.

[19] Eaton, D. L., & Gallagher, E. P. (1994). Mechanisms of aflatoxin carcinogenesis. *Annu Rev Pharmacol Toxicol*, 34, 275-301.

[20] Ferlay, J, Shin, H. R., Bray, F, et al. (2010). Establishment of world-wide burden of cancer in 2008. GLOBOCAN 2008. *Int J Cancer*, 127, 2893-2917.

[21] Gemechu-Hatewu, M., Platt, K. L., Oesch, F., et al. (1997). Metabolic activation of aflatoxin B_1 to aflatoxin B_1-8,9-epoxide in woodchucks undergoing chronic active hepatitis. *Int J Cancer*, 73, 587-591.

[22] Gouas, D., Shi, H., & Hainaup, P. (2009). The aflatoxin-induced TP53 muation at codon 249(R249S): biomarker of exposure, early detection and target for therapy. *Cancer Lett*, 286, 29-37.

[23] Groopman, J. D., Hall, A. J., Whittle, W. H., et al. (1992). Molecular dosimetry of aflatoxin-N^7 guanine in human urine obtained in The Gambia. *Cancer Epidemiol Biomarkers Prev*, 1, 221-227.

[24] Groopman, J. D., Kensler, T. W., & Wild, C. P. (2008). Protective interventions to prevent aflatoxin-induced carcinogenesis in developing countries. *Annu Rev Public Health*, 28, 187-203.

[25] Groopman, J. D., Wang, B. J., & Scholl, P. (1996). Molecular biomarkers of aflatoxins: from adducts to gene mutations to human liver cancer. *Can J Physiol Pharmacol*, 74, 203-209.

[26] Groopman, J. D., Scholl, P., Wang, J. S., et al. (1996). Epidemiology of human aflatoxin exposures and their relationship to liver cancer. *Progr Clin Biol Res*, 395, 211-222.

[27] Gursoy-Yuzugullu, O., Yuzullu, H., Yilmaz, M., et al. (2011). Aflatoxin genotoxicity is associated with a defective DNA damage response by bypassing p53 activation. *Liv Internat*, 31, 561-571.

[28] Hsu, I. C., Metcalf, R. A., Sun, T., et al. (1991). Mutational hotspot in the p53 gene in human hepatocellular carinomas. *Nature (Lond)*, 350, 427-428.

[29] Hussain, S. P., Aquilar, F, Amstad, P, et al. (1994). Oxyradical-induced mutagenesis of hotspot codons 248 and 249 of the human p53 gene. *Oncogene*, 9, 2277-2281.

[30] Hussain, S. P., Schwank, S. J., Staib, F, et al. (2007). TP53 mutations and hepatocellular carcinoma: insights into the etiology and pathogenesis of liver cancer. *Oncogene*, 26, 2166-2176.

[31] IARC. (1987). Overall evaluations of Carcinogenicity: An Updating of IARC Monographs. IARC Press, Lyon., 1-42.

[32] IARC. (1993). Some naturally occurring substances: Food Items and Constituents. Heterocyclic Aromatic Amines and Mycotoxins. IARC Press, Lyon.

[33] IARC . (2002). Monographs on the evaluation of Carcinogenic Risks to Humans. *International Agency for Research on Cancer: Lyon*, France, 82, 59.

[34] IARC. (2002). Some Traditional Herbal Medicines, Some Mycotoxins, Naphthalene and Styrene. IARC Press, Lyon.

[35] Ishima, H., & Bartsch, H. (1994). Chronic infection and inflammatory process as cancer risk factors: possible role for nitric oxide in carcinogenesis. *Mutation Res*, 305, 253-264.

[36] Jackson, P. E., Kuang, S. Y., Wang, G. B., et al. (2003). Prospective detection of codon 249 mutations in plasma of hepatocellular carcinoma patients. *Carcinogenesis*, 24, 1657-1663.

[37] Jemal, A., Bray, F., Center, MM, et al. (2011). Global cancer statistics. *CA Cancer J Clin*, 61, 69-90.

[38] Jia, L., Wang, X. W., & Harris, C. C. (1999). Hepatitis B virus x protein inhibits nucleotide excision repair. *Int J Cancer*, 80, 875-879.

[39] Kamdem, L. K., Meineke, I., Godtel-Armbrust, U., et al. (2006). Dominant contribution of P450 3A4 to the hepatic carcinogenic activation of aflatoxin B_1. *Chem Res Toxicol*, 19, 577-586.

[40] Kensler, T. W., Qian, G. S., Chen, J. G., et al. (2003). Translational strategies for cancer prevention in liver. *Nat Rev Cancer*, 3, 321-329.

[41] Kew, M. C. (2007). Hepatocellular carcinoma in Africa. In: About Cancer in Africa. *Paris: National Cancer Institute of France (INCa)*, 221-241.

[42] Kew, M. C. (2010). Epidemiology of hepatitis B virus infection, hepatocellular carcinoma and hepatitis B virus-induced hepatocellular carcinoma. *Pathologie Biologie (Paris)*, 58, 273-277.

[43] Kew, M. C. (2011). Hepatitis B virus x gene in the pathogenesis of hepatitis B virus-induced hepatocellular carcinoma. *J Gastroenterol Hepatol*, (1), 144-152.

[44] Kirby, G., Chemin, I., Montesana, R., et al. (1994). Induction of specific cytochrome p450 sinvolved in aflatoxin metabolism in hepatitis B transgenic mice. *Mol Carcinogenesis*, 11, 74-80.

[45] Kirk, G. D., Lesi, O. A., Mendy, M., et al. (2005). 249 (Ser) TP53 mutation in plasma DNA, hepatitis B virus infection, and risk of hepatocellular carcinoma. *Oncogene*, 24, 5858-5867.

[46] Leong, Y. H., Rosma, M. A., Latiff, A. A., et al. (2012). Association of serum aflatoxin B_1 adduct level with socio-demographic factors in a study in Malaysia and aflatoxins intake from nuts and nut products in Malaysia. *Int J Hyg Environ Health*, 215, 368-372.

[47] Lian, M., Liu, Y., Yu, S. Z., et al. (2006). Hepatitis B virus x gene and cyanobacterial toxins promote hepato-tumorigenesis in mice. *World J Gastroenterol*, 12, 3065-3072.

[48] Li, Y., Su, J., Qin, C., et al. (1999). Synergistic effect of hepatitis B virus and aflatoxin B_1 in hepatocarcinogenesis in tree shrews. *Ann Acad Med Singapore*, 28, 68-71.

[49] Liu, Y., Chang, C. C., Marsh, G. M., et al. (2012). Population attributable risk of aflatoxin-related liver cancer: a systematic review and meta-analysis. Eur J Cancer Mar 8th. Epub ahead of print.

[50] Liu, R. H., Jacob, J. R., Hotchkiss, J. H., et al. (1994). Woodchuck hepatitis virus surface antigen induces nitric oxide synthesis in hepatocytes: possible role in hepatocarcinogenesis. *Carcinogenesis* , 15, 2875-2877.

[51] Liu, Y., & Wu, F. (2010). Global burden of aflatoxin-induced hepatocellular carcinoma: a risk assessment. *Environ Health Perspect*, 118, 818-824.

[52] Liu, Z. M., Li, L. Q., Peng, M. H., et al. (2008). Hepatitis B virus infection contributes to oxidative stress in a population exposed to aflatoxin B1 and high-risk for hepatocellular carcinoma. *Cancer Lett*, 263, 745-749.

[53] Liu, S. N., Lin, T. M., Chen, C. J., et al. (1988). A case-control study of primary hepatocellular carcinoma. *Cancer*, 62, 2051-2055.

[54] Lunn, R. M., Zhang, Y. J., Wang, L. Y., et al. (1997). p53 mutations, chronic hepatitis B virus infection, and aflatoxin exposure in hepatocellular carcinoma in Taiwan. *Cancer Res*, 57, 3471-3477.

[55] Madden, C. R., Finegold, M. J., & Slagel, B. L. (2002). Altered DNA mutation spectrum in aflatoxin treated transgenic mice that express the hepatitis B virus X protein. *J Virol*, 76, 11770-11774.

[56] McGlynn, K. A., Rosvold, E. A., Lustbader, E. D., et al. (1995). Susceptibility to hepatocellular carcinoma is associated with genetic variation in the enzymatic detoxification of aflatoxin B1. *Proc Natl Acad Sci*, 92, 2384-2387.

[57] Ming, L., Thorgeirsson, S., Gail, M. H., et al. (2002). Dominant role of hepatitis B virus and co-factor role of aflatoxin in hepatocarcinogenesis in Qidong, China. *Hepatology*, 36, 1214-1220.

[58] Montesano, R, Hainaut, P, & Wild, C. P. (1997). Hepatocellular carcinoma; from gene to public health. *J Natl Cancer Inst*, 89, 1844-1851.

[59] Omer, R. E., Kuijsten, A., Kadaru, M. Y. K., et al. (2004). Population attributable risk of dietary aflatoxins and hepatitis B virus infection with respect to hepatocellular carcinoma. *Nutr Cancer*, 48, 15-21.

[60] Ozturk, M., Bressac, B., Pusieux, A., et al. (1991). A p53 mutational hotspot in primary liver cancer is geographically localised to high aflatoxin areas of the world. *Lancet*, 338, 260-265.

[61] Peers, F. G., Gilman, G. A., & Linsell, C. A. (1976). Dietary aflatoxins and human liver cancer. A study in Swaziland. 1976, *Int J Cancer*, 17-167.

[62] Pineau, P., Marchio, A., Battiston, C., et al. (2008). Chromosome instability in human hepatocellular carcinoma depends upon p53 status and aflatoxin exposure. *Mutat Res*, 653, 6-13.

[63] Plymoth, A., Vivani, S., & Hainaut, P. (2008). Control of hepatocellular carcinoma through hepatitis B vaccination in areas of high endomicity: perspectives for global liver cancer prevention. *Cancer Lett*, 286, 15-21.

[64] Qian, G. S., Ross, R. K., Yu, M. C., et al. (1994). A follow-up study of urinary markers of aflatoxin exposure and liver cancer risk in Shanghai, People's republic of China. *Epidemiol Biomarkers Prev*, 3, 3-10.

[65] Ross, R. K., Yuan, J. M., Yu, M. C., et al. (1992). Urinary aflatoxin biomarkers and risk of hepatocellular carcinoma. *Lancet*, 339, 943-946.

[66] Sancar, A., Lindsey-Boltz, L. A., Unsal-Kacmaz, K., et al. (2004). Molecular mechanisms of mammalian DNA repair and the DNA damage checkpoints. *Annu Rev Biochem*, 73, 39-85.

[67] Sell, S., Hunt, J. M., Dunsford, H. A., et al. (1991). Synergy between hepatitis B virus expression and chemical hepatocarcinogenesis in transgenic mice. *Cancer Res*, 511, 1278-1285.

[68] Sheu, J. C., Huang, G. T., Lee, P. H., et al. (1992). Mutations of p53 in hepatocellular carcinoma in Taiwan. *Cancer Res*, 52, 6089-6100.

[69] Stern, M. C., Umbach, D. M., Yu, M. C., et al. (2001). Hepatitis B, aflatoxin B1 p53 codon 249 mutation in hepatocellular carcinoma from Guanxi, People's Republic of China, and a meta-analysis of existing studies. *Cancer Epidemiol Biomarkers Prev*, 10, 617-625.

[70] Strosnider, H, Baumgartner-Azziz, F, Banzinger, M, et al. (2006). Workgroup report: Public health strategies for reducing aflatoxin exposure in developing countries. *Environ Health Perspect*, 114, 1898-1903.

[71] Sudaikin, D. I. (2003). Dietary aflatoxin exposure and chemoprevention of cancer: a clinical review. *J Toxicol Clin Toxicol*, 41, 195-204.

[72] Sun, Z., Lu, P., Gail, M. H., et al. (1999). Increased risk of hepatocellular carcinoma in male hepatitis B surface antigen carriers with chronic hepatitis who have detectable uruinary aflatoxin metabolite M1. *Hepatology*, 30, 379-383.

[73] Sun, CA, Wang, LY, Chen, CJ, et al. (2001). Genetic polymorphisms of glutathione S-transferases M1 and T1 associated with susceptibility to aflatoxin-related hepatocarcinogenesis among chronic hepatitis B carriers: a nested case-control study in Taiwan. *Carcinogenesis* , 22, 1289-1294.

[74] Sun, C. A., Wu, D. M., Wang, L. Y., et al. (2002). Determinants of formation of aflatoxin-albumin : a seven township study in Taiwan. *Br J Cancer*, 87, 966-970.

[75] Tennant, B., Horbuckle, W. E., Yeager, A. E., et al. (1990). Effects of aflatoxin B_1 on experimental woodchuck hepatitis virus infection and hepatocellular carcinomas. In: Hollinger FB, Lemon SM, Margolis H, editors. Viral Hepatitis and Liver Disease. Baltimore Williams and Wilkens , 599-600.

[76] Turner, P. C., Mendy, M., Whittle, H., et al. (2000). Hepatitis B infection and aflatoxin levels in Gambian children. *Trop Med Int Health*, 5, 837-841.

[77] Villar, S., Le Roux-Goglin, E., Gouas, D. A., et al. (2011). Seasonal variation in TP53 R249S-mutated serum DNA with aflatoxina exposure and hepatitis B virus infection. *Environ Health Perspect*, 119, 1635-1640.

[78] Wang, J. S., & Groopman, J. D. (1999). DNA damage by mycotoxins. *Mutat Res*, 424, 167-181.

[79] Wang, L. Y., Hatch, M., Chen, C. J., et al. (1996). Aflatoxin exposure and risk of hepatocellular carcinoma. *Int J Cancer*, 67, 620-630.

[80] Wang, J. S., Qian, G. S., Zarba, A., et al. (1996). Temporal patterns of aflatoxin-albumin adducts in hepatitis B surface antigen positive and negative residents of Davin, Qidong County, People's Republic of China. *Cancer Epidemiol Biomarkers Prev*, 5, 253-261.

[81] Wild, C. P., Fortuin, M., Donato, F., et al. (1993). Aflatoxin, liver enzymes and hepatitis B virus infection in Gambian children. *Cancer Epidemiol Biomarkers Prev*, 2, 555-561.

[82] Wild, C. P., & Gong, Y. Y. (2010). Mycotoxins and human disease: a largely ignored global health issue. *Carcinogenesis*, 31, 71-82.

[83] Wild, C. P., Hasegawa, R., Barraud, L., et al. (1996). Aflatoxin-albumin adducts: a basis for comparative carcinogenesis between animals and humans. *Cancer Epidemiol Biomarkers Prev*, 5, 179-189.

[84] Wild, C. P., & Montesano, R. (2009). A model of interaction: aflatoxin B1 and hepatitis viruses in liver cancer aetiology. *Cancer Lett*, 286, 22-28.

[85] Wild, C. P., Pionneau, F. A., Montesano, R., et al. (1987). Aflatoxin detected in human breast milk by radioimmunoassay. *Int J Cancer*, 40, 328-333.

[86] Wild, C. P., Shrestha, S. M., Anwar, W. A., et al. (1992). Field studies of aflatoxin exposure, metabolism and induction of genetic alterations in relation to HBV infection and hepatocellular carcinomas in The Gambia and Thailand. *Toxicol Lett*, 455 -461.

[87] Wild, C. P., & Turner, P. C. (2002). The toxicity of aflatoxins as a basis for public health decisions. *Mutagenesis*, 17, 471-481.

[88] Williams, J. H., Phillips, T. D., Jolly, P. E., et al. (2004). Human aflatoxicosis in developing countries: review of toxicology, exposure, potential health consequences, and interventions. *Am J Clin Clin Nutr*, 80, 10106-1022.

[89] Wogan, G. N. (1997). Aflatoxin exposure as a risk factor in the etiology of hepatocellular carcinoma. In:Liver Cancer. Okuda K, Tabor E (eds). New York Churchill Livingstone , S 1-S8.

[90] Wogan, G. N. (2000). Impact of chemicals on liver cancer risk. *Semin Cancer Biol*, 10, 201-210.

[91] Wu, H. C., Wang, Q., Yang, H. I., et al. (2009). Aflatoxin B1 exposure, hepatitis B virus infection and hepatocellular carcinoma in Taiwan. *Cancer Epidemiol Biomarkers Prev.*, 18, 846-853.

[92] Yan, R. Q. (1989). A study of primary liver cancer in tree shrews induced by human hepatitis B virus and aflatoxin B1. *Zhongua Bing Li Xue Za Zhi*, 18, 19-22.

[93] Yeh, F. S., Mo, C. C., Yen, R. C., et al. (1989). Risk factors for hepatocellular carcinoma in Guanxi, People's republic of China. *Nati Cancer Inst Monog*, 69, 47-48.

[94] Yu, M. W., Lien, J. P., Chiu, Y. H., et al. (1997). Effect of aflatoxin metabolism and DNA adduct formation on hepatocellular carcinoma among chronic hepatitis B carriers in Taiwan. *J Hepatol*, 27, 320-330.

[95] Zhang, Y. J., Ahsan, H., Chen, Y., et al. (2002). High frequency of promoter hypermethylation of RASSFIA and p16 and its relationship to aflatoxin B_1-DNA adducts in human hepatocellular carcinoma. *Mol Carcinog*, 35, 85-92.

Section 4

Detection and Analysis

Chapter 14

Characteristics of Mycotoxin Analysis Tools for Tomorrow

Luis Miguel Contreras-Medina,
Alejandro Espinosa-Calderon, Carlos Duarte-Galvan,
Arturo Alfonso Fernandez-Jaramillo,
Rafael Francisco Muñoz-Huerta,
Jesus Roberto Millan-Almaraz,
Ramon Gerardo Guevara-Gonzalez and
Irineo Torres-Pacheco

Additional information is available at the end of the chapter

1. Introduction

Aflatoxins difuranocoumarin derivatives are produced by fungi *Aspergillus flavus, Aspergillus parasiticus and Aspergillus nomius* [1] and they are part of the group of mycotoxins. From the twenty metabolites that have been formed endogenously in animals, aflatoxins B_1, B_2, G_1 and G_2 (AFB_1, AFB_2, AFG_1 and AFG_2) are the most common and the most toxic. The names of aflatoxins B_1, B_2, G_1, and G_2 are based on their florescence characteristics. Aflatoxin B_1 and B_2 show strong blue fluorescence under UV light, whereas aflatoxins G_1 and G_2 exhibit greenish yellow fluorescence [2]. All the aflatoxins have been classified as carcinogenic compounds for humans, but AFB_1 has been tagged as the most dangerous, highly toxic, immunosuppressive, mutagenic, and teratogenic compound and its effects have been identified as well. Also, malabsorption syndrome and reduction in bone strength may occur due to AFs consumption. Aflatoxins not only have adverse effects on human health but also cause serious economic losses when tons of foods have to be dropped or destroyed for being contaminated with AFs.

To ensure food's safety, the maximum level of aflatoxins in food has been set by international organizations. For each kind of aflatoxin a minimum quantity of concentration is allowed, for instance, European Commission Regulation 2010/165/EC established limits of 8 and 15 μg/kg for AFB_1 and total AFs respectively. Several methods have been developed to determine AFs in foods, for instance: immunoassays techniques [3], Thin layer chromatography (TLC) [4]. High-Performance liquid chromatography (HPLC) with fluorescence detection [5]. Not long ago, analytical methods based on clean-up with immunoaffinity column and HPLC with postcolumn derivatization and fluorescence detection have gained much popularity. Even though, several works have been reported to determine AFs in foods by using these methods, only few validation studies are available which comply with certain regulations. There are immunochemical methods which are based principally on enzyme-linked immunosorbent assay (ELISA) that has a good sensitivity; speed and simplicity; however these kinds of instruments are expensive. An alternative of improving the disadvantages of the previous methods are trying to be solved by biosensors which are devices that enable identification and quantification of aflatoxins. Exists a variety of biosensors that base their performance in several principles, those are: optical, optoelectronic, electrochemical, piezoelectric, DNA and combined. In the same way, there are other methods not as common as the previous methods but they have a wide utility as well. The most important are those that base their principle on electrochemistry, spectroscopy and fluorescence.

The chapter has two main proposes. First, to give general description of the most common methods used for quantifying aflatoxin concentrations. And second, to give a perspective about the tendencies in the development of systems, based on the so far used methods, which could be employed in the near future to detect and quantify aflatoxins in food.

2. Chromatography methods

Chromatography is one of the most common methods for quantifying aflatoxins. This method started with Gas chromatography (GC). However, technology advancements allow the development of new chromatography-based techniques. Examples of these improvements are Liquid Chromatography (LC), Thin Layer Chromatography (TLC) and High-Performance Liquid Chromatography (HPLC). The quantification of aflatoxins using chromatography relies principally on fluorescence detection depending of the compounds under analysis. So that, nowadays there are several works employing a variety of fluorescence detection in order to improve the sensibility of these techniques. In the same manner, UV visible (Vis) wavelength spectrum has been used to improve the detection and quantification of aflatoxins. Others methods employed to accomplish the chromatographic quantification of AFs are array of diodes and refraction index.

Before the chromatographic analysis, the toxic compounds must be extracted to remove the interfering particles; such extraction is commonly done by a solvent in a clean-up step that regularly uses an immunoaffinity column (IAC). This procedure increases the sensitivity and diminishes the necessary sample quantity in the analysis. Other system used to quantify single and multiple aflatoxins is the mass spectrometer which is coupled commonly with a HPLC system. This section explains the most common methods based on chromatographic principle and the steps before and after for accomplishing the analysis with better assessments.

2.1. Gas Chromatography (GC)

The instrumentation of gas chromatography comprises well defined components that accomplish specific functions of the overall process. GC almost reaches the complete development of technological level in 50 years. The Figure 1 shows the principal components that constitute a Gas Chromatography System.

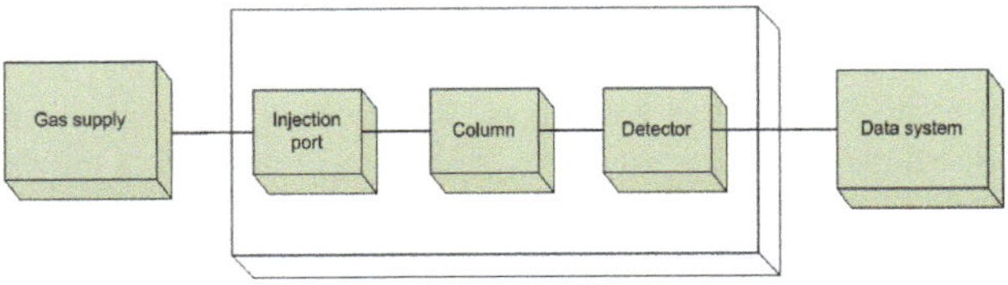

Figure 1. Block diagram of a gas chromatograph.

Gas supply means to move the sample through the column; the possible gases to choice are restricted and the most commonly used are nitrogen and helium. It is also necessary to control the gas flow because it can have impact on the separating performance. Tramps can be purchased to reduce or remove hydrocarbons and oxygen in the carrier gas. The chromatographic process starts when the sample is introduced into the column, ideally without disrupting flows in the column. Therefore, the deliberation of the sample into the column should be controlled, reproducible and rapid. The GC include an oven which is an important component in this process, because the vapor state must be maintained thought the GC separation, therefore, a good control of temperature must be kept. Another important component of the gas chromatography is the detector which has been evolved through the years. Nowadays, the mass spectrometer (MS) promises to be the most suitable method to be coupled with GC.

2.2. Liquid Chromatography (LC)

The principle of liquid chromatography is the separation process which is based on the distribution between two phases. The sample is propelled by a liquid which percolates a solid stationary phase. Thus a variety of stationary phases can be used in liquid chromatographic systems. The liquid chromatographic process and the separation of the sample may be achieved, both, in low and high pressure systems. And the correct selection of the separation mode stationary phase and mobile phase may be straight (normal) phase, reversed phase and size-exclusion (SEC) or ion-exchange (IEC) liquid chromatography respectively.

2.3. Thin-Layer Chromatography (TLC)

Thin-layer chromatography is a very commonly used technique in syntactic chemistry. This technique identifies compounds by determining the purity and progress of a reaction. Such reaction is fast and only requires a small quantity of the compounds. In TLC the mobile phase is liquid and the stationary process is a solid adsorbent. Several factors determine the efficiency of a chromatographic separation. The adsorbent should show a maximum of selectivity toward the substances that are being separated so that the differences in rate of elution will be large. For the separation of any mixture, some adsorbents may be too strongly adsorbing or too weakly adsorbing.

2.4. High-Performance Liquid Chromatography (HPLC)

HPLC is now the most common used chromatographic technique for a detection of a wide diversity of mycotoxins, especially for aflatoxins [6]. The analysis sample cleanup can be performed by liquid-liquid partitioning, solid phase extraction (SPE), column chromatography, immunoaffinity clean-up (IAC) columns, and multifunctional clean columns [7]. Recently the utility of the IAC columns has become very popular because of its high selectivity. IAC columns can be used for sample preparation before HPLC analysis either in off-line or in-line mode [8]. While in the off-line immunoaffinity cleanup the purification step is done separately by an expert, the IAC column is directly coupled to the HPLC system in the in-line immnunoaffinity cleanup. A chromatographic process can be defined as separation technique which involves mass-transfer between stationary and mobile phase. HPLC utilizes a liquid mobile phase to separate the components of a mixture. The stationary phase can be a liquid or a solid.

2.5. Combined methods

Nowadays there are combinations of the aforementioned methods with pre-process techniques. Such methods are able to detect in a better way, the concentration of aflatoxins in a solution. Immunoaffinity column sample clean-up followed by a normal or reverse phase of HPLC separation with fluorometric detection is mostly used for quantitative determination of AFM_1 because of the characteristics of specificity, high sensitivity and simplicity of operation [9].

There are several works that employ IAC combined with HPLC and fluorometric detection for detecting and quantifying precisely concentrations of AFM_1 [10]. In the reference [11] the authors employ IAC, HPLC and an optimized photo-derivatization to assess the concentration of mycotoxins of airborne from a house in Dalian, china. In [12] is employed, in the same manner, IAC and an HPLC equipped with fluorescence detector to determine the quantity of aflatoxins AFM_1 in milk used for preparing market milk, yogurt and infant formula products in South Korea. As can be seen, the HPLC process commonly needs a clean-up process with immunoaffinity columns before detection. Such sample preparations are multistage, expensive and time-consuming. The combination of GC with MS (GC-MS) for the analysis of aflatoxins can provide definitive, qualitative and quantitative results, but it requires a derivatization step, which lengthens the analysis time and may compromise analyte recoveries [13]. Post-column derivatization is a version of chromatography where the components that were separated eluting from the column are derivatized prior to entering the detector. The derivatization process is generally carried out during the process, during the transfer of the sample components from the column to the detector. The derivatization may also be accomplished before the sample enters into the column or the planar medium, thus it is called pre-column derivatization.

In the work presented by [14], aflatoxin B_1 was detected in animal liver (pig, chicken, turkey, beef, calf) and chicken eggs by a process consisting of sample immunoaffinity column clean-up and liquid chromatography with post column bromination in kobra cell and florescence detection, which was introduced and validated. The validation process was done based on the decision 2002/657/EC established by European commission. Figure 2 shows an exemplification of the common steps included in the detection and quantification of aflatoxins used nowadays. These steps include the processing of the sample before introduced it into the HPLC system. The most commonly pre-process technique is the IAC which permits to purify and decreases the quantity of sample needed in the experiment. The other problem that IAC tries to diminish is the non homogenization of the sample. HPLC is the most common based-chromatography technique used at this time; nevertheless, it needs of the steps shown in Figure 2 to work better. The principal advantages of IAC are the effectiveness and specificity in the purification of the extraction that provide the economic use of solvents and the improved chromatographic performance achieved with samples [15]. Also IAC can be used to analyze commodities that contain different aflatoxins. The fluorescence of aflatoxins make possible the use of a fluorometers to detect and quantify the concentration of aflatoxins in a sample, however, sometimes it is necessary to improve this property. Derivatization processes have been employed to improve the fluorescence of aflatoxins and by consequence the sensitivity of the system. Years ago, the quantification of the concentration of aflatoxins was accomplished by comparison of the sample and authentic standards using visual estimation of fluorescence of the separated spot long wavelength UV radiation [15]. A great advance in the detection and quantification of aflatoxins is the introduction of mass spectrometry as a viable detector system. The advantage of coupling mass spectrometer to LC is that, it allows improving the detection limits. For aflatoxins quantification, a number of instruments have been used including single quadrapole, triple quadrapole and lines ion single quadrapole and linear ion trap instrument.

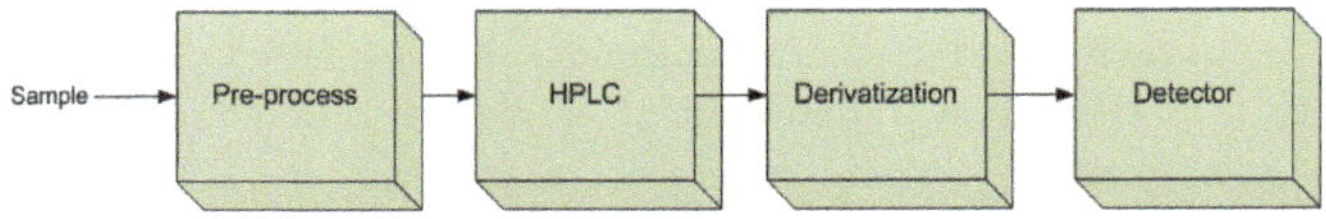

Figure 2. Common steps followed actually to quantify aflatoxins using HPLC.

Chromatography has been one of the most common methods used for the detection and quantification of aflatoxins. This technique has been evolving through the years from GC to HPLC which nowadays is the most used chromatography-based technique employed for aflatoxin detection. TLC and LC are methods that have been going displaced by HPLC because of its sensitivity, specificity and facility of operation. At the present time, HPLC by itself is not enough and it is necessary to employ pre-process and derivatization techniques that, jointly with detectors, improve the purity of the sample and the fluorescence property of the aflatoxins. In consequence, the detectors, which generally take into account the fluorescence of the aflatoxins, accomplish better quantification and offer more sensitivity. The most common method for pre-processing the sample is IAC that allow having high specificity in the selection of aflatoxin in samples contaminated with more compounds. So that, it is clear that the tendency of the technology is principally to improve the stage of pre-processes and derivatizations in order to achieve a more precise quantification. The more sophisticated the technology of pre-process and derivatization the more high specificity and sensitivity the method gets.

3. Immunoassays for aflatoxins detection

Among aflatoxin detection methods, there are those that base their operation on antibody-antigen reactions (Ab-Ag), known as immunoassays. Different kind of Aflatoxin molecules (AF) can be considered as antigens from the immunological point of view, so that it is possible to develop antibodies against them.

Most of immunological the methods are based on enzyme-linked immunosorbent assay (ELISA), which require less expensive instruments, have good sensitivity, speed and simplicity. However, ELISA kits are expensive especially for third-world countries [16], so several studies have focused on developing less expensive methods, without losing the benefits they offer. Besides, other alternatives will have some advantages over ELISA, as the use of magnetic droplets together with RT-PCR (Reverse Transcription Polymerase Chain Reaction), which has sensitivity to 1000 times greater than ELISA [17].

3.1. Antibody-Antigen reactions

Immunoassays are based on antibody-antigen reactions, in which one of reactants is marked and the other is immobilized on a platform. There are several kinds of molecule markers, which may be radioisotope, enzyme, fluorescent compound and colloidal Au. Because of small size of AFs, they are bounded to proteins so they can be captured with antibodies (Abs).

There are three type of antibodies used in immunoassays: polyclonal, monoclonal and recombinant. Polyclonal Abs (pAbs) is produced by rabbit, horse or goat blood immunized with protein-AF conjugate. This type of Abs is low-cost for preparation and easily produced. Monoclonal Abs (mAbs) is produced from positive hybridomas, which are usually produced by fusing murine myeloma cells and spleen cells from immunized mice. In [16], the authors used an indirect-competitive ELISA to detect AFB_1 by using a platform coated with monoclonal antibody. Recombinant Abs (rAbs) is produced by cloning the functional gene of some Abs and transmitted it into a prokaryotic or eukaryotic genetically-modified organism to hybridoma or spleen cells with or without immunization.

3.2. Competitive and non-competitive assay

There are two types of immunoassays: competitive and non-competitive assays. Competitive assays in turn are divided into two types: indirect and direct assays. In an indirect competitive assay, aflatoxins are immobilized by a protein-aflatoxin conjugate (Ag). The set is exposed to a buffer with the tested sample. Antibodies are released into the buffer; some of them will bind to the immobilized conjugate, while the remainder will join the analytes in the buffer. After a while, it is released a second group of enzyme-labeled antibodies, or other fluorescent-core kind of signal material; these are joined to the first antibodies, which in turn are attached to the protein-conjugate immobilized aflatoxins.

The process of a direct competitive assay does not require a second labeled antibody. For this type of analysis, either aflatoxins (aflatoxin protein conjugate) or specific antibodies can be immobilized. The complementary component to that immobilized is marked and added to the sample. A competitive reaction occurs between the Ab-Ag, so that some marked components remain adhered to the immobilized one, while others adhere to those present in the sample.

For non-competitive assays, Abs are immobilized. When immobilized Abs make contact with the sample molecules and AF bind to them because Abs are attached to a sensitive surface, the amount of analyte bounded by Abs will result in an electrical or optical variation. However, sandwich format is preferred for this kind of assay, in which the sample is mixed previously with Bovine Serum Albumine (BSA) because AF molecules are small. Previously it was carried out a treating of the sample with Bovine Serum Albumine (BSA) (carrier protein). BSA binds to the AFs and the conjugate is captured by immobilized Abs.

3.3. Enzyme-Linked Immunoabsorbent Assays (ELISAs)

Any type of assay involving Ab-Ag reaction, where one of the reactants is conjugated with an enzyme, is considered as an ELISA. Amplification and visualization of Ab-Ag interaction are achieved by this enzyme conjugation. ELISA is the most used immunoassays used in food-aflatoxin detection.

Antibodies or antigens are immobilized on a solid-phase matrix by linking them, either through adsorption or covalently. Reactants are usually adsorbed on to the wells of 96- or 384- microtitre plate of polystyrene, where adsorption is characterized by a strong hydrophobic binding and slow dissociation rate. After this coating process, the residual protein-binding capacity of solid matrix is blocked by exposing it to an excess of unrelated protein (e.g. gelatin or bovine serum albumin "BSA"). The next step is the addition of a test solution, which may be serum with an unknown concentration of antibodies against the immobilized antigen. After incubation and washing, binding of specific antibodies is visualized by the addition of antiimmunoglobulin-enzyme conjugate followed by a substrate, generating a colored product when hydrolysed. This change of color is proportional to the amount of antibodies bounded and may be recorded visually or spectrophotometrically. In case of an antigen measurement, the process is the same but may be done by using competitive- or sandwich-type assays. When using microarray format, ELISA may detect other toxins, such as AFs in a sample [18].

3.4. Recent advances

ELISA has been modified by using electrochemical techniques. Antibodies or antigens may be immobilized on an electrode with a free and enzyme-conjugate. So in a competitive assay, some enzyme-conjugate will bind to the electrode, and enzyme density can be shown by current produced from the catalytic oxidation reaction of the enzyme with the substrates. In a non-competitive assay, the formation of an Ab-Ag complex generates a barrier of direct electrical communication between the immobilized enzyme and the electrode surface.

Some authors have reported the use of electrochemical sensors. In [19] developed a sensor based on enzymatic silver deposition amplification to detect AFB_1 in rice. A linear sweep voltage was done in order to read the sensor response. In [20] the authors proposed the use of electrical impedance spectroscopy and free-labeled molecules.

Optical ELISAs often uses surface plasmon resonance (SPR). They are similar to electrochemical sensors, but in SPR, Ab or Ag is immobilized in an optical-sensitive surface. As AFB_1 changes, the angle of Spectral Power Distribution (SPD) varies. A combination of competitive-direct ELISA and an immunochromatographic assay was done by [21], in order to increase its sensitivity.

In recent years, some articles have developed modifications on ELISAs (e.g. with the using of nano particles). In [22] the authors refined the ELISA process for aflatoxin detection by using anti-AFB_1 single chain fragment variables, in order to detect only free AFB_1 instead of an AFB_1-protein conjugate. In the references [23] developed an ultra-sensitive ELISA by coupling a micro plate ELISA with sensitive magnetic particles. An important feature of this hybrid system is its small column size, high capture efficiency and lower cost over other reported materials.

A combination of a competitive direct ELISA and gold nanoparticle immunochromatographic strip was done by [21], with a detection limit of 1.0 ng/ml for AFM_1 in milk. Immunochromatographic assay (ICA) is rapid and simple, and can be carried out by untrained personnel without using electronic devices. However, this type of assay has low selectivity, so in [24], an improved ICA by using a new monoclonal antibody against AFB_1 was developed.

4. Biosensors for aflatoxins

Aflatoxins are harmful organisms. Their toxicity is due to their capacity to covalent binding DNA and proteins. The most acutely and chronically toxic aflatoxin is the B_1. The legal limits set for AFB_1 or for total aflatoxins vary from country to country [25]. The detection and quantification of aflatoxins are the first steps in the challenging task of controlling such organisms.

The rearing of livestock, the storage of grains, and the stock of their derivatives are daily life activities which are susceptible to be infected with pests and diseases. Such infections may cause human death and economical losses.

Every topic exposed in this chapter is so vast that could be by itself a single chapter. The aims of this chapter are to give a general overview of all the existing methods for the measurement and quantification of aflatoxins; to signalize their principles of operation; and to expose their tendencies.

Biosensors are multidisciplinary tools with an enormous potential in detection and quantification of aflatoxins. Thus, such devices have a huge impact in healthcare, food management, agronomical economy and bio-defense [26].

4.1. Biosensors

The International Union of Pure and Applied Chemistry (IUPAC) define biosensor as:

"A device that uses specific biochemical reactions mediated by isolated enzymes, immunosystems, tissues, organelles or whole cells to detect chemical compounds usually by electrical, thermal or optical signals".

4.2. Application of biosensors

Dr. Leland C. Clark established the concept of a biosensor as a biological sensing element whose change its properties when reacting biochemically with a specific compound or analyte [27]. Such reaction is converted into an electronic signal for its quantification. Dr. Clark developed a glucose oxidase enzyme electrode for detecting glucose.

There are two different approaches which can be carried out by biosensors.

a. The enzyme metabolizes the analyte, thus the analyte can be determined through the measuring of the enzymatic product.

b. The analyte inhibits the enzyme, thus the decrease of the enzymatic product formation is correlated with the analyte concentration. This case is called "biosensor based on enzyme inhibition".

Biosensors are tools basically conformed of a substrate (silicon, glass or polymers). Common polymers are: polymethyl methacrylate, polydimethyl siloxane, etc. The substrate is often coated with a conductive layer like: polysilicon, silicon dioxide, silicon nitrite, gold, and metal oxides. The specific recognition elements include: antigens, antibodies, nucleic acids, whole cells, proteins, enzymes, DNA/RNA probes, and phage-derived biomolecular recognition probes. The changes in these elements are detected via optical, electrochemical, calorimetric, acoustic, piezoelectric (quartz crystal, potassium sodium tartrate, lithium niobate), magnetic, and micromechanical transducers [28].

4.3. Biosensor based on optical techniques

Optical sensors are analytical tools that satisfy requirements as accuracy, precision and specificity in the selection of the analyte, allowing *in vivo* or *in vitro* investigations. Optical techniques provide a large realm of possibilities based on properties such as absorbance, reflectance and luminescence of single elements or groups of analytes [29].

Among the optical techniques used in biosensors it can be found: non linear optics (based on surface plasmon resonance) [30], resonant mirror, fiber-optics [31], complementary metal oxide semiconductors, fluorescence/phosphorescence [32], reflectance, light scattering, chemiluminescence, and refractive index [33].

Such advantages, plus their easy operation and wide detection capacity, have made of optical biosensors useful tools for the detection of dangerous organisms as aflatoxins.

4.4. Electrochemical biosensors

The first biosensor based on cholinesterase (ChE) inhibition for detection of nerve agents was developed by G.Guilbaut in 1962 [34]. Since then, many other enzymes have been used in biosensor for detecting and quantifying a huge realm of parameters.

Other important enzyme used in biosensors is the acetylcholinesterase (AchE). The principal biological role of AchE is the termination of the nervous impulse transmission at cholinergic synapses by rapid hydrolysis of the neurotransmitter acetylcholine [35]. The AFB_1 inhibits AChE by binding at the peripheral site, located at the entrance of the active site (at the tryptophane residue) [36]. Even though, there other enzymes, as butyrylcholinesterase (BChE) that are also used for detection of AFB_1, AChE is preferred because it is more sensitive than BChE for this purpose [37].

Biosensors based on the amperometric method allow the detection of low aflatoxin concentration. For example, acetylcholinesterase (AchE) is measured using a choline oxidase amperometric biosensor [38]. In this example the decrease in the amperometric activity of AchE has a direct relationship with the quantity of aflatoxins. This method is commonly used when the aflatoxin concentration is too low, and it cannot be detected with the spectrophotometric method.

4.5. Biosensors that combine techniques

Commonly, polymerase chain reaction (PCR) has been used to accurately detect low numbers of different pathogens with multiple sets of primers. But, important disadvantages of PCR are: the inhibition of the polymerase enzyme by the contaminants from the sample; difficulties in quantification; false positives resulting from the detection of naked nucleic acids; and non-viable microorganisms or contamination of samples in the laboratory. Biosensors are useful tools that provide a rapidly detection of the presence and amount of microorganisms in any given environments [26]. Thus, the mixture of different techniques might overcome the exposed problem. For example, in [39] there was a decrease of contaminants by coupling PCR with a piezoelectric biosensor.

There are biosensors that combine biological and physical/physicochemical transducers (SPR, piezoelectric, acoustic, and amperometric biosensors). The related problems for these biosensors are: chemical/physical stability of the transducers in the biological samples, the difficulty in production of highly specific antibodies, poor signal, etc. Such problems are often overcome by: coating the surface to make the transducer compatible with the biological samples; using of highly specific monoclonal antibodies; and incorporating amplification steps to generate stronger signals [26].

Latest researches on nanomaterials, such as carbon nanotubes, metal nanoparticles, nanowires, nanocomposite and nanostructurated materials reveal to be a key points in the design of the near future biosensing systems with applications in aflatoxin detection [40].

The aforementioned methods to quantify aflatoxins present several disadvantages, for instance those based on chromatography, however they have laborious and time-consuming process [41]. Therefore, a pathway to improve AFs detection is through biosensors. This term was first used by Cammaann in 1977 [42], who defined it as a device that enables the identification and quantification of the interest sample (e.g. water, air, food, solutions, among others). Nevertheless, the main characteristic in a biosensor is the biological recognition element that is capable to create a response of interest. Such element can be an antibody, an antigen or an enzyme [43].

There are many kinds of biosensors applied to detection aflatoxins, however they majorly work in conjunction with immunochemical methods. Such junctions are based on the high affinity of antigen-antibody interaction and have the aim of increasing the sensitivity and decreasing the detection time of the toxic element [41].

4.6. Immunochemical

These kinds of sensors use mainly immunological receptor units such as antibodies or antigens, and detection methods as optic effects (e.g. fluorescence and plasmon resonance), electrochemical, or acoustical readout [44]. The majorly of these techniques are comprised of three main steps: First, the extraction of the aflatoxin from de complex mixtures of materials in which it is found; then, the purification of the sample for removing pollutants; and finally, the detection and quantification of the toxins [45].

The main challenges of these types of biosensors are the design and construction of prototypes which minimize their handling. Besides, they must use the best immunochemical techniques, with the aim to generate automated sensors that replace the existing large, complex, cumbersome, and chemical laboratory analysis systems. Such immunochemical biosensors would offer the benefit of an increasingly developing of modular design that would permit the rapid substitution of other reagents to detect different toxics with the same platforms [45].

In [45] is reported a biosensor that it is based in the property of fluorescence. This fluorescence system consists on an arc lamp that generates a microsecond flash and a lens that focuses in the radiation into the sample. Such sample was previously treated, with process shown on the Figure 3 which in turn shows the three main steps before the antigen detection with the automated process placed in the arrows. Then the detection consists in using a filter which allows the passing of UV radiation, around 365 nm. This wavelength excites the fluorescence of aflatoxins.

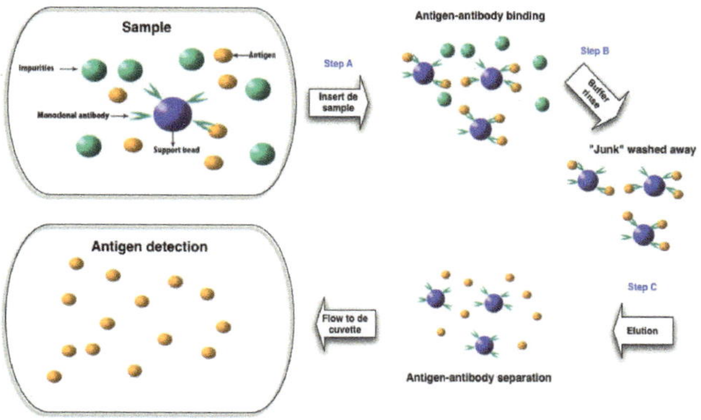

Figure 3. immunochemical-based capture, purification and detection process modify by [45].

After the excitation it is necessary the monitoring of the fluorescence response. This monitoring is carried out by a second lens that captures some of the light emitted by the sample through a photomultiplier tube and a filter centered around 455 nm. This is the wavelength where these AFs fluorescence. This device detects concentrations from 0.1 o 50 ppb in less than 2 minutes using a sample volume of 1 ml [45].

Another method used in conjunction with immunological techniques is named optical waveguide lightmode spectroscopy (OWLS). This technique is based in the precise measurement of the resonance angle of polarized laser light, diffracted by a grating in coupled into a thin waveguide. The incoupling resonance effect is very sensitive; such effect depends on the optical parameters of the sensor and the refractive index of the covering sample medium. The intensity effect response is carried out by a photodiode with the aim of determining the refractive index of the resonance incloupling angle detected with high accuracy [31].

There is another versatile technique named self-assambly structures that are considered as promising noble nanoscale systems with a several numbers of applications (solar cells, data storage, and biosensors). With this process it is possible to create biomarkers to exploit the absence of ligands on these nanoparticles surface that enhances the possibility of working better with molecules [46].

The self-assembly nanoparticles of nickel and gold are widely used for biosensing applications due to their biocompability, high surface to volume ratio, strong adsorption, fast electron transfer, enhanced sensitivity, high selectivity, and large detection range [47].

4.7. Centralized testing of DNA

Due to the necessity of creating simpler and more user-friendly methods for aflatoxins detection, it has been developed centralized testing of DNA. This method allows the early detection of genes associated with human diseases [48]. In this case, it is interesting to denote that the biosynthesis of aflatoxins has been extensively studied, and more than 25 genes arranged in a 70-kbp gene cluster were identified [49].

DNA biosensor has given out rapid and accurate measurements of aflatoxins in milk or dairy products [41]. The novel contribution of such system is its measurement technique based on electrochemical impedance spectroscopy (EIS) to analyze compounds that have restricted catalytic interaction activity such as aflatoxins.

The EIS method in recent years has become a powerful tool for evaluating many biochemical and biophysical processes. The biosensor's characterization and fabrication can be generated through EIS. Moreover, with the interaction between enzyme–substrate, biomolecule which have no reaction sequence after binding (such as antigen–antibodies), and DNA, among others, charge transfer changes occurred after affinity interactions can also be monitored with EIS.

To identify the aflatoxin M_{1A}, it is necessary the ss-HSDNA, which was specifically bounded to this aflatoxin. It is necessary to immobilize the ss-HSDNA on gold electrodes with the help of cysteamine and gold nanoparticles. The differences between before and after binding of aflatoxin M_1 to the HSDNA probe can be analyzed with a cyclic voltammetry and IES. An aflatoxin M_1 calibration curve was prepared by considering the differences in electron transfer resistances before and after aflatoxin M_1 binding [41].

However, in the most of the cases, the sample needs a pretreatment. For milk case, in order to remove the milk fat, the sample was centrifuged. Then three completely separated phases were obtained. The layer at the top was the fat; the cream was at the center; and the fat free milk was at the bottom. This last phase was used for the experiment as a sample, in order to avoid any possibly negative effect of fat on the EIS.

4.8. Piezoelectric biosensors

Biosensors based on piezoelectric effect are commonly used for aflatoxins detection because they have the property of providing sensitive measurements in air as well as in liquids. This kind of biosensors, based on piezoelectric quartz crystal (PQC), is usually combined with most of the above mentioned methods, like immunosensors with cells, bacteria, proteins (including antibody or antigen), DNA and so on [50].

Between the different existing kinds of immunosensors, the PQC has been extensively applied to biorecognition sensing due to its advantages of cost-effectiveness, direct detection, experimental simplicity, and real-time output. The principle of these sensors is based in the fact that the quartz is used as transducer, its resonance frequency changes with the change in the mass, according to the Sauerbrey equation.

In [49] the author reports a DNA-based piezoelectric biosensor with the aim of detecting a PCR-amplified 248-bp fragment of the aflD gene of *A. flavus and A. parasiticus* involved in the conversion from norsolorinic acid to averantin. Such biosensor was used for the analysis of DNA fragments coming from the amplification of DNA extracted from reference strains of *A. parasiticus.* Originally it was designed with the objective of researching about the influence of different parameters, such as amplicon concentration, dilution, and PCR specificity on the biosensor's response.

An important point of these kinds of devices is that the crystal only can be used for 25 measurements without losing sensitivity; this is because the devices that work with mechanical effects are majorly affected with the use. This is the reason for coupling the PCR protocol and the DNA piezoelectric biosensor. After its characterization with synthetic oligonucleotides, the piezoelectric-DNA biosensor led to the clear identification and quantification of contaminated feed samples with aflatoxins [49].

4.9. Optoelectronic

The principle of the optical waveguide light-mode spectroscopy (OWLS) technique is the measurement of the resonance angle of polarized laser light, diffracted by a grating and incoupled into a thin waveguide. Incoupling resonance occurs at very precise angles depending on the optical parameters of the sensor chips and the complex refractive index of the covering sample medium. The intensity of the incoupled light guided within the waveguide layer by multiple internal reflections is measured with a photodiode [31].The refractive index is determined from the resonance incoupling angle detected at high precision. Such index allows the determination of layer thickness and coverage of the adsorbed or bound material with high sensitivity. This method allows the construction of both chemical and biosensors. Therefore, it can be applied for direct sensing of various types of biomolecules.

Other optical based biosensor uses a high-tech semiconductor material–silicon for the efficient accuracy registration of narrow spectral bands or specific wavelengths. This biosensor is used for detecting and quantifying aflatoxins that are commonly found in a variety of agricultural products.

Based on the above mentioned techniques it was developed a structure with two oppositely directed potential barriers, the total current conditioned by these barriers depended on both, the external voltage and the wavelength of the absorbed radiation. A modification in these parameters resulted in the obtaining of high-accuracy data of aflatoxins contaminants in food and provender in natural conditions [51].

Detection and identification of harmful organisms, such as aflatoxins, in a cost and time effective way is a challenge for the researchers. Biosensors have proved to be useful tools for detection and quantification of such organisms. These sensors have advantages such as: fast response, relative easiness of use, a huge realm of applications, and flexibility for combination of techniques. Such advantages are derived from the involvement of multidisciplinary research activities. But, even though the vast research on biosensors, it is still needed the injection of economical funds to locate them in the commercial market, and impulse their use in real applications.

The research on biosensor has the aim to develop, at low cost an analytical approach simpler and faster. Being an alternative improving the classical techniques. It is necessary that the improvement of the processes is focused in autonomous measurement, in order to avoid, as much as possible, the human error. The mostly classical measurements are linked to the laboratory; the biosensors must be the way to create embedded systems with the aim to detect the aflatoxins *in vivo*. Because of this, it is necessary to automate the whole process including the pretreatment of the sample to generate more efficient systems and manageable and better aflatoxin detection biosensors.

5. Miscellaneous methods

5.1. Electrochemical methods for aflatoxins determination

Aflatoxins measurement usually implies complex, expensive and slow methods. However, this determination can be carried out taking into account the response of aflatoxins to determinate electrical stimulus. These methods are called electrochemical, where immunosensors are applied to determine the presence of aflatoxins in a sample. Usually, these sensors are composed by two screen-printed-electrodes (SPE), the first one is made of graphite, platinum, or gold; and it is known as working, active, or measuring electrode. The second electrode is the reference and is commonly made of Ag/AgCl. In general, this technique involves two basic steps. In the first, the immunosensor working electrode is coated with an antibody; after an incubated time, the sample that contains the aflatoxins is added to this electrode, while the left one reacts for a determinate time; finally, a conjugated of aflatoxins and enzymes is added to the electrode, it is then when the competitive reaction begins. In this reaction, free aflatoxins compete to link to antibodies present in the working electrode. After a stabilization time, the measuring electrode is removed from the sample and rinsed with a buffer solution. The second step implies to apply an electrical potential (commonly 100 mV) to the electrode, which changes its electrical conductivity according with the aflatoxins concentration. After sampling the electrode; an increase or reduction in the electric current flow will appear according with the concentration of aflatoxins in the sample. This technique has received improvements; disposable immunosensors have been reached for measurement of aflatoxins M_1 (AFM_1) directly in milk following a simple centrifugation step but without dilution or other pretreatment steps. Exhibiting a good working range, comparable to the ones obtained in buffer; linearity between 30 and 240 ng/ml making it useful for AFM_1 monitoring in milk (maximum acceptable level of AFM_1 in milk is 50 ppt) [52]. It is easy to notice that electrochemical techniques offer some advantages over traditional methods for aflatoxins determination, among which it can be found: reliability, low cost, *in-situ* measurements, fast processes, and easier methodology than common chromatography techniques through a similar performance

Other improvements to this methodology involve the analysis of thermal stability given that the conductivity properties of materials also change with temperature variations and not only for the aflatoxins concentration in the electrode. SPEs with platinum as substrate for the working electrode have been used to achieve long-term stability. Probes have shown that this type of electrodes maintain a good biorecognition affinity for antibodies on its layer and a decrease in the detected signal of less than 10% after two weeks inside a refrigerator (5 °C) and less than 22% at laboratory temperature (25 °C), values that allow partial usability for practical assaying [37]. Using this type of electrodes, a voltage of 50 mV, and a stabilization time of 1 minute are suggested to begin current measurements. Limits detection of 2.4 ppb has been reached in real capsicum spice samples, producing good correlations comparing with data from HPLC with fluorescence detector.

Working range for electrochemical immunnosensors from 0.1 to 10 ng/ml with a detection limit of 0.06 ng/ml has been achieved by using gold electrodes and enzymatic silver deposition amplification. In this procedure, an aflatoxin B_1-bovine serum albumin (AFB_1-SBA) conjugated is immobilized on the measuring electrode (gold electrode). An indirect competitive format between the selected analyte in solution and the AFB_1-BSA on the electrode is performed. After the competition step, monoclonal antibody against AFB_1 was bounded to the electrode and then conjugated to a secondary antibody-alkaline phosphatase (ALP) conjugated. The ALP could catalyze the substrate, ascorbic acid 2-phosphate, into ascorbic acid, and the latter could reduce silver ions in solution to metal silver deposited onto de electrode surface. Finally, the metallic silver deposited onto the electrode was determined by linear sweep voltametry (LSV). The peak current for this immunosensor exhibited a negative linear correlation to AFB_1 concentration [53].

As it can be noticed, electrochemical sensors and biosensors have, in some cases, the advantage of rapidity and sensitivity over the traditional techniques. Electrochemical sensors based on acetycholinesterase (AChE) inhibition by aflatoxins have been rapidly applied due to detection limits of 2 ppb. As reported by [53], the AFB_1 determination can be based on AChE inhibition, while the AChE residual activity is determined by using a choline oxidase amperometric biosensor coupled with AChE enzyme in solution. The amperometric detection of AChE activity is based on a second enzyme, cholesterol oxidease (ChOx), providing a consecutive conversion of the native substrate (acetylcholine) to an electrochemically active H_2O_2. Finally this component is measured at the screen-printed electrode previously modified with Prussian Blue (PB) at a potential of -0.05V versus screen-printed internal silver pseudo reference electrode. The linear working range was assessed to be 10-60 ppb.

Single electrode immunosensors have proved to be a reliable alternative to complex methods for aflatoxins determination. However, devices with multiple electrodes have been developed to offer the possibility to combine the high sensitivity of electrochemical SPE-based immunosensors with the favourable characteristics of high throughput ELISA procedures. An analytical immunosensor array, based on a microtiter plate coupled to a multichannel electrochemical detection system using the intermittent pulse amperometry technique is presented for detection of aflatoxins B_1 [54].

The device is composed by 96-well screen-printed microplated, their thick-film carbon sensors was modified according with a competitive indirect enzyme-linked immunoassay (ELISA) format for aflatoxins detection. Spectrophotometry and electrochemical procedures were both applied to determinate the reliability of the proposed system. The principal advantage of the aforementioned system is the possibility to separately apply the amperometric to each of the 96 sensing electrodes. The applied potential is +400 mV with a pulse of 1 ms and a selected frequency of 50 Hz. This immunoassay was applied for analysis of corn samples. AFB_1 could be measured at a level of 30 pg/ml and with a working range between

0.05 and 2 ng/mL. Aflatoxin AFM_1 was also quantified by this method. The suitability of the immunosensor for the direct analysis of the toxin in milk was assessed. AFM_1 was correctly measured with a working range of 5-250 pg/ml and a detection limit of 1 pg/ml was achieved. For this experiment, the intermittent pulse amperometry parameters were adjusted to -100mV with a pulse width of 10 ms and a 5 Hz frequency [55].

Variation of electrochemical immunosensors appeared recently to determinate aflatoxins through detection of a specific DNA [56]. The detection technique was optimized applying DNA sequences from *Aspergillus* gene *aflR* that codes a biochemical pathway of aflatoxins B_1 production. Then, voltametric detection of the specific *Aspergillus* DNA sequence is based on hybridization of adsorbed target DNA with a biotinylated probe and subsequent binding with streptavidin alkaline phosphatase conjugated. Then, the modified electrode surface of carbon paste electrode is incubated in a buffer solution with an electrochemically inactive substrate (1-naphthyl phosphate). Alkaline phosphatase converts 1-naphthyl phosphate into 1-naphthol, which is determinate by the selected voltammetric technique. The optimize procedure is capable to distinguish potentially aflatoxigenic fungi from other *Aspergillus* species.

5.2. Spectroscopy techniques

Spectroscopy techniques have been popularized because they present fast, low-cost and non-destructive analytical methods suitable to work with solid and liquid samples. This method involves the study of the interrelationships between the spectral characteristics of objects and their biophysical attributes, specifically, the interaction with radiated energy as a function of its wavelength or frequency [57].

In the particular case of aflatoxins, different studies have been carried out to determinate the wavelength in which these substances respond to radiant energy. The different spectroscopy systems available in the market have the facility to scan a sample over a determinate wavelength range and acquire the spectral data in different modes as reflectance, absorbance, or transmittance. The procedure to detect aflatoxins in a sample is quite similar to the aforementioned methods. The sample preparation implies extraction and clean up. However, some authors use the sample without any preparation. The samples are scanned with a spectrophotometer commonly over a wavelength range from 250 nm to 2500 nm at different steps (2 nm steps can be reached). Finally the results are shown in a graph of wavelength against reflectance or absorbance.

Near infrared spectroscopy (NIRS) is an excellent method for a rapid and low cost detection of aflatoxins in cereals [58]. Aflatoxin B_1 was successfully measured in maize and barley by applying grating and Fourier transform NIR spectroscopy instruments with multivariable statistical methods on intact, non-milled samples. This technique quantifies aflatoxins in order of 20 ppb. Variations to this method imply the use of horizontal attenuated total reflectance technique for determination of aflatoxin B_1, B_2, G_1 and G_2 in groundnut. The mid-band infrared attenuated total reflectance (ATR) spectra were obtained with a Fourier transform spectrometer equipped with a horizontal ATR accessory. This variant in the method generates rapid and substantial spectra of aflatoxins with a minimum sample size (>2 mL) and chemicals [59]. Other authors have incorporated a bundle reflectance fiber-optic probe to

NIRS system. Here, the fiber-optic probe is immersed in the sample without any previous treatment or manipulation of the samples. Then, NIR spectra are recorded direct from the fiber. This combination of technologies has proved to quantify aflatoxin B_1, ocharatoxin A and total aflatoxins in paprika successfully [60].

5.3. Fluorescence methods

Aflatoxins have a native luminescence due to their oxygenated pentaherocyclic structure. Thus, most analytical and microbiological methods for detection and quantification of aflatoxins are based on this feature. There are a number of microbiological methods that can be used for the direct visual detection of aflatoxin-producing *Aspergillus* strains. The aim of these procedures is to increase the production of aflatoxins and elicit at bright blue or blue-green fluorescent areas surrounding colonies under UV radiation. Complex agar media containing different additives to increase the production of aflatoxins have been implemented for this purpose. The addition of a methylated derivative of of β-CD plus sodium deoxycholate (NaDC) to yeast extract agar (YES) was found to be suitable for the identification of aflatoxigenic *Aspergillus* strains. This was achieved through the visualization of a beige ring surrounding the colonies. When this ring was examined under UV light, it exhibited blue fluorescence. Furthermore, it was observed that aflatoxigenic colonies grown in such environment also emitted room temperature phosphorescence (RTP), when examined in the dark, following excitation with a UV light lamp [61]. The main problem with this technique is related with the disturbance due to the background emission origination from matrix constituents, this because the emission maxima depends on the solvent and the pH. This problematic was addressed and solved by applying two-photon excitation conditions [62].

6. Conclusions

More than 300 micotoxins are discovered. They are toxic metabolites of a variety of fungi growing in a wide range of food and animal feedstuffs. Of all micotoxins, the aflatoxins are the major concerns as they are mutagenic, carcinogenic, teratogenic and immunosuppressive compounds. Consumption even at very low concentration may cause serious health problems. For the aforementioned reasons, it is important to develop new methodologies and systems able to quantify the aflatoxins concentrations that satisfy the restrictions proposed by the organizations in charge of control this compounds. To do this, several techniques have been employed such as: chromatography, immunological methods, biosensors and others methods. Through the paper can be noticed that almost all techniques need to combine efforts to accomplish precise quantifications. These combinations have depended greatly of technology development during the last years. In the case of chromatography, if the methods of pre-process, derivatization and detections improve their capabilities to achieve their functions, it can be developed new systems with higher sensitivity and portability than the so far developed systems.

In the case of immunological methods, there are several research papers reporting advances in the development and improvement of immunological techniques for detection of aflatoxins. Most of them are based on ELISA, although there are other techniques such as ICA and real-time PCR that have been used for this purpose; the objective of these studies is to achieve the development of rapid, simple, highly sensitive and low-cost techniques.

Future aflatoxins detection methods shall be guided by biosensors with mixed techniques, which have already proved their contribution, and utility in sensing and detection technology. Such sensors might be also used in biosecurity brigades along international borders. Biosensors may play a major role in this field as they provide rapid and specific detection compared to other techniques. A barrier that shall be overcome is the production of biosensors for harsh environments. Research on materials, techniques and working parameters need to be made to solve such problems. Portability is another obstacle to be defeated. The use of biosensors in small laboratories and the agricultural industry will increase as biosensors become more portable.

Tendencies in the development of new methods for quantifying the aflatoxins suggest a continuous combination among the different techniques. The combination of different techniques allows increasing the sensibility, portability and rapidness of analysis.

Acknowledgements

Authors give thanks to Consejo Nacional de Ciencia y Tecnología (CONACyT), in Mexico, for its financial support through the scholarships with Registration Numbers: 201401 (LMCM), 239421 (AEC), 231946 (CDG), 226888 (AAFJ), 209021 (RFMH).

Author details

Luis Miguel Contreras-Medina[1*], Alejandro Espinosa-Calderon[1], Carlos Duarte-Galvan[1], Arturo Alfonso Fernandez-Jaramillo[1], Rafael Francisco Muñoz-Huerta[1], Jesus Roberto Millan-Almaraz[2], Ramon Gerardo Guevara-Gonzalez[1] and Irineo Torres-Pacheco[1]

*Address all correspondence to: mcontreras@hspdigital.org

1 Biosystems Engineering CA, Postgraduate Study Division, Engineering Faculty, Autonomous University of Querétaro, Querétaro, Qro., México

2 Faculty of Physics and Mathematics, Autonomous University of Sinaloa, Culiacán, Sinaloa , México

References

[1] Juan, C., Zinedine, A., Molto, J. C., Idrissi, L., & Mañes, J. (2008). Aflatoxins levels in dried fruits and nuts from Rabat-Salé area, Morocco. *Food Control*, 9(9), 849-853.

[2] Baltaci, C., Iyasoglu, H., & Yuksel, F. (2011). Single-Laboratory Validation for the determination of Aflatoxins B1, B2, G1 and G2 in food based on immunoaffinity column and liquid chromatography with postcolumn derivatization and fluorescence detection. *Food Analytical Methods*, DOI s12161-012-9417-3.

[3] Lee, N. A., Wang, S., Allan, R. D., & Kennedy, I. R. (2004). A rapid aflatoxin B1 ELISA: development and validation with reduced matrix effects for peanuts, corn, pistachio and soyabeans. *Journal of Agricultural Food Chemistry*, 52(10), 2746-2755.

[4] Stroka, J., Van Otterdijk, R., & Anklam, E. (2000). Immunoaffinity column clean-up prior to thin-layer chromatography for the determination of aflatoxins in various food matrices. *Journal of Chromatography A*, 904(2), 251-256.

[5] Bacaloni, A., Cavaliere, C., Cucci, F., & Patrizia, F. (2008). Determination of aflatoxins in hazelnuts by various sample preparation methods and liquid chromatography-tandem mass spectrometry. *Journal of Chromatography A*, 1179(2), 182-189.

[6] De Rijk, T. C., Arzandeh, S., Kleijnen, H. C. H., Zomer, P., Van der Weg, G., & Mol, J. G. J. (2011). Aflatoxins determination using in-line immunoaffinity chromatography in foods. *Food Control*, 26(1), 42-48.

[7] Rahmani, A., Jinap, S., & Soleimany, F. (2009). Qualitative and quantitative analysis of mycotoxins. *Comprehensive Reviews in Food Science and Food Safety*, 8(3), 202-251.

[8] Cichna, M., Markl, P., Knopp, D., & Nuesser, R. (2001). On-line coupling of sol--gel-generated immunoaffinity columns with high-performance liquid chromatography. *Journal of Chromatography A*, 919(1), 51-58.

[9] Muscarella, M., Lo, Magro. S., Palermo, C., & Centonze, D. (2007). Validation according to European Commission Decision '02/657/EC of a confirmatory method for aflatoxin M1 in milk based on immunoaffinity columns and high performance liquid chromatography with fluorescence detection. *Analytica chimica acta*, 594(2), 257-264.

[10] Lee, J. E., Kwak, B. M., Ahn, J. H., & Jeon, T. H. (2009). Occurrence of aflatoxin M1 in raw milk in South Korea using an immnoaffinity column and liquid chromatography. *Food Control*, 20(2), 136-138.

[11] Wang, Y., Chai, T., Lu, G., Quan, C., Duan, H., Yao, M., Zucker, B., & Schlenker, G. (2008). Simultaneous detection of airborne Aflatoxin, Ochratoxin and Zearalenone in poultry house by immunoaffinity clean-up and high-performance liquid chromatography. *Environmental Research*, 107(2), 139-144.

[12] Lee, J. E., Kwak, B. M., Ahn, J. H., & Jeong, T. H. (2010). Ocurrence of aflatoxins M_1 in raw milk from South Korea winter season using an immunoaffinity column and liquid chromatography. *Food Control*, 20(2), 136-138.

[13] Wang, H., Xiangjuan, Z., Liu, Y., Yang, H., & Guo, Q. (2011). Simultaneous determination of chloramphenicol and aflatoxin M1 Residues in Milk by triple quadrupole liquid chromatography-Tandem Mass spectrometry. *Journal of agricultural and food chemistry*, 59(8), 3532-3538.

[14] Tavcar-Kalcher, G., Vrtac, K., Pestevsek, U., & Vengust, A. (2007). Validation of the procedure for the determination of aflatoxina B_1 in animal liver using immunoaffinity columns and liquid chromatography with postcolumn derivatisation and fluorescente detection. *Food control*, 18(4), 333-337.

[15] Shepard, GS. (2009). Aflatoxin analysis at the beginning of the twenty-first century. *Analytical and bioanalytical Chemistry*, 395(5), 1215-1224.

[16] Rossi, N. C., Takabayashi, C. R., Ono, M. A., Saito, G. H., Itano, E. N., Kawamura, O., Hirooka, E. Y., & Ono, E. Y. S. (2012). Immunoassay based on monoclonal antibody for aflatoxin detection in poultry feed. *Food Chemistry*, 132(4), 2211-2216.

[17] Babu, D., & Muriana, P. M. (2011). Immunomagnetic bead-based recovery and real time quantitative PCR (RT iq-PCR) for sensitive quantification of aflatoxin B(1). *Journal Microbiological Methods*, 86(2), 188-194.

[18] Lamberti, I., Tanzarella, C., Solinas, I., Padula, C., & Mosiello, L. (2009). An antibody-based microarray assay for the simultaneous detection of aflatoxin B1 and fumonisin B1. *Mycotoxin Research*, 25(4), 193-200.

[19] Tan, Y., Chu, X., Shen, G. L., & Yu, R. Q. (2009). A signal-amplified electrochemical immunosensor for aflatoxin B1 determination in rice. *Analytical biochemistry*, 387(1), 82-86.

[20] Wu, N. N., Cao, L., Yan, P. S., & Wang, M. H. (2011). Study on the Novel Label Free Impedimetric Immunosensor for Aflatoxin B1 Based on Poly(o-phenylenediamine) Film Modified Gold Electrode. *Advanced Materials Research*, 322-385.

[21] Wang, JH, Liu, BH, Hsu, YT, & Yu, FY. (2011). Sensitive competitive direct enzyme-linked immunosorbent assay and gold nanoparticle immunochromatographic strip for detecting aflatoxin M1 in milk. *Food Control*, 22(6), 964-969.

[22] Yang, L., Ding, H., Gu, Z., Zhao, J., Chen, H., Tian, F., Chen, Y. Q., Zhang, H., & Chen, W. (2009). Selection of Single Chain Fragment Variables with Direct Coating of Aflatoxin B1 to Enzyme-Linked Immunosorbent Assay Plates. *Journal of Agricultural and Food Chemistry*, 57(19), 8927-8932.

[23] Kanungo, L., Pal, S., & Bhand, S. (2011). Miniaturised hybrid immunoassay for high sensitivity analysis of aflatoxin M1 in milk. *Biosensors and Bioelectronics*, 26(5), 2601-2606.

[24] Zhang, D., Li, P., Yang, Y., Zhang, Q., Zhang, W., Xiao, Z., & Ding, X. (2011). A high selective immunochromatographic assay for rapid detection of aflatoxin B_1. *Talanta,* 85(1), 736-742.

[25] Delmulle, B. S., De Saeger, S. M., Sibanda, L., Barna-Vetro, I., & Van Peteghem, C. H. (2005). Development of an immunoassay-based lateral flow dipstick for the rapid detection of aflatoxin B1 in pig feed. Journal of Agricultural Food Chemistry; , 53(9), 3364-3368.

[26] Nayak, M., Kotian, A., Marathe, S., & Chakravortty, D. (2009). Detection of microorganisms using biosensors-A smarter way towards detection techniques. *Biosensors and Bioelectronics,* 25(4), 661-667.

[27] Guilbault, G. G., Kramer, D. N., & Cannon, P. L. (1962). Electrochemical determination of organophosphorus compounds. *Analitical Chemistry,* 34(11), 1437-1439.

[28] Velasco-Garcia, M. N. (2009). Optical biosensors for probing at the cellular level: A review of recent progress and future prospects. *Seminars in Cell Development Biology,* 20(1), 27-33.

[29] Espinosa-Calderon, A., Torres-Pacheco, I., Padilla-Medina, J., Osornio-Rios, R., de Jesus-Troncoso, Romero. R., Villaseñor-Mora, C., Rico-Garcia, E., & Guevara-Gonzalez, R. (2011). Description of photosynthesis measurement methods in green plants involving optical techniques, advantages and limitations. *African Journal of Agricultural Research,* 6(12), 2638-2647.

[30] Puiu, M., Istrate, O., Rotariu, L., & Bala, C. (2012). Kinetic approach of aflatoxin B1-acetylcholinesterase interaction: A tool for developing surfaceplasmonresonance biosensors. *Analytical Biochemistry,* 421(2), 587-594.

[31] Adanyi, N., Levkovets, I. A., Rodriguez-Gil, S., Ronald, A., Váradi, M., & Szendro, I. (2007). Development of immunosensor based on OWLS technique for determining Aflatoxin B1 and Ochratoxin A. *Biosensors and. Bioelectronics,* 22(6), 797-802.

[32] Hu, H., Garcia-Uribe, A., Deng, Y., & Zou, J. (2012). Layer-by-layer assembled smectite-polymer nanocomposite film for rapid fluorometric detection of aflatoxin $B_{1.}$ *Sensors and Actuators B: Chemical,* 166-205.

[33] Pohanka, M., Jun, D., Kalasz, H., & Kuca, K. (2008). Cholinesterase biosensor construction-a review. *Protein and Peptide Letters,* 15(8), 795-798.

[34] Guilbeau, E. J., Clark, L. C., Pizziconi, V. B., Schultz, J. S., & Towe, B. C. (1987). Biosensors in artificial organs. *ASAIO Journal,* 33(4), 834.

[35] Arduini, F., Errico, I., Amine, A., Micheli, L., Palleschi, G., & Moscone, D. (2007). Enzymatic spectrophotometric method for aflatoxin B detection based on acetylcholinesterase inhibition. *Analytical Chemistry,* 79(9), 3409-3415.

[36] Hansmann, T, Sanson, B, Stojan, J, Weik, M, Marty, JL, & Fournier, D. (2009). Kinetic insight into the mechanism of cholinesterase inhibition by aflatoxin B1 to develop biosensors. *Biosensors and Bioelectronics*, 24(7), 2119-2124.

[37] Pohanka, M, Malir, F, Roubal, T, & Kuca, K. (2008). Detection of Aflatoxins in Capsicum Spice Using an Electrochemical Immunosensor. *Analytical Letters*, 41(13), 2344-2353.

[38] Ben, Rejeb I, Arduini, F, Arvinte, A, Amine, A, Gargouri, M, Micheli, L, Bala, C, Moscone, D, & Palleschi, G. (2009). Development of a bio-electrochemical assay for AFB1 detection in olive oil. *Biosensors and Bioelectronics*, 24(7), 1962-1968.

[39] Tombelli, S., Mascini, M., Sacco, C., & Turner, A. P. F. (2000). A DNA piezoelectric biosensor assay coupled with a polymerase chain reaction for bacterial toxicity determination in environmental samples. *Analytica Chimica Acta*, 418(1), 1-9.

[40] Pérez-López, B., & Merkoçi, A. (2011). Nanomaterials based biosensors for food analysis applications. *Trends in Food Science & Technology*, 22(11), 625-639.

[41] Dinçkaya, E., Kınık, O., Sezgintürk, M. K., Altuğ, C., & Akkoca, A. (2011). Development of an impedimetric aflatoxin M1 biosensor based on a DNA probe and gold nanoparticles. *Biosensors and Bioelectronics*, 26(9), 3806-3811.

[42] Cammann, K. (1977). Biosensors based on ion-selective electrodes. *Fresenius Zeitschrift für Analytische Chemie*, 287(1), 1-9.

[43] Alkire, R. C., Kolb, D. M., & Lipkowski, J. (2011). Bioelectrochemistry: fundamental, aplications and recent developments. *Germany: Wiley-VCH*.

[44] Simone, S. M., & Schäferling, M. (2009). Toxin immunosensors and sensor arrays for food quality control. Bioanalytical Reviews; , 1(1), 73-104.

[45] Carlson, M. A., Bargeron, C. B., Benson, R. C., Fraser, A. B., Groopman, J. D., Ko, H. W., Phillips, T. E., Strickland, P. T., & Velky, J. T. (1999). Development of an Automated Handheld Immunoaffinity Fluorometric Biosensor. *Johns Hopkins APL technical digest*, 20(3), 373.

[46] Kalita, P., Singh, J., Singh, M. K., Solanki, P. R., Sumana, G., & Malhotra, BD. (2012). Ring like self assembled Ni nanoparticles based biosensor for food toxin detection. *Applied physics letters*, 100(9), 093702.

[47] Nie, Z., Petukhova, P., & Kumacheva, E. (2010). Properties and emerging applications of self-assembled structures made from inorganic nanoparticles. *Nature Nanotechnology*, 5(1), 15-25.

[48] Marrazza, G., Chianella, I., & Mascini, M. (1999). Disposable DNA electrochemical biosensors for environmental monitoring. *Analytica Chimica acta*, 387(3), 297-307.

[49] Tombelli, S., Mascini, M., Scherm, B., Battacone, G., & Migheli, Q. (2009). DNA biosensors for the detection of aflatoxin producing Aspergillus flavus and A. Parasiticus. Monatshefte für Chemie / Chemical Monthly , 140(8), 901-907.

[50] Jin, X., Jin, X., Chen, L., Jiang, J., Shen, G., & Yu, R. (2009). Piezoelectric immunosensor with gold nanoparticles enhanced competitive immunoreaction technique for quantification of aflatoxin B1. *Biosensors and Bioelectronics*, 24(8), 2580-2585.

[51] Khudaverdyan, S, Petrosyan, O, Dokholyan, J, Khudaverdyan, D, & Tsaturyan, S. (2012). Modeling of a New Type of an Optoelectronic Biosensor for the Monitoring of the Environment and the Food Products. Optoelectronic Biosensor for Monitoring Environment. *Technological Innovations in Sensing and Detection 179 of Chemical, Biological, Radiological, Nuclear Threats and Ecological Terrorism*, DOI 10.1007/978-94-007-2488-4_17.

[52] Micheli, L., Grecco, R., Badea, M., Moscome, D., & Palleschi, G. (2005). An electrochemical immunosensor for aflatoxin M1 determination in milk using screen-printed electrodes. *Biosensors and Bioelectronics*, 21(4), 588-596.

[53] Ben Rejeb, I, Arduini, F, Arvinte, A, Amine, A, Gargouri, M, Micheli, L, Bala, C, Moscone, D, & Palleschi, G. (2009). Development of a bio-electrochemical assay for AFB1 detection in olive oil. *Biosensors and Bioelectronics*, 24(7), 1962-1968.

[54] Piermarini, S., Micheli, L., Ammida, N. H. S., Palleschi, G., & Moscone, D. (2007). Electrochemical immunosensor array using a 96-weel screen-printed microplate for aflatoxin B_1 detection. *Biosensors and Bioelectronics*, 22(7), 1434-1440.

[55] Neagu, D., Perrino, S., Micheli, L., Palleschi, G., & Mosocone, D. (2009). Aflatoxin M_1 determination and stability study in milk samples using a screen-printed 96-well electrochemical microplate. *Internatioanal Dairy Journal*, 19(12), 753-758.

[56] Snevajsova, P., Tison, L., Brozkova, I., Vytrasova, J., Metelka, R., & Vytras, K. (2010). Carbon paste electrode for voltametric detection of specific DNA sequence from potentially aflatoxigenic Aspergillus species. *Electrochemistry Communications*, 12(1), 106-109.

[57] Roggo, Y., Chalus, P., Maurer, L., Lema-Martinez, C., Edmond, A., & Jent, N. (2007). A review of near infrared spectroscopy and chemometrics in pharmaceutical technologies. 44(3), 683-700.

[58] Fernández-Ibáñez, V., Soldado, A., Martínez-Fernández, A., & De la Roza-Delgado, B. (2009). Application of near infrared spectroscopy for rapid detection of aflatoxin B1 in maize and barley as analytical quality assessment. *Food Chemistry*, 113(2), 629-634.

[59] Mirghani, M. E. S., Chen, Man. Y. B., Jinap, S., Baharin, BS, & Bakar, J. (2001). A new method for determining aflatoxins in groundnut and groundnut cake using fourier transform infrared spectroscopy with attenuated total reflectance. *Journal of the american oil chemists' society*, 78(10), 985-992.

[60] Hernández-Hierro, J. M., García-Villanova, R. J., & González-Martín, I. (2008). Potential of near infrared spectroscopy for the analysis of mycotoxins applied to naturally

contaminated red paprika founding the spanish market. *Analytica chimica acta*, 622(1), 189-194.

[61] Rojas-Durán, T. R., Fente, Vázquez. B. I., Franco, C. M., Sanz-Medel, A., & Cepeda, A. (2007). Study of a room temperature phosporescence phenomenon to allow the detection of aflatoxigenic strains in culture media. *International Journal of food microbiology*, 115(2), 149-158.

[62] Rasch, C., Böttcher, M., & Kumke, M. (2010). Determination of aflatoxin B1 in alcoholic beverages: comparison of one-and two-photon-induced fluorescence. *Analytical and Bioanalytical Chemistry*, 397(1), 87-92.

Chapter 15

Lateral Flow Immunoassays for Aflatoxins B and G and for Aflatoxin M_1

Laura Anfossi, Claudio Baggiani,
Cristina Giovannoli and Gianfranco Giraudi

Additional information is available at the end of the chapter

1. Introduction

Aflatoxins (AFs), secondary metabolites produced by *Aspergillus Flavus* and *Aspergillusparasiticus*, are a numerous group of chemically related compounds characterised by high toxicity. Among these, aflatoxin B_1 (AFB_1) is the most potent known carcinogen for liver and, together with aflatoxins B_2 (AFB_2), G_1 (AFG_1) and G_2 (AFG_2) is the most frequently found and the most toxic of the group [1]. Therefore, maximum residue levels (MRLs) for AFB_1 and for the sum of the four $AFB_1 + AFB_2 + AFG_1 + AFG_2$ (total aflatoxins) in food and feed have been set by the European Union [2-4] and all over the world [5-7].

The occurrence of aflatoxins (AFs) has been widely reported in a variety of crops (including maize, wheat, barley, rice, groundnuts, nuts, pistachios, cottonseed, and spices) which can be infected pre-, during and post-harvest. Moreover, due to the relative stability of AFs to thermal and chemical stresses, they are found on commodities despite the elimination of mould, after long periods of storage, and also after the transformation of raw materials; therefore the presence of AFs has also been ascertained in commodities such as composite feed, flour, bakery products, and roasted peanuts.

In addition, products of the animal metabolism of aflatoxins could retain toxicity, such as in the case of AFB_1, which, once ingested, is rapidly absorbed and transformed into a hydroxylated metabolite. The latter is secreted into the milk and thus has been named aflatoxin M_1 (AFM_1). The hepatotoxicity and carcinogenic effects of AFM_1 have also been demonstrated and IARC have included this toxin in the group I human carcinogens as well as the parent AFB_1[1]. Milk and derived products can consequently also be implicated in the spreading of aflatoxins. Therefore, most countries have also set up MRLs of AFM_1 in milk, which varies from the 50 ng kg^{-1} established by the EU to the 500 ng kg^{-1} established by US FDA [2, 8]. More restrictive MRLs have been decided by the EU for the presence of AFM_1 in baby food [2].

A part from safety issue, food contamination caused by AFs also strongly affects economic interests; so much effortis devoted to the development of analytical methods for detecting these contaminants. Newly developed methods of analysis are intended both for screening purposes (rapid, economic and simple methods) and for the accurate, reproducible and sensitive quantification by confirmatory methods.

Numerous chromatographic methods to detect AFs in foods have been developed, coupled to fluorescent or mass spectrometric detection [9-11]. Likewise, several methods for aflatoxin M_1 determination in milk based on high-performance liquid chromatography associated to fluorescence or mass spectrometric detection have been developed [12-13]. However, chromatographic techniques are mainly used in confirmatory analyses and are usually not applied to routine controls owing to the necessity to use expensive equipment and extensive clean-up steps.

The first rapid methods of analysis for AFs were based on Thin Layer Chromatography [14]; this technique is still used today even though in a significant lesser extent compared to methods based on the use of antibodies. Immunochemical methods of analysis are widely employed as screening methods for measuring AFs in food and feed [9, 14-18] and also for AFM_1 quantification in milk and dairy products [19-21] thanks to their rapidity, selectivity and sensitivity. Several ELISA kits are commercially available, whose performances are generally adequate to meet legal requirements, and are routinely employed for aflatoxin monitoring. Some of these methods have also been validated [17-18]. However, even immunoassays need to be run in a laboratory, use a minimum of equipment and occasionally require some sample treatments, which may also involve the use of hazardous chemicals. Instead, affordable monitoring of food contaminants requires the highest-through put and more economical methods of detection and, possibly, little or no sample treatment, user-friendliness, employment of non-hazardous chemicals, in situ applicability. Additional requisites in aflatoxin detection would be low detection limits (especially for aflatoxin M_1) and adaptability to very differing commodities (for aflatoxins B and G).

Several innovative strategies have been proposed for the rapid, qualitative, semi-quantitative or quantitative detection of aflatoxins, also based on the use of specific antibodies without constraints of classical immunoassays [22]. For example, an interesting qualitative approach has been described for the detection of AFM_1 in milk [23-24]. The proposed method is based on a flow-through immunoassay with visual detection. Main advantages are represented by the high sensitivity and by the on site applicability of the assay which does not require any equipment for the treatment of the sample, norfor the analysis. In addition, it allowed the possibility of obtaining sample pre-concentration and/or clean-up in the same device used for the analysis [25]. Nevertheless, this method implies several subsequent steps to be carried out, thus limiting simplicity and rapidness of use. Very recently, the same approach has also been demonstrated for the multi-detection of different mycotoxins, thus increasing its potentiality of utilization [26].

Numerous immunosensors have been described [27] as well, and research is constantly evolving in this area, particularly for the development immunosensors for the selective determination of AFB_1 [28-32] and for AFM_1 detection [33-35].

In parallel, strategies aimed at avoiding the use of antibodies in the development of rapid methods for aflatoxin detection have also been reported, such as those based on the preparation of polymers with molecular recognition properties towards AFB_1as capture systems [36-37] or those based on the exploitation of its natural fluorescence for the detection [38]. A combination of the surface plasmon resonance phenomenon and fluorescence has been exploited in the work of Wang et al and permitteda very sensitive determination of AFM_1, though the proposed assay took almost an hour to be accomplished and couldn't be considered as a truly rapid method [39]. A fancy and cunning approach for the rapid quantification of AFB_1 have been described in the work of Arduini et al, who exploited the inhibiting effect of the toxin towards the enzyme acetylcholinesterase. The measurement of the enzymatic activity was demonstrated to directly allow AFB_1 quantification in 3 minutes and within the 10-60 $\mu g\ l^{-1}$ range [40].

Among the rapid methods for screening of food contaminants, the'lateral flow immunoassay" (LFIA) (also known as immunochromato graphic assayorimmuno-colloid gold immunoassay, ICG) has recently attracted the interest of researchers and industry. This technology has long been known in medical fields for diagnosing blood infections and failure of internal organs, disclosing drug abuse or ascertaining pregnancy and combines a series of benefits, including extreme simplicity, rapidity, and cost effectiveness [41]. These features make it ideally suited for screening large number of samples, for being conducted by non-trained personnel and practically everywhere, thus enabling the effective possibility of food safety assessment at all stages of food and feed production.

2. Lateral flow immunoassays for aflatoxins

Since the early 2000's, scientific papers and commercial devices aimed at measuring mycotoxinsin food and feed have appeared, and recentlya certain amount of literature on this topic has become available, including comprehensive reviews [42-44]. In particular, some LFIAs for the qualitative and semi-quantitative detection of aflatoxins in food and feed have been described and will be discussed below. At the same time, commercial LFDs for the detection of aflatoxinsin various commodities have become available and some of them have also been validated by USDA-GIPSA [45].

2.1. Principle of the method

As aflatoxins are low-molecular-mass compounds, immunoassays in competitive formats should be conceived to measure them. The same principles and reagents as in the microwell-type immunoassays could be applied, except for the fact that,in LFIA, the separation of bound and unbound antibody sites is obtained by means of the lateral flow on a suitable support (the membrane). The liquid flow transports immunoreagents along the membrane where they encounter their partners in spatially confined zones of the membrane itself where immuno reactions take place.

Besides the porous membrane which assures the flow, lateral flow devices (LFDs) usually include: an absorbent pad (positioned at the top of the membrane to increase the volume of the flowing liquid), a sample pad (to assure contact between the liquid sample and the membrane), and a rigid backing (Figure 1). A release pad can be added, whose role is to adsorb labelled antibodies in such a way that they are included in the device itself.

The membrane is almost exclusively made of nitrocellulose (NC); sample and adsorbent pads are usually made of cellulose, although sample pads could also be made of glass fibre or other materials and sometimes are soaked with proteins and/or surfactants for special applications. Release pads are usually glass fibre pads. Lines are traced on the NC membrane by means of dedicated dispensers which enables the dispensing of small volumes (typically few μl per cm) with high reproducibility.

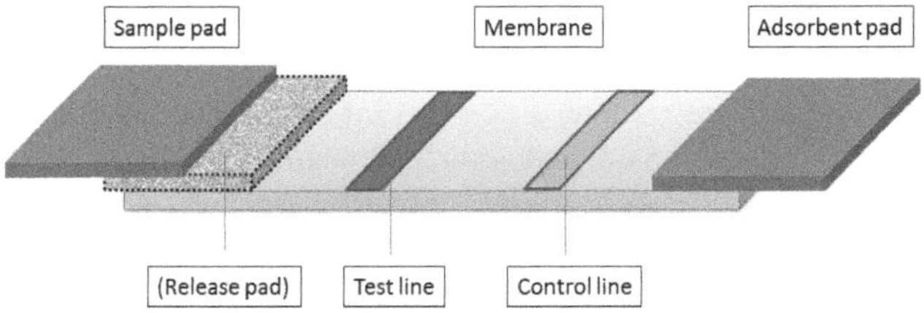

Figure 1. Schematic of a lateral flow device in the dipstick format.

The simplest LFD is a dipstick, which is dipped directly into the sample solution. Labelled antibodies can be added to the sample as a concentrated suspension or provided in a lyophilized form to be re-suspended by the sample itself. Alternatively, the labelled antibody can be pre-adsorbed onto the releasing pad, which partially overlaps the membrane. The liquid sample itself causes the re-suspension of the pre-adsorbed labelled antibodies during the assay. The sample pad is added in such a way that it overlaps the membrane or the releasing pad. Its role is the reduction of matrix interference by filtration alone or combined with some chemical action by means soaked reagents.

Besides the most popular dipstick format, LFDs exist in which the strip is inserted into a rigid plastic cassette provided with a sample well and a reading window. The main advantage of these housings is the guarantee of a reproducible compression of all components in the overlapping zones, which assures faster and more reproducible flows.

With few exceptions, the indirect competitive format, in which the antigen (a protein conjugate of the target toxin) is coated on the membrane and the antibody is labelled, is strongly preferred in the development of LFIA for AFs. Antibody labelling can be obtain by using virtually whatever nanoparticles that have a spectroscopically detectable property, such as, for example, coloured or fluorescent nanoparticles. However, gold nanoparticles (GNPs) are generally employed, with few exceptions, because of some characteristics which make them particularly suitable for the purpose. First, the conjugation of antibodies with GNPs is very easily obtained by simply mixing the two components at a proper pH (at or above the pI of the antibodies). The preparation and characterisation of stable colloidal solution of GNPs also follows well-established, easy protocols and a wide literature is available on this topic. The surface plasmon resonance of GNPs determines an intense colour of colloidal gold, which varies from orange to pink depending on particle dimensions and on surface overlay, therefore coloured nanoparticles can be prepared and the colour nuance can be use to monitor preparation and conjugation to antibodies.

The principles of the indirect competitive LFIAs which exploit GNP-labelled antibodies have been widely described and are schematized in Figure 2 and 3.

Briefly, the labelled specific antibody is suspended in the liquid sample and flows through the membrane where it first encounters the antigen coated in a zone indicated as "Test line" (T-line). In the absence of the target compound (negative sample, Figure 2), labelled antibodies bind to the coated antigen and are focused on the T-line, so that a visible (detectable) line is formed.

Usually, a second so-called "Control line" (C-line) follows and is constituted by secondary anti-species antibodies which capture any excess of specific antibodies.

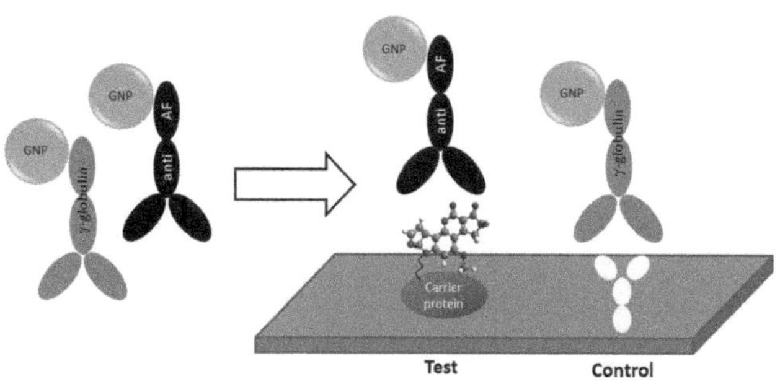

Figure 2. A lateral flow immunoassay in the indirect formatwith GNP-labelled antibodies for a negative sample (no AF is present). The Test line is made by a protein conjugate of the target toxin, while the Control line is constituted of anti-species antibodies. Anti-aflatoxin antibodies mixed together with non-specific γ-globulins (both GNP-labelled) move along the membrane. Anti-AF antibodies bind the antigen coated in the Test zone and are focused, thus forming a visible (detectable) line. Non-specific γ-globulins pass the Test line and are captured by the anti-species antibodies in the Control line where they are focused and form a second visible (detectable) line.

The appearance of a C-line can be regarded simply as the confirmation of the correct development of the assay (reagents and materials integrity) or else can be exploited to calculate the T/C signal ratio with the aim of normalizing strip-to-strip variations [46] or can also be regarded as an internal standard to which the intensity of the T-line is compared to determine positivity/negativity [47-48].

When the target is present above the lower detectable concentration level (positive sample, Figure 3), binding of labelled antibodies to the coated antigen is inhibited, resulting in a non-visible (undetectable) T-line.

Interpretation of assay results depends on the presence and intensity of both Test and Control lines as schematized in Figure 4.

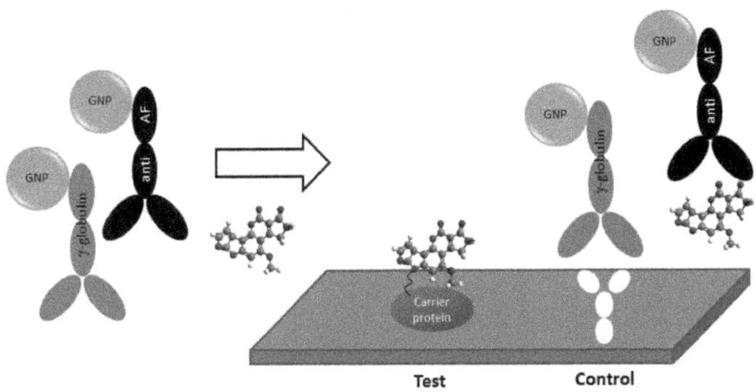

Figure 3. A lateral flow immunoassay in the indirect format for a positive sample (AF above the detectable limit). GNP-labelled anti-aflatoxin antibodies and non-specific γ-globulins move along the membrane. Anti-AF antibodies bind the toxin in the sample and the interaction with coated antigen is thus inhibited. Non-specific γ-globulins pass the T-line and are captured by the anti-species antibodies in the Control line where they are focused. Therefore, a single line (C-line) appears on the membrane.

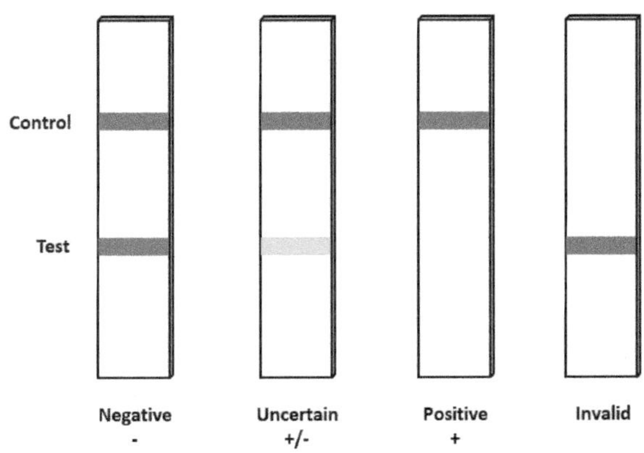

Figure 4. Assay result interpretation. Two intense lines: valid test, negative sample (target toxin below the detection limit of the method); intense C-line and fading T-line: valid test, the amount of the target toxin is near to the detection limit; intense C- line: test valid, positive sample (target toxin above the detection limit); intense or fading T-line: invalid test.

2.2. LFIAs for aflatoxins B and G

The first LFIA aimed at measuring any one of aflatoxins appeared in the scientific literature ten years ago and was one of the first reported lateral flow assays for food contaminants. The authors described a simplified device formed by aNC membrane on which the T-line had been traced upon by dispensing antibodies towards AFB_1. The signal reporters were liposomes, which were tagged with AFB_1 and encapsulated a visible dye. The tagged liposomes flowed along the membrane where encountered the coated anti-AFB_1 antibodies and were captured, thus determining the appearance of a coloured T-line due to the focalization of the encapsulated dye. If some AFB_1 was present in the sample, the binding of the tagged liposomes to the coated antibodies was inhibited and the colour of the T-line faded. The absolute limit of detection of such a device was 18 ng of AFB_1 and the test could be completed in a total of 12 minutes, including sample preparation [49].

Apart from this early approach, following papers described more usual LFDs based on the use of GNPs as antibody labels. In 2005, Delmulle and co-workers reported the development of a dipstick which allowed authors to detect AFB_1in pig feed. The visual detection limit (VDL) was set at 5 $\mu g\ kg^{-1}$ and the analysis could be completed in 10 minutes [50]. In the same year, the group of Xiulianal so described the preparation of GNP-labelled antibodies towards AFB_1 and their exploitation in a visual LFIA [51]. The application of the developed dipstick to measure AFB_1 in rice, corn, and wheat was reported in a following paper of the same group [52]. The described LFD showed a VDL of 2.5 $\mu g\ l^{-1}$ in buffer, which became 0.05 $\mu g\ l^{-1}$ when the colour intensity of lines was determined by means of a photometric reader. Therefore, a sensitive quantification of the target toxin (limit of detection, LOD, 2 $\mu g\ kg^{-1}$ in food) could be demonstrated; moreover, accuracy of the developed assay was confirmed on 60 samples through comparison with ELISA.

A visual LFIA for detecting AFB_1 was also described by papers of Shim et al [53-54]. The developed LFD was shown to cross-react to some extent to other major aflatoxins (AFB_2, AFG_1, and AFG_2) but not to differing mycotoxins (such as ochratoxin A, citrinine, patulin, zearalenone, and T-2 toxin). Nevertheless, it was applied for selectively measuring the sole AFB_1in rice, barley and feed. VDLs of 5-10 $\mu g\ l^{-1}$(rice, barley) and 10-20 $\mu g\ l^{-1}$(feed) were obtained and the proposed method showed agreeing results towards HPLC analysis on up to 172 food and feed samples. The same group also published results obtained with a multi-analyte device aimed at contemporary measuring AFB_1 and ochratoxin A in feed. The described method allowed the simultaneous detection of the two toxins which could be completed in 15 minutes and showed a VDL of 10 $\mu g\ kg^{-1}$for AFB_1. Method validation by means of ELISA and HPLC confirmatory analyses was also reported [55].

Although regulations prescribe the simultaneous determination of AFB1, AFB_2, AFG_1, and AFG_2 beside AFB_1 quantification, most papers described LFIA selective towards AFB_1.To meet the need of measuring all the four major AFs our group developed a quantitative LFIA for total aflatoxin determination in corn samples. The assay could be completed in 10 minutes, showed a LOD of 10 $\mu g\ l^{-1}$ and was validated through comparison with HPLC on 25 samples. In addition, an aqueous extracting medium was also optimized and proven to allow reliable quantification of total aflatoxin [56]. Except in this case, AFs were always extracted in methanol/water (typically 70/30 or 80/20 v/v) followed by dilution of the extract before LFIA analysis to reduce the proportion of the organic solvent, which is hardly compatible with materials composing LFDs. However, a methanol amount lower than 15-20% has been demonstrated by most authors to be compatible with LFD materials and further more not to affect immunoassay performance.

Most recent contributes to the topic are due to the group of Zhang and co-workers, who described two LFDs, the first highly selective towards AFB_1 and the second able to measure total aflatoxins [57-58]. Both devices have been applied to visually detect target toxins in peanuts (the highly selective one could also be exploited to detect AFB_1 in pu-erh tea, vegetable oil and feed). Both methods allowed reliable results (agreeing with HPLC determination) to be obtain in 15 minutes. In addition, the LFIA aimed at measuring total AFs was extremely sensitive, with VDL in peanut extracts as low as 0.03, 0.06, 0.12, and 0.25 μg $l^{-1}$for AFB_1, AFB_2, AFG_1, and AFG_2, respectively.

In addition to papers aimed at describing actually functioning devices for measuring AFs, those targets have often been chosen as system models for the development of original devices which exploited non-traditional signal reporters to label antibodies. Besides the above mentioned approach of Ho and Wauchope, based on the use of dye-encapsulating liposomes, Liao and Li described a visual device which exploited nanoparticles with a silver core and a gold shell as the reporters in the construction of a LFD for AFB_1. The toxin was determined in cereals and nuts and performances were compared to those of a GNP-based LFIA and to results obtained through a classic microwell-based immunoassay. The authors demonstrated that the newly developed LFD was comparable to the GNP-LFD in terms of stability of components and reproducibility of signals. On the other hand, it allowed a great enhancement in sensitivity so that values as low as 0.1 μg $l^{-1}$$AFB_1$ could be measured [59].

With the expectation of increasing the useful signal, therefore being able to reduce immunore agents for the benefits of the competition, magnetic nanogold microspheres with a Fe_2O_3 core and a shell of multiple GNPs have also been proposed. The magnetic core of particles allowed authors to simplify separation steps during the labelling of antibodies and their micro- dimensions to enhance colour during the test itself. A three-fold increase in sensitivity was stated for the visual detection of AFB_2 compared to the use of simple gold colloid nanoparticles [60].

2.2.1. Application of LFIA for aflatoxins B and G in food analysis

A major concern in the development of LFDs for aflatoxins is the unpredictable effects due to food components co-extracted from the sample beyond the target and which affect not only the antigen-antibody interaction on which the immunoassay is based, but also the mechanics of the device itself.

Some authors experienced the apparently inexplicable failure of recovery experiments conducted on fortified materials and the incongruity of results attained for artificially and naturally contaminated samples, which necessitate matrix-matched calibrations and recommended the use of naturally contaminated samples blended in varying proportions with blank samples as calibrators [56, 61-63]. Matrix components not only interfere in defining appropriate standards for calibration but also determine requirement of distinct devices to be developed for individual foods.

Despite the fact that the some authors reported calibration by using standard AFs diluted in buffers (to which methanol is added in limited proportions, as discussed above) and stated no interference from matrix given a limited dilution of sample extracts, the application of LFDs for the effective AF B and G detection in food remains the bottleneck in the development of new LFIAs. This taskis also made particularly complex by the multiplicity andvariety of matrices to be considered in aflatoxin B and G analysis.

2.3. LFIAs for aflatoxin M_1

The development of LFIAs for AFM_1 is one of the most challenging goals in this field of research because of the extreme sensitivity required by legislation for this contaminant (particularly in the European Union).

The first paper dealing with the subject reported a validation study on a commercial device which was conceived for meet US regulations and did not described any preparation protocols and methods. The ROSA Charm Aflatoxin M_1™ aimed at quantitatively measuring AFM_1 in milk was validated as the result of an inter-laboratory trial, which involved 21 participants, at four levels above and two below the declared LOD of the assay (400 ng l^{-1}) [64]. Less than 5% of false negative (n=83) and no false positive below 300 ng l^{-1} were found. For contaminations between 350 and 450 ng l^{-1} false positivity increased from 21 to 93%.

More recently, Wang et al first described the development of a LFD for the detection of AFM_1 [65]. The cut-off level (0.5-1 μg l^{-1}) is just above the eligible value required by the US regulation [8] and far beyond the more severe limits imposed by the European Union for this contaminant [2]. However, it is an effectively sensitive and rapid assay, provided that the whole analytical procedure can be completed in 10 minutes, as no sample treatment is required.

A visual device has also been developed by Zhang et al which showed a VDL for $AFM_1$of 0.3 μg l^{-1} [66]. Although the sensitivity improvement respect to the work of Wang et al, the obtained VDL remains far away from the detectability demand imposed by EU MRLs for this contaminant.

3. Development of a highly sensitive LFIA for measuring AFM_1 in milk

With the aim of producingasystem sensitive enough to reach the limits imposed by European regulations, we developed a competitive lateral flow immunoassay which exploited rabbit polyclonal antibodies towards AFM_1that had been previously employed in the development of a sensitive ELISA [19]. A classic device, including a NC membrane (onto which the two lines of reagents had been immobilized), cellulose sample and adsorbent pads, and a glass fibre release pad (on which GNP-labelled antibodies are pre-adsorbed) was conceived.

The method was designed to be a competitive LFIA, in which the Test line comprised an AFM_1 conjugate (competitor) and the Control line was composed of anti-rabbit IgG antibodies. GNP-labelled anti-AFM_1 antibodies were furnished as pre-adsorbed in a release pad. When re-suspended by the sample, flowed across the membrane where first encountered the T-line and bound to the immobilized AFM_1 conjugate. A red colour became visible at the T-line, due to the focusing of nanoparticles. If some AFM_1 was present in the sample, it competed with the immobilized AFM_1-BSA for binding to the GNP-labelled antibodies, resulting in a reduction of the T-line intensity. The anti-rabbit IgG antibodies on the Control line captured any excess GNP-labelled antibodies to produce a C-line as a visible confirmation of particle flow. Signal intensities of the two lines were read by a portable scanner connected to a laptop and processed by dedicated software, which acquires images, determines colour intensity, interpolates values on a memorized standard curve and returns the concentration of the analyte in the sample.

Since the methodin development was a competitive immunoassay, its sensitivity was influenced by several well-known factors, such as antibody dilution and competitor concentration, provided that a definite antiserum was used. Additional factors that could be considered were: the chemical structure of the hapten (actually, the use of heterologous competitors had been shown to improve sensitivity [67]), the structure of the antigen used as the competitor in the assay (as far as the nature of the carrier-protein and the degree of conjugation between the hapten and the carrier-protein itself were considered); the specific response of the reporter used to label the antibody; the extent of antibody labelling (moles of reporter per mole of antibody). In effect, the work of Byzovaet al [68] firstly reported the effect of varying some of the described factors on LFIA performances and, in particular, showed that the diminishing of the molar substitution ratio (SR) between the hapten and the carrier-protein in the preparation of the competitor significantly improved as say sensitivity. The same authors also studied the binding capacity of different anti-species antibodies (which were used to trace the C-line) concluding, in this case, that no evident differences could be observed.

The need of developing a very high sensitive assay for determining AFM_1 in milk at levels of regulatory concern according to EU regulation [2], forced us to investigate further in these directions and to question other established practices, such as the assumption that the labelling of antibodies should be conducted in such a way to obtain a complete coating of GNP surfaces.

Therefore, the effects of varying: the competitor (use of homologous or heterologous hapten; nature of the carrier-protein and hapten-to-protein molar ratio) and the reporter (extent of antibody labelling)on sensitivity were studied and optimized.

3.1. Materials and methods

3.1.1. LFD preparation

Gold colloids with mean diameter of about 40 nm were prepared using the sodium citrate method as previously described [46]. The saturation concentration of the anti-AFM_1 antibody for conjugation with gold nanoparticles was determined according to Horisbergand Rosset [69]. GNP-antibody conjugation was carried out using an amount of antibodies which is the half the saturation concentration and was carried out as follows: 100 μl of a 0.5 mg ml^{-1} anti-AFM_1antibodies in borate buffer was added to 10 mL of pH-adjusted colloidal gold solution. After 30′ incubation at room temperature, 1 ml of borate buffer containing 1% of BSA was added. The mixture was centrifuged and the pellet was washed twice by re-suspension in borate buffer with 0.1% BSA added. Finally, the pellet was re-suspended in borate buffer supplied with 1% BSA, 0.25% Tween 20, 2% sucrose, and 0.02% sodium azide and stored at 4°C until use.

Release pads were previously treated with borate buffer supplied with 1% BSA, 0.25% Tween 20, 2% sucrose, and 0.02% sodium azide. After drying, gold-labelled antibodies were distributed near the lower edge of the pads and left to dry.

Test and Control lines were spotted upon a NC membrane as follows: the AFM_1-protein conjugate (SR 4) at 0.3 mg/ml was the capture reagent, and the goat anti-rabbit IgG antibodies (2 mg/ml) formed the C-line. Then, the membrane was dried. Strips were composed as follows: from the top; the adsorbent pad, the NC membrane, the release pad and the sample pad were pasted, in sequence, with 1-2 mm overlap. Release pad was positioned so that the line of GNP-labelled antibodies was on the opposite site from the edge of the membrane. The prepared membrane was cut into strips of 5 mm, which were inserted into rigid plastic cassettes. Cassettes were stored in plastic bags containing silica at room temperature until use.

3.1.2. Lateral flow immunoassay procedure

Pasteurized milk samples were purchased in large stores, and raw milk samples were obtained from farms. Whole and semi-skimmed milk (1 ml) were centrifuged for 2 min at 6000 rpm. The upper fat layer was discharged, 500 μl of the underlying serum was transferred into a tube and 25 μl of 10% Tween 20 was added. The mixture was immediately used in the lateral flow assay.

The test was carried out by placing 100 μl of sample into the sample well. After 15 minutes of incubation at 37°C, the cassette was placed above a mobile scanner connected to a laptop. The Skannex 3.0 software (SkannexAS,Hoenefoss, Norway) was used to acquire and process images. Calibration curves were obtained by plotting the ratio between the intensity of the test (T) and the control line (C) [46] against the log of AFM_1 concentration. For each experiment, a calibration curve was determined by a nonlinear regression analysis of the data using the four-parameter logistic equation [70]. For the construction of the standard curve and for recovery experiments blank milk samples that did not show any detectable residues of the target when analysed by a reference ELISA (LOD 5 ng l^{-1}) [19] were fortified with appropriate amounts of an AFM_1 standard solution.

3.2. Optimization of the LFIAs

3.2.1. Effect of varying the hapten, the AFM_1-protein substitution ratio and the carrier protein in the T line

The polyclonal antiserum used in this work had showncertain cross-reactivity towards aflatoxin B_1 (about 35% when measured by means of the ELISA); therefore a competitor synthesized by using a hapten derived from this toxin was considered as a "heterologous" competitor respect to AFM_1 protein conjugates (which were homologous to the immunogen). Therefore, three conjugates of AFM_1 with Bovine Serum Albumin (AFM_1-BSA) conjugates which varied in the hapten-to-protein ratio, one conjugate of $AFM_1$with ovalbumin (AFM_1 –OVA) and one conjugate between $AFB_1$and BSA (AFB_1-BSA) were evaluated as potential competitors to be immobilized in the Test line (Table 1). Each conjugate was dispensed on the membrane at the same rate and volume (1µl/cm), however the concentration was varied to obtain an absolute signal of about 20-25 arbitrary units in the T-line when the strip were read by means of the software. AFM_1 standard solutions (0-10-100-1000 ng l^{-1}) prepared in a blank pasteurized whole milk were measured in triplicate and IC_{50} values were compared (Table 1). The AFB_1 conjugate qualitatively behaved as the AFM_1 conjugate with a similar SR, except for the absolute signal, which was less intense at the same concentration of dispensing. Interestingly, the decrease of the amount of AFM_1 per mole of protein strongly influenced the sensitivity of the assay. Indeed, the reducing of the substitution ratio (SR) from about 22 to about 4 allowed an improvement of nearly 10-folds in the IC_{50} to be obtained. This result is in good agreement with the observation of Byzova and co-workers [68] and with expectations based on the experience with competitive immunoassays in other formats (such as for example in ELISA). In parallel, the absolute signal decreased and forced to increase the amount of antigen to be dispensed. Nevertheless, the advantage of reducing the hapten density strongly predominated over the increase of the absolute antigen concentration.

Conjugate	SR	Dispensing concentration (mg ml^{-1})	IC_{50} (µg l^{-1})
AFM_1-BSA	4	0.8	0.2
AFM_1-BSA	15	0.4	1.1
AFM_1-BSA	22	0.2	1.7
AFM_1-OVA	10	0.8	0.6
AFB_1-BSA	24	0.4	1.6

Table 1. Effect of varying the competitor to be used in the Test line of the LFD. Protein conjugates were dispensed onto the membrane at different concentrations to reach an absolute signal comprises between 20 and 25 arbitrary units on the T-line. SR represents the molar substitution ratio between the toxin and the protein which had been estimated by spectrophotometric measurements.

On the contrary, the substitution of the bovine serum albumin (which had been used to prepare the immunogen) with ovalbumin as the carrier-protein seemed irrelevant. In fact, antibodies binding the BSA used as the immunogenic carrier-protein are saturated in the preparation of the gold labelled- antibody. This preparation involves the GNP overcoating with exceeding amount of BSA to prevent aggregation; however, the inhibition of further non-specific binding to BSA of antigens could also be attained.

3.2.2. Labelling of antibodies with gold nanoparticles

Optimization of LFIA usually involves checkerboard titrations where the amounts of antibodies and of the competitor are varied to achieve the lower limit of detection and the maximum slope of the calibration curve. Varying the amount of antibodies is exclusively intended as diluting the colloid of GNPs coated with antibodies themselves. The parameter used to measure this dilution is the optical density (OD) of the gold colloid, assuming that GNPs surface had been saturated with antibodies; a typical protocol prescribes that the saturation amount of antibodies, intended as the amount that prevent GNP aggregation, has to be determined firstly and this stabilizing amount or, more usually, a small excess of antibodies, has to be conjugated to GNPs to prepare the signal reporter. Nevertheless, contrarily to this generally accepted assumption, Laycock et al reported a huge increase in sensitivity due to the reduction of antibodies coated onto GNPs in comparison to the stabilizing amount [47].

Therefore, besides studying the effect of varying GNP-labelled antibody (intended as varying the OD under saturating conditions); we considered that dilution of antibodies to favour competitive conditions would also be achieved by reducing the number of molecules of antibody bound per GNP at a fixed OD value. Consequent risk of GNP aggregation, due to incomplete shielding of the superficial GNP charges, could be efficiently prevented by the further addition of exceeding amount of other proteins, such as for example BSA, which is particularly effective in this purpose.The variation of the amount of GNP-labelled antibodies dispensed at different ODs (3 and 6) under saturating conditions, apparently did not directly influence the sensitivity of the LFIA (data not shown) Nevertheless, the increasing of the OD allows the development of more intense absolute signals, which in turn means that the amount of competitor could be decreased in the T-line therefore improving detectability.

To study antibody dilution intended as the reduction of antibody amount per GNP, different amount of antibodies were reacted with portions of the same GNP colloid as follows: saturation amount (Ab_{SAT}), excess of antibody ($Ab/ Ab_{SAT} = 1.5$), defect of antibody ($Ab/ Ab_{SAT} = 0.7$), and half the saturation amount ($Ab/ Ab_{SAT} = 0.5$). The four GNP-antibody preparations were dispensed onto release pads at OD 3 and applied to strips where the AFM1-BSA with SR of 22 and a concentration of 0.2 mg/ml had been traced upon to form the T-line. AFM1 calibrators prepared in milk were run onto these strips in triplicate. Resulting curves are show in in Figure 4. Besides a significant signal reduction, a certain improvement in sensitivity was observed when the amount of antibody was lowered from saturating conditions (IC50 = 1.71 ± 0.01) to its half (IC50 = 0.99 ± 0.01); however detectability was influenced in a considerably lesser extent respect than when modifying the nature of the competitor (i.e.: the SR of the conjugate used to obtain the T-line), as discussed above.

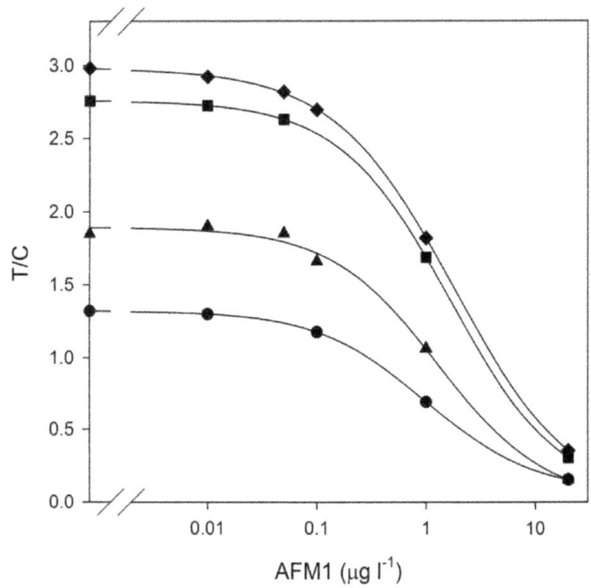

Figure 5. Effect of the amount of antibodies coated onto the GNPs (Ab) compared to the amount needed for saturating GNP surface (Ab_{SAT})for varying Ab/ Ab_{SAT}: 0.5 (•), 0.7 (▲), 1 (•), 1.5 (•).GNP-antibodies dispensed at OD 3; T-line: AFM_1-BSA conjugate (0.2 mgml^{-1}, SR=22).

3.3. AFM_1 detection in milk by the developed LFIA

Protein and fat contents of milk may influence test results in various ways: the sample flow can be altered (for example fat content strongly affectsviscosity) and any of the milk components can give specific or non-specific interactions with immunoreagents involved in the assay. In fact, we observed that casein determined a strong signal depression of both the Test and Control lines. With the aim of developing a unique system that could be used on milk samples undergone to different thermal treatments, i.e.: with different levels of protein denaturation (raw, pasteurized, UHT milk) and with variable fat content (whole, semi-skimmed, skimmed milk), samples were standardized by a rapid centrifugation stepto allow the removal of the fat layer and by adding Tween 20 to control protein interferences.

After development (15′ at 37°C), strips were scanned. Dedicated software acquires and processed images and the signal, intended as the T/C ratio, was plotted against the logarithm of AFM_1 concentration to carry out calibration. As previously observed in the development of

LFIA for other mycotoxin [56, 61-63], matrix-matched calibration should be carried out to fit experimental results obtained on milk samples. Therefore, a pasteurized whole milk sample in which AFM_1 was found out undetectable when analysed by the reference ELISA kit was used to prepare diluted calibrators. A typical calibration curve is depicted in Figure 5. A LOD (calculated as the average of the blank minus 3 standard deviations from the average) and IC_{50} of 20 ng l^{-1} and 102 ± 19 ng l^{-1} were estimated, respectively.

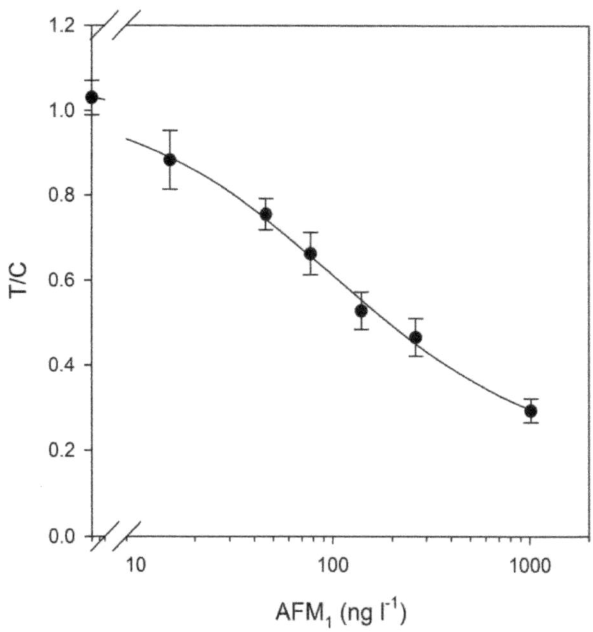

Figure 6. A typical calibration curve for AFM_1 measurement in milk by the developed LFIA. Error bars represent SD of 3 replicates.

Accuracy of the developed LFIA was evaluated on different kind of milk samples (Table 2). Milk samples were purchased on the market and were found undetectable according to the developed LFIA. Therefore, accuracy was evaluated on samples fortified at two levels (50 and 100 ng l^{-1}). Acceptable results were obtained, although a slight overestimation or underestimation were observed for the raw and the UHT samples, respectively, which can be attributed to the fact that calibration was carried out in pasteurized milk.

The intra- and inter-day precision was evaluated at 3 levels of fortification (0-50-100 ng l^{-1}). RSD values were generally high (above 30%) which makes reliability of quantification questionable.

Milk sample	AFM_1 measured by ELISA (ng l^{-1})	Fortification level (ng l^{-1})	Estimated AFM_1± SD (ng l^{-1})	Recovery (%)
raw	17.8	0	<LOD	
		50	78.4 ± 6.2	121
		100	153.2 ± 14.1	135
whole 1	< LOD	0	<LOD	
		50	40.0 ± 2.0	80
		100	121.5 ± 9.8	122
whole 2	16.0	0	<LOD	
		50	79.0 ± 8.6	126
		100	125.5 ± 11.0	126
skimmed	15.7	0	34.6 ± 1.2	
		50	74.4± 4.0	117
		100	113.0 ± 20.5	97
UHT	<LOD	0	<LOD	
		50	46.8 ± 5.3	94
		100	87.5 ± 10.8	88

Table 2. Recovery of AFM_1 determination from artificially contaminated milk samples undergone to various thermal treatments and with different fat content as determined by the developed LFIA. Recovery was calculated as follows: (estimated AFM_1 for the fortified sample – estimated AFM_1 for the non fortified sample) / fortification level *100

3.4. Intra-laboratory validation of the semi-quantitative LFIA

The objective of analytical methods such as those based on the LFIA technology is the parting between samples surely complying with legislation in force and samples which do not comply. However, a further category of samples should be considered and is represented by those samples in which the toxin content is close to the legal limit which because of measure uncertainty cannot be classified as compliant or noncompliant (Figure 4). These "uncertain samples" should be submitted to further controls before entering the transformation chain. In the case of milk, rejection is more often the fate of such uncertain samples (as for noncompliant samples), because the perishable nature of milk discourages time-consuming investigations. Therefore, the purpose of the work could become the development of a very rapid screening method which allowed the semi-quantitation of AFM_1 in milk in such a way to permit the discrimination between compliant and noncompliant samples. The instrumental quantification of coloured lines and their correlation with a calibration curve, in this context, could be regarded as a way to limit subjectivity in the interpretation of results and to improve detectability [52, 44] rather than going into the direction of factual quantitative measurements.

To achieve the useful ability to discriminate compliant from noncompliant samples, a proper cut-off value should be established. The eligible EU MRL value (i.e.: 50 ng l^{-1}) would be expected to be at tain able given the high sensitivity of the developed LFIA. Nevertheless, the definition of a cut-off level should consider precision and technical limitations of the method, besides sensitivity. Moreover, the calibration curve being a continuously descending curve characterized by a finite slope, the definition of a single-point cut-off value is less appropriate than the identification of an indicator range of analyte concentrations within which uncertain or "non-attributable" results (neither "compliant" nor "noncompliant") fall [44].

As regard precision, European legislation for screening methods of analysis defines as appropriate a relative uncertainty of 47% of the MRL and as acceptable even 94% for AFM_1 based on the application of Horwitz equation [71]. Accepting the more restrictive criterion, this means that any screening methods should be able to discriminate between AFM_1 content less than 26.5 ng l^{-1} (negative sample) and AFM_1 content over 73.5 ng l^{-1} (positive sample). Samples that have AFM_1 content close to the thres hold limit should thus be defined as uncertain because precision did not allow to reliably attributing them to one or another group.

In spite of this, it should be noted that a "non-attributable" judgement would determine rejection of the sample with a considerable economic damage, as discussed above. Therefore, the minimum number of non-attributable results would be expected for a worth while method and this number obviously depends on the combination of accuracy and precision of the method itself. To indicate the capability of a qualitative/semi-quantitative method to produce the lowest score of non-attributable results, for a defined uncertainty interval, we introduced a new parameter indicated as "efficiency" of the method, defined as the ability of the method itself to detect truly non-attributable as non-attributable. Efficiency was thus calculated as the number of truly non-attributable tests divided by the sum of known non-attributable samples, in strict analogy with "sensitivity" and "selectivity" of qualitative and semi-quantitative as says, which are defined as the rate of truly positive e and truly negative test results, respectively [50, 60]. The more efficient the assay, the highest the score of useful results (samples certainly attributed as compliant or noncompliant).

The ability of the developed LFIA to correctly attribute to each of the groups milk samples found on the market was thus assessed; in particular, negative (compliant) samples were defined as those in which AFM_1 content was below 30 ng l^{-1}, positive (noncompliant) samples those in which AFM_1 content was above 70 ng l^{-1} and uncertain (non-attributable) those having an AFM_1 content between 30 and 70 ng l^{-1}. Since all tested samples were always contaminated below 30 ng l^{-1} as established by the reference ELISA, positive samples were generated through fortification at 50 and 100 ng l^{-1}. Results of this evaluation, together with the definition of sensitivity, selectivity and efficiency, are reported in Table 3.

It can be observed from data that the definition of an indicator range instead of a cut-off level allowed us to avoid occurrence of false compliant and false noncompliant. Incorrect attribution occurred in 15% of samples (6/40), though 3 of them would represent a minor issue being assigned as non-attributable rather than noncompliant, which anyhow mean that samples would be discarded. The efficiency is relatively low, however it could still be considered acceptable.

Parameter	Definition	Calculated as	Value (%)
Sensitivity	truly positive / known positive	tp / (tp + fn + fup)	81.3
Selectivity	truly negative / known negative	tn / (tn + fp + fun)	100.0
Efficiency	truly uncertain / known uncertain	tu / (tu + fun + fup)	62.5
False compliant rate	false negative / known negative	fn / (tn + fn + fun)	0
False noncompliant rate	false positive / known positive	fp / (tp + fp + fup)	0
False non-attributable rate	false uncertain / known uncertain	fu / (tu + fun + fup)	37.5

Table 3. Evaluation of LFIAs performances on 40 milk samples: 16 negatives, 16 positives and 8 uncertain. The AFM_1 reference content was determined by an ELISA kit [19]. Abbreviations used: tp, truly positive (AFM_1 below 30 ng l^{-1}); tn, truly negative (AFM_1 above 70 ng l^{-1}); tu, truly uncertain (AFM_1 between 30 and 70 ng l^{-1}); fn, false negative; fp, false positive; fun, false uncertain and known to be negative; fup, false uncertain and known to be positive.

Finally, the stability of the overall device at room temperature was evaluated as the possibility of correctly measuring samples contaminated at low (<30 ng l^{-1}) and high levels (> 70 ng l^{-1}) and by using calibration curves carried out with freshly prepared strips; nevertheless, it could not be confirmed for periods longer than a month.

4. Conclusions

Despite LFIAs still being regarded in some ways as an emerging and incoming technology for food safety monitoring, there are several examples of fully developed devices described in the literature and also available as commercial kits for detecting a variety of natural and xenobiotic contaminants. Annual updates of state-of-the-art techniques underline the growing interest in the field and the increasing relevance of this technology over more established screening techniques. Not with standing the research is conditioned by the attainment of effectively functioning devices, often at the expense of true innovation, except in a few rare cases.

The literature concerning lateral flow immunoassays for aflatoxins is stilllimited, partly because the subject is very recent; indeed, the first published work on this topic dates back to just adecade ago. From this pioneering approach, several papers have been published which describes devices mainly aimed at measuring aflatoxin B_1. The use of LFDs for aflatoxin determination in nuts has also been demonstrated, even if the principal application is represented by their use to monitor aflatoxin contamination in cereals and derived products. This can be explained by the fact that research in this field is strongly driven by industry and by the prevalent economic impact of cereals in comparison to other commodities potentially affected by aflatoxin contamination.

The development of reliable devices for AFM_1 detection, conversely, suffers the extreme sensitivity required to analytical methods aimed at measuring such a contaminant. Very few papers have been published which describe LFIAs for $AFM_1$and none actually meet those requirements, despite the high interest in obtaining adequate systems for the rapid and on site monitoring of this toxin.

In this paper, we demonstrated that modifying the format of the classic lateral flow assay (such as tailoring the toxin conjugate, used as the competitor in the T-line, and the antibody labelling procedure)a greatdetect ability improvement could be obtained. The estimated LOD of the developed semi-quantitative LFIA was one order of magnitude lower than previously published LFIAs for AFM_1, therefore allowed us to effectively discriminate between compliant and noncompliant samples at a level required by the most severe legislation in force. Matrix-matched calibration was necessary to level results obtained on milk samples, however, various matrices (undergone to different thermal treatment and with differing fat contents) could be analysed after a very rapid and easy sample treatment, which involves 2′ centrifugation followed by the addition of a small volume of a concentrated solution of a surfactant.

Author details

Laura Anfossi*, Claudio Baggiani, Cristina Giovannoli and Gianfranco Giraudi

*Address all correspondence to: laura.anfossi@unito.it

Department of Chemistry, University of Turin, Giraudi

References

[1] International Agency for Research on Cancer (IARC). (2002). Evaluation of carcinogenic risks in humans. 82, *Lyon (France):IARC*, 171-274.

[2] European Commission. (2006). Commission Regulation (EC) No 1881/2006 of 19 December 2006 setting maximum levels for certain contaminants in foodstuffs. *Official Journal of the European Community*, L364, 5-24.

[3] European Commission. (2010). Commission Regulation (EU) No 165/2010 of 26 February 2010 amending Regulation (EC) No 1881/2006 setting maximum levels for certain contaminants in foodstuffs as regards aflatoxins. *Official Journal of the European Community*, L050, 8-12.

[4] European Commission. (2003). Commission Directive 2003/100/EC of 31 October 2003 amending Annex I to Directive 2002/32/EC of the European Parliament and of

the Council on undesirable substances in animal feed. *Official Journal of the European Community*, L285, 33-37.

[5] U.S. Food and Drug Administration. (2012). FDA Regulatory Guidance for Toxinsand Contaminants. http://www.ngfa.org/files/misc/Guidance_for_Toxins.pdf, accessed June.

[6] Canadian Food Inspection Agency. (2012). Fact sheet- Mycotoxins. http://www.inspection.gc.ca/english/anima/feebet/pol/mycoe.shtml, accessed 12 June.

[7] European Mycotoxins Awareness Network. (2012). Mycotoxins Legislation Worldwide. (last updated February)., http://services.leatherheadfood.com/eman/FactSheet.aspx?ID=79, accessed 25 June 2012.

[8] U.S. Food and Drug Administration. (2012). Guidance for Industry: Action Levels for Poisonous or Deleterious Substances in Human Food and Animal Feed. http://www.fda.gov/Food/GuidanceComplianceRegulatoryInformation/GuidanceDocuments/ChemicalContaminantsandPesticides/ucm077969.htm, accessed 13 June.

[9] Shephard, G. S. (2009). Aflatoxin analysis at the beginning of the twenty-first century. *Analytical and Bioanalytical Chemistry*, 395-1215.

[10] Köppen, R., Koch, M., Siegel, D., Merkel, S., Maul, R., et al. (2010). Determination of mycotoxins in foods: current state of analytical methods and limitations. *Applied Microbiology and Biotechnology*, 86-1595.

[11] Sforza, S., Dall'Asta, C., & Marchelli, R. (2006). Recent advances in mycotoxin determination in food and feed by hyphenated chromatographic techniques/mass spectrometry. *Mass Spectrometry Review*, 25-54.

[12] Muscarella, M., Lo, Magro. S., Palermo, C., & Centonze, D. (2007). Validation according to European Commission Decision 2002/657/EC of a confirmatory method for aflatoxin M_1 in milk based on immunoaffinity columns and high performance liquid chromatography with fluorescence detection. *Analitica Chimica Acta*, 594(2), 257-264.

[13] Chen, W.L, Hsu, T.F, & Chen, C.Y. (2011). Measurement of aflatoxin M_1 in milk by ultra-high-performance liquid chromatography/tandem mass spectrometry. *Journal of AOAC International*, 94(3), 872-877.

[14] Betina, V. (1985). Thin-layer chromatography of mycotoxins. *Journal of Chromatography*, 334(3), 211-76.

[15] Reiter, E., Zentek, J., & Razzazi, E. (2009). Review on sample preparation strategies and methods used for the analysis of aflatoxins in food and feed. *Molecular Nutrition and Food Research*, 53(4), 508-524.

[16] Li, P., Zhang, Q., & Zhang, W. (2009). Immunoassays for aflatoxins. *Trends in Analytical Chemistry;*, 28(9), 1115-1126.

[17] Trombley, A., Fan, T., & La Budde, R. (2011). Aflatoxin plate kit Performance Tested Method 081003. *Journal of AOAC International*, 94(5), 1519-1530.

[18] Lupo, A., Roebuck, C., Dutcher, M., Kennedy, J., & Abouzied, M. (2010). Validation study of a rapid ELISA for detection of aflatoxin in corn Performance Tested Method 050901. *Journal of AOAC International*, 93(2), 587-599.

[19] Magliulo, M., Mirasoli, M., Simoni, P., Lelli, R., Portanti, O., & Roda, A. (2005). Development and validation of an ultrasensitive chemiluminescent enzyme immunoassay for aflatoxin M_1 in milk. *Journal of Agricultural and Food Chemistry*, 53(9), 3300-3305.

[20] Anfossi, L, Calderara, M, Baggiani, C, Giovannoli, C, Arletti, E, & Giraudi, G. (2008). Development and application of solvent-free extraction for the detection of aflatoxin M1 in dairy products by enzyme immunoassay. *Journal of Agricultural and Food Chemistry*, 56(6), 1852-1857.

[21] Guan, D., Li, P., Cui, Y., Zhang, Q., & Zhang, W. (2011). A competitive immunoassay with a surrogate calibrator curve for aflatoxin M_1 in milk. *Analytica Chimica Acta;*, 703(1), 64-9.

[22] Zheng, M. Z., Richard, J. L., & Binder, J. (2006). A review of rapid methods for the analysis of mycotoxins. *Mycopathologia*, 161-261.

[23] Sibanda, L., De Saeger, S., & Van Peteghem, C. (1999). Development of a portable field immunoassay for the detection of aflatoxin M_1 in milk. *International Journal of Food Microbiology*, 48(3), 203-209.

[24] Goryacheva, I. Y., Karagusheva, M. A., Van Peteghem, C., Sibanda, L., & De Saeger, S. (2009). Immunoaffinity pre-concentration combined with on-column visual detection as a tool for rapid aflatoxin M_1 screening in milk. *Food Control*, 20(9), 802-806.

[25] Goryacheva, I. Y., De Saeger, S., Nesterenko, I. S., Eremin, S. A., & Van Peteghem, C. (2007). Rapid all-in-one three-step immunoassay for non-instrumental detection of ochratoxin A in high-coloured herbs and spices. *Talanta*, 72(3), 1230-1234.

[26] Ediage, E. N., Di Mavungu, J. D., Goryacheva, I. Y., Van Peteghem, C., & De Saeger, S. (2012). Multiplex flow-through immunoassay formats for screening of mycotoxins in a variety of food matrices. *Analytical and Bioanalytical Chemistry*, 403-265.

[27] Ricci, F., Volpe, G., Micheli, L., & Palleschi, G. (2007). A review on novel developments and applications of immunosensors in food analysis. *Analytica Chimica Acta*, 605-111.

[28] Tang, D., Zhong, Z., Niessner, R., & Knopp, D. (2009). Multifunctional magnetic bead-based electrochemical immunoassay for the detection of aflatoxin B_1 in food. *Analyst*, 134(8), 1554-60.

[29] Zaijun, L., Zhongyun, W., Xiulan, S., Yinjun, F., & Peipei, C. (2010). A sensitive and highly stable electrochemical impedance immunosensor based on the formation of silica gel-ionic liquid biocompatible film on the glassy carbon electrode for the determination of aflatoxin B_1 in bee pollen. *Talanta*, 80(5), 1632-1637.

[30] Tan, Y., Chu, X., Shen, G. L., & Yu, R. Q. (2009). A signal-amplified electrochemical immunosensor for aflatoxin B_1 determination in rice. *Analytical Biochemistry;*, 387(1), 82-86.

[31] Pohanka, M, Malir, F, Roubal, T, & Kuca, K. (2008). Detection of Aflatoxins in Capsicum Spice Using an Electrochemical Immunosensor. *Analytical Letters*, 41(13), 2344-2353.

[32] Jin, X., Jin, X., Chen, L., Jiang, J., Shen, G., & Yu, R. (2009). Piezoelectric immunosensor with gold nanoparticles enhanced competitive immunoreaction technique for quantification of aflatoxin B_1. *Biosensensors and Bioelectronics*, 24(8), 2580-2585.

[33] Micheli, L., Grecco, R., Badea, M., Moscone, D., & Palleschi, G. (2005). An electrochemical immunosensor for aflatoxin M_1 determination in milk using screen-printed electrodes. *Biosensensors and Bioelectronics*, 21(4), 588-596.

[34] Parker, C. O., Lanyon, Y. H., Manning, M., Arrigan, D. W., & Tothill, I. E. (2009). Electrochemical immunochip sensor for aflatoxin M_1 detection. *Analytical Chemistry*, 81(13), 5291-5298.

[35] Kanungo, L., Pal, S., & Bhand, S. (2011). Miniaturised hybrid immunoassay for high sensitivity analysis of aflatoxin M_1 in milk. *Biosensensors and Bioelectronics*, 26(5), 2601-2606.

[36] Piletska, E., Karim, K., Coker, R., & Piletsky, S. (2010). Development of the custom polymeric materials specific for aflatoxin B_1 and ochratoxin A for application with the ToxiQuant T1 sensor tool. *Journal of Chromatography A*, 1217(16), 2543-2547.

[37] Baggiani, C., Anfossi, L., & Giovannoli, C. (2008). Molecular imprinted polymers as synthetic receptors for the analysis of myco- and phyco-toxins. *Analyst*, 133(6), 719-730.

[38] Molina-García, L., Fernández-de Córdova, M. L., & Ruiz-Medina, A. (2012). Indirect determination of aflatoxin B_1 in beer via a multi-commuted optical sensor Food Additives & Contaminants. *Part A: Chemistry, Analysis, Control, Exposure & Risk Assessment*, 29(3), 392-402.

[39] Wang, Y., Dostálek, J., & Knoll, W. (2009). Long range surface plasmon-enhanced fluorescence spectroscopy for the detection of aflatoxin M1 in milk. *Biosensensors and Bioelectronics*, 24(7), 2264-7.

[40] Arduini, F., Errico, I., Amine, A., Micheli, L., Palleschi, G., & Moscone, D. (2007). Enzymatic Spectrophotometric Method for Aflatoxin B Detection Based on Acetylcholinesterase Inhibition. *Analytical Chemistry*, 79-3409.

[41] Posthuma-Trumpie, G. A., Korf, J., & Van Amerongen, A. (2009). Lateral flow (immuno)assay: its strengths, weakness, opportunities and threats. A literature survey. *Analytical and Bioanalytical Chemistry*, 393-569.

[42] Krska, R., & Molinelli, A. (2009). Rapid test strips for analysis of mycotoxins in food and feed. *Analytical and Bioanalytical Chemitry*, 393-67.

[43] Ngom, B., Guo, Y., Wang, X., & Bi, D. (2010). Development and application of lateral flow test strip technology for detection of infectious agents and chemical contaminants: a review. *Analytical and Bioanalytical Chemitry*, 397-1113.

[44] Anfossi, L., Baggiani, C., Giovannoli, C., D'Arco, G., & Giraudi, G. (2012). Lateral-flow immunoassays for mycotoxins and phycotoxins: a review. *Analytical and Bioanalytical Chemistry*, 1-14, DOI s00216-012-6033-4.

[45] U.S. Department of Agriculture. (2012). GIPSA Performance Verified Rapid Test Kits for Analysis of Mycotoxins. http://www.gipsa.usda.gov/fgis/tech-servsup/metheqp/testkits.pdf, accessed June.

[46] Anfossi, L., Calderara, M., Baggiani, C., Giovanoli, C., Arletti, E., & Giraudi, G. (2010). Development and application of a quantitative lateral flow immunoassay for fumonisins in maize. *Analytica Chimica Acta;*, 682-104.

[47] Laycock, M. V, Donovan, M. A, & Easy, D. J. (2010). Sensitivity of lateral flow tests to mixtures of saxitoxins and applications to shellfish and phytoplankton monitoring. *Toxicon*, 55-597.

[48] Komano, A., Maruko, H., Sekiguchi, H., & Seto, Y. (2011). Detection of saxitoxin in counterterrorism using a commercial lateral flow immunoassay kit. *Forensic Toxicology*, 2938-43.

[49] Ho, J. A, & Wauchope, R. D. (2002). A Strip Liposome Immunoassay for Aflatoxin B_1. *Analytical Chemistry;*, 74-1493.

[50] Delmulle, B. S., De Saeger, S. M., Sibanda, L., Barna-Vetro, I., & Van Peteghem, C. H. (2005). Development of an Immunoassay-Based Lateral Flow Dipstick for the Rapid Detection of Aflatoxin B_1 in Pig Feed. *Journal of Agricultural and Food Chemistry*, 53-3364.

[51] Xiulan, S., Xiaolian, Z., Jiana, T., Zhoub, J., & Chu, F. S. (2005). Preparation of gold-labeled antibody probe and its use in immunochromatography assay for detection of aflatoxin B_1. *International Journal of Food Microbiology*, 99-185.

[52] Xiulan, S., Xiaolian, Z., Jian, T., Xiaohong, G., Jun, Z., & Chu, F. S. (2006). Development of an immunochromatographic assay for detection of aflatoxin B_1 in foods. *Food Control*, 17-256.

[53] Shim, W. B, Yang, Z. Y, Kim, J. S, Kim, J. Y, Kang, S. J, Woo, G. J, Chung, Y. C, Eremin, S. A, & Chung, D. H. (2007). Development of immunochromatography strip-test using nanocolloidal gold-antibody probe for the rapid detection of aflatoxin B_1 in grain and feed samples. *Journal of Microbiology and Biotechnology*, 171629-1637.

[54] Shim, W. B, Kim, J. S, Kim, J. Y, Choi, J. G, Je, J. H, Kuzmina, N. S, Eremin, S. A, & Chung, D. H. (2008). Determination of aflatoxin B_1 in rice, barley, and feed by non-

instrumental immunochromatographic strip-test and high sensitive ELISA. *Food Science and Biotechnology;*, 17623-630.

[55] Shim, W. B., Kim, G., Ryu, H. J., Nam, M., & Chung, D. H. (2009). Development of One-step Simultaneous Immunochromatographic Assay for Rapid Analysis of Aflatoxin B_1 and Ochratoxin A. *Food Science and Biotechnology*, 18(3), 641-648.

[56] Anfossi, L., D'Arco, G., Calderara, M., Baggiani, C., Giovannoli, C., & Giraudi, G. (2011). Development of a quantitative lateral flow immunoassay for the detection of aflatoxins in maize. *Food Additives & Contaminants: Part A: Chemistry, Analysis, Control, Exposure & Risk Assessment*, 28(2), 226-234.

[57] Zhang, D., Li, P., Yang, Y., Zhang, Q., Zhang, W., Xiao, Z., & Ding, X. (2011). A high selective immunochromatographic assay for rapid detection of aflatoxin B_1. *Talanta*, 85-736.

[58] Zhang, D., Li, P., Zhang, Q., & Zhang, W. (2011). Ultrasensitive nanogold probe-based immunochromato graphic assay for simultaneous detection of total aflatoxins in peanuts. *Biosensors and Bioelectronics*, 26-2877.

[59] Liao, J. Y., & Li, H. (2010). Lateral flow immunodipstick for visual detection of aflatoxin B_1 in food using immuno-nanoparticles composed of a silver core and a gold shell. *Microchimica Acta*, 171(3-4), 289-295.

[60] Tang, D., Sauceda, J. C., Lin, Z., Ott, S., Basova, E., Goryacheva, I., Biselli, S., Lin, J., Niessner, R., & Knopp, D. (2009). Magnetic nanogold microspheres-based lateral-flow immunodipstick for rapid detection of aflatoxin B_2 in food. *Biosensors and Bioelectronics*, 25-514.

[61] Anfossi, L, D'Arco, G, Baggiani, C, Giovannoli, C, & Giraudi, G. (2011). A lateral flow immunoassay for measuring ochratoxin a: development of a single system for maize, wheat and durum wheat. *Food Control*, 22(12), 1965-1970.

[62] Molinelli, A., Grossalber, K., Führer, M., Baumgartner, S., Sulyok, M., & Krska, R. (2008). Development of Qualitative and Semiquantitative Immunoassay-Based Rapid Strip Tests for the Detection of T-2 Toxin in Wheat and Oat. *Journal of Agricultural and Food Chemistry*, 56-2589.

[63] Molinelli, A., Grossalber, K., & Krska, R. (2009). A rapid lateral flow test for the determination of total type B fumonisins in maize. *Analytical and Bioanalytical Chemistry*, 395-1309.

[64] Salter, R., Douglas, D., Tess, M., Markovsky, B., & Saul, S. J. (2006). Interlaboratory study of the Charm ROSA Safe Level Aflatoxin M_1 Quantitative lateral flow test for raw bovine milk. *Journal of AOAC International*, 89(5), 1327-34.

[65] Wang, J. J, Liu, B. H, Hsu, Y. T, & Yu, F. Y. (2011). Sensitive competitive direct enzyme-linked immunosorbent assay and gold nanoparticle immunochromatographic strip for detecting aflatoxin M_1 in milk. *Food Control*, 22-964.

[66] Zhang, D., Li, P., Zhang, Q., Yang, Y., Zhang, W., Guan, D., & Ding, X. (2012). Extract-free immunochromatographic assay for on-site tests of aflatoxin M_1 in milk Analytical Methods. *Accepted Manuscript*, DOI:10.1039/C2AY25205H.

[67] Liu, Y. H., Xie, R., Guo, Y. R., Zhu, G. N., & Tang, F. B. (2012). Comparison of homologous and heterologous formats in nanocolloidal gold-based immunoassays for parathion residue determination. *Journal of Environmental and Science and Health Part. B, Pesticides, Food contaminants and agricultural wastes*, 47(5), 475-83.

[68] Byzova, N. A, Zvereva, E. A, Zherdev, A. V, Eremin, S. A, Sveshnikov, P. G, & Dzantiev, B. B. (2011). Pretreatment-free immunochromatographic assay for the detection of streptomycin and its application to the control of milk and dairy products. *Analytica Chimica Acta;*, 701(2), 209-217.

[69] Horsiberger, M., & Rosset, J. (1977). Colloidal gold, a useful marker for transmission and scanning electron microscopy. *Journal of Histochemistry and Cytochemistry*, 25-295.

[70] Findlay, J. W., Smith, W. C., Lee, J. W., Nordblom, G. D., Das, I., De Silva, B. S., Khan, M. N., & Bowsher, R. R. (2000). Validation of immunoassays for bioanalysis: a pharmaceutical industry perspective. *Journal of Pharmaceutical and Biomedical Analysis*, 21-1249.

[71] European Commission. (2006). Commission Regulation (EC) No 401/2006 of 23 February 2006 laying down the methods of sampling and analysis for the official control of the levels of mycotoxins in foodstuffs. *Official Journal of the European Community*, L070, 12-34.

Section 5

Risk Assessment, Economics and Trade

Chapter 16

Aflatoxin and Peanut Production Risk and Net Incomes

Cynthia Bley N'Dede, Curtis M. Jolly,
Davo Simplice Vodouhe and Pauline E. Jolly

Additional information is available at the end of the chapter

1. Introduction

Aflatoxins (AFs) are naturally occurring mycotoxins and they are produced by species of *Aspergillus fungi*, namely *A. flavus* and *A. parasiicus*. AF contamination of peanut affects the quantity and quality produced and marketed. AF contaminated peanut is tainted and cannot be marketed and must be thrown away. Awuah et al. (2009) stated that about 5 to 15 percent of peanut in Ghana was discarded during sorting. This reduces the supply of peanut marketed at the farm level. Lower quality peanut is less attractive to buyers who offer a lower price for AF contaminated peanut. Hence it is expected that AF will lower farmer revenue and increase production and marketing risks.

The toxic effects of AF on human and animal health constitute one of the major factors for establishing regulations regarding acceptable levels of AFs in food. These regulations require pre-harvest and post-harvest control, such as appropriate drying, sorting and storage structures. Therefore, implementing food safety standards can be costly. According to Mitchell (2003), government regulations increase production costs which generally cause the supply curve of a firm to shift to the left: from S_0 to S_1 (Figure 1). Hence, if consumers are aware of the AF problem and its consequences, they will be willing to pay a higher price for a safer food supply.

Although, consumers throughout the world desire a safe food supply, not all consumers are willing to pay a higher price for safer foods. Furthermore, if they perceive the product as unsafe they may be willing to buy less of the product. This will lead to the following conclusion: Risk of AF negatively affects demand. Both (cost for sorting peanut and perception of lower quality) result in a decrease of the firm's revenue (Figure 1).

Since AF contamination of peanut is both a pre-harvest and a post-harvest problem, factors that affect production of the mycotoxin will also be discussed along these lines. AF production may occur during pre-harvest. In tropical countries, humidity, high temperature and rainfall are some of the factors encouraging fungal growth and AF production. Moisture content exceeding a safe range of 8 to 12% may contribute to fungal growth (Schatzki and Haddon, 2002; Diener and Davis, 1967). At harvest, moisture content in peanut is generally high and can lead to development of aflatoxigenic fungi. ICRISAT (2008) recommended drying of peanut immediately after harvest down to 8% moisture in order to avoid the production of AF when the crop is stored. In Benin, the most commonly used drying method for peanut is sun drying. Farmers spread the nuts on a wooden or concrete floor usually for one to three days. Paz et al. (1989) reported that delayed drying could lead to a rapid increase in AF from 14.0 ppb at harvest to 93.8 ppb, if maize is not dried for 5 days after harvest. This was confirmed by Hell et al. (2003) who found that post-harvest contamination with AF in Benin increased when harvesting took more than 5 days and drying was delayed.

Another important post-harvest factor affecting AF contamination is storage condition. Grain crops may be attacked by fungi in the field which can then develop rapidly during storage when conditions are suitable for producing mycotoxins (Turner et al., 2005). During the survey, it was reported that the most common storage system for unshelled peanuts used by farmers across all regions was polyester bags in storage houses or rooms. In addition, farmers were questioned about the time of storage of their products. Most of them reported that they can store peanut for up to six months depending on market conditions. However, this time can be expanded to 8 to 12 months for other market participants. The relationship between the length of storage and the level of AF was, therefore, examined to assess the risk associated with storage time. Sorting of peanuts includes the elimination of broken, shriveled, discolored nuts or nuts burrowed by insect. The removal of contaminated nuts from a pile of nuts can reduce contamination level to less than 4ppb. In larger households the young children usually are assigned the tasks for sorting the nuts (Awuah et al., 2009). The risk of contamination by AF is an important food safety hazard for field crops (Dolman, 2003). In order to protect consumers from health risks, regulatory limits have been imposed on field crops intended for use as food and feed, and have significant impacts on world export market. The World Health Organization (WHO) has set a maximum level for AF at 20 ppb in human foods and 100 ppb in animal feed (WHO, 1998). Likewise, the Food and Drug Administration (F.D.A) set a tolerance limit for peanut at 15 ppb for human (FDA, 1978). The European Union (E.U) has set stricter standards: any food products for human consumption with a concentration of AF greater than 4 ppb cannot be marketed. These standards are bound to affect the production and marketing of peanut as efforts to reduce contamination result in lower grain supply at a higher cost. This means that decisions lo reduce levels of AF in peanut may affect producers net income.

The main objective of this paper was to examine the effects of AF on peanut production in Benin. The first hypothesis states that AF contamination of peanut affects quality and quantity marketed. Based on survey answers and visual observations peanut were sorted. We examined the peanut that were discolored, broken, punctured and discarded them. The removal of bad peanut from the lot leaves us with lower quality of marketable peanut. The second hypothesis is that AF contamination of peanut influences selling price; we examined buyers' responses to test the influence of AF on selling price. Our third hypothesis is that AF contamination influences labor costs and reduces net returns from peanut. We examined sorting, labor requirement, labor costs from survey data, and enterprise budgets. We also examined how sorting affects financial and marketing risks with capital budgets and risk analysis.

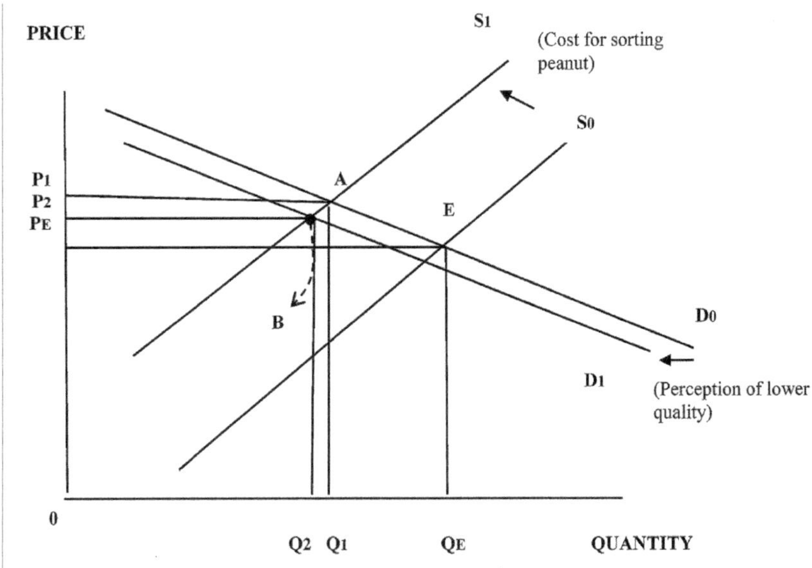

Figure 1. Market impact of food safety regulation

We can further represent the market equilibrium by these equations:

$$S_1 = \alpha + \beta_1 P - \theta_1 \tag{1}$$

$$D_1 = \alpha - \beta_1 P - \theta_2 \tag{2}$$

S_1 and D_1 are supply and demand respectively,

P represents the market price,

θ_1 and θ_2 represent both risk factors of AF.

2. Material and methods

2.1. Market participant identification, data and sample collection

We conducted surveys on AF contamination in peanut in three agro-ecological zones of Benin. Kandi (North), Savalou (South-east) and Abomey-Bohicon (South) were selected on the basis of their climatic conditions and levels of peanut production (Figure 2). A total of 30 farmers were selected in each of the three peanut producing regions of Benin during the period of May to July 2007. Peanut farmers were identified through the assistance of agricultural officers in the Ministry of Food and Agriculture (MoFA) and through the help of personnel from the University of Abomey-Calavi, Republic of Benin.

During the visits to each farm household, farming practices related to grain storage and handling were observed and documented using information on the survey instrument. Questionnaires were administrated to farmers by trained interviewers. Primary data collected included information on demographic and socioeconomic status, farming, post-harvest handling, storing and sorting practices, scheduled production activities, production level, household revenues and consumption frequency of peanut.

Peanut samples were collected at different post-harvest points and under different storage conditions. The levels of infestation of AF contaminated peanut under farmers' storage and marketing conditions were determined. It has been noted that AF levels vary along the marketing chain and the level of AF is more pronounced during storage and processing (Awuah et al., 2009).

Samples of 0.800 Kg (= 1.764 pounds) of peanut were taken in the fields and markets from each farmer. These samples were divided into two groups: sorted clean and rejected (bad) nuts. Bad nuts were the ones with discoloration and holes.

2.2. Determination of aflatoxin (AF) level

Assessment of AF levels was undertaken using the VICAM technique. VICAM is an AF test that produces numerical results using monoclonal antibody-based affinity chromatography. The test can isolate AF β1, β2, ğ1, and ğ2 from feeds, foods, grains, and nuts, and from dairy products.

This test involved observation of post-harvest and handling of peanut, collection of data on management issues related to grain storage and handling, collection of peanut samples, and testing them to detect AF levels. Each farmer's farm or business selected for the study was visited. The frequency of levels of AF found in peanut was used to estimate the probabilities of occurrence of AF.

2.3. Statistical analysis

Data collected during the survey were entered into an EXCEL spreadsheet and analyzed using SAS software package version 9.1. These data were used to develop enterprise budgets for producing peanut. The costs of each business activity were estimated based on the data collected in the survey. Furthermore, a simulation of the risk of AF contamination on farmers' income from the production, storage and trading of peanut was done using the @RISK software.

Figure 2. Map of Benin showing research sites

3. General assumptions

Regional effects: The survey was carried out in three different agro-ecological zones: Kandi, the northern region has one growing season starting at the end of May to September, with a temperature ranging from 28 to 45°C and a low rainfall averaging 800 to 900 mm. However, Savalou and Abomey-Bohicon both have two growing seasons (April to July and September to November) with higher rainfall between 1,300 and 1,500 mm, and temperature from 25 to 35°C (Setamou et al., 1997). Because of its dry climate, Kandi is the most productive region and is also the least prone to AF production.

In the southern regions, AF production is due to the high rainfall and high temperature. Under these tropical conditions, development of fungi and AF proliferation are facilitated. A higher concentration of AF in Abomey-Bohicon than in Savalou and Kandi.

Decision on drying, sorting and storing peanut: Based on previous studies, drying, sorting and storing methods are reported as the most important factors that encourage AF production. Farmers are recommended to dry peanut immediately after harvest, importantly to bring the moisture level of less than 8% (ICRISAT, 2008).

Sorting is considered as one of the ultimate solutions for the AF problem. This method has been reported as a post-harvest intervention strategy successful in reducing AF levels in peanut. An essential question was how much farmers or market participants will lose if they decide to sort peanut. In case the decision was "no sorting", not only quantity is affected but also labor cost. Hence, based on the answers obtained during the survey, the probability to throw away some nuts was estimated at one to five percent of quantity produced if farmers decided to sort them. However, if not, the risk of fungal growth and from nuts (molded or contaminated with AF) will increase.

Long-term storage in warm environment results in *Aspergillus* growth and increased in AF contamination. Previous research has yet to suggest a safe period in which peanut can be stored. We assumed, therefore, that after two months, with a risk of having bad nuts (mold, insect damage, and AF contamination), the percentage of rejection will be one percent and will increase by one percent more after each of two months. This percentage is applied on the quantity harvested as the percentage representing the loss in quantity if the storage length exceeds two months. This period (two months) was chosen based on survey reports.

Enterprise budgets: Budget analyses are used to evaluate the profitability of peanut enterprises in the short run. Costs and returns were estimated for each region. Most information used to develop each enterprise budget was obtained during the survey. Data such as seed quantity, seed price, quantity of peanut harvested, peanut selling price, material and equipment, labor hours and costs were obtained from the survey. They are the averages for the various size farms. All lands included in the budgets are treated as owned by farmers. Material and equipment are the same in each region and are depreciated according to the useful life, using the straight-line method. Costs for repairs and maintenance are assumed to be $1.00 for a one-hectare farm. Labor costs include land preparation, planting, harvest, drying, sorting, bagging, and transport costs. Labor costs and hour of use vary depending on the farm size.

Risk analysis: Parameters such as price of output, inputs and quantity are manipulated to examine how changes in parameters affect peanut production and revenues. A total of 5,000 iterations of the model are executed to generate all probability distributions that are used to establish stochastic dominance. All parameters used to develop the risks models are presented in Appendix 1.

Here we assume that net returns from peanut sales are affected by the costs of production and post-harvest handling. Hence, we use the formula:

$p_i{*}N.R = [p_i{*}(P_p{*}Q_p) - (p_i{*}Cost)]$

where,

p_i is the probability of the occurrence,

N.R is net return,

P_p is the price for peanut,

Q_p is the quantity for peanut, and

Cost is the cost of production; cost includes seed quantity and price, equipment, cost of pre-harvest, harvest, sorting, storage, bagging, winnowing…

$Cost = p_1{*}\beta_1$ drying cost + $p_2{*}\beta_2$ storing cost + $p_3{*}\beta_3$ sorting cost +…+ $p_n{*}\beta_n$ costs of n

Stepwise least squares regression is conducted between the collected input distribution values and the selected output values. The assumption is that there is a relationship between each input and output. The output of the stepwise regression is expressed in the form of a tornado chart.

Tornado chart is used to show the influence an input distribution has on the change in value of the output. Its main use is to enable the researcher to determine which variable contributes more to the output. It is also used for model diagnostic.

Therefore, the coefficient for any of the variables is standardized and will vary from -1.0 to +1.0. Variables contributing zero to the cost will be eliminated. Variation in cost, each year will be kept and their importance to cost will be explained.

4. Results

4.1. Demographics and socio-economic results

Socio-demographic information, knowledge of AF on peanut and farming practices were collected during the survey. Age of the respondents ranged from 35 to 55 years old, and over 55 years old (Table 1). Peanut production is done mostly by men in Kandi (63.3%), Savalou (100%), and in Abomey-Bohicon (54.4%). A large number of peanut producers in Benin have not received any formal education, and have never heard of AF contamination of peanut.

Most respondents have no formal education. A large number is found in Kandi with 43.3% (13) literates. Over 9 respondents who received a formal education in Savalou, only 3.3% (1) continued to secondary school in Abomey-Bohicon, 36.7% (11) had primary education and only 6.7% (2) attended secondary school (Table 1).

	Kandi		**Savalou**		**Abomey-Bohicon**	
	Number	%	Number	%	Number	%
Age groups						
Under 35	8	26.7	14	46.7	8	26.7
36-55	20	66.7	9	30.0	38	43.3
over 55	2	6.7	7	23.3	8	26.7
Gender						
Female	11	36.7	0	0	5	45.6
Male	19	63.3	30	100	25	54.4
Education						
No formal education	17	56.7	23	76.7	19	63.3
Primary school	13	43.3	7	23.3	11	36.7
Secondary school	5	16.7	1	3.3	2	6.7
Years of experience						
0-15	9	30.0	16	53.3	7	23.3
16-30	14	46.7	6	20.0	7	23.3
Over 30	7	23.3	8	26.7	16	53.3
Land tenure						
Owner	26	86.7	18	60.0	5	16.7
Renter	4	13.3	5	16.7	25	83.3
Income levels (month)						
$0-$350.14	10	33.3	13	43.3	18	60.0
$350.14 - $700.28	3	10.0	15	50.0	9	30.0
$700.28 - $1,400.56	10	33.3	2	6.7	2	6.7
Over $1,400.56	7	23.3	0	0.0	1	3.3

Table 1. Socio-demographics characteristics of peanut producers in Kandi, Savalou and Abomey-Bohicon.

Years of experience were divided into 3 groups: less than 15 years (group one), between 15 and 30 years (group two) and over 30 years (group three). In Kandi, most farmers belong to the second group (46.7%); In Savalou, the majority (53.3%) is in group 1. More than half of the respondents in Abomey (53.3%) have been farming for at least 30 years.

The majority of the respondents own their land in Kandi (86.7%) and Savalou (60%), while in Abomey a large percent (83.3%) rent land to produce peanut.

Income levels for most farmers in Kandi (33.3%) are less than $350.14, and between $700.28 and $1,400.56. Half of the producers in Savalou an income level between to generate an income level to$350.14 and $700.28, while in Abomey, approximately 60.0% earn less than $350.14.

Aflatoxin Knowledge and Identification: Very few respondents know about AF contamination of food. As Kaaya and Warren (2005) reported, a large number of producers, traders and even consumers are not aware of food contamination with AF. When respondents were asked about the criteria used to identify AF contaminated peanut, some of them reported that they could identify spoiled or contaminated crops by the color or the shape; common colors are black, brown, white dust and greenish. Respondents suspect also any nut that are broken or attacked by insects to be contaminated by AF.

When asked if they had ever been sick from ingestion of AF contaminated peanut, most of the respondents' answers were negative. There is no report of diseases related to AF; however, it was reported that important consumption of peanut could affect consumers' health (Table 2). About 27.78% (Kandi), 43.33% (Savalou) and 47.78% (Abomey) of respondents reported that they were affected by diseases such as malaria, diarrhea and coughing, due to a large and frequent consumption of peanut. This may show limited knowledge of the health effects of consumption of AF contaminated peanut.

Of 90 farmers interviewed in Benin, about 95.6% dry peanut immediately after harvesting, and only 10% sort peanut before selling. However, the remaining farmers explained that not only it is time consuming to sort peanut but also, it reduces peanut quantity by 5 percent on average. Nevertheless, when peanut samples were tested for AF, results indicated that 91.5% of the samples tested were below the European standard (4 ppb), and only 8.5% were above that limit.

During the survey, a number of respondents (78%) stated that they store their products for approximately 2 to 6 months or longer if market price is not favorable. In the northern region (Kandi), this period can exceed 6 months (up to 12 months) because there is only one growing season each year.

Aflatoxin level: Distribution of AF levels for farmers samples are shown in Table 3. Based on European standards, we observe that a large number of the samples tested (91.5%) have a concentration level of less than 4 parts per billion (ppb). About 93.2% of the samples have a level less than the tolerance limit (15 ppb) set by the Food and Drug Administration (FDA). Based on WHO standards, the majority of the samples (96.6%) were safe for consumption, while 3.4% exceeded 20 ppb. In addition, most of the samples (98.3%) were less than the permissible level in animal feed (100 ppb).

Region (N = 90 per region)	Yes		No	
	Number	%	Number	%
Producers report sickness related to aflatoxin in three regions of Benin, 2007				
Kandi	25	27.8	65	72.2
Savalou	39	43.3	51	56.7
Abomey	43	47.8	47	52.2
Characteristics				
Dry peanut after harvesting	86	95.6	4	4.4
Sort peanut	9	10.0	81	90.0
Consume bad* grains	0	0.0	90	100.0
Give bad* grains to your animal	10	10.0	80	90.0

* Bad: discolored or contaminated

- Source: survey data

Table 2. Producers report sickness related to aflatoxin in three regions of Benin, and their characteristics

Aflatoxin limit	European standards 4 ppb	USA standards 15 ppb	WHO standards 20 ppb	Animal standards 100 ppb
Less than	91.5	93.2	96.6	98.3
Greater than	8.5	6.8	3.4	1.7

- Source: analysis of marketed peanut

Table 3. Distribution of aflatoxin level for farmers based on standards (%).

4.2. Enterprise budget

Enterprise budgets for each region studied are shown in table 4. AF contamination reduces farmers' net returns. Peanut production is more profitable in Kandi than in the other regions. Table 4 shows that net returns above total expenses are $1,626.54, $1,294.26, and $802.62 in Kandi, Savalou and Abomey-Bohicon, repectively. Estimated costs and returns budgets for sorting are also reported in table 4. Results show that there is a decrease in revenue and returns when farmers decide to sort peanut to improve quality. In addition, there is a decrease in yield (5%) and an increase in labor cost due to sorting, which in turn reduce farmers revenue and net returns. Previous studies conducted on the relationship between AF contamination and environmental conditions showed that high levels of AF are found in regions with warm and humid climates (Dohlman, 2003; Farombi, 2006). The present study demonstrates that farmers in the most humid area (Abomey-Bohicon) generate lower net returns ($783.09).

-Sell unshelled and per bag of 100 kg	-Sell in local markets (price: $0.42/kg).
-Straight line method for depreciation	-Kandi (Plant in May and harvest in September)
-Savalou (Plant in April and Harvest in August)	-Abomey (Plant in March and Harvest in July)

	Kandi		Savalou		Abomey	
	Not sorted	Sorted	Not sorted	Sorted	Not sorted	Sorted
Yield (Kg)	4,500	4,455	3,600	3,564	2,400	2,376
Revenue ($)	1,890.75	1,871.85	1,512.60	1,497.48	1,008.39	998.31
Labor costs ($)	174.69	184.14	174.69	184.14	174.69	184.14
Total variable costs ($)	259.29	268.74	213.93	223.38	200.85	210.30
Total fixed costs ($)	4.92	4.92	4.41	4.41	4.92	4.92
Income above variable costs ($)	1,631.46	1,603.11	1,298.70	1,274.10	807.57	788.04
Net returns ($)	1,626.54	1,598.16	1,294.26	1,269.69	802.62	783.09
Break-even price ($/kg)	0.06	0.06	0.06	0.06	0.08	0.09
NPV (6%)	5,714	5,614	4,536	4,450	2,811	2,743
PI (6%)	1,060.00	1,042.42	1,134.02	1,112.44	522.46	509.75
IRR	95.13	93.47	10.08	98.79	46.88	45.74

- Material and Equipment are physical resources used in the farm operation. They are renewed every year.

Table 4. Estimated annual costs and returns budget for a large size farm (3ha) in each region, assuming that there is no change in price when farmers sort peanut and using the following peanut production practices

Table 5 summarizes the costs and returns generated by farmers after 6 months of storage. Following Hell et al. (2000) and Kaaya and Kyamuhangire (2006) reports, who indicated that duration of storage positively influences fungal growth and AF production in food crops, this paper hypothesized that peanut stored for more than 6 months have a negative effect on farmers net returns. Since consumers may perceive that peanut quality will deteriorate during storage, due to AF contamination, they might lower price. Results show that AF growth increases with the length of storage and lowers revenue from peanut production, due to lower peanut quality.

The assumption in this table is that there is a decrease in price by five percent, due to peanut quality. After 6 months of storage, significant differences are observed in product quality and on farmers' income. Hence, net returns per hectare above all expenses are reduced.

	Kandi	Savalou	Abomey
Revenue ($)	1,796.22	1,440.60	957.98
Income above variable costs ($)	1,536.90	1,226.27	757.14
Net returns ($)	1,531.90	1,222.27	752.34

- Assuming that after 6 months, peanut quality worsens resulting in a lower price.
- Peanut price decrease from $0.42 to $0.40 per kg.
- Results in this table are compared to the results in Table 4.

Table 5. Storage impact in each agro-ecological region (large farms 3 ha).

4.3. Risk analysis

Table 6 displays the results for the risk analysis. As farmers sort their stored product, we assume that an increase in peanut price of 15, 10 and 5 percent is offered over the storage period. Assumptions are shown in table 6.1, table 6.2 and table 6.3. These tables report the simulated effects of change in price and storage duration on farmers' costs and returns. We observe a significant relationship between net returns and price, and also a negative relationship between net returns and sorting when farmers sort their peanut. The longer peanut is stored, the smaller is the final quantity due to fungal and AF production; however, for each region, as price increases by 5%, 10%, and 15%, revenue and net returns also increase. Overall, to improve quality of stored peanut, farmers sort peanut which results in an increase in labor cost, a decrease in yield and higher net returns. This finding confirms that sorting causes economic losses to peanut producers who want to improve peanut quality. Drying has also a positive impact on farmers' revenue and net returns, which shows that farmers have to dry peanut efficiently before selling their products. Further, as storage period exceeds 6 months, the enterprise becomes less profitable. It is, therefore, more profitable and less risky, to increase selling price to cover cost of sorting; however, it is more risky for farmers to sort peanut 6 months after harvesting than to sort at harvest.

Storage time	Change in price	Price	Quantity	Revenue	Net returns
(months)	%	($/kg)	(kg)	($)	($)
0-2	15	0.48319	4,455.00	2,152.62	1888.38
2-4	10	0.46219	4,410.45	2,038.45	1774.21
4-6	5	0.44118	4,366.35	1926.33	1662.09
6-8	-5	0.39916	4,322.68	1725.43	1461.19
8-10	-10	0.37815	4,279.46	1618.28	1354.04
10-12	-15	0.35714	4,236.66	1513.09	1248.85

Table 6.1. Sensitivity analysis for peanut budget by changing price and the effect on revenue, and net returns (Kandi)

Storage time	Change in price	Price	Quantity	Revenue	Net returns
(months)	%	($/kg)	(kg)	($)	($)
0-2	15	0.48	3564.00	1722.10	1503.77
2-4	10	0.46	3528.36	1630.76	1412.43
4-6	5	0.44	3493.08	1541.07	1322.74
6-8	-5	0.40	3458.15	1380.35	1162.02
8-10	-10	0.38	3423.56	1294.63	1076.30
10-12	-15	0.36	3389.33	1210.47	992.14

Table 6.2. Sensitivity analysis for peanut budget by changing price and the effect on revenue and net returns (Savalou)

Storage time	Change in price	Price	Quantity	Revenue	Net returns
(months)	%	($/kg)	(kg)	($)	($)
0-2	15	0.48	2376.00	1148.06	942.42
2-4	10	0.46	2352.24	1087.17	881.53
4-6	5	0.44	2328.72	1027.38	821.73
6-8	-5	0.40	2305.43	920.23	714.59
8-10	-10	0.38	2282.38	863.08	657.44
10-12	-15	0.36	2259.55	806.98	601.34

Table 6.3. Sensitivity analysis for peanut budget by changing price and the effect on revenue and net returns (Abomey)

Table 6. Sensitivity analysis for large farms (3 ha) gross margins, assuming that price varies through sorting and storage.

Results are also confirmed further. Figure 3 presents the tornado graphs for net returns for farmers who sort peanut before marketing. Price is the most important variable in the regression analysis. Drying has also a positive impact on farmers' revenue and net returns, which shows that farmers have to dry peanut efficiently before selling their products. However, there is a negative relationship between sorting and net returns. It is evident that when farmers sort peanut, it negatively affects net returns. Similarly, coefficients for storage (-0.03) and other labor variables like harvesting (-0.001) have a negative influence on net returns for each region.

In addition, based on the assumptions used to develop the sensitivity analysis of the NPVs in Table 6, risk is incorporated in NPV at different price levels and at different storage times. Figure 4 shows that NPV for farmers who sort peanut and sell at the normal price is smaller than those who sell sorted peanut at a higher price (5%). It is, therefore, more profitable and less risky, to increase selling price to cover cost of sorting. Tornado graphs above show that there is a significant relationship between price and NPV. As price goes up due to sorting, the NPV also increases; for instance, with a probability of 80%, NPV is 15.24% smaller when farmers sort immediately at harvest (Figure 5). Sorting peanut stored for 6 months is more risky than when farmers sort at harvest;

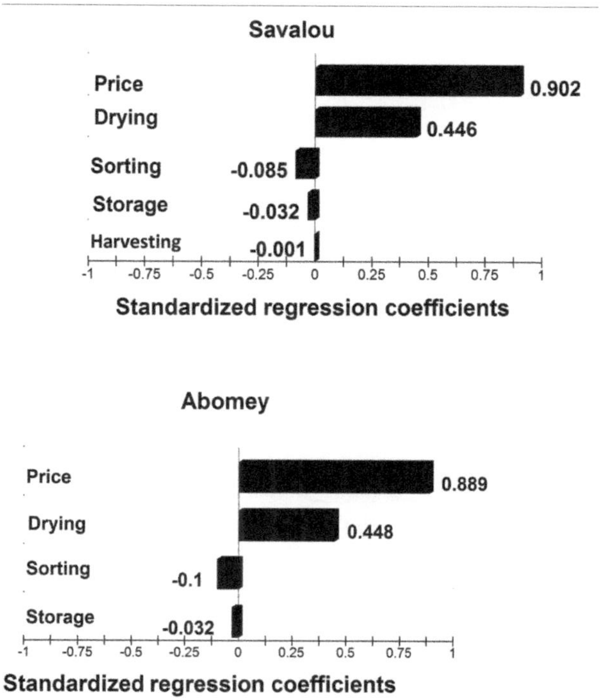

Figure 3. Tornado graphs of the net returns of peanut production in each region, assuming that peanut is sorted before marketing.

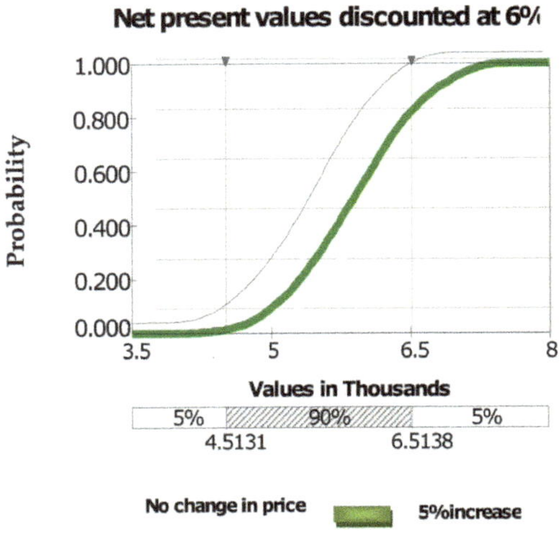

Figure 4. Cumulative probability distribution of the net present value for sorted and non-sorted peanut at varying prices according storage time (no change, 5% increase).

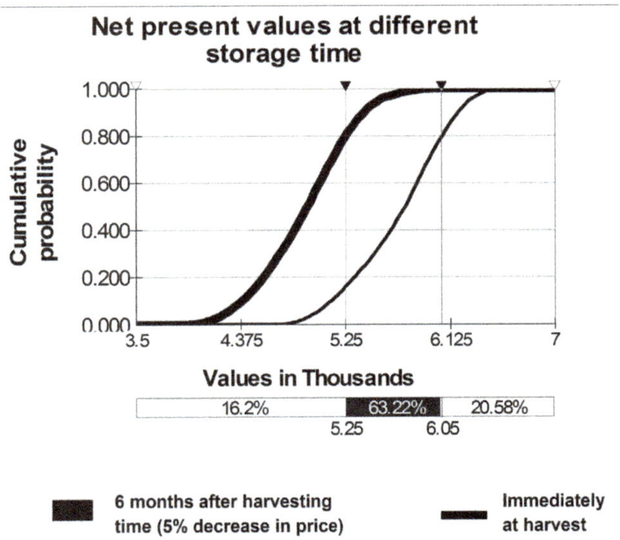

Figure 5. Cumulative probability distribution of the net present value for stored peanut at harvest and six months later.

5. Conclusion

This study compares the costs and returns of peanut production in three agro-ecological zones of Benin. Findings demonstrate that AF is affected by pre-harvest and post-harvest factors. During the survey, most farmers stated that drying of peanut was done immediately after harvest. However, sorting was practiced only by few respondents. In many studies, sorting has been suggested as an efficient method to control AF development in peanut. In addition, another factor that needs to be highlighted is storage condition. Growth of storage fungi followed by AF production is also determined by storage structure and storage length. Plastic bags or other synthetic bags used mostly by farmers during storage promote increases in humidity, and hence, increase in AF levels. Since AF contamination in storage is dependent on the storage system, the solution would be to sort peanut during storage.

Results from enterprise budgets show that AF reduces farmers' net returns. Sorting of peanut results in higher labor costs and smaller net returns than the costs and returns generated when farmers do not sort peanut. Net returns per hectare after sorting peanut were reduced to $532.7 in Kandi, $423.2 in Savalou, and $261.03 in Abomey-Bohicon. Net returns were higher for Kandi which is the most productive region.

Results also demonstrate that AF increases with length of storage and lowers revenue from peanut production. After 6 months of storage, farmers' net revenues decrease due to lower peanut quality. It is evident that storage conditions have a significant impact on AF development. Moreover, in the risk analysis results, we note a significant relationship between net returns and price, and also a negative relationship between net returns and sorting when farmers sort peanut. This finding confirms that sorting causes economic losses to peanut producers who want to improve quality. Hence to compensate for their losses due to costs of sorting, producers have to increase price to cover at least their variable costs.

Although investigations in this study indicate that it is more profitable for farmers to sell peanut immediately after harvest than to store it, the solution would be to improve farming practices and management, storage conditions, increase price in order to improve peanut quality and minimize risk of losses from AF. Improvements of quality and higher prices are obtainable with government legislations, and consumer and producer education.

Appendix 1. Definition of parameters (inputs) used for risk models

Parameters	Unit	Risk function
Price (selling)	$	RiskTriang (0.32,0.42, 0.51)
Drying: No (0), Yes (1)	-	RiskDiscrete ({0,1},{0.044,0.956})
Sorting : No (0), Yes (1)	-	RiskDiscrete ({0,1},{0.04,0.96})

Parameters	Unit	Risk function
Storage	Month	RiskTriang (0, 2, 4)
Pre-harvest cost	$	RiskTriang (50, 52.5, 55)
Harvest cost	$	RiskTriang (1, 2, 3.5)
Drying costs	$	RiskTriang (0, 1.58, 3)
Sorting cost	$	RiskTriang (0, 3.15, 6.5)
Bagging cost	$	RiskTriang (0.5, 1.4, 2.5)
Transportation cost	$	RiskTriang (0.1, 0.5, 1.5)

Acknowledgement

This study was funded by the Peanut Collaborative Research Program, USAID Grant No. LAG-G-00-96-90013-00, and supported by the University of Georgia, Auburn University and the University of the Republic of Benin. The authors would like to thank the Ministry of Agriculture, Breeding and Fisheries of Benin for its support.

Author details

Cynthia Bley N'Dede[1], Curtis M. Jolly[1*], Davo Simplice Vodouhe[2] and Pauline E. Jolly[3]

1 Department of Agricultural Economics and Rural Sociology, Alabama Experiment Station, Auburn University, USA

2 The University of the Republic of Benin, Republic of Benin

3 University of Alabama, Birmingham, USA

References

[1] Awuah, R. T., Fialor, S. C., Binns, A. D., Kagochi, J., & Jolly, C. M. (2009). Factors Influencing Market Participants Decision to Sort Groundnuts along the Marketing Chain in Ghana. *Peanut Science*, 36, 68-76.

[2] Diener, U. L. N. D., & Davis, . (1967). Limiting Temperature and Relative Humidity for Growth and Production of Aflatoxin and Free Fatty Acids by *Aspergillus flavus* in Sterile Peanuts". *Journal of the American Oil Chemist's Society* 44(4):, 259 EOF-63 EOF.

[3] Dohlman, E. (2003). Mycotoxin Hazards and Regulations: Impacts on Food and Animal Feed Crop Trade," International Trade and Food Safety: Economic Theory and Case Studies, Jean Buzby (editor), Agricultural Economic Report 828. USDA, ERS.

[4] FAO (Food and AgricultureOrganization).(1997). Worldwide Regulations for Mycotoxins 1995: A compendium. FAO Food and Nutrition Paper". Rome, Italy.(64)

[5] Food and Drug Administration.(1978). Assessment of Estimated Risk Resulting from Aflatoxins in Consumer Peanut Products and Other Food Commodities". Bureau of Foods, Washington, D.C.

[6] Farombi, E. O. (2006). Aflatoxin Contamination of Foods in Developing Countries: Implications for Hepatocellular Carcinoma and Chemopreventive Strategies.". *African Journal of Biotechnology*, 5(1), 001-014.

[7] Hell, K., Cardwell, K. F., Setamou, M., & Poehling, H. M. (2000). The Influence of Storage Practices on Aflatoxin Contamination in Maize in four Agroecological Zones of Benin, West Africa. *Journal of Stored Products Research*, 36, 365-382.

[8] Hell, K., Cardwell, K. F., & Poehling, H. M. (2003). Relationship between Management Practices, Fungal Infection and Aflatoxin for stored in Benin." . *Journal of Phytopatology*, 151, 690-698.

[9] ICRISAT(2008). VIth International Conference on Legume Genentics and Genomics, Hyderbad India.

[10] Kaaya, A. N., & Warren, H. L. (2005). A Review of Past and Present Research on Aflatoxin in Uganda. *African Journal of Food Agriculture Nutrition and Development (AJFAND)* , 5(1), 1-18.

[11] Kaaya, A. N., & Kyamuhangire, W. (2006). The Effect of Storage Time and Agroecological Zone on Mould Incidence and Aflatoxin Contamination of Maize from Traders in Uganda. *International Journal of Food Microbiology*, 217 EOF-223 EOF.

[12] Mitchell, Lorraine. (2003). Economic Theory and Conceptual Relationships between Food Safety and International Trade. " Chapter 2 in Buzby, J.C., editor International Trade and Food Safety: Economic Theory and Case Studies. USDA/ERS Agricultural Economic Report # 848.

[13] Paz, R. R., Tiongson, R. L., Dayanghirang, D. D., & Rodriguez, A. C. (1989). 12th ASEAN Seminar on Grain Postharvest Technology Phase I". In Proceedings o f the 12th ASEAN Seminar on Grain Postharvest Technology" edited by ASEAN Grain Postharvest Programme, Bangkok, Thailand, 1991, August, Surabaya, Indonesia., 4-8.

[14] Setamou, M., Cardwell, K. F., Schulthess, F., & Hell, K. (1997). *Aspergillus flavus* Infection and Aflatoxin Contamination of Pre-harvest Maize in Benin", Plant Health Man-

agement Division, International Institute of Tropical Agriculture (IITA), 08 BP 932 Tri Postal, Cotonou, Republic of Benin. , 81(11), 1323-1327.

[15] Schatzki, T. F., & Haddon, W. F. (2002). Rapid, Non-destructive Selection of Peanuts for High Aflatoxin Content by Soaking and Tandem Mass Spectrometry". *Journal of Agricultural and Food Chemistry*, 50(10), 3062-3069.

[16] Turner, P., A., Sylla, Y., Gong, M., Diallo, A.Sutcliffe, A.., Hall, C, & Wild, . (2005). "Reduction in Exposure to Carcinogenic Aflatoxins by the Postharvest Intervention Measures in West Africa: A Community-based Intervention Study". *Lancet.* 365:1950-1956.

[17] World Health Organization (WHO) (1998). . Food additives series 40: Safety evaluations of certain food additives *Aflatoxins*. WHO: Geneva, 359-468.

Chapter 17

Aflatoxins: Risk, Exposure and Remediation

Antonello Santini and Alberto Ritieni

Additional information is available at the end of the chapter

1. Introduction

Aflatoxins family includes a great number of lipophilic molecules produced by aerobic microscopic fungi belonging to the genus *Aspergillus*. The chapter describes their chemical structure, chemical and physical properties, and aspects related to their presence in food and commodities. Aflatoxins presence in food is considered a real and severe risk to consumers for their toxicity. Aflatoxins levels and frequency of foods natural contamination as reported in the scientific literature are briefly analyzed. Focus is given to the different foodstuffs that may be at risk of contamination by *Aspergillus* and the subsequent accumulation of aflatoxins in the food chain. Bioavailability and bioaccessibility of aflatoxins will be discussed considering that these unwanted molecules can be assumed by the humans with the diet. Bioaccessibility, that deals with the fraction of micro-nutrients released from the food matrix during digestion and gastro-intestinal available for absorption, will be discussed with reference to aflatoxins bioaccessibility of during the digestion process, considering the relationships between the food matrix and its influences on aflatoxins fate. Bioavailability of the aflatoxins assumed from the diet depends on their stability during digestion, since they are released from the food matrix (bioaccessibility) and on the efficiency of their passage through the gastro-intestinal mucosa. The term bioavailability includes the concepts of availability to the absorption, metabolism, distribution of nutrients to tissues and bioactivity and indicates the fraction of micro-nutrients absorbed by the body and the speed with which these molecules are absorbed and made available at their site of action. Despite of the practical difficulties in measuring the distribution and bioactivity of aflatoxins on a specific human body organ, the bioavailability is the fraction of an oral dose of a compound or precursor of an active metabolite that reaches the bloodstream. Bioaccessibility includes the entire sequence of events that take place during the digestion of food material that can be assimilated by the body through the epithelial cells of the gastro-intestinal mucosa. Aflatoxins are often present in very small amounts or in traces and, for this reason, a part of the chapter addresses the advanced new chromatographic and spectrometric methods described in the literature and applied to research, that can reveal, even in trace amounts, aflatoxins in biological fluids as free form or as by-products, e.g. non-covalent adducts.

2. Structure and chemistry of aflatoxins

Aflatoxins were isolated and characterized after the Turkey X desease, that caused the death of more than 100.000 turkey poultries due to the intake of a contaminated peanut meal produced in South America starting from contaminated raw material (Blout, 1961; Goldblatt, 1969).

The most important aflatoxins, among the about 13 compounds so far identified, are the aflatoxin B_1 and B_2, the aflatoxin G_1 and G_2 and the aflatoxin metabolic byproducts M_1 and M_2. The four major aflatoxins are called B_1, B_2, G_1, and G_2 based on their fluorescence under UV light (blue or green) and relative chromatographic mobility during thin-layer chromatography. Figure 1 shows the chemical structures of the main aflatoxins. Their chemical structure incorporates dihydrofuran and tetrahydrofuran moieties coupled to a substituted coumarin. They are produced by a polyketide pathway by many strains of *Aspergillus flavus* and *Aspergillus parasiticus*; in particular, *Aspergillus flavus* is a common contaminant in agriculture. *Aspergillus bombycis, Aspergillus ochraceoroseus, Aspergillus nomius*, and *Aspergillus pseudotamari* are also aflatoxin-producing species, but they are encountered less frequently (Goto, Wicklow, Ito, 1996; Klich, Mullaney, Daly, Cary, 2000; Peterson, Ito, Horn, Goto, 2001). Table 1 gives some relevant chemical properties of these compounds. Aflatoxin B_1 is considered the most toxic and is produced, together with aflatoxin B_2 by both *Aspergillus flavus* and *Aspergillus parasiticus*. Aflatoxin G_1 and G_2 are produced exclusively by *Aspergillus parasiticus*. While the presence of *Aspergillus* spp. in food products does not always indicate harmful levels of aflatoxins are also present, it does imply a significant risk in consumption. Aflatoxins M_1 and M_2 were originally discovered in the milk of cows which fed on moldy grain. Aflatoxin M_1 has been observed also in the fermentation broth of *Aspergillus parasiticus*. These compounds are products of a conversion process in the animal's liver that try to make these molecules more hydrophilic to be easily excreted from body via the kidney. Aflatoxin M_1 is a metabolite of aflatoxin B_1 in humans and animals where exposure at ng levels can come from mother's milk. Similarly, aflatoxin M_2 is a metabolite of aflatoxin B_2 in milk of cattle fed on contaminated food (Tara, 2005). Other metabolites can derive from these main ones, like Aflatoxicol, that forms by biological reduction of aflatoxin B_1 (Pawlowski, Schoenhard, Lee, Libbey, Loveland, Sinnhuber, 1977). The levels considered safe for these compounds are reported in Table 2. Aflatoxin B_1 is the most potent natural carcinogen known, and is probably also the most studied aflatoxin being often the major aflatoxin produced by toxigenic strains (Squire, R. A. 1981). For this reason, it is also the best studied: in a large percentage of the papers published the term aflatoxin can be assumed to refer to aflatoxin B_1. However, many other aflatoxins (e.g., P_1. Q_1, B_{2a}, and G_{2a}) have been described, especially as mammalian biotransformation products of the major metabolites (Heathcote, Hibbert, 1978).

Aflatoxin	MW (g/ mol)	Formula	Melting point (°C)	IUPAC name
B_1	312.28	$C_{17}H_{12}O_6$	268–269	2,3,6a,9a-tetrahydro-4-methoxycyclopenta(c)furo(3',2': 4,5)furo(2,3-h)(1)benzo-pyran-1,11-dione
B_2	314.29	$C_{17}H_{14}O_6$	286–289	2,3,6aa,8,9,9aa-Hexahydro-4-methoxycyclopenta(c)furo(2',3': 4,5)furo(2,3-h)chromene-1,11-dione
G_1	328.28	$C_{17}H_{12}O_7$	244-246	7AR,cis)3,4,7a,10a-tetrahydro-5-methoxy-1H,12H-furo(3',2': 4,5)furo(2,3-h)pyrano(3,4-c)chromene-1,12-dione
G_2	330.29	$C_{17}H_{14}O_7$	237–240	1H,12H-furo(3',2':4,5)furo(2,3-h)pyrano(3,4-c) (1)benzopyran-1,12-dione
M_1	328.28	$C_{17}H_{12}O_7$	299	(6AR-cis)-2,3,6a,9a-tetrahydro-9a-hydroxy-4-methoxycyclopenta(c)furo(3',2':4,5)furo(2,3-h) (1)benzopyran-1,11-dione
M_2	330.29	$C_{17}H_{14}O_7$	293	2,3,6a,8,9,9a-Hexahydro-9a-hydroxy-4-methoxycyclopenta(c)furo(3',2':4,5)furo(2,3-h)(1) benzopyran-1,11-dione

Table 1. Chemical relevant data for main aflatoxins (O'Neil, Smith, Heckelman, 2001).

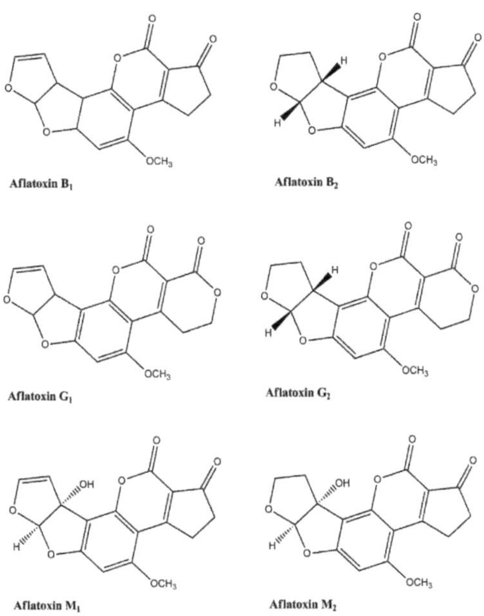

Figure 1. Chemical structures of the main aflatoxins.

μg/kg	Food
20	Food addressed to humans consumption. Corn and grains for animal feeds.
100	Corn and grains for breeding beef cattle, breeding swine, or poultry.
200	Corn and grains intended for swine.
300	Corn and grains for finishing beef cattle, swine, poultry.

Table 2. Aflatoxins levels limits generally considered as safe.

3. Biosynthesis of aflatoxins

Many relevant aspects of aflatoxins biosynthesis and molecular biology have been studied and extensively described. The first step in the biosynthetic pathway is considered the production of norsolorinic acid, an anthraquinone precursor, by a type II polyketide synthase. A series of about 15 post-polyketide synthase steps follows, yielding increasingly toxigenic metabolites (Bennett, Chang, Bhatnagar, 1997; Cleveland, Bhatnagar, 1992; Hicks, Shimizu, Keller, 2002; Payne, Brown, 1998; Townsend, 1997; Trail, Mahanti, Linz, 1995). Sterigmatocystin, a related dihydrofuran toxin, mutagenic and tumorigenic but less potent than aflatoxin (Berry, 1988), is a late metabolite in the aflatoxin pathway, and is also produced as a final biosynthetic product by a number of species like *Aspergillus versicolor* and *Aspergillus nidulans*. Analysis of the molecular genetics of sterigmatocystin biosynthesis in the genetically tractable species *Aspergillus nidulans* has provided a useful model system. The genes for the sterigmatocystin gene cluster from *Aspergillus nidulans* have been cloned and sequenced (Brown, Yu, Kelkar, Fernandes, Nesbitt, Keller, Adams, Leonard, 1996). Cognate genes for aflatoxins pathway enzymes from *Aspergillus flavus* and *Aspergillus parasiticus* show high sequence similarity to the sterigmatocystin pathway genes (Payne, Brown, 1998; Yu, Chang, Bhatnagar, Cleveland, 2000; Yu, Woloshuk, Bhatnagar, Cleveland, 2000). Genes organization for *Aspergillus flavus, Aspergillus nidulans,* and *Aspergillus parasiticus* sterigmatocystin-aflatoxin pathway has been studied as reported by Cary et al. (Cary, Chang, Bhatnagar, 2001) and Hicks et al. (Hicks, Shimizu, Keller, 2002).

Aspergillus oryzae and *Aspergillus sojae*, species that are widely used in Asian food fermentations such as soy sauce, miso, and sake, are closely related to the aflatoxigenic species *Aspergillus flavus* and *Aspergillus parasiticus*. Although these food fungi have never been shown to produce aflatoxin (Wei, Jong. 1986), they contain homologues of several aflatoxin biosynthesis pathway genes (Klich, Yu, Chang, Mullaney, Bhatnagar, Cleveland, 1995). Deletions and other genetic defects have led to silencing of the aflatoxin pathway in both *Aspergillus oryzae* and *Aspergillus sojae* (Takahashi, Chang, Matsushima, Abe, Bhatnagar, Cleveland, Koyama, 2002; Watson, Fuller, Jeens, Archer, 1999; Bennett, Klich, 2003).

4. Frequency and levels of contamination in food

Aflatoxins have received greater attention than any other mycotoxins because of their demonstrated potent carcinogenic effect in susceptible laboratory animals and their acute toxicological effects in humans. Many countries have attempted to limit exposure to aflatoxins by imposing regulatory limits on commodities to be used as food and feed. The two species of *Aspergillus* fungi, aflatoxin producing, are especially found in areas with hot and humid climate. Since aflatoxins are known to be genotoxic and carcinogenic, exposure through food should be kept as low as possible. Aflatoxins have been also associated with various diseases, such as aflatoxicosis. Aflatoxin B_1 is the most common in food, and has the most potent genotoxic and carcinogenic effects. Aflatoxin M_1 is a major metabolite of aflatoxin B_1 in humans and animals, which may be present in milk from animals fed with aflatoxin B_1 contaminated feed. Aflatoxins can occur in foods, such as groundnuts, treenuts, maize, rice, figs, grapes, raisins, and other dried foods, spices and crude vegetable oils, and cocoa beans, as a result of fungal contamination before and after harvest. The biosynthesis and the occurrence of aflatoxins is influenced by environmental factors; consequently the extent of contamination varies with geographic location, agricultural and agronomic practices. The susceptibility of commodities to fungal invasion during preharvest, storage, and/or processing periods is also important to assess the possible contamination.

5. Aflatoxins in food and commodities

From the mycological perspective, there are great qualitative and quantitative differences in the toxigenic abilities displayed by different strains within each aflatoxigenic species. For example, only about half of *Aspergillus flavus* strains produce aflatoxins (Klich, Pitt, 1988), while those that do may produce more than 10^6 μg/kg (Cotty, Bayman, Egel, Elias, 1994). Many substrates support growth and aflatoxin production by aflatoxigenic molds. Natural contamination of cereals, figs, oilseeds, nuts, tobacco, and a long list of other commodities is a common occurrence (Detroy, Lillehoj, Ciegler, 1971; Diener, Cole, Sanders, Payne, Lee, Klich, 1987).

Crops can be contaminated with aflatoxins in the field before harvest (Diener, Cole, Sanders, Payne, Lee, Klich, 1987; Klich, 1987). Even more problematic is the fate of crops stored under conditions that favor mold growth. The most relevant variables to keep under control during the storage are considered the moisture content of the substrate and the relative humidity of the surroundings (Detroy, Lillehoj, Ciegler, 1971; Wilson, Payne, 1994). There are many side implications of aflatoxins contamination. Aflatoxin contamination has been linked to increased mortality in farm animals and thus significantly lowers the value of grains as an animal feed and as an export commodity (Smith, Moss, 1985). Milk products can also be an indirect source of information on aflatoxins presence in the diet, and considering the broad diffusion of these products mainly addressed to infants, children, and people affected by health conditions, the risk associated to aflatoxins M_1 and M_2 is relevant. When cows assume aflatoxin-contaminated feed, they metabolically biotransform aflatoxin B_1 into a hydroxylated form, namely aflatoxin M_1, as a detoxification way for animal exposed to aflatoxins B_1 or B_2 (Van Egmond, 1989).

6. Occurence

Aflatoxins often occur in crops in the field before harvest so frequently that they are considered mycotoxins originating from the field compared to other mycotoxins that are commonly found in post-harvesting of field crops. Postharvest contamination can occur if crop drying is delayed and during crop storage if water is present in the amount required for the mold growth. Insect or rodent presence can facilitate mold onset on stored commodities. Aflatoxins have been also detected in milk, cheese, corn, peanuts, cottonseed, nuts, almonds, figs, grape berries, spices, and a variety of other foods and feeds. Milk, eggs and meat products are contaminated sometimes due to the consumption by the animal of aflatoxin contaminated feed, and are a clear example of carry-over. A few years after the discovery of mycotoxins, scientific understanding of the carry-over phenomenon raised immediately the interest of scientists and put focus on the risk related to food contaminated by molds. The commodities with the highest risk of aflatoxin contamination are corn, peanuts, and cottonseed. Corn is probably the commodity of greatest worldwide concern, because it is grown in climates that are likely to have perennial contamination with aflatoxins. Corn is the staple food of many countries, and, also for some population corn represents the main ingredient of the diet. It is usually named as single-food with all nutritional and unwanted contaminants related to its consumption. Corn can be used to produce flour and starch products and this links back to the problem statement such as aflatoxins is a likely toxin to be found in foodstuff. However, procedures used in the processing of corn help to reduce contamination of the resulting food product. This is because although aflatoxins are stable to moderately stable in most food processes, they are unstable in processes such as those used in making tortillas that employ alkaline conditions or oxidizing steps. Aflatoxin-contaminated corn and cottonseed meal in dairy rations have resulted in aflatoxin M_1 contaminated milk and milk products, including non-fat dry milk, cheese, ice creams and yogurts. Even in the case of the butter, during its production due to its chemical lipid rich compostion, the accumulation and concentration of any aflatoxin M_1 present in milk is usually involved.

7. Aflatoxins toxicity

Aflatoxins, and especially aflatoxin B_1, are associated with both toxicity and carcinogenicity in human and animal populations. The International Agency for Research on Cancer has classified aflatoxin B_1 as a group I carcinogen (International Agency for Research on Cancer, 1982).

In particular, aflatoxin B_1 is considered by medicine doctors and toxicologists as the most hepatocarcinogenic compound not produced by human activites but produced by a life organism (Newberne, Butler, 1969; Shank, Bhamarapravati, Gordon, Wogan, 1972; Peers, Linsell, 1973; Eaton, Groopman, 1994). The diseases caused by aflatoxin consumption are loosely called aflatoxicoses. Acute aflatoxicosis results in death; chronic aflatoxicosis results in cancer, immune suppression, and other "slow" pathological conditions (Hsieh, 1988). The liver is the primary target organ, with liver damage occurring when poultry, fish, rodents, and non human primates are fed aflatoxin B_1 contaminated foodstuff. This data is not unexpected because the liver is a lipophilic organ and all compounds carried by blood stream, i.e. drugs, contaminants, mycotoxins etc., are stored and concentrated in the hepatocytes that, with a long exposure time, may transform themselves in a cancer cell line. There are substantial differences in species susceptibility. Moreover, within a given species, the magnitude of the response is influenced by age, sex, weight, diet, exposure to infectious agents, and the presence of other mycotoxins and pharmacologically active substances. Thousands of studies on aflatoxin toxicity have been conducted, mostly on laboratory models or agricultural important species (Cullen, Newberne, 1994; Eaton, Groopman, 1994; Newberne, Butler, 1969).

Cytochrome P450 enzymes convert aflatoxins to the reactive 8,9-epoxide form (also known as aflatoxin-2,3 epoxide), which is capable of binding to both DNA and proteins (Eaton, Groopman, 1994). The reactive aflatoxin epoxide binds to the N^7 position of guanines. Moreover, aflatoxin B_1-DNA adducts can result in GC to TA transversions. A reactive glutathione *S*-transferase system found in the cytosol and microsomes catalyzes the conjugation of activated aflatoxins with reduced glutathione, leading to the excretion of aflatoxins (Raj, Prasanna, Mage, Lotlikar, 1986). Variation in the level of the glutathione transferase system as well as variations in the cytochrome P450 system are considered contributor to the differences observed in interspecies aflatoxin susceptibility (Eaton, Ramsdel, 1992; Eaton, Groopman, 1994).

Considering the differences exhisting in aflatoxin susceptibility in test animals, it has been proven not easy to extrapolate the possible effects of aflatoxins to humans. Acute toxicity of aflatoxins in humans however represent a serious threat.

In 1974 it has been reported in India an outbreak of hepatitis and 100 cases of death attributed to the consumption is heavily aflatoxins contaminated maize, causing an aflatoxins intake of 2 to 6 mg per day (Krishnamachari, Bhat, Nagarajan, Tilnak, 1975). Based on these data, it has been estimated that the acute lethal dose (LD) for adults is approximately 10 to 20 mg of aflatoxins (Pitt, 2000). Aflatoxins have been in years associated to various health conditions and are considered a poison. For example it has been associated kwashiorkor, a severe malnutrition disease, to a form of pediatric aflatoxicosis (Hendrickse, 1997). Aflatoxins, according to reported studies non completely assessed, could be involved in Reye's syndrome, an encephalopathy, and to fatty degeneration of some target organs in children and adolescents (Hayes, 1980).

Exposure to aflatoxins in the diet is considered an important risk factor for the development of primary hepatocellular carcinoma, particularly in individuals already exposed to hepatitis B. There are also observed nonhepatic effects of aflatoxin B_1 as reported by Coulombe (Coulombe, 1994). Several epidemiological studies have linked liver cancer incidence to estimated aflatoxin consumption in the diet (Peers, Linsell, 1973; Van Rensburg, Cook-Mazaffari, van Schalkwyk, van der Watt, Vincent, Purchase, 1985; Li, Yoshizawa, Kawamura, Luo, Li, 2001) even if the long term quantification of individual exposure to aflatoxins is difficult. The incidence of liver cancer varies widely from country to country, but it is one of the most common occurring in China, the Philippines, Thailand, and many African countries. The presence of hepatitis B virus infection, an important risk factor for primary liver cancer, complicates many of the epidemiological studies. In one case-control study involving more than 18.000 urine samples collected over 3.5 years in Shanghai, China, aflatoxin exposure alone yielded a relative risk of about 2; hepatitis B virus antigen alone yielded a relative risk of about 5; combined exposure to aflatoxin and hepatitis B yielded a relative risk of about 60 (Ross, Yuan, Yu, Wogan, Qian, Tu, Groopman, Gao, Henderson, 1992).

Using molecular epidemiology, it is possible to asses a link exhisting between putative carcinogens and specific cancers. Biomonitoring of aflatoxins can be done by analyzing for the presence of aflatoxin metabolites in blood, milk, and urine. In addition, excreted DNA adducts and blood protein adducts can also be monitored (Sabbioni, Sepai, 1994). The aflatoxin B_1-N^7-guanine adduct is considered a reliable urinary biomarker for aflatoxin exposure but reflects only recent exposure. Many studies have shown that carcinogenic potency is highly correlated with the extent of total DNA adducts formed *in vivo* (Eaton, Gallagher, 1994; Eaton, Groopman, 1994).

Inactivation of the p53 tumor suppressor gene may be important in the development of primary hepatocellular carcinoma. Studies of liver cancer patients in Africa and China have shown that a mutation in the p53 tumor suppressor gene at codon 249 is associated with a G-to-T transversion (Bressac, Kew, Wands, Ozturk, 1991; Hsu, Metcalf, Sun, Welsh, Wang, Harris, 1991). It is known that the reactive aflatoxin epoxide binds to the N^7 position of guanines. Moreover, aflatoxin B_1-DNA adducts can result in GC to TA inversion. The specific mutation in codon 249 of the p53 gene has been called the first example of a "carcinogen-specific" biomarker that remains fixed in the tumor tissue (Eaton, Gallagher, 1994).

There is also considerable evidence associating aflatoxin with neoplasms in extrahepatic tissues, particularly the lungs. For example, one early epidemiological study of Dutch peanut processing workers exposed to dust contaminated with aflatoxin B_1 showed a correlation between both respiratory cancer and total cancer in the exposed group compared with unexposed cohorts (Hayes, van Nienwenhuise, Raatgever, Ten Kate, 1984). Exposition even indirect to aflatoxins can result in a severe health issue: Deger (Deger, 1976) reported for example that dust from scrapings of chromatographic plates from aflatoxin analyses contributed to causing cancer in two young adults.

In developed countries, sufficient amounts of food combined with regulations that monitor aflatoxin levels protect human populations from significant aflatoxins ingestion. However, in countries where populations are facing starvation or where regulations are either not enforced or nonexistent, routine ingestion of aflatoxin may occur (Cotty, Bayman, Egel, Elias, 1994). Worldwide, liver cancer incidence rates are 2 to 10 times higher in developing countries than in developed countries (Henry, Bosch, Troxell, Bolger, 1999). A joint Food and Agriculture Organization/World Health Organization/United Nations Environment Programme Conference report stated that "in developing countries, where food supplies are already limited, drastic legal measure may lead to lack of food and to excessive prices. It must be remembered that people living in these countries cannot exercise the option of starving to death today in order to live a better life tomorrow" (Henry, Bosch,. Troxell, Bolger, 1999).

8. Monitoring techniques for assessing human exposure to aflatoxins

In the last few years, new technologies have been developed that more accurately monitor individual exposures to aflatoxins. Particular attention has been paid to the analysis of aflatoxin DNA adducts and albumin adducts as surrogates for genotoxicity in people. Autrup et al. (Autrup, Bradly, Shamsuddin, Wakhisi, Wasunna, 1983) proposed for the first time the use of synchronous fluorescence spectroscopy for the measurement of aflatoxin DNA adducts in urine. Urine samples collected after exposure to alfatoxins were found to contain 2,3-dihydroxy-2-(N7-guanyl)-3-hydroxyaflatoxin B_1, trivially known as aflatoxin B-Gual. Wild et al. used highly sensitive immunoassays to quantitate aflatoxins in human body fluids (Wild, Umbenhauer, Chapot, Montesano, 1986).

An enzyme linked immunosorbent assay (ELISA) was used to quantitate aflatoxin B_1 in a range from 0.01 ng/mL to 10 ng/mL, and was validated in human urine samples. Using this method, aflatoxin-DNA adduct excretion into urine was found to be positively correlated with dietary intake, and the major aflatoxin B_1-DNA adduct excreted in urine was shown to be an appropriate dosimeter for monitoring aflatoxin dietary exposure. Several epidemiological studies have found positive association between aflatoxin B_1 dietary exposure and an increased risk of human liver cancer (Sudakin, 2003; Zhu, Zhang, Hu, Xiao, Chen, Xu, Fremy, Chu, 1987; Groopman, Donahue,1988; Bean, Yourtee, 1989). Cytochrome P-450 enzymes further convert aflatoxins to different metabolites (Eaton, Ramsdell, Neal, 1994), e.g. aflatoxin B_1 is converted to metabolites like aflatoxin B_1-epoxide and the hydroxylated aflatoxins M_1, P_1 and Q_1. The hydroxylated metabolites form glucuronide and sulfate conjugates that can be enzymatically hydrolysed by b-glucuronidase and sulfatase (Wei, Marshall, Hsieh, 1985).

The European Union (EU) introduced measures to minimise the presence of aflatoxins in different foodstuffs. Maximum levels of aflatoxins are laid down in Commission Regulation (EC) No. 1881/2006. In an opinion adopted in January 2007, the European Food Safety Authority (EFSA) scientific Panel on contaminants in the food chain (CONTAM), concluded that increasing the current EU maximum levels of 4 μg/kg total aflatoxins in nuts to 8 or 10 μg/kg total afatoxins would have had minor effects on the estimated dietary exposure, cancer risk and calculated margin of exposure. The Panel also concluded that exposure to aflatoxins from all food sources should be kept as low as reasonably achievable because aflatoxins are genotoxic and carcinogenic. In June 2009 the European Commission asked EFSA to assess the effect on public health of an increase of the maximum level for total aflatoxins from 4 μg/kg to 10 μg/kg allowed for tree nuts other than almonds, hazelnuts and pistachios (e.g. Brazil nuts and cashews). This would facilitate the enforcement of the maximum levels, in particular regarding commercially available mixtures of nuts. The Panel concluded that public health would not be adversely affected by increasing the levels for total aflatoxins from 4 μg/kg to 8 or 10 μg/kg. However, the Panel reiterated its previous conclusions regarding the importance of reducing the number of highly contaminated foods reaching the market.

9. Bioavailability

In human health risk assessment, ingestion of contaminated food is considered a major route of exposure to many contaminants either caused by industrial or environmental contamination or as result of production processes. The total amount of an ingested contaminant (intake) does not always reflect the amount that is available to the body. Only a certain amount of the contaminant will be bioavailable (Versantvoort, Oomen, Van de Kamp, Rompelberg, Sips, 2005). Bioavailability is a term used to describe the proportion of the ingested contaminant in food that reaches the systemic circulation and then the organ or the apparatus. Studies in animals and humans show that oral bioavailability of compounds from food can be significantly different depending on the food source (food product), food processing or food preparation (Wienk, Marx, Beynen, 1999; van het Hof, West, Weststrate, Hautvast, 2000). As a consequence, the intake of a contaminant in food matrix A can lead to toxicity whereas the intake of the same amount of contaminant in food matrix B will not exert a toxic effects. Thus, a better insight in the effect of the matrix on the oral bioavailability of a contaminant will lead to a more accurate health risk assessment (Versantvoort, Oomen, Van de Kamp, Rompelberg, Sips, 2005).

Oral bioavailability of a compound can be seen as the resultant of three processes, namely the release of the compound from its matrix into digestive juice in the gastrointestinal tract (bioaccessibility); the transport across the intestinal epithelium into the vena Portae (intestinal transport); and the degradation of the compound in the liver and intestine (metabolism).

Release of the contaminant from the ingested product in the gastrointestinal tract is a prerequisite for uptake and bioavailability of a contaminant in the body. The oral bioavailability of the contaminant can be reduced subsequently by partial transport of the contaminant across the intestinal epithelium, or by degradation of the contaminant. Thus, determination of the bioaccessibility of a contaminant from its matrix can be seen as an indicator for the maximal oral bioavailability of the contaminant. Quantification of bioavailability and bioaccessibility of a compound from a certain matrix is difficult and often hampered by complex processes comprising digestion. The last decade there is an increasing interest in the use of *in vitro* methodologies to study the human oral bioavailability of compounds from the food chain (Minekus, Marteau, Havenaar, Huis, 1995; Glahn, Wien, Van Campen, Miller, 1996; Garrett, Failla, Sarama, 1999; Ruby, Schoof, Brattin, Goldade, Post, Harnois, Mosby, Casteel, Berti, Carpenter, Edwards, Cragin, Chappell, 1999; Oomen, Hack, Minekus, Zeijdner, Cornelis, Schoeters, Verstraete, Wiele, Wragg, Rompelberg, Sips, Wijnen, 2002).

Most of the *in vitro* digestion models simulate in a simplified manner the digestion processes in mouth, stomach and small intestine, in order to enable investigation of the bioaccessibility of compounds from their matrix during transit in the gastrointestinal tract.

Extensive studies involving animal models have indicated that the primary site for absorption of aflatoxin is the small intestine, in particular the duodenum (Wogan, Edwards, Shank, 1967; Ramos, Hernandez, 1996). *Lactobacillus spp.* has previously proven to be capable to survive at the gastrointestinal tract after oral intake (Taranto, Medici, Perdigon, Ruiz-Holgado, Valdez, 2000; Valeur, Engel, Carbajal, Connolly, Ladefoged, 2004); therefore, it is probable that mycotoxins were in contact with bacteria in the intestinal lumen, which then favored aflatoxin B_1 binding by bacteria prior to its natu ral process of absorption.

It has been reported that the binding process might be dependent on the environmental pH (Bolognani, Rumney, Rowland, 1997) and that the presence of bile salts could produce significant effects in the aflatoxin B_1 binding ability of the bacteria (Hernandez-Mendoza, Garcia, Steele, 2009). These two factors are closely related during the normal digestive process and its relationship varies along the small intestine (Low, 1990). Hence, the difference on aflatoxin binding ability of *Lactobacillus spp.* observed at the different portions of the intestine could be influenced by conditions prevailing in each region of the gastrointestinal tract.

Once the aflatoxin B_1 has been absorbed at intestinal level, it proceeds to the bloodstream and binds with plasma proteins especially albumin to form aflatoxin B_1-albumin adduct (Verma, 2004). The average half-life of albumin (approximately 20 days in humans) allows accumulation of adducts after chronic exposure to the toxin (Chapot, Wild, 1991). According to this, the amount of adducts present in blood samples of rats treated only with aflatoxin B_1 represent the cumulative dose of aflatoxin intake over the experimental period, which indicates that the reduction of aflatoxin B_1-Lys adduct observed in animals treated with aflatoxin plus bacteria was originated by the ability of *Lactobacillus spp.*to bind aflatoxin B_1 inside the intestinal lumen, thus avoiding its passage into the bloodstream. In a related work (Gratz, Täubel, Juvonen, Viluksela, Turner, 2006) no significant differences were found in the amounts of aflatoxin B_1-Lys adduct present in animals receiving *Lactobacillus rhamnosus* GG daily for 3 d before and 3 d after a single oral dose of aflatoxin B_1 compared with those receiving only the mycotoxin. Other reports suggested that probiotics are less capable of binding aflatoxin B_1 in the presence of mucus and are more susceptible to interfere factors in the intestinal tract, which may explain the behavior observed in the levels of adduct (Gratz, Mykkänen, Ouwehand, Juvonen, Salminen, 2004; Gratz, Täubel, Juvonen, Viluksela, Turner, 2006). This effect could have been surmounted by the numbers of bacteria implanted before oral dose of aflatoxin B_1, and the constant administration of probiotic bacteria during the experimental period (Gratz, Mykkänen, Ouwehand, Juvonen, Salminen, 2004).

In agreement with earlier reports (Ward, Sontag, Weisburger, Brown, 1975; Maurice, Bodine, Rehrer, 1983), body weight gain was not adversely affected. However, there was a reduction in feed intake in rats receiving only aflatoxin B_1. This effect could be induced by the dose of aflatoxin received, since it has been reported that aflatoxin B_1 induces reduction of food intake in some animal species, including rats and birds, in a dose-dependent manner (Maurice, Bodine, Rehrer, 1983). In addition, toxicological studies in rats have shown that aflatoxin

B_1 consumption may produce a significant decrease of serum leptin levels (Abdel-Wahhab, Ahmed, Hagazi, 2006). Leptin concentration is usually associated with the high levels of cortisol and interleukin-6, which act together to influence the feeding response (Barber, McMillan, Wallace, Ross, Preston, 2004). *Lactobacillus reuteri* might have contributed to reduce the aflatoxin B_1 absorption in bacteria-treated rats and thus diminish its effect on leptin levels in blood serum.

The volume of the stomach is considered an important parameter for oral dosing in experimental animals. For rats, maximum oral dosage volume recommended is 10 mL kg^{-1} of body weight; for a 200 g rat this would mean a dosing volume of 2 mL (McConnell, Basit, Murdan, 2008). Therefore, it is possible that the volume supplied (every third day) by oral gavage of aflatoxin and/or bacteria over the experiment, had partially met the basic water needs of the rats, which may explain the observed reduction in water consumption at the end of the experimental period (21 days).

A world-wide-accepted method for protecting animals against mycotoxicosis is the use of adsorbent materials. An effective adsorbent is one that tightly binds the mycotoxin in contaminated feed without dissociating in the gastrointestinal tract of the animal. The toxin–adsorbent complex passes then through the gastrointestinal tract without absorption and is eliminated via the faeces. In other words, the bioavailability of the mycotoxin is reduced as less mycotoxin is absorbed because it is bound to the adsorbent, i.e. lower bioaccessibility. Therefore, these adsorbents can be used to evaluate the use of the in vitro digestion model as indicator for the in vivo bioavailability. The following materials, representative for different classes of adsorbents, have been selected: an aluminosilicate (HSCAS), which is a common anticaking additive in animal feeds to reduce mycotoxicosis in animals; activated charcoal, which is used in humans and animals as an antidote against poisoning; cholestyramine is an anion exchange resin and binds bile acids in the gastrointestinal tract and it has been used for over 20 years in the clinic for reduction of lowdensit y lipoproteins and cholesterol.

The effect of chlorophyllin on intestinal transport of aflatoxin B_1 was studied by measurement of the transport of aflatoxin B_1 with the intestinal Caco-2 cells. The rate at which compounds are transported across the Caco-2 cells, which is expressed as a permeability coefficient, is correlated with absorption in humans (Artursson, Karlsson, 1991).

Transport of 5ng/mL aflatoxin B_1 across Caco-2 cells revealed that after 4h, 25±6% aflatoxin B_1 was transported across Caco-2 cells into the basolateral compartment. Addition of chlorophyllin (1 mg/mL) greatly reduced (>20-fold) the transport of aflatoxin B_1 to only 1±1%. From this transport, a permeability coefficient can be calculated for aflatoxin B_1 of $9x10^{-6}$ cm/s in absence, and $0.4x10^{-6}$ cm/s in presence of chlorophyllin. When we compare these transport rates with the S-shaped correlation found for absorption of compounds in humans, the permeability coefficient of aflatoxin B_1 alone ($9x10^{-6}$ cm/s) corresponds with high absorption in humans whereas the permeability coefficient of aflatoxin B_1 in presence of chlorophyllin ($0.4x10^{-6}$ cm/s) indicates an intermediate absorption. Thus, these data are in accordance with the human intervention study on chlorophyllin and aflatoxin B_1, where a 50%

reduction in excretion of aflatoxin metabolites in urine was found in presence of chlorophyllin (Versantvoort, Oomen, Van de Kamp, Rompelberg, Sips, 2005).

Furthermore, these results provide evidence for the hypothesis that chlorophyllin reduces the absorption of aflatoxin B_1 in humans.

In risk assessment, a dose proportional relationship between contamination level and bioavailability is taken as basic assumption. This assumption simplifies risk assessment, since it can be assumed that regardless the level of contamination, a constant percentage of the contaminant will be bioavailable. The extreme sensitivity of turkeys to the toxic effects of aflatoxin B_1, a condition associated with a combination of efficient CYP-mediated activation and deficient GST-mediated detoxification of aflatoxin B_1 (Klein, Buckner, Kelly, Coulombe, 2000), makes turkeys an excellent model in which to study various chemopreventives. We have recently shown that the observed chemopreventive properties of BHT in turkeys is due, at least in part, to its ability to inhibit hepatic conversion of aflatoxin B_1 to the exo-aflatoxin B_1-8,9-epoxide (AFBO) in vivo and in vitro (Guarisco, Hall, Coulombe, 2008).

Determining the outcome of inhibition of hepatic aflatoxin B_1 bioactivation in whole animals is relevant to veterinary medicine and to food safety. Dietary butylated hydroxytoluene (BHT) can reduce aflatoxin B_1 bioavailability, as demonstrated by serum concentrations of radiolabel which were reduced at every time interval after aflatoxin B_1 administration. Among the possible explanations for reduced bioavailability is high first pass elimination prior to absorption into the blood, and/or an attenuation of mucosal aflatoxin B_1 absorption. However, since no quantitative difference in the biliary elimination of aflatoxin B_1 or its metabolites was identified, any change due to increased first pass effect would have to result in increased elimination by non-biliary pathways.

The observed reduction in hepatic aflatoxin B_1–DNA adducts in BHT fed animals is consistent with the fact that this antioxidant is a competitive inhibitor of hepatic in vitro CYP1A5-mediated aflatoxin B_1 epoxidation to AFBO. Because of the critical role of AFBO and subsequent adduct formation in aflatoxicosis (as well as longer-term consequences such as tumor formation), a reduction in hepatic aflatoxin B_1–DNA adducts would be expected to have a positive effect on the overall health of aflatoxin B_1-exposed turkeys.

10. Bioaccessibility

Bioaccessibility has been defined as the fraction of a compound that is released from its matrix in the gastrointestinal tract and thus becomes available for intestinal absorption i.e. enters the blood stream (Benito, Miller, 1998). Bioaccessibility includes the entire sequence of events that take place during the digestive transformation of food into material that can be assimilated by the body, the absorption/assimilation into the cells of the intestinal epithelium, and lastly, the presystemic metabolism (both intestinal and hepatic). Bioaccessibility analyses can be approached using general experimental techniques (there are systematic techniques common to all types of foods) that can be adapted to all types of claims regarding nutritional content.

In vivo, as soon as a compound is released from its matrix in the chyme, the compound can be transported across the intestinal epithelium into the body thereby keeping the compound concentration low in the chyme.

Different analytical approaches can be applied to measure bioaccessibility of nutrients and bioactive compounds: *in vivo* and *in vitro* studies both present strengths and drawbacks. Within *in vivo* studies, balance studies and tissue concentration are two strategies that allow determination of the absorbed amount of nutrients, bioactive compounds, or their metabolites. Balance studies determine the absorbed amount by measuring the difference between the fed and excreted amounts of the nutrient or bioactive compound. Tissue concentration consists of monitoring the increase in plasma/serum concentration of the nutrient or bioactive compound. These approaches have been applied these approaches have been used with both animals and humans to determine absorption of carbohydrates, minerals, vitamins, phytochemicals, and different compounds (Benito, Miller, 1998; Hallberg, 1991). *In vivo* human studies are the criterion standard approach to determine bioaccessibility of food nutrients or bioactive compounds, although some experimental approaches are ethically and technically unaffordable.

Digestion and absorption involve several different steps, and each one could cause an effect on the nutrient or bioactive compound so that a detailed picture is not obtained with the balance and bioassay studies. *In vitro* studies have been developed to simulate the physiologic conditions and the sequence of events that occur during digestion in the human gastrointestinal tract. In a first step, an in vitro gastrointestinal method is applied to the food, mirroring the physiochemical conditions that take place during human digestion, considering the three areas of the human digestive system (mouth, stomach, and intestine).

The main features of the in vitro gastrointestinal methods are temperature, shaking or agitation, and the chemical and enzymatic composition of saliva, gastric juice, duodenal juice, and bile juice (Wittsiepe, Schrey, Hack, Selenka, Wilhelm, 2001). When physical processes that occur in vivo are not reproduced (shear, mixing, hydration, changes in conditions over time, peristalsis), the in vitro gastrointestinal model is defined as a static or biochemical model. The dynamic models mimic the *in vivo* physical processes so that they take into account new variables, such as changes on viscosity of the digesta, particle size reduction, diffusion, and partitioning of nutrients. Several examples of *in vitro* gastrointestinal static and dynamic models have been described (Rotard, Christmann, Knoth, Mailahn, 1995; Arcand, Mainville, Farnworth, 2007). During the application of the in vitro gastrointestinal method, food nutrients or bioactive compounds can be monitored to determine whether they are affected by digestion conditions (pH, enzymes) or if interactions with other food components (fiber, sucrose polyester, fat replacers) take place, which could affect efficiency of digestion. The final processed material of the experimental procedure is a digesta or intestinal preparation.

To analyze the lipophilic content that has been effectively incorporated to mixed micelles, the micellar fraction can be isolated from that processed material by the application of an ultracentrifugation protocol (Hernell, Staggers, Carey, 1999). In the digestion model, the compounds are not removed from the chyme during digestion and therefore, bioaccesibility may be underestimated when saturation of the compound occurs in the chyme. Thus, one factor potentially affecting bioaccessibility is the level of contamination.

The bioaccessibility of aflatoxin B_1 in chime has been determined from nine peanut slurries ranging from 0.6 to 14 μg/kg aflatoxin B_1 (contamination level in peanuts 1.5-36 μg/kg). Aflatoxin B_1 was almost completely mobilised from the peanut slurries during digestion evidencing a mean bioaccessibility of 94%. The concentration of aflatoxin B_1 in chyme of the two highest contaminated peanut slurries was higher than those in the calibration curve and a smaller volume of chyme was used for analysis (Versantvoort, Oomen, Van de Kamp, Rompelberg, Sips, 2005).

The amount of aflatoxin B_1 released from the peanut slurries into the chyme is dose proportional to the contamination level in the peanut slurries. These results showed no saturation of the chyme with aflatoxin B_1. The amount of food in the digestion model was varied in order to study whether release of the contaminant from its food matrix was linearly, i.e. whether bioaccessibility was independent from the amount of food in the model. Application of 0.5 g and 4.5 g peanut slurry in the in vitro digestion model corresponds to the consumption of approximately 10 and 100 g peanuts, respectively. The bioaccessibility of some bioactive compounds can be influenced by the food composition as observed by Versantvoort et al. (Versantvoort, Oomen, Van de Kamp, Rompelberg, Sips, 2005) that studied the effects of different food components on the bioaccessibility of aflatoxin B_1 from peanut slurry considering an average meal. Bioaccessibilities of aflatoxin B_1 (108±11%) from 6 g food-mix (4.5g standard meal + 0.5g peanut slurry + 1g buckwheat) were compared to the bioaccessibility of aflatoxin B_1 from 0.5g corresponding peanut slurry (83±18%) showing that the bioaccessibility of aflatoxin B_1 did not vary significantly.

11. Advanced analysis of aflatoxins in biological fluids

11.1. Sampling and sample preparation

Sampling and sample preparation remain a considerable source of error in the analytical identification of aflatoxins. Thus, systematic approaches to sampling, sample preparation, and analysis are absolutely necessary to determine aflatoxins at the parts-per-billion level. In this regard, specific plans have been developed and tested rigorously for some commodities such as corn, peanuts, and tree nuts; sampling plans for some other commodities have been modeled after them. A common feature of all sampling plans is that the entire primary sample must be ground and mixed so that the analytical test portion has the same concentration of toxin as the original sample. Methods of sampling and analysis for the official control of mycotoxins, including aflatoxins, are laid down in Commission Regulation No 401/2006. This ensures that the same sampling criteria intended for the control of mycotoxin content in food are applied to the same products by the competent authorities throughout the EU and that certain performance criteria, such as recovery and precision, are fulfilled. In 2008, the Codex Alimentarius set a maximum level of 10 μg/kg total aflatoxins in ready-to-eat almonds, hazelnuts, and pistachios at a level higher than that currently in force in the EU (4 μg/kg total aflatoxins).

11.2. Solid-phase extraction

All analytical procedures include three steps: extraction, purification, and determination. The most significant recent improvement in the purification step is the use of solid-phase extraction. Extracts are extracts are cleaned up before instrumental analysis (thin layer or liquid chromatography) to remove coextracted materials that often interfere with the determination of target analytes.

11.3. Thin-layer chromatography

Thin layer chromatography (TLC), also known as flat bed chromatography or planar chromatography is one of the most widely used separation techniques in aflatoxin analysis. Since 1990, it has been considered the AOAC official method and the method of choice to identify and quantitate aflatoxins at levels as low as 1 ng/g. The TLC method is also used to screen and corroborate findings by newer, more rapid techniques.

11.4. Liquid chromatograph

Liquid chromatography (LC) is similar to TLC in many respects, including analyte application, stationary phase, and mobile phase. Liquid chromatography and TLC complement each other. For an analyst to use TLC for preliminary work to optimize LC separation conditions is not unusual. Liquid chromatography methods for the determination of aflatoxins in foods include normal-phase LC (NPLC), reversed-phase LC (RPLC) with pre- or before-column derivatization (BCD), RPLC followed by postcolumn derivatization (PCD), and RPLC with electrochemical detection.

11.5. Immunochemical methods

Thin layer chromatography and LC methods for determining aflatoxins in food are laborious and time consuming. Often, these techniques require knowledge and experience of chromatographic techniques to solve separation and and interference problems. Through advances in biotechnology, highly specific antibody-based tests are now commercially available that can identify and measure aflatoxins in food in less than 10 minutes. These tests are based on the affinities of the monoclonal or polyclonal antibodies for aflatoxins. The three types of immunochemical methods are radioimmunoassay (RIA), enzyme-linked immunosorbent assay (ELISA), and immunoaffinity column assay (ICA). These are mostly chemical methods of detection but still provide an insight into the immunochemical methods such as ELISA and RIA which can used to detect aflatoxins in foods, such as flour and starch products.

12. Aflatoxins identity assessment

Although analytical methods might consist of different extraction, clean-up, and quantitation steps, the results of the analyses by such methods should be similar when the methods are applied properly. Since the reliability of the quantitative data is not in question, the problem still to be solved is the confirmation of identity of the aflatoxins. The confirmation techniques used involve either chemical derivatization or mass spectrometry (MS).

Different analytical methods have been reported in the literature in order to facilitate the investigation of the role of ingested aflatoxins in small volumes of human sera (Grio, Jose, Frenich, Martinez Vidal, Luis, Romero-Gonzalez, 2010; Yuanjing, Yi, Huiming, Bingnan, Haicheng, Fanli, Miaomiao, Wei, Wendong, 2010). Aflatoxin B_1 has been extracted from 1 mL or less of human sera spiked with a known concentration of aflatoxin B_1 and analyzed using high-performance liquid chromatography (HPLC) as the detection system. Several methods have been used to analyze feed, foods and bodyfluids, human and animal plasma, serum, milk, etc. (Santini, Ferracane, Meca, Ritieni, 2009; Rampone, Piccinelli, Aliberti, Rastrelli, 2009; Monbaliu, Van Poucke, Detavernier, Dumoulin, Van De Velde, Schoeters, Van Dyck, Averkieva, Van Peteghem, De Saeger, 2010). The ELISA (Zhu, Zhang, Hu, Xiao, Chen, Xu, Fremy, Chu, 1987) or radioimmunoassay (RIA) methods (Groopman, Donahue, 1988; Tang, Pang, 2009; Li, Zhang, Zang, 2009) allow the quantification of the total amount of aflatoxins, and results are expressed in term of aflatoxin B_1 equivalents. Both methods however involve the use of specific antibodies not commercially available. Recently immunoensors (Sun, Yan, Tang, Zhang, 2012) and biosensor have been proposed for the analyses of mycotoxins in different matrices (Campàs, Garibo, Prieto-Simón, 2012).

The International Agency for Research on Cancer has classified aflatoxin B_1 as a human carcinogen and aflatoxins B_2, G_1 and G_2 as possible nephrotoxic natural compounds and carcinogenic to humans (IARC, 1993; Commission Regulation (EC) No. 1525/98, 1998). Due to carryover in food and feed they are considered nowadays to have the most severe impact of all mycotoxins on human health. Maximum residue levels have been set down to the • g/mL range in a wide variety of agricultural commodities, food, feed and milk, e.g. 0.01 mg/kg of aflatoxin M_1 in milk for infants (Groopman, Donahue, Zhu, Chen, Wogan,1985).

Methods like liquid chromatography-mass spectroscopy (LC/MS) have been repeatedly used for structural elucidation in metabolism on aflatoxin containing analytes and specific matrices but only a limited number of quantitative methods have been published to determine the more common aflatoxins present in food (Papp, Otta, Zaray, Mincsovics, 2002; Biselli, Hartig, Wegener, Hummert, 2004; Biselli, Hartig, Wegener, Hummert, 2005; Sorensen, Elbaek, 2005; Kokkonen, Jestoi, Rizzo, 2005) milk, (Sorensen, Elbaek, 2005) cheese, (Cavaliere, Foglia, Pastorini, Samperi, Lagana, 2006) herbs, (Ventura, Gomez, Anaya, Diaz, Broto, Agut, Comellas, 2004) urine, (Scholl, Musser, Groopman, 1997; Walton, Egner, Scholl, Walker, Kensler, Groopman, 2001; Egner, Yu, Johnson, Nathasingh, Groopman, Kensler, Roebuck, 2003; Wang-Buhler, Lee, Chung, Stevens, Tseng, Hseu, Hu, Westerfield, Yang, Miranda, Buhler, 2005) airborne dust (Kussak, Nilsson, Andersso, Langridge, 1995) and cigarette smoke (Edinboro, Karnes, 2005).

LC/MS has been used as a confirmation technique for the already well established, reliable and robust LC-FL methodology (Kussak, Nilsson, Andersson, Langridge, 1995; Abbas, Williams, Windham, Pringle, Xie, Shier, 2002; Blesa, Soriano, Molto, Marin, Manes, 2003; Abbas, Cartwright, Xie, Shier, 2006) and has also been used to confirm positive results of TLC and ELISA based screening analyses. All the aflatoxins exhibit good ESI ionisation efficiency in the positive ion mode with abundant protonated molecules $[MH]^+$ and sodium adduct ions (Blesa, Soriano, Molto, Marin, Manes, 2003; Ventura, Gomez, Anaya, Diaz, Broto, Agut, Comellas, 2004; Kussak, Nilsson, Andersson, Langridge, 1995) and typically, for aflatoxins B_1, B_2, G_1 and G_2, the formation of sodium adduct ions can easily be suppressed by the addition of ammonium ions to the mobile phase leading to a better mass spectroscopy (MS) sensitivity (Cavaliere, Foglia, Pastorini, Samperi, Lagana, 2006). Reports about the utility of atmospheric pressure chemical ionization (APCI) interfaces and ionization efficiencies in this mode seem to be highly dependent on the aflatoxin studied and the APCI interface geometry (Abbas, Williams, Windham, Pringle, Xie, Shier, 2002; Abbas, Cartwright, Xie, Shier, 2006).

This method has been proved to be more sensitive for the simultaneous determination of aflatoxins B_1, B_2, G_1, G_2, M_1, M_2, and moreover smaller sample volumes of serum can be used for the analysis. Aflatoxins are in free equilibrium with the albumin combined form and it is reported in the literature the effect of pH and/or serum concentration of fatty acids on the formation of the adducts. Moreover, a recent study showed that green tea polyphenols might modulate the formation of the adducts between aflatoxin B_1 and albumin (Tang, Tang, Xu, Luo, Huang, Yu, Zhang, Gao, Cox, Wang, 2008).

Advanced spectrometric methods, such as LC-MS/MS, permit quantification and recognition of the free aflatoxins in the sera with fewer problems on recovery, sensitivity and chemical identification (Santini, Ferracane, Meca, Ritieni, 2009; Huang, Zheng, Zengxuan, Yongjiang, Yiping, 2010) evaluating the aflatoxin exposure directly from their free forms.

13. Decontamination, detoxification, exposition

Aflatoxin produced by *Aspergillus flavus*, *A. parasiticus* and *A. nomius* crops contamination, is a worldwide food safety concern. Several strategies, including chemical, physical and biological control methods have been investigated to manage these potent toxic secondary metabolites in foods. Among them, biological control seems nowadays to be the most promising approach for the aflatoxins control. From the food safety point of view, fermentation with microorganisms, a technique quite commonly used in food production (e.g. fermentation with lactic acid bacteria, alcoholic fermentation, conventional fermentation of the protein from vegetables as common in South Asia, etc.) should be preferred. In optimal conditions, this procedure can result in a mycotoxin-free food or feed.

The reaction of aflatoxins to various physical conditions and reagents have been studied extensively because of the possible application of such reactions to the detoxification of aflatoxins contaminated material. Aflatoxins in dry state are stable to heat up to the melting point. However, in the presence of moisture and at elevated temperatures there is destruction of aflatoxin and this can occur with aflatoxin in oilseed meals, roasted peanuts or in aqueous solution at pH 7. Although the reaction products have not been examined in detail it seems likely that such treatment leads to opening of the lactone ring with the possibility of decarboxylation at elevated temperatures. At a temperature of about 100°C, ring opening followed by decarboxylation occurs, and reaction may proceed further, leading to the loss of the methoxy group from the aromatic ring.

In alkali solution reversible hydrolysis of the lactone moiety occurs. Recyclization has been observed after acidification of a basic aflatoxin containing solution.

In the presence of acids, aflatoxin B_1 and G_1 are converted in to aflatoxin B_{2A} and G_{2A} due to acid-catalyzed addition of water to the double bond in the furan ring. In the presence of acetic anhydride and hydrochloric acid the reation proceeds further to give the acetoxy derivative. Similar adducts of aflatoxin B_1 and G_1 are formed with formic acid-thionyl chloride, acetic acid-thionyl chloride and trifluoroacetic acid.

Many oxidizing agents, e.g. sodium hypochlorite, potassium permanganate, chlorine, hydrogen peroxide, ozone and sodium perborate react with aflatoxin and change the aflatoxin molecule in some way as indicated by the loss of fluorescence. The mechanisms of these reactions are uncertain and the reaction products remain unidentified in most cases. Reduction of aflatoxin B_1 and B_2 with sodium borohydride yielded aflatoxin RB_1 and RB_2, respectively. These arise as a result of opening of the lactone ring followed by reduction of the acid group and reduction of the keto group in the cyclopentene ring. Hydrogenation of aflatoxin B_1 and G_1 yields aflatoxin B_2 and G_2 respectively. Further reduction of aflatoxin B_1 using 3 moles of hydrogen yields tetrahydroxyaflatoxin.

Food and feed contaminated with mycotoxins pose a severe health risk to animals and they may cause big economical losses due to the lower efficacy of animal husbandry and crop performances.

In addition, directly or indirectly (carry on through animal products) contaminated foods may also pose a health risk to humans. For this reason it is understandable that many research has been addressed in an attempt to salvage mycotoxin contaminated commodities and to avert health risks associated with the toxins.

Relevant basic criteria to be followed when a decontamination strategy is assessed have been suggested (Scott, 1990; Pomeranz, Bechtel, Sauer, Seitz, 1990):

- the mycotoxin must be inactivated (destroyed) by transformation to non-toxic compounds;
- fungal spores and mycelia should be destroyed, so that new toxins are not produced;
- the food or feed material should retain its nutritive value and remain palatable for consumers;
- the physical properties of raw material should not change significantly;

- it must be economically feasible (the cost of decontamination should be less than the value of contaminated commodity).

The main three possibilities to avoid any possible harmful effects of contamination of food and feed caused by mycotoxins habe been described by Halàsz et al. (Halasz, Lasztity, Abonyi, Bata, 2009):

- prevention of contamination;
- decontamination of mycotoxin-containing food and feed;
- inhibition of absorption of mycotoxin in consumed food in the digestive tract.

Although the different methods used at present are to some extent successful, they have big disadvantages with limited efficacy and possible losses of important nutrients and normally with high costs. It is a common opinion that the best solution for decontamination should be detoxification by biodegradation, giving a possibility for removal of mycotoxins under mild conditions without using harmful chemicals without significant losses in nutritive value and palatability of decontaminated food and feed. One of the most frequently used strategies for biodegradation of mycotoxins includes isolation of microorganisms able to degrade the given mycotoxin and treatment of food or feed in an appropriate fermentation process.

Thousand of microorganisms habe been screened for their ability to degrade aflatoxins from solutions (Lillehoj, Ciegler, Hall, 1967; Ciegler, Lillehoj, Peterson, Hall, 1996). As a result it was found that only one bacterium, the *Flavobacterium aurantiacum* B-184, was able to eliminate aflatoxin from solutions and uptake of the mycotoxin by the cells was influenced by pH and temperature.

Another interesting result was that an high concentration populations of the cells, more than 10^{11} per mL, is more useful to remove the aflatoxin from solutions than lower cell concentrations. Large populations of heat inactivated cells were also shown to bind some aflatoxin from solution, which was easily recovered by washing with water (Line, Brackett, 1967). The ability of *Flavobacterium aurantiacum* B-184 to remove aflatoxins from foods was demonstrated in milk, vegetable oil, corn, peanut, peanut butter and peanut milk (Hao, Brackett, 1988; Hao, Brackett, 1989; Line, Brackett, 1995). To assess the exact fate of the aflatoxin B_1 treated with *Flavobacterium aurantiacum*, Line et al. used radio-labeled carbon (C_{14}) aflatoxin B_1 and detected the formed radioactive carbon dioxide confirming this way the biodegradation pathway of aflatoxin (Line, Brackett, Wilkinson, 1994).

It should be noted that the interest of the biological approach to degrade aflatoxin is increasing since the consumers prefer this tool to chemical treatments used on food and feed to eliminate aflatoxins.

Nevertheless, one of the big obstacle to the developing of biological approaches is the bright pigmentation associated with the bacterium treatment, that hampers the applicability for food and feed. Microorganisms that are able to degrade aflatoxin B_1 include *Corynebacterium rubrum, Aspergillus niger, Trichoderma viride, Mucor ambiguus, Dactylium dendroides, Mucor griseocyanus, Absidia repens, Helminthosporium sativum, Mucor alternans, Rhizopus archisus, Rhizo-*

pus oryzae, Rhizopus solonifer and a protozoan *Tetrahymena pyriformis* (Doyle, Applebaum, Brackett, Marth, 1982; Karlovsky, 1999).

Recently, a growing interest can be observed concerning the use of *Rhodococci* for aflatoxins degradation: these microorganisms have a wide-range ability to degrade compounds like aflatoxins (Alberts, Engelbrecht, Steyn, Holzapfel, van Zyl, 2006; Teniola, Addo, Brost, Farber, Jany, Alberts, Van Zyl, Steyn, Holzapfel, 2005). Teniola et al. (Teniola, Addo, Brost, Farber, Jany, Alberts, Van Zyl, Steyn, Holzapfel, 2005) reported the degradation of aflatoxin B_1 using liquid cultures of *Rhodococcus erythropolis* and the analysis of the intracellular extracts separated from *Rhodococcus erythropolis* liquid cultures suggested that a cascade of enzymatic reactions with loss of fluorescence (the intact aflatoxin is a fluorescent compound and degradation results in loss of fluorescence in time) occurred. Aflatoxin B_1 is probably degraded by the same enzymes (biphenyl-dioxygenases, dihydro-diol-dehydrogenases, and hydrolases) that are involved in catabolic pathways of polychlorinated biphenyls. Knowledge of gene coding for these enzymes may be helpful in development and production of new effective enzyme preparations for degradation of aflatoxins. The role of trace metal ions in microbial aflatoxin B_1 degradation has been studied studied by Souza et al. (Souza, Brackett, 1998) who found that copper and zinc ions may inhibit the degradation of aflatoxin B_1 by *Flavobacterium aurantiacum*. This effect is probably connected with an influence on the enzyme system involved in the degradation process. Peltonen et al. (Peltonen, El-Nezami, Salminen, Ahokas, 2000) and El Nezami et al. (El-Nezami, Kankaanpaa, Salminen, Ahokas, 1998) studied the ability of dairy strains of lactic acid bacteria to bind Aflatoxin B_1. It has been observed that *Lactobacillus rhamnosus* can significantly remove Aflatoxin B_1 compared with other strains. Removal was observed as very rapid, with 80% of toxin removed within the first 60 min of treatment.

Several bacterial species, such as *Bacillus, Lactobacilli, Pseudomonas, Ralstonia* and *Burkholderia* spp., have shown ability to inhibit fungal growth and production of aflatoxins by *Aspergillus* spp. in laboratory tests. Palumbo et al. (Palumbo, Baker, Mahoney, 2006) reported that a number of *Bacillus, Pseudomonas, Ralstonia* and *Burkholderia* strains could completely inhibit *A. flavus* growth. *B.subtilis* and *P. solanacearum* strains isolated from maize soil were also able to inhibit aflatoxin accumulation. In most cases, although these strains were highly effective against fungal growth and against the produced toxins in laboratory conditions, they did not give good efficacies on field. This could be attributed to the difficulty to bring the bacterial cells to the *Aspergillus* infection sites on commodities under field conditions. Saprophytic yeast species, such as *Candida krusei* and *Pichia anomala,* have revealed promising efficacy as biocontrol agents for aflatoxins decontamination (Yin, Yan, Jiang, Ma, 2008). In a similar way to bacterial agents, these yeast strains were able to significantly inhibit *Aspergillus* growth and resultant toxins in laboratory conditions. Shetty et al. (Shetty, Hald, Jespersen, 2007) observed that the ability of *S. cerevisiae* to bind aflatoxin B_1 was strain specific with 7 strains binding 10-20%, 8 strains binding 20-40% and 3 strains binding more than 40% of the added aflatoxin B_1. Though the yeasts are considered to be potential biocontrol agents for the aflatoxins management, further experiments conducted on field are necessary to test their efficacies in reducing aflatoxin contamination in real on field situations.

Many reports exists on the use of physically separated yeast cell walls obtained from brewery as feed additive in poultry diet resulting in amelioration of aflatoxins toxic effects (Shetty, Jespersen, 2003; (Santin, Paulillo, Maiorka, Okada Nakaghi, Macari, Fischer da Silva, Alessi, 2003). When dried, yeast and yeast cell walls have been added to rat-ration along with aflatoxin B_1, and a significant reduction in the toxicity has been observed (Baptista et al., 2004). In an *in vitro* study with the cell wall material, there was a dose dependent binding of as much as 77% (w/w) and modified mannan-oligosaccharides derived from the *S. cerevisiae* cell resulted in as much as 95% (w/w) binding (Girish and Devegowda, 2006).

14. Conclusions

Total quality of food is the main goal to reach and a mission both for food industry and for the world Government Institutions. Quality means also safety, and this aspect is the most relevant goal to pursue and achieve. Consumers have often prejudicials ideas about risks associated to food and feed; nowadays based on the information available, they tend to have a quite large knowledge about genetically modified microorganisms, phytopharmaceutical origin active principles, heavy metals contamination or unbalanced dietary habit. These are the main issues for the majority of the people. However, the perceived risk related to mycotoxin occurrence in food is very neglected and underestimated.

Aflatoxins are a serious problem for human health, and it is not possible to evaluate this threat without paying great attention to the exposure to these compounds. The frequency and level of mycotoxin presence in the food chain are grown up in the last decades, probably due to the changed global weather conditions, to the market globalisation, and to the worldwide deployment of mold. The development of new analytical methods, more sensitive and more specific to evaluate aflatoxins presence, ensures the managment of the risk and, consequently, could allow to guarantee the safety of food from aflatoxin contamination. It is not possible however to completely avoid aflatoxins contamination in the food chain since the colonisation by molds and their mycotoxins biosynthesis are not under the full humans control due to many different biological, genetic and biochemical reasons.

Consumers, together with safety food agencies and with the worldwide research, must maximize efforts to reduce the global aflatoxin exposure. Focus should not be given only to Aflatoxin B_1 for which the neologism ALARA where this acronym say "As Low As Reasonably Achievable" has been considered applicable.

It can be noted as final remark that, in general, consumers do not appreciate any chemical procedure, e.g. the use of alkali or acid solutions. The same results to control unwanted aflatoxins presence are achievable using ammonium or different physical or chemical approaches. Another aspect regards the develop of a biological protocols that use microrganism generally recognised as safe for food (GRAS), a procedure that is considered much more acceptable by the consumers. Nevertheless, aflatoxins fate should be determined considering also the toxicological bioactivities of aflatoxins byproducts, like their many metabolites. It is important to determine these aspects before to propose new microrganisms able to detoxify aflatoxins without causeing any unwanted side effect, e.g. changes in sensorial and technological properties of foods. In addition, any new method should be economically convenient if compared with any actually used procedure, especially for food industry that may hamper for these economic reasons.

Safe food is a non-negotiable topic both for ethic reasons and for economic aspects. The social costs linked to an increase of health conditions like liver diseases, or the problems connected to crop destruction, withdrawal of food from the shelves, etc., can be more expensive than a preventive actions to reduce aflatoxin presence in the food chain.

Author details

Antonello Santini[1*] and Alberto Ritieni[2]

*Address all correspondence to: asantini@unina.it

1 Department of Food Science, University of Napoli "Federico II", Portici, Napoli, Italy

2 Department of Pharmaceutical and Toxicological Chemistry, University of Napoli "Federico II", Napoli, Italy

References

[1] Abbas, H. K., Cartwright, R. D., Xie, W., & Shier, W. T. (2006). Aflatoxin and fumonisin contamination of corn (maize, Zea mays) hybrids in Arkansas. Crop Prot. , 25(1), 1-9.

[2] Abbas, H. K., Williams, W. P., Windham, G. L., Pringle, H. C., Xie, W., & Shier, W. T. (2002). Aflatoxin and Fumonisin Contamination of Commercial Corn (Zea mays) Hybrids in Mississippi. J. Agric. Food Chem. , 50, 5246-5254.

[3] Abdel-Wahhab, M. A., Ahmed, H. H., & Hagazi, M. M. (2006). Prevention of aflatoxin B1-initiated hepatotoxicity in rat by marine algae extracts. J. Appl. Toxicol. , 26, 229-238.

[4] Alberts, J. F., Engelbrecht, Y., Steyn, P. S., Holzapfel, W. H., & van Zyl, W. H. (2006). Biological degradation of aflatoxin B1 by Rhodococcus erythropolis cultures. Int.J.Food Microbiol. , 109, 121-126.

[5] Arcand, Y., Mainville, I., & Farnworth, E. R. (2007). A dynamic model that simulates the human upper gastrointestinal tract for the study of probiotics. Int. J. Food Microbiol. , 99, 287-96.

[6] Artursson, P., , P., & Karlsson, J. (1991). Correlation between oral drug absorption in humans and apparent drug permeability coefficients in human intestinal epithelial (Caco-2) cells. *Biochemical and Biophysical Research Communications*, 880 EOF-885 EOF.

[7] Autrup, H., Bradly, K. A., Shamsuddin, A. K. M., Wakhisi, J., & Wasunna, A. (1983). Detection of putative adduct with fluorescence characteristics identical to 2,3-dihydro-2-(7′-guanyl)-3-hydroxyaflatoxin B_1 in human urine collected in Murang'a district, Kenya. Carcinogesis, , 4(9), 1193-1195.

[8] Barber, M. D., Mc Millan, D. C., Wallace, A. M., Ross, J. A., & Preston, T. (2004). The response of leptin, interleukin-6 and fat oxidation to feeding in weight-losing patients with pancreatic cancer. Br. J. Cancer. , 90, 1129-1132.

[9] Benito, P., & Miller, D. (1998). Iron absorption and bioavailability: an updated review. Nutr. Res. , 18, 581-603.

[10] Bennett, J. W., & Klich, M. (2003). Mycotoxins. Clinical Microbiology Reviews. , 16(3), 497-516.

[11] Bennett, J. W., Chang, P. K., & Bhatnagar, D. (1997). One gene to whole pathway: the role of norsolorinic acid in aflatoxin research. Adv. Appl. Microbiol. , 45, 1-15.

[12] Berry, C. L. (1988). The pathology of mycotoxins. J. Pathol. , 154, 301-311.

[13] Biselli, S., Hartig, L., Wegener, H., & Hummert, C. (2004). Analysis of Fusarium Toxins using LC/MS-MS: Application to Various Food and Feed Matrices. Recent Applications in LC-GC Eur. , 17, 25-31.

[14] Biselli, S., Hartig, L., Wegener, H., & Hummert, C. (2005). Analysis of Fusarium Toxins Using LC-MS-MS: Application to Various Food and Feed Matrices. Spectroscopy. , 20(2), 20-28.

[15] Blesa, J., Soriano, J. M., Molto, J., Marin, C., , R., & Manes, J. J. (2003). Determination of aflatoxins in peanuts by matrix solid-phase dispersion and liquid chromatography. J. Chromatogr. A. , 1011, 49-54.

[16] Blout, W. P. (1961). Turkey "X" disease. Turkeys 9:52, 55-58.

[17] Bolognani, F., Rumney, C. J., & Rowland, R. (1997). Influence of carcinogen binding by lactic acid producing bacteria on tissue distribution and in vivo mutagenicity of dietary carcinogens. Food Chem. Toxicol. , 35, 535-545.

[18] Bressac, B., Kew, M., Wands, J., & Ozturk, M. (1991). Selective G to T mutations of gene in hepatocellular carcinoma from southern Africa. Nature 350:429-431., 53.

[19] Brown, D. H., Yu, J. H., Kelkar, H. S., Fernandes, M., Nesbitt, T. C., Keller, N. P., Adams, T. H., & Leonard, T. J. (1996). Twenty-five co-regulated transcripts define a sterigmatocystin gene cluster in Aspergillus nidulans. Proc. Natl. Acad. Sci. USA , 93, 1418-1422.

[20] Campàs, M., Diana, Garibo. M., & Prieto-Simón, B. (2012). Novel nanobiotechnological concepts in electrochemical biosensors for the analysis of toxins. Analyst. , 137, 1055-1067.

[21] Cary, J. W., Chang, P. K., & Bhatnagar, D. (2001). Clustered metabolic pathway genes in filamentous fungi, In Khachatourians G. G. and Arora D. K. (Ed.), Applied mycology and biotechnology, Vol. I. Agriculture and food production. Elsevier Science, Amsterdam, The Netherlands., 165-198.

[22] Cavaliere, C., Foglia, P., Pastorini, E., Samperi, R., & Lagana, A. (2006). Liquid chromatography/tandem mass spectrometric confirmatory method for determining aflatoxin M1 in cow milk: Comparison between electrospray and atmospheric pressure photoionization sources. J. Chromatogr. A. 1101(1-2):69-78.

[23] Chapot, B., & Wild, C. P. (1991). ELISA for quantification of aflatoxin-albumin adducts and their application to human exposure assessment. In: Van-Warhol, M., Velzen, D., Bullock, G.R. (Eds.), Techniques in Diagnostic Pathology (New York: Academic Press, , 2, 135-155.

[24] Ciegler, A., Lillehoj, B., Peterson, R. E., & Hall, H. H. (1966). Microbial detoxification of aflatoxin. Appl. Microbiol. , 14, 934-939.

[25] Cleveland, T. E., & Bhatnagar, D. (1992). Molecular strategies for reducing aflatoxin levels in crops before harvest, In D. Bhatnagar and T. E. Cleveland (ed.), Molecular approaches to improving food quality and safety. Van Nostrand Reinhold, New York (USA)., 205-228.

[26] Commission, Regulation. . E. C., No, 1525/9., amending, Regulation. . E. C., No, 184/., setting, maximum., residue, levels., for, certain., contaminants, in., & foodstuffs, Official. J. Eur. Commun. L 201/43, (1998).

[27] Cotty, P. J., Bayman, P., Egel, D. S., & Elias, K. S. (1994). Agriculture, aflatoxins and Aspergillus, In Powell, K. A., Renwick, A., Peberdy, J. F. (Ed.). The genus Aspergillus. Plenum Press, New York (USA)., 1-27.

[28] Coulombe, R. A., & Jr , . (1994). Nonhepatic effects and Biotransformations of aflatoxin B1, In Eaton, D.L., Groopman, J.D. (Eds.). The Toxicology of Aflatoxins: Human Health, Veterinary and Agricultural Significance. Academic Press, Orlando (USA)., 89-110.

[29] Cullen, J. M., & Newberne, P. N. (1994). Acute heptoxotoxicity of aflatoxins. In Eaton D. L. and Groopman J. J. (Ed.). The toxicity of aflatoxins. Human health, veterinary, and agricultural significance. Academic Press, San Diego (USA)., 3-26.

[30] Deger, G. E. (1976). Aflatoxin-human colon carcinogenesis? Ann. Intern. Med. , 85, 204-205.

[31] Detroy, R. W., Lillehoj, E. B., & Ciegler, A. (1971). Aflatoxin and related compounds, In Ciegler, A., Kadis, S., Ajl S.J. (Ed.). Microbial toxins. vol. VI: fungal toxins. Academic Press, New York (USA)., 3-178.

[32] Diener, U. L., Cole, R. J., Sanders, T. H., Payne, G. A., Lee, L. S., & Klich, M. A. (1987). Epidemiology of aflatoxin formation by Aspergillus flavus. Annu. Rev. Phytopathol. , 25, 249-270.

[33] Doyle, M. P., Applebaum, R. S., Brackett, R. E., & Marth, H. (1982). Physical, Chemical and Biological Degradation of Mycotoxins. Foods and Agricultural Commodities. J. of Food Protection. , 45, 964-971.

[34] Eaton, D. L., & Gallagher, E. P. (1994). Mechanisms of aflatoxin carcinogenesis. Annu. Rev. Pharmacol. Toxicol. , 34, 135-172.

[35] Eaton, D. L., Groopman, J. D. ., & Ed., . (1994). The toxicology of aflatoxins: human health, veterinary, and agricultural significance. Academic Press, San Diego (USA).

[36] Eaton, D. L., & Ramsdel, H. S. (1992). Species and diet related differences in aflatoxin biotransformation, In Bhatnagar, D., Lillehoj, E.B., Arora, D.K. (Ed.), Handbook of applied mycology, mycotoxins in ecological systems. Marcel Dekker, Inc., New York (USA)., 5, 157-182.

[37] Eaton, D. L., Ramsdell, H. S., & Neal, G. E. (1994). In Eaton, D.L., Groopman, J.D. (Editors). The Toxicology of Aflatoxins. Academic Press, San Diego, CA (USA)., 45.

[38] Edinboro, L. E., & Karnes, H. T. (2005). Determination of aflatoxin B_1 in sidestream cigarette smoke by immunoaffinity column extraction coupled with liquid chromatography/mass spectrometry. J. Chromatogr. A. 1083 (1-2), 127-132.

[39] Egner, P. A., Yu, X., Johnson, J. K., Nathasingh, C. K., Groopman, J. D., Kensler, T. W., & Roebuck, B. D. (2003). Identification of Aflatoxin M_1-N^7-Guanine in Liver and Urine of Tree Shrews and Rats Following Administration of Aflatoxin $B_{1.}$ Chem. Res. Toxicol. , 16, 1174-1180.

[40] El -Nezami, H., Kankaanpaa, P., Salminen, S., & Ahokas, J. (1998). Ability of dairy strains of lactic acid bacteria to bind a common food carcinogen, aflatoxin B1. Food Chem.Toxicol. , 36, 321-326.

[41] Garrett, D. A., Failla, M. L., & Sarama, R. J. (1999). Development of an in vitro digestion method to assess carotenoid bioavailability from meals. J. Agric. Food Chem. , 47, 4301-4309.

[42] Girish, C. K., & Devegowda, G. (2006). Efficacy of Glucomannan-containing Yeast Product (Mycosorb) and Hydrated Sodium Calcium Aluminosilicate in Preventing the Individual and Combined Toxicity of Aflatoxin and T-2 Toxin in Commercial Broilers. Asian-Australian J. Anim. Sci. , 19, 877-883.

[43] Glahn, R. P., Wien, E. M., Van Campen, D. R., & Miller, D. D. (1996). Caco-2 cell iron uptake from meat and casein digests parallels in vivo studies: use of a novel in vitro method for rapid estimation of iron bioavailability. J. Nutr. , 126, 332-339.

[44] Goldblatt, L., & (ed., . (1969). Aflatoxin. scientific background, control, and implications. Academic Press, New York, N.Y.

[45] Goto, T., Wicklow, D. T., & Ito, Y. (1996). Aflatoxin and cyclopiazonic acid production by a sclerotium-producing Aspergillus tamarii strain. Appl. Environ. Microbiol., 62, 4036-4038.

[46] Gratz, S., Mykkänen, H., Ouwehand, A. C., Juvonen, R., & Salminen, S. (2004). Intestinal mucus alters the ability of probiotic bacteria to bind aflatoxin B1 in vitro. Appl. Environ. Microbiol., 70, 6306-6308.

[47] Gratz, S., Täubel, M., Juvonen, R. O., Viluksela, M., & Turner, P. C. (2006). Lactobacillus rhamnosus strain GG modulates intestinal absorption, fecal excretion, and toxicity of aflatoxin B1 in rats. Appl. Environ. Microbiol., 72, 7398-7400.

[48] Grio, L., Jose, S., Frenich, G., Martinez, Vidal. A., Luis, J., & Romero-Gonzalez, R. (2010). J. Sep. Sci., 33, 502-510.

[49] Groopman, J. D., Donahue, K., Zhu, J., Chen, J., & Wogan, G. N. (1985). Aflatoxin metabolism in humans: Detection of metabolites and nucleic adducts in urine by affinity chromatography. Proc. Natl. Acad. Sci. USA., 82, 6492-6496.

[50] Guarisco, J. A., Hall, J. O., Coulombe, R. A., & Jr , . (2008). Mechanisms of butylated hydroxytoluene chemoprevention of aflatoxicosis-inhibition of aflatoxin B1 metabolism. Toxicol. Appl. Pharmacol., 227, 339-346.

[51] Halasz, A., Lasztity, R., Abonyi, T., & Bata, A. (2009). Decontamination of Mycotoxin-Containing Food and Feed by Biodegradation. Food Reviews International. , 25, 284-298.

[52] Hallberg, L. (1981). Bioavailability of dietary iron in man. Ann. Rev. Nutr., 1, 123-47.

[53] Hao, D. Y. Y., & Brackett, R. E. (1988). Removal of aflatoxin B1 from peanut milk inoculated with Flavobacterium aurantiacum J.Food Sci., 53, 1384-1386.

[54] Hao, D. Y. Y., & Brackett, R. E. (1989). Growth and survival of Flavobacterium aurantiacum in peanut milk. J.Food Protect., 52, 165-168.

[55] Hayes, A. W. (1980). Mycotoxins: a review of biological effects and their role in human diseases. Clin. Toxicol., 17, 45-83.

[56] Hayes, R. B., van Nienwenhuise, J. P., Raatgever, J. W., & Ten, Kate. F. J. W. (1984). Aflatoxin exposure in the industrial setting: an epidemiological study of mortality. Food Chem. Toxicol., 22, 39-43.

[57] Heathcote, J. G., & Hibbert, J. R. (1978). Aflatoxins: chemical and biological aspects. Elsevier Scientific Publishing Company, Amsterdam, The Netherlands.

[58] Hendrickse, R. G. (1997). Of sick turkeys, kwashiorkor, malaria, perinatal mortality, heroin addicts and food poisoning: research on the influence of aflatoxins on child health in the tropics. Ann. Trop. Med. Parasitol., 91, 787-793.

[59] Henry, S. H., Bosch, F. X., Troxell, T. C., & Bolger, P. M. (1999). Reducing liver cancer-global control of aflatoxin. Science., 286, 2453-2454.

[60] Henry, S. H., Bosch, F. X., & Bowers, J. C. (2002). Aflatoxin, hepatitis and worldwide liver cancer risks, In DeVries, J.W., Trucksess, M.W., Jackson, L.S. (Ed.). Mycotoxins and food safety. Kluwer Academic/Plenum Publications., New York, N.Y. (USA)., 229-320.

[61] Hernandez-Mendoza, A., Garcia, H. S., & Steele, J. L. (2009). Screening of Lactobacillus casei strains for their ability to bind aflatoxin B1. Food Chem. Toxicol. , 47, 1064-1068.

[62] Hernandez-Mendoza, A., Guzman-de-Peña, D., & Garcia, H. S. (2009). Key role of teichoic acids on Aflatoxin B1 binding by probiotic bacteria. J. Appl. Microbiol. , 107, 395-403.

[63] Hernell, O., Staggers, J. E., & Carey, M. C. (1999). Physical-chemical behavior of dietary and biliary lipids during intestinal digestion and absorption: phase analysis and aggregation states of luminal lipids during duodenal fat digestion in healthy adult human beings. Biochem. , 29, 2041-56.

[64] Hicks, J. K., Shimizu, K., & Keller, N. P. (2002). Genetics and biosynthesis of aflatoxins and sterigmatocystin, In F. Kempken (ed.), The mycota, vol. XI. Agricultural applications. Springer-Verlag, Berlin, Germany., 55-69.

[65] Hsieh, D. (1988). Potential human health hazards of mycotoxins, In Natori, S., Hashimoto, K., Ueno, Y. (Ed.), Mycotoxins and phytotoxins. Third Joint Food and Agriculture Organization W.H.O. United Nations Program International Conference of Mycotoxins. Elsevier, Amsterdam, The Netherlands., 69-80.

[66] Hsu, I. C., Metcalf, R. A., Sun, T., Welsh, J. A., Wang, N. J., & Harris, C. C. (1991). Mutational hotspot in the gene in human hepatocellular carcinomas. Nature 350:377-378., 53.

[67] Huang, B., Zheng, H., Zengxuan, C., Yongjiang, W., & Yiping, R. (2012). Simultaneous determination of aflatoxins B1, B2, G1, G2, M1 and M2 in peanuts and their derivative products by ultra-high-performance liquid chromatography-tandem mass spectrometry. Anal. Chim. Acta. , 662, 62-68.

[68] IARC Monographs on the Evaluation of Carcinogenic Risks to Humans. (1993). International Agency for Research on Cancer, Lyon, France. , 489.

[69] International Agency for Research on Cancer. (1982). The evaluation of the carcinogenic risk of chemicals to humans. IARC Monograph Supplement 4. International Agency for Research on Cancer, Lyon, France.

[70] Karlovsky, P. (1999). Biological detoxification of fungal toxins and its use in plant breeding, feed and food production. Nat. Toxins. , 7, 1-23.

[71] Klein, P. J., Buckner, R., Kelly, J., Coulombe Jr, , & , R. A. (2000). Biochemical basis for the extreme sensitivity of turkeys to aflatoxin B1. Toxicol. Appl. Pharmacol. , 165, 45-52.

[72] Klich, M. A. (1987). Relation of plant water potential at flowering to subsequent cottonseed infection by Aspergillus flavus. Phytopathology , 77, 739-741.

[73] Klich, M. A., Mullaney, E. J., Daly, C. B., & Cary, J. W. (2000). Molecular and physiological aspects of aflatoxin and sterigmatocystin biosynthesis by A. tamarii and A. ochraceoroseus. Appl. Microbiol. Biotechnol. , 53, 605-609.

[74] Klich, M. A., & Pitt, J. I. (1988). Differentiation of Aspergillus flavus from Aspergillus parasiticus and other closely related species. Trans. Br. Mycol. Soc. , 91, 99-108.

[75] Klich, M. A., Yu, J., Chang, P. K., Mullaney, E. J., Bhatnagar, D., & Cleveland, T. E. (1995). Hybridization of genes involved in aflatoxin biosynthesis to DNA of aflatoxigenic and nonaflatoxigenic aspergilli. Appl. Microbiol. Biotechnol. , 44, 439-443.

[76] Kokkonen, M., Jestoi, M., & Rizzo, A. (2005). Determination of selected mycotoxins in mould cheeses with liquid chromatography coupled to tandem with mass spectrometry. Food Addit. Contam. , 22, 449-456.

[77] Krishnamachari, K. A. V. R., Bhat, R. V., Nagarajan, V., & Tilnak, T. M. G. (1975). Hepatitis due to aflatoxicosis. An outbreak in Western India. Lancet. , 1, 1061-1063.

[78] Kussak, A., Nilsson, C. A., Andersson, B., & Langridge, J. (1995). Determination of aflatoxins in dust and urine by liquid chromatography/electrospray ionization tandem mass spectrometry. Rapid Commun. Mass Spectrom. , 9(13), 1234-1237.

[79] Li, F. Q., Yoshizawa, T., Kawamura, S., Luo, S. Y., & Li, Y. W. (2001). Aflatoxins and fumonisins in corn from the high-incidence area for human hepatocellular carcinoma in Guangxi, China. J. Agric. Food Chem. , 49, 4122-4126.

[80] Lillehoj, E. B., Ciegler, A., & Hall, H. H. (1967). Aflatoxin B1 uptake by Flavobacterium aurantiacum and resulting toxic effects. J. Bacteriology. , 93, 464-471.

[81] Line, E. J., Brackett, R. E., & Wilkinson, R. E. (1994). Evidence for degradation of aflatoxin B1 by Flavobacterium aurantiacum. J. Food Protect. , 57, 788-791.

[82] Line, J. E., & Brackett, R. E. (1995). Factors affecting aflatoxin B1 removal by Flavobacterium aurantiacum. J. Food Protect. , 58, 91-94.

[83] Line, J. E., & Brackett, R. E. (1995). Role of toxin concentration and second carbon source in microbial transformation of aflatoxin B1. J. of Food Protect. , 58, 1042-1044.

[84] Low, A. G. (1990). Nutritional regulation of gastric secretion, digestion and emptying. Nutr. Res. Rev. , 3, 229-252.

[85] Maurice, D. V., Bodine, A. B., & Rehrer, N. J. (1983). Metabolic effects of low aflatoxin B1 levels on broiler chicks. Appl. Environ. Microbiol. , 45, 980-984.

[86] Mc Connell, E. L., Basit, A. W., & Murdan, S. (2008). Measurements of rat and mouse gastrointestinal pH, fluid and lymphoid tissue, and implications for in-vivo experiments. J. Pharm. Pharmacol. , 60, 63-70.

[87] Minekus, M., Marteau, P., Havenaar, R., & Huis, . (1995). In Veld, J.H.J. A multicompartimental dynamic computer-controlled model simulating the stomach and small intestine. ATLA , 23, 197-209.

[88] Monbaliu, S., Van Poucke, C., Detavernier, C., Dumoulin, F., van De Velde, M., Schoeters, E., van Dyck, S., Averkieva, O., Van Peteghem, C., & De Saeger, S. (2010). J. Agric. Food Chem. , 58, 66-75.

[89] Newberne, P. M., & Butler, W. H. (1969). Acute and chronic effect of aflatoxin B_1 on the liver of domestic and laboratory animals: a review. Cancer Res. , 29, 236-250.

[90] Pawlowski, N. E., Schoenhard, G. L., Lee, D. J., Libbey, L. M., Loveland, P. M., & Sinnhuber, R. O. (1977). Reduction of aflatoxin B1 to aflatoxicol. J. Agric. Food Chem. 25(2), , 437-438.

[91] Oomen, A. G., Hack, A., Minekus, M., Zeijdner, E., Cornelis, C., Schoeters, G., Verstraete, W., Wiele, T. V., Wragg, J., Rompelberg, C. J. M., Sips, A. J. A. M., & Wijnen, J. V. (2002). Comparison of five in vitro digestion models to study the bioaccessibility of soil contaminants. Environ. Sci. Technol. , 36, 3326-3334.

[92] O'Neil, M. J., Smith, A., & Heckelman, P. E. (2001). The Merck Index, 13th Ed., Whitehouse Station, NJ, Merck & Co., (USA) , 34-35.

[93] Palumbo, J. D., Baker, J. L., & Mahoney, N. E. (2006). Isolation of bacterial antagonists of Aspergillus flavus from almonds. Microb Ecol. , 52(1), 45-52.

[94] Payne, G., & Brown, M. P. (1998). Genetics and physiology of aflatoxin biosynthesis. Annu. Rev. Plant Pathol. , 36, 329-362.

[95] Peers, F. G., & Linsell, C. A. (1973). Dietary aflatoxins and human liver cancer-a population study based in Kenya. Br. J. Cancer. , 27, 473-484.

[96] Peltonen, K. D., El -Nezami, H. S., Salminen, S. J., & Ahokas, J. T. (2000). Binding of aflatoxin B1 by probiotic bacteria. J. Sci. Food Agric. , 80, 1942-1945.

[97] Peterson, S. W., Ito, Y., Horn, B. W., & Goto, T. (2001). Aspergillus bombycis a new aflatoxigenic species and genetic variation in its sibling species A. nomius. Mycologia. , 93, 689-703.

[98] Pitt, J. I. (2000). Toxigenic fungi: which are important? Med. Mycol. , 38(1), 17-22.

[99] Pomeranz, Y., Bechtel, D. B., Sauer, D. B., & Seitz, L. M. (1990). Fusarium headblight / scab/ in cereal grain. In Advances in Cereal Science and Technology, Pomeranz, Y.; Ed.; AACC: St.Paul, 373-73., 10

[100] Raj, H. G., Prasanna, H. R., Mage, N., & Lotlikar, P. D. (1986). Effect of purified rat and hamster hepatic glutathione S-transferases on the microsome mediated binding of aflatoxin B_1 to DNA. Cancer Lett. , 33, 1-9.

[101] Ramos, A. J., & Hernández, E. (1996). In situ absortion of aflatoxins in rat small intestine. Mycopathologia. , 134, 27-30.

[102] Rampone, L., Piccinelli, A. L., Aliberti, L., & Rastrelli, L. (2009). Application of pressurized liquid extraction in the analysis of aflatoxins B1, B2, G1 and G2 in nuts. J. Sep. Sci. , 32, 3837-3843.

[103] Ross, R. K., Yuan, J. M., Yu, M. C., Wogan, G. N., Qian, G. S., Tu, J. T., Groopman, J., Gao, Y. T., & Henderson, B. E. (1992). Urinary aflatoxin biomarkers and risk of hepatocellular carcinoma. Lancet. , 339, 1413-1414.

[104] Rotard, W., Christmann, W., Knoth, W., & Mailahn, W. (1995). Bestimmung der resorptionsverfügbaren PCDD/PCDF aus Kieselrot. UWSF-Z Umweltchem Ökotox. , 7, 3-9.

[105] Ruby, M. V., Schoof, R., Brattin, W., Goldade, M., Post, G., Harnois, M., Mosby, D. E., Casteel, S. W., Berti, W., Carpenter, M., Edwards, D., Cragin, D., & Chappell, W. (1999). Advances in evaluating the oral bioavailability of inorganics in soil for use in human health risk assessment. Environ. Sci. Technol. , 33, 3697-3705.

[106] Sabbioni, G., & Sepai, O. (1994). Determination of human exposure to aflatoxins, In Sinha, K.K., Bhatnagar, K. (Ed.), Mycotoxins in agriculture and food safety. Marcel Dekker, Inc., New York (USA)., 183-226.

[107] Santin, E., Paulillo, A. C., Maiorka, A., Okada, Nakaghi. L. S., Macari, M., Fischer, da., Silva, A. V., & Alessi, A. C. (2003). Evaluation of the Efficacy of Saccharomyces cerevisiae Cell Wall to Ameliorate the Toxic Effects of Aflatoxin in Broilers. International J. Poultry Sci. , 2(5), 341-344.

[108] Santini, A., Ferracane, R., Meca, G., & Ritieni, A. (2009). Overview of analytical methods for beauvericin and fusaproliferin in food matrices. Anal. Bioanal. Chem. , 395, 1253-1258.

[109] Scholl, P. F., Musser, S. M., & Groopman, J. D. (1997). Synthesis and Characterization of Aflatoxin B_1 Mercapturic Acids and Their Identification in Rat Urine. Chem. Res. Toxicol. , 10, 1144-1151.

[110] Scott, P. M. (1990). Trichothecenes in grain. Cereal Foods World, , 35, 661-665.

[111] Shank, R. C., Bhamarapravati, N., Gordon, J. E., & Wogan, G. N. (1972). Dietary aflatoxins and human liver cancer. IV. Incidence of primary liver cancer in two municipal populations in Thailand. Food Cosmet. Toxicol. , 10, 171-179.

[112] Shetty, P. H., & Jespersen, L. (2003). Trends in Food Science & Technology Saccharomyces cerevisiae and lactic acid bacteria as potential mycotoxin decontaminating agents. International Journal of Poultry Science. , 2(5), 341-344.

[113] Shetty, P. H., Hald, B., & Jespersen, L. (2007). Surface binding of aflatoxin B_1 by Saccharomyces cerevisiae strains with potential decontaminating abilities in indigenous fermented foods. Int. J. Food Microbiol. , 113(1), 41-6.

[114] Smith, J. E., & Moss, M. O. (1985). Mycotoxins. Formation, analyses and significance. John Wiley and Sons Ed., Chichester, United Kingdom.

[115] Sorensen, L. K., & Elbaek, T. H. (2005). Determination of mycotoxins in bovine milk by liquid chromatography tandem mass spectrometry. J. Chromatogr. B. , 820(2), 183-196.

[116] Souza, D. H., & Brackett, R. E. (1998). The role of trace metal ions in aflatoxin B1 degradation by Flavobacterium aurantiacum J.Food Protect. , 61, 1666-1669.

[117] Squire, R. A. (1981). Ranking animal carcinogens: a proposed regulatory approach. Science. , 214, 877-880.

[118] Sudakin, D. L. (2003). Dietary Aflatoxin Exposure and Chemoprevention of Cancer: A Clinical Review. Clinical Toxicology. , 41(2), 195-204.

[119] Sun, X., Yan, L., Tang, Y., & Zhang, Y. (2012). A rapid and specific immunosensor for the detection of aflatoxigenic Aspergilli. European Food Research and Technology. , 234(6), 1013-1021.

[120] Takahashi, T., Chang, P. K., Matsushima, K., Yu, J., Abe, K., Bhatnagar, D., Cleveland, T. E., & Koyama, Y. (2002). Nonfunctionality of Aspergillus sojae aflR in a strain of Aspergillus parasiticus with a disrupted aflR gene. Appl. Environ. Microbiol. , 68, 3737-3743.

[121] Tang, L., Tang, M., Xu, L., Luo, H., Huang, T., Yu, J., Zhang, L., Gao, W., Cox, S. B., & Wang, J. S. (2008). Modulation of aflatoxin biomarkers in human blood and urine by green tea polyphenols intervention. Carcinogenesis. , 29(2), 411-417.

[122] Taranto, M. P., Medici, M., Perdigon, G., Ruiz-Holgado, A. P., & Valdez, G. F. (2000). Effect of Lactobacillus reuteri on the prevention of hypercholesterolemia in mice. J. Dairy Sci. , 83, 401-403.

[123] Teniola, O. D., Addo, P. A., Brost, I. M., Farber, P., Jany, K. D., Alberts, J. F., van Zyl, W. H., Steyn, P. S., & Holzapfel, W. H. (2005). Degradation of aflatoxin B1 by cell-free extracts of Rhodococcus erythropolis and Mycobacterium fluoranthenivorans sp. Nov. Int. J. Food Microbiol. , 105, 111-117.

[124] Townsend, C. A. (1997). Progress towards a biosynthetic rationale of the aflatoxin pathway. Pure Appl. Chem. , 58, 227-238.

[125] Trail, F., Mahanti, N., & Linz, J. (1995). Molecular biology of aflatoxin biosynthesis. Microbiology , 141, 755-765.

[126] Valeur, N., Engel, P., Carbajal, N., Connolly, E., & Ladefoged, K. (2004). Colonization and immunomodulation by Lactobacillus reuteri ATCC 55730 in the human gastrointestinal tract. Appl. Environ. Microbiol. , 70, 1176-1181.

[127] van Egmond, H. P. (1989). Aflatoxin M_1: occurrence, toxicity, regulation, In H. P. van Egmond (ed.), Mycotoxins in dairy products. Elsevier Applied Science, London (UK)., 11-55.

[128] van het, Hof. K. H., West, C. E., Weststrate, J. A., & Hautvast, J. G. A. J. (2000). Dietary factors that affect the bioavailability of carotenoids. J. Nutr. , 130, 503-506.

[129] van Rensburg, S. J., Cook-Mazaffari, P., van Schalkwyk, D. J., van der Watt, J. J., Vincent, T. J., & Purchase, I. F. (1985). Heptatocellular carcinoma and dietary aflatoxin in Mozambique and Transkei. Br. J. Cancer , 51, 713-720.

[130] Ventura, M., Gomez, A., Anaya, I., Diaz, J., Broto, F., Agut, M., & Comellas, L. (2004). Determination of aflatoxins B1, G1, B2 and G2 in medicinal herbs by liquid chromatography-tandem mass spectrometry. J. Chromatogr. A. , 1048, 25-29.

[131] Verma, R. J. (2004). Aflatoxin cause DNA damage. Int. J. Hum. Genet. , 4, 231-236.

[132] Versantvoort, C. H. M., Oomen, A. G., Van de Kamp, E., Rompelberg, C. J. M., & Sips, A. J. A. M. (2005). Applicability of an in vitro digestion model in assessing the bioaccessibility of mycotoxins from food Food and Chemical Toxicology. , 43(1), 31-40.

[133] Walton, M., Egner, P., Scholl, P. F., Walker, J., Kensler, T. W., & Groopman, J. D. (2001). Liquid Chromatography Electrospray-Mass Spectrometry of Urinary Aflatoxin Biomarkers: Characterization and Application to Dosimetry and Chemoprevention in Rats. Chem. Res. Toxicol. , 14(7), 919-926.

[134] Wang-Buhler, J. L., Lee, S. J., Chung, W. G., Stevens, J. F., Tseng, H. P., Hseu, T. H., Hu, C. H., Westerfield, M., Yang, Y. H., Miranda, C. L., & Buhler, D. R. (2005). CYP2K6 from zebrafish (Danio rerio): Cloning, mapping, developmental/tissue expression, and aflatoxin B1 activation by baculovirus expressed enzyme. Comp. Biochem. Physiol., Part C Toxicol. And Pharmacol. , 140(2), 207-219.

[135] Ward, J. M., Sontag, J. M., Weisburger, E. K., & Brown, C. A. (1975). Effect of life time exposure to aflatoxin B1 in rats. J. Natl. Cancer Inst. , 55, 107-113.

[136] Watson, A. J., Fuller, L. J., Jeens, D. J., & Archer, D. B. (1999). Homologues of aflatoxin biosynthesis genes and sequence of aflR in Aspergillus oryzae and Aspergillus sojae. Appl. Environ. Microbiol. , 65, 307-310.

[137] Wei, C. I., Marshall, M. R., & Hsieh, D. P. H. (1985). Characterization of water-soluble glucuronide and sulphate conjugates of aflatoxin B_1. Urinary excretion in monkey, rat and mouse. Food Chem. Toxicol. , 23, 809-819.

[138] Wei, D. L., & Jong, S. C. (1986). Production of aflatoxins by strains of the Aspergillus flavus group maintained in ATCC. Mycopathologia. , 93, 19-24.

[139] Wienk, K. J. H., Marx, J. J. M., & Beynen, A. C. (1999). The concept of iron bioavailability and its assessment. Eur. J. Nutr. , 38, 51-75.

[140] Wild, C. P., Umbenhauer, D., Chapot, B., & Montesano, R. (1986). Monitoring of individual human exposure to aflatoxins (AF) and N-nitrosamines (NNO) by immunoassays. J. Cell. Biochem. , 30(2), 171-179.

[141] Wilson, D. M., & Payne, G. A. (1994). Factors affecting Aspergillus flavus group infection and aflatoxin contamination of crops, In Eaton D. L. and Groopman J. D.

(Ed.). The toxicology of aflatoxins. Human health, veterinary and agricultural significance. Academic Press. San Diego (USA)., 309-325.

[142] Wittsiepe, J., Schrey, P., Hack, A., Selenka, F., & Wilhelm, M. (2001). Comparison of different digestive tract models for estimating bioaccessibility of polychlorinated dibenzo-p-dioxins and dibenzofurans (PCDD/F) from red slag 'Kieselrot'. Int. J. Hyg. Environ. Health. , 203, 263-73.

[143] Wogan, G. N., Edwards, G. S., & Shank, R. C. (1967). Excretion and tissue distribution of radioactivity from aflatoxin B1- ^{14}C in rats. Cancer Res. , 27, 1729-1736.

[144] Yin, Y. N., Yan, L. Y., Jiang, J. H., & , Z. H. (2008). Biological control of aflatoxin contamination of crops. J. Zhejiang Univ. Sci. B. (10):787-92.

[145] Yu, J., Chang, P. K., Bhatnagar, D., & Cleveland, T. E. (2000). Genes encoding cytochrome and monooxygenase enzymes define one end of the aflatoxin pathway gene cluster in Aspergillus parasiticus. Appl. Microbiol. Biotechnol. 53:583-590., 450.

[146] Yu, J., Woloshuk, C. P., Bhatnagar, D., & Cleveland, T. E. (2000). Cloning and characterization of avfA and omtB genes involved in aflatoxin biosynthesis in three Aspergillus species. Gene , 248, 157-167.

[147] Yuanjing, F., Yi, Y., Huiming, X., Bingnan, P., Haicheng, G., Fanli, M., Miaomiao, Y., Wei, C., & Wendong, H. (2010). Faming Zhuanli Shenqing Gongkai Shuomingshu. , 1, 1-9.

[148] Zhu, J. Q., Zhang, L. S., Hu, X., Xiao, Y., Chen, J. S., Xu, Y. C., Fremy, J., & Chu, F. S. (1987). Correlation of Dietary Aflatoxin B_1 Levels with Excretion of Aflatoxin M_1 in Human Urine. Cancer Res. , 47(7), 1848-1852.

Permissions

The contributors of this book come from diverse backgrounds, making this book a truly international effort. This book will bring forth new frontiers with its revolutionizing research information and detailed analysis of the nascent developments around the world.

We would like to thank Mehdi Razzaghi-Abyaneh, for lending his expertise to make the book truly unique. He has played a crucial role in the development of this book. Without his invaluable contribution this book wouldn't have been possible. He has made vital efforts to compile up to date information on the varied aspects of this subject to make this book a valuable addition to the collection of many professionals and students.

This book was conceptualized with the vision of imparting up-to-date information and advanced data in this field. To ensure the same, a matchless editorial board was set up. Every individual on the board went through rigorous rounds of assessment to prove their worth. After which they invested a large part of their time researching and compiling the most relevant data for our readers. Conferences and sessions were held from time to time between the editorial board and the contributing authors to present the data in the most comprehensible form. The editorial team has worked tirelessly to provide valuable and valid information to help people across the globe.

Every chapter published in this book has been scrutinized by our experts. Their significance has been extensively debated. The topics covered herein carry significant findings which will fuel the growth of the discipline. They may even be implemented as practical applications or may be referred to as a beginning point for another development. Chapters in this book were first published by InTech; hereby published with permission under the Creative Commons Attribution License or equivalent.

The editorial board has been involved in producing this book since its inception. They have spent rigorous hours researching and exploring the diverse topics which have resulted in the successful publishing of this book. They have passed on their knowledge of decades through this book. To expedite this challenging task, the publisher supported the team at every step. A small team of assistant editors was also appointed to further simplify the editing procedure and attain best results for the readers.

Our editorial team has been hand-picked from every corner of the world. Their multi-ethnicity adds dynamic inputs to the discussions which result in innovative outcomes. These outcomes are then further discussed with the researchers and contributors who give their valuable feedback and opinion regarding the same. The feedback is then collaborated with the researches and they are edited in a comprehensive manner to aid the understanding of the subject.

Apart from the editorial board, the designing team has also invested a significant amount of their time in understanding the subject and creating the most relevant covers. They scrutinized every image to scout for the most suitable representation of the subject and create an appropriate cover for the book.

The publishing team has been involved in this book since its early stages. They were actively engaged in every process, be it collecting the data, connecting with the contributors or procuring relevant information. The team has been an ardent support to the editorial, designing and production team. Their endless efforts to recruit the best for this project, has resulted in the accomplishment of this book. They are a veteran in the field of academics and their pool of knowledge is as vast as their experience in printing. Their expertise and guidance has proved useful at every step. Their uncompromising quality standards have made this book an exceptional effort. Their encouragement from time to time has been an inspiration for everyone.

The publisher and the editorial board hope that this book will prove to be a valuable piece of knowledge for researchers, students, practitioners and scholars across the globe.

List of Contributors

Masoomeh Shams-Ghahfarokhi
Department of Mycology, Faculty of Medical Sciences, Tarbiat Modares University, Tehran 14115-331 Iran

Sanaz Kalantari and Mehdi Razzaghi-Abyaneh
Department of Mycology, Pasteur Institute of Iran, Tehran 13164, Iran
Robert L. Brown, Deepak Bhatnagar and Thomas E. Cleveland USDA-ARS, Southern Regional Research Center, New Orleans, LA, USA

Zhi-Yuan Chen
Department of Plant Pathology and Crop Physiology, Louisiana State University Agricultural Center, Baton Rouge, LA, USA

Abebe Menkir
International Institute of Tropical Agriculture, Ibadan, Nigeria

Amos Emitati Alakonya
Institute of Biotechnology Research, Jomo Kenyatta University of Agriculture and Technology, P. O. Box 62000-00200, Nairobi

Ethel Oranga Monda
Department of Plant and Microbial Sciences, Kenyatta University, P. O Box 43844 -00100, Nairobi

Alma Virginia Lara-Sagahón and Eva Guadalupe Lizárraga-Paulín
Laboratorio de Biotecnología, Facultad de Estudios Superiores Cuautitlán, Campo 1, UNAM. Cuautitlán Izcalli, Estado de México, México

Ernesto Moreno-Martínez and Susana Patricia Miranda-Castro
Unidad de Investigación en Granos y Semillas, Facultad de Estudios Superiores Cuautitlán, CAT, UNAM. Cuautitlán Izcalli, Estado de México, México

Irineo Torres-Pacheco
CA Ingeniería de Biosistemas, División de Investigación y Posgrado, Facultad de Ingeniería, Universidad Autónoma de Querétaro, Cerro de las Campanas s/n, 76010, Querétaro, Qro., México

Carlos Augusto Fernandes Oliveira, Fernanda Bovo, Carlos Humberto Corassin and Alessandra Vincenzi Jager
Faculty of Animal Science and Food Engineering, University of São Paulo, Brazil

Kasa Ravindranadha Reddy
Research and Development Department, SRI Biotech Laboratories Ltd, India

N.K.S. Gowda and H.V.L.N. Swamy
National Institute of Animal Nutrition and Physiology, Bangalore, India

P. Mahajan
Alltech Biotechnology Private Ltd., Bangalore, India

Ayhan Filazi
Department of Pharmacology and Toxicology, Faculty of Veterinary Medicine, Ankara University, Turkey

Ufuk Tansel Sireli
Department of Food Hygiene and Technology, Faculty of Veterinary Medicine, Ankara University, Turkey

Vivian Feddern , Fernando de C. Tavernari, Helenice Mazzuco, Anildo Cunha Jr., Everton L. Krabbe and Gerson N. Scheuermann
Embrapa Suínos e Aves, Concórdia, SC, Brazil

Giniani C. Dors
Instituto Federal Catarinense (IFC) - Campus Concórdia, Concórdia, SC, Brazil

Suthep Ruangwises
Department of Veterinary Public Health, Faculty of Veterinary Science, Chulalongkorn University, Bangkok, Thailand

Piyawat Saipan
Department of Veterinary Public Health, Faculty of Veterinary Medicine, Khon Kaen University, Khon Kaen, Thailand

Nongluck Ruangwises
Department of Pharmaceutical Chemistry, Faculty of Pharmacy, Mahidol University, Bangkok, Thailand

Ariane M. Kluczkovski and Augusto Kluczkovski Junior
Federal University of Amazonas, Faculty of Pharmaceutical Sciences, Manaus, Am, Brazil

Godfrey S. Bbosa, A. Lubega, Jasper Ogwal-Okeng and William W. Anokbonggo
Department of Pharmacology and Therapeutics, Makerere University College of Health sciences, Kampala, Uganda

David Kitya
Department of Surgery, Mbarara University of Science & Technology Medical School, Mbarara, Uganda

David B. Kyegombe
Department of Pharmacology and Toxicology, Kampala International University, School of Health Sciences, Ishaka Campus, Busyenyi, Uganda

Tahereh Ziglari
Herman Ostrow School of Dentistry, University of Southern California, Los Angeles, California, USA

Abdolamir Allameh
Department of Biochemistry, Faculty of Medical Sciences, Tarbiat Modares University, Tehran, I.R. Iran

Michael C. Kew
Department of Medicine, Faculty of Health Sciences, University of Cape Town, Department of Medicine, Faculty of Health Sciences, University of the Witwatersrand, Johannesburg, South Africa

Luis Miguel Contreras-Medina, Alejandro Espinosa-Calderon, Carlos Duarte-Galvan, Arturo Alfonso Fernandez-Jaramillo, Rafael Francisco Muñoz-Huerta,, Ramon Gerardo Guevara-Gonzalez and Irineo Torres-Pacheco
Biosystems Engineering CA, Postgraduate Study Division, Engineering Faculty, Autonomous University of Querétaro, Querétaro, Qro., México

Jesus Roberto Millan-Almaraz
Faculty of Physics and Mathematics, Autonomous University of Sinaloa, Culiacán, Sinaloa, México

Laura Anfossi, Claudio Baggiani, Cristina Giovannoli and Gianfranco Giraudi
Department of Chemistry, University of Turin, Giraudi

Cynthia Bley N'Dede and Curtis M. Jolly
Department of Agricultural Economics and Rural Sociology, Alabama Experiment Station, Auburn University, USA

Davo Simplice Vodouhe
The University of the Republic of Benin, Republic of Benin

Pauline E. Jolly
University of Alabama, Birmingham, USA

Antonello Santini
Department of Food Science, University of Napoli "Federico II", Portici, Napoli, Italy

Alberto Ritieni
Department of Pharmaceutical and Toxicological Chemistry, University of Napoli "Federico II", Napoli, Italy